CHEMISTRY AND BIOLOGY OF PEPTIDES

PROCEEDINGS OF THE THIRD AMERICAN PEPTIDE SYMPOSIUM

T,

CHEMISTRY AND BIOLOGY OF PEPTIDES

PROCEEDINGS OF THE THIRD AMERICAN PEPTIDE SYMPOSIUM,

3d, BOSTON, 1972.

Edited by

Johannes Meienhofer
The Children's Cancer Research Foundation and
Department of Biological Chemistry, Harvard
Medical School, Boston, Massachusetts 02115

ann arbor science PUBLISHERS INC.
P.O. BOX 1425 ● ANN ARBOR, MICHIGAN 48106

PREFACE

The Third American Peptide Symposium in Boston on June
19-23, 1972, was attended by three hundred scientists including over sixty colleagues from other countries. The
meeting was held at the Jimmy Fund Auditorium of the
Children's Cancer Research Foundation. The keen interest
in this Symposium was due to an unprecedented upsurge of
scientific and industrial interest and involvement in peptides in the past few years.

The increasing interaction of traditional peptide
chemistry with physically and biologically oriented disciplines was reflected in this Symposium by a series of
invited lectures. NMR in the conformational analysis of
polypeptides was reviewed by F. A. Bovey. Properties of
antamanide were presented by Th. Wieland. R. A. Bradshaw
elaborated on a comparison between nerve growth factor and
insulin. Angiotensin was discussed in terms of its physiological roles by M. J. Peach and its conformations by P.
Fromageot. K. Hofmann reported on the progress toward the
synthesis of ribonuclease T_1. N. Izumiya illustrated the
problems involved in synthesizing this protein by the
solid-phase technique. The significance of the hypothalamic
hormones in physiology and medicine was elucidated by R.
Guillemin.* The precision and sensitivity of amino acid
analysis was reviewed by S. Moore, and S. Udenfriend revealed a fluorometric assay in the picomole range. E. Gross
structural relationships in peptides containing α, β unsaturated amino acids. Many thanks are extended to the invited
speakers for their outstanding presentations and the efforts
in preparing the manuscripts.

---

*Admirably presented by R. Burgus when hurricane Agnes
prevented Dr. R. Guillemin's travel from Washington, D.C.
to Boston.

The large number of contributions presented during the
ten sessions of the meeting placed high demands on the individual presiding chairmen. Their skillful and excellent
performance is gratefully acknowledged. Nine papers on
structure and conformation emphasized the usefulness of
NMR and especially the high potential of carbon-13 NMR.
Studies on oxytocin, deamino lysine-vasopressin, proline-
containing cyclic hexapeptides, and on model aromatic
dipeptides, tetra- and pentapeptides were reported. Solvent
exposure of peptide protons was determined by NMR. Particularly fruitful proved to be conformational analysis of
cyclic peptides, and this was again part of the lively
forum discussion. Discourses on synthetic problems, side-
reactions during cyclization and chemical behavior alter-
nated with those on biological aspects, in particular,
antamanide, neurohypophyseal hormones, valinomycin and
actinomycin. Nineteen contributions dealt with *solid-phase
synthesis*. Methodical studies on modified solid supports,
improved protecting groups, insoluble coupling reagent and
active esters, and the use of oxidation-reduction condensation and *o*-nitrophenyl esters were reported followed by
accounts on problems and difficulties. Long-chain solid-
phase syntheses with target sequences as lysozyme or
ribonuclease T_1 apparently experienced many difficulties.
Improved solid-phase syntheses of LH-RH/FSH-RH and encephalitogenic peptides and preparations of valine-gramicidin
B and C and of semisynthetic noncovalent protein (nuclease)
complexes were presented. *Synthetic advances* by conventional methods in solution and *progress in synthetic
procedures* were covered by sixteen papers. Complex syntheses of ribonuclease T_1, a porcine gastric inhibitory
polypeptide and porcine proinsulin are very close to completion. Single crystals were obtained of $(Pro-Pro-Gly)_{10}$
and human ACTH was synthesized according to the corrected
structure. Syntheses of hypothalamic hormones are currently
pursued in many laboratories. Several novel amine protecting groups were disclosed. Amide protection, new catalytic
agents, problems of racemization and of cystine peptide
syntheses continue to be of interest. Broad coverage was
given to biological aspects. Ten presentations in the
session on *biologically active peptides* dealt with antamanide,
scotophobin, tuftsin, nerve growth factor, vasopressins,
biosynthetic actinomycins, fibrinogen peptides, bradykinin-
potentiators, a polypeptide from the lung and a L-glutamine
antimetabolite. An entire session with thirteen papers was
devoted to *angiotensin*, its pharmacology, its *in vivo*
generation, and its conformation as well as structure-activity

relationships and syntheses of inhibitors. *Hormonal messengers*, such as hypothalamic factors, β-lipotropin, cyclic AMP, and gastrointestinal hormones, and studies on structure and function of ACTH were presented in six papers. The eleven papers in the session on *analytical techniques* were concerned with amino acid analysis, fluorometric assay in the picomole range, sequence analysis by (a) mass spectrometry of hydrolysis mixtures, (b) thermal degradation, and (c) automated Edman degradation. Structure elucidation of viomycin and a human plasma alipoprotein were also covered. Probable future trends of the field were examined in a *forecast* at the end of the meeting.

The discussions were very lively and stimulating. The difficulties of preserving meeting discussions in print cannot easily be resolved. At the end of each chapter I have attempted to summarize various points of general interest raised during the discussions. The already large volume of the book required that these comments be brief. They were taken from the tape recordings without trying to identify individual discussants.

I am very indebted to all members of the Program Committee for their many valuable suggestions during the organization of the program and for their devoted help in editing manuscripts. R. Walter undertook the difficult task of editing the forum discussion on cyclic peptides.

Generous financial support to the Symposium fund has greatly assisted organizing the meeting. We wish to thank the following sponsors for their contributions: Abbott Laboratories; Armour Pharmaceutical Company; Ayerst Research Laboratories; Beckman Instruments, Inc.; Children's Cancer Research Foundation, Inc.; CIBA-GEIGY; Hoffmann-La Roche Inc.; Merck and Company, Inc.; Norwich Pharmacal Company; Pierce Chemical Company; Sandoz Pharmaceuticals; Schering Corporation; G. D. Searle and Company; and The Dow Chemical Company.

The Planning Committee entered into an appointive three-term membership system, and the terms of office of G. W. Anderson, M. Bodanszky, M. Goodman, R. B. Merrifield, M. A. Ondetti, and B. Weinstein expired. Their services are deeply appreciated. The new committee appointed Dr. R. Walter to organize the fourth symposium in New York.

J. Meienhofer   October 1972

THIRD AMERICAN PEPTIDE SYMPOSIUM

Boston, Massachusetts
June 19 - 23, 1972

HONORARY PRESIDENTS

Leonidas Zervas and Vincent du Vigneaud

HOST & ORGANIZING INSTITUTION

Children's Cancer Research Foundation, Inc.

PROGRAM COMMITTEE

J. Meienhofer, Chairman

E. Gross   R. B. Merrifield
R. Hirschmann   J. T. Potts, Jr.
E. C. Jorgensen   J. Ramachandran
D. S. Kemp   R. R. Smeby
S. Lande   R. Walter

PLANNING COMMITTEE

(1972 - 1975)

E. R. Blout   S. Lande
F. M. Bumpus   J. Meienhofer
E. Gross   J. Ramachandran
R. Hirschmann   R. R. Smeby
K. D. Kopple   R. Walter

TABLE OF CONTENTS

SECTION VI
BIOLOGICALLY ACTIVE PEPTIDES

xiii

SECTION VIII
HORMONAL MESSENGERS

SECTION IX
ANALYTICAL TECHNIQUES

SECTION X
FORECAST

ABBREVIATIONS

The abbreviations used in this book are listed below. For
amino acid residues and several protecting groups they are
those recommended by IUPAC-IUB Commission on Biochemical
Nomenclature in Biochemistry $\underline{5}$, 1445, 2485 (1966); $\underline{6}$, 322
(1967); J. Biol. Chem. $\underline{241}$, 2491 (1966); $\underline{247}$, 977 (1972).

A I	Angiotensin I
A II	Angiotensin II
A II'	Angiotensin II amide
	([Asn[1]]-Angiotensin II)
AA	Antamanide
Abu	L-α-Aminobutyric acid
Ac	Acetyl
Acm	Acetamidomethyl
AcOH, HOAc	Acetic acid
Acpc	1-Aminocyclopentanecarboxylic acid
	(Cycloleucine)
ACTH	Adrenocorticotropin
$(Alk)_3N$	Trialkylamine, t-amine
apoLp-Gln II	Apolipoprotein II (Gln C-terminal)
AVP	Arginine-vasopressin
Boc	t-Butyloxycarbonyl
Bpoc	2-(p-Biphenylyl)isopropyloxycarbonyl
BPTI	Bovine pancreatic trypsin inhibitor
m-BrBzl	m-Bromobenzyl
o-BrZ	o-Bromobenzyloxycarbonyl
p-BrZ	p-Bromobenzyloxycarbonyl
Bu^tOCOCl	Isobutyl chloroformate
Bu^t	t-Butyl
Bz	Benzoyl
Bzl	Benzyl

CCD	Countercurrent Distribution
CD	Circular dichroism
Cha	β-Cyclohexylalanine
CHO	Formyl
4-CH₃OBzl (MeOBzl)	4-Methyloxybenzyl
2,6-Cl₂Bzl	2,6-Dichlorobenzyl
2,4-Cl₂Z	2,4-Dichlorobenzyloxycarbonyl
2,6-Cl₂Z	2,6-Dichlorobenzyloxycarbonyl
3,4-Cl₂Z	3,4-Dichlorobenzyloxycarbonyl
CM	Carboxymethyl (cellulose, Sephadex)
CP-B	Carboxypeptidase B
CT	Cobrotoxin
cyclic AMP	cyclic Adenosine-3',5'-monophosphate
Cys(Cm)	S-Carboxymethylcysteine
DCC, DCCI	Dicyclohexylcarbodiimide
DCHA	Dicyclohexylamine
DEAE	Diethylaminoethyl
DFP	Diisopropyl fluorophosphate
Dipmoc	Diisopropylmethyloxycarbonyl
DLVP	1-Desamino lysine-vasopressin, (1-(β-mercaptopropionyl]-lysine-vasopressin
DMF	Dimethylformamide
DMSO	Dimethylsulfoxide
Dnp	2,4-Dinitrophenyl
Dns	Dansyl, 1-dimethylaminonaphthalene-5-sulfonyl
DTP	2,2'-Dithiopyridine
Ec	Ethylcarbamyl
EDAC	N-Ethyl-N'-(3-dimethylaminopropyl) carbodiimide
EDTA	Ethylenediamine tetraacetic acid
EEDQ	N-Ethyloxycarbonyl-2-ethyloxy-1,2-dihydroquinoline
Et₃N, NEt₃	Triethylamine
EtOH	Ethanol
FDNB	1-Fluoro-2,4-dinitrobenzene
FSH	Follicle-stimulating hormone
FSH-RH	Follicle-stimulating hormone-releasing hormone
<Glu	Pyroglutamic acid, pyrrolid-2-one-5-carboxylic acid
GRF, GRH	Growth hormone releasing factor

GS–MS	Gas chromatography – mass spectrometry
GU	Goldblatt unit
HAA	Perhydro-antamanide
HDL	High density lipoprotein
HFiPA	Hexafluoro-2-propanol
HGH	Human growth hormone
HOAc, AcOH	Acetic Acid
HOBt	1-Hydroxybenzotriazole
HONSu, HOSu	N-Hydroxysuccinimide
HPT	Hexamethyl phosphoric acid triamide
Hyp	Hydroxyproline
iNoc	Isonicotinyloxycarbonyl
IR	Infrared spectroscopy
$i.v.$	Intravenous (injection)
LAP	Leucine aminopeptidase
LH	Luteinizing hormone
LH–RH, LRF	Luteinizing hormone – releasing hormone
LPH	Lipotropic hormone
LVP	Lysine-vasopressin
MA	Carbonic acid mixed anhydride (method)
Mbh	4,4'-Dimethyloxybenzhydryl
McBoc	1-Methylcyclobutyloxycarbonyl
Me	Methyl
βMe	β-Mercaptoethanol
3,4-Me$_2$Bzl	3,4-Dimethylbenzyl
MeIle	N-Methylisoleucine
MeLeu	N-Methylleucine
MeOBzl, 4-CH$_3$OBzl	p-Methyloxybenzyl
MeOH	Methanol
MePhe	N-Methylphenylalanine
MRF	Melanotropin releasing factor
MRIF	Melanotropin-release inhibiting factor
MSH	Melanocyte-stimulating hormone, melanotropin
NCA	N-Carboxyanhydride
Ncps	2-Nitro-4-carboxyphenylsulfenyl
NGF	Nerve growth factor
NHPH	Neurohypophyseal hormone
N^{π}-MeHis	$pros$ Methylhistidine (N nearer to C^{β})
N^{τ}-MeHis	$tele$ Methylhistidine (N away from C^{β})
NMR	Nuclear magnetic resonance spectroscopy

Nps	*o*-Nitrophenylsulfenyl
Nps-ACTH	*o*-Nitrophenylsulfenyl adrenocortico-tropin
Nva	Norvaline
OBt	1-Hydroxybenzotriazole ester
OBzl	Benzyl ester
OBu^t	*t*-Butyl ester
OEt	Ethyl ester
ONb	*p*-Nitrobenzyl ester
ONo	*o*-Nitrophenyl ester
ONp	*p*-Nitrophenyl ester
ONSu	*N*-Hydroxysuccinimide ester
OPcp	Pentachlorophenyl ester
OPfp	Pentafluorophenyl ester
OPsp	*N*-5-(Polystyryl-4-methyloxycarbonyl)-imino-4-oximino-1,3-dimethyl-2-pyrazoline ester
ORD	Optical rotatory dispersion spectroscopy
OTcp, Ocp	2,4,5-Trichlorophenyl ester
PDE	3',5'-cyclic Adenosine monophosphate phosphodiesterase
PGC	Pyrolysis - gas chromatography method
Pht	Phthalyl
PI	Proinsulin
PLV-2	[Phe^2]-Lysine-vasopressin
PMA	Phenylmercuri acetate
PMR	Proton magnetic resonance spectroscopy
PRF	Prolactin releasing factor
Pro(4Br)	4-Bromo-L-proline
REMA	Repetitive excess mixed anhydride (method)
RNase T$_1$	Ribonuclease T$_1$
SPS, SPPS	Solid-phase peptide synthesis
TFA	Trifluoroacetic acid
Tfa	Trifluoroacetyl
tlc	Thin-layer chromatography
tle	Thin-layer electrophoresis
TMS	Trimethylsilane
Tms	Trimethylsilyl
Tos	Tosyl,*p*-toluenesulfonyl
TPCK	Tosylphenylalanine chloromethylketone
TPP	Triphenylphosphine

TRH, TRF	Thyrotropin – releasing hormone
Trt	Trityl, triphenylmethyl
TSH	Phyroid-stimulating hormone, Thyrotropin
Tyr(Ome)	*O*-Methyltyrosine
UV	Ultraviolet spectroscopy
Z	Benzyloxycarbonyl
Ztf	2,2,2-Trifluoro-*N*-benzyloxycarbonyl-aminoethyl

SECTION I

STRUCTURE AND CONFORMATION

Session Chairmen

Manfred Rothe and Frank R. N. Gurd

NMR IN THE CONFORMATIONAL ANALYSIS OF POLYPEPTIDES,
ESPECIALLY CYCLIC POLYPEPTIDES

F. A. Bovey. Bell Laboratories, Murray Hill, New
Jersey 07974

SUMMARY--The application of proton and C-13 NMR to the study
of the conformations of polypeptides, particularly cyclic
polypeptides, is discussed. Proton NMR is considerably
further developed for this purpose at the present time than
C-13 NMR. Main chain conformations can be deduced from the
J coupling of the α- and NH protons, together with information concerning internal hydrogen bonds from measurement of
the NH resonance positions as a function of temperature,
from observations of *cis-trans* isomerization of X-Pro peptide
bonds, and from energy calculations. In C-13 spectroscopy,
no coupling information is available and one must rely on
chemical shifts, at least some of which (particularly in
proline residues) appear to be sensitive to conformation.
The use of NMR is illustrated for cyclic hexapeptides, for
antamanide, and for oxytocin and its open chain precursor
peptides.

INTRODUCTION

OF THE SPECTROSCOPIC METHODS useful for the study of polypeptide conformations, high resolution NMR has recently
emerged as one of the most powerful. Many studies in our
laboratory and in others have amply demonstrated that proton
NMR (pmr) can provide large numbers of spectral parameters,
i.e. chemical shifts and spin-spin couplings. If they could
be fully interpreted, these data would give a fairly complete
conformational picture of these molecules. As yet, such a

3

complete interpretation is usually not possible but nevertheless certain features of the spectra can be made to
yield information concerning the main-chain conformations
of polypeptides and oligopeptides. Of the latter, cyclic
polypeptides, both synthetic and naturally occurring,
have been the objectives of particularly intensive investigation in our laboratory[1] and in many others.

Limited C-13 NMR (cmr) data have also been recently
published for several amino acids,[2] di- and tripeptides,[2]
a cyclic polypeptide (gramicidin S),[3] a linear polypeptide
(poly-γ-benzyl-L-glutamate),[4] a cyclic depsipeptide
(valinomycin)[5] and a protein (ribonuclease A).[6] C-13
studies are in course of publication from our laboratory
on oxytocin and its precursor peptides, valinomycin and
its K+ complex,[8] antamanide,[9] and antamanide-sodium
complex.[10] I will mention certain aspects of this work
today. In addition, C-13 investigations of synthetic
cyclic hexapeptides, carried out in collaboration with
Professor Blout's group at Harvard, are in course of preparation for publication. This work will be discussed in
part in this presentation, and also in a separate paper by
Dr. Deber. We may note also several other C-13 studies to
be presented in this meeting, including one by Smith *et al.*
on both oxytocin and deamino lysine vasopressin and their
precursor peptides.

The greater part of the 15 years since natural abundance
C-13 spectroscopy was first described[11,12] has been a period
of extensive data collecting with relatively little interpretation in terms of structure or conformation. One reason
for this state of affairs is the very large range of C-13
chemical shifts,   over 350 ppm, or roughly 30 times the
range usually observed for protons. This is in itself
attractive, since it promises fine discrimination, but it
has tended to baffle fundamental understanding, although
there is a considerable body of empirical correlations.
Another reason is that $^{13}C-^1H$ J couplings have not as yet
proved particularly useful in conformational analysis, as
$^1H-^1H$ couplings are, and in any case are banished from
present day C-13 spectra by double irradiation in order to
improve the signal-to-noise ratio. Thus, in polypeptide
C-13 spectra we have only chemical shifts to help us. I
will show, however, that these may be helpful and in fact
can supply data which are not provided,   or at best more
ambiguously provided, in proton spectra.

Proton NMR

The proton parameters which can be most incisively interpreted in terms of main chain conformation, i.e. rotation angles, are as follows:

1. *The vicinal coupling of the C_α proton and the NH proton*, termed $J_{N\alpha}$, which can be interpreted in terms of a Karplus-like relationship to give the rotation angle ϕ:

$J_{N\alpha}$ is measured as the spacing of the NH multiplets, usually observed in the $1.5 - 2.5$ τ range and appearing as doublets for all residues except glycine. An appropriate relationship has been found to be:

$$J_{N\alpha} = \begin{cases} 8.5 \cos^2\phi' & (\ 0° \le \phi' \le \ 90°) \\ 9.5 \cos^2\phi' & (90° \le \phi' \le 180°) \end{cases} \tag{1}$$

where ϕ' is the *dihedral* angle, not ϕ as defined by the 1966[13] or 1970[14] angle conventions.

2. *The temperature coefficient of the NH chemical shift* gives information concerning the participation of these protons in hydrogen bonds. Upfield shifts with increasing temperature are expected for protons capable of forming hydrogen bonds and are attributed to the breaking of an increasing fraction of such bonds. This dependence should be small for intramolecular hydrogen bonds but substantial for those capable of forming only external hydrogen bonds to solvent. The method is very useful but is not always straightforward. For example, the occasional observation of *negative* slopes instills caution.

Observations of Cis-Trans Peptide Bond Isomerism. It is well known that because of the planar nature of the peptide bond, the angle ω must be either 0° (*trans*) or 180° (*cis*), major departures from these values being costly in terms of energy. The *cis* structure can normally be adopted only by N-substituted residues, i.e. prolines or (more rarely) sarcosine. The energy barrier separating the *cis* and *trans* conformations of X-Pro bonds is of the order

of 20 kcal. The nmr correlation of these familiar facts
is the appearance of separate *cis* and *trans* spectra when
both forms are present, owing to the relatively slow
equilibration between them. Since most biological polypeptides contain proline and the more intriguing conformational behavior of synthetic polypeptides often depends
on the presence of proline residues, this is an observation
of major importance. Strong emphasis has been placed on
proline in much of our work, both in pmr and cmr.

4. *Peptide NH Exchange Rate.* The NH proton exchange
rate with hydroxylic solvents can be monitored analytically
by observing NH peak intensity reductions in deuterated
solvents, usually D_2O. More rarely, fast exchange rates
can be measured in H_2O by peak broadening. It is a little
difficult to know what to say about this type of study,
which is, of course, an old and well established one for
proteins, using isotopic tracers. It can be helpful when
marked retardation of rate can be correlated with internal
hydrogen bond formation, as observed, for example, by
Stern, Gibbons, and Craig in their pioneering study of
gramidicin S.[15] Like others, we have found it useful, but
we have also found it to be erratic; for some compounds,
rates have been observed to vary inexplicably by orders of
magnitude between different preparations of the same polypeptide solution. We now treat results of such measurements
with great reserve.

5. *Energy Calculations.* For simpler polypeptides,
particularly if cyclic, NMR alone may be sufficient to yield
a conformational structure fairly unambiguously. For more
complex polypeptides, the NMR data alone, although very rich,
cannot be completely and unambiguously interpreted in terms
of conformation. *Ab initio* energy calculations are virtually
impossible because of the large number of possible conformations. However, by incorporating NMR data into the energy
calculations, it is found that self-consistent structures
(or groups of structures) can be arrived at. This is done
by eliminating from consideration all structures having one
or more residues which do not correspond to the observed
$J_{N\alpha}$ or to a low conformational energy.

Let us now look at a few applications of these methods.
The use of criteria 1, 3 and 4 is well illustrated in the
study of gramicidin S by Stern, Gibbons, and Craig.[15] We
shall consider here two simpler hexapeptides:

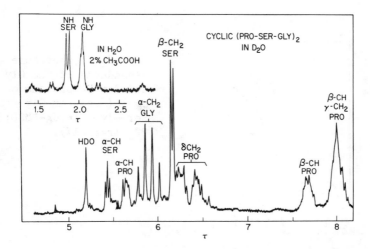

c-(PSG)$_2$

c-(SPG)$_2$

These compounds illustrate well the typical features of the NMR spectra of cyclic polypeptides and turn out to exhibit some surprising conformational behavior; the conclusions drawn by NMR are fully supported by energy calculations.[16]   In Figure 1 are shown the spectra of

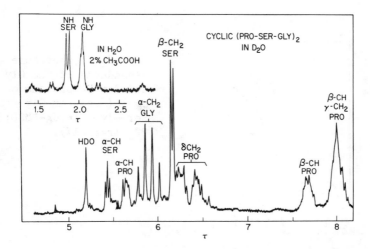

Figure 1:   220 MHz spectrum of *cyclo*(-Pro-Ser-Gly-Pro-Ser-Gly-) in D$_2$O and (upper left) in H$_2$O-CH$_3$COOH, 98:2 by vol.; 22°.   The H$_2$O solution is made acidic to retard exchange of NH protons with solvent (ref. 17).

cyclo (-Pro-Ser-Gly-Pro-Ser-Gly-), abbreviated *c*-(PSG)$_2$, in
D$_2$O and (inset) in H$_2$O.[17] From the spacings of the Gly NH
"triplet" (actually a quartet from coupling to two Gly β
protons) and the Ser NH doublet, $J_{N\alpha}$ is determined and
limits for the φ angles for these residues are established.
(There is, of course, no NH resonance for Pro residues,
but this is not important as in this case φ is fixed at
ca. 120° by the pyrrolidine ring.) In D$_2$O, the NH protons
are exchanged for deuterium and these couplings cannot be
measured, but the α-proton spectrum is now simpler to
analyze. The assignments of the glycine and serine α-
protons are shown in Figure 1; the former appear as an AB
quartet in the asymmetric environment of the serine and
proline residues; the latter is effectively a triplet,
although actually the X part of an ABX system. The proline
ring protons give a complex spectrum, the assignments being
based on extensive previous study.[18-21]

In d$_6$-dimethylsulfoxide (d$_6$-DMSO), the spectrum (not
shown) of *c*-(PSG)$_2$ is essentially similar to Figure 1; in
this solvent, the NH protons are not exchanged and appear
in the same relative positions as in Figure 2 (inset).

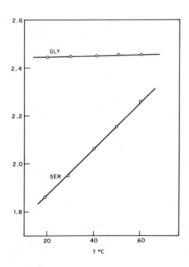

Figure 2: Temperature
dependence of the Gly and
and Ser NH protons of
cyclo(-Pro-Ser-Gly-Pro-
Ser-Gly-) in d$_6$-DMSO;
22° (ref. 17).

In Figure 2, the glycine and serine NH peak positions are plotted vs. temperature. The results strongly suggest that the glycine peptide protons participate in internal hydrogen bonds whereas those of the serine residues are externally hydrogen bonded; the relatively greater de-shielding of the latter probably reflects principally the strong hydrogen bond acceptor properties of DMSO.

The observation of a single spectrum for each pair of like residues in c-(PSG)$_2$ demonstrates C_2 symmetry. The evidence for glycine NH hydrogen bonding together with the values of $J_{N\alpha}$ for the glycine (*ca*. 4.0 and 5.0 Hz) and serine (8.5 Hz) residues are compatible only with the conformations shown in Figure 3 in both water and DMSO.

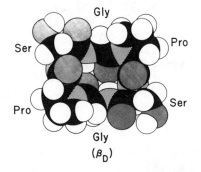 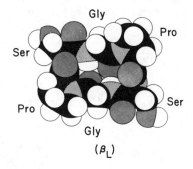

Figure 3: The β_D and β_L conformations of *cyclo*(-Pro-Ser-Gly-Pro-Ser-Gly-). Angles: all $\omega = 0°$;
β_D: $\phi_{Pro} = 290°$; $(\phi,\psi)_{Ser} = 240°, 210°$; $(\phi,\psi)_{Gly} = 330°, 60°$.
β_L: $\phi_{Pro} = 290°$; $(\phi,\psi)_{Ser} = 240°, 210°$; $(\phi,\psi)_{Gly} = 30°, 0°$.

In these structures, all peptide bonds are *trans* and the glycine and serine residues are in somewhat distorted anti-parallel β-type conformations. That designated β_L has glycine angles appropriate for an L-residue β-structure, while in the β_D conformation the glycine residues have angles appropriate to a D-residue β-structure. Calculations[16] show these conformations to be of nearly equal energy, while the middle-range values of $J_{N\alpha}$ for both the glycine protons indicate an averaging of these couplings corresponding to rapid equilibration of the two forms. The two hydrogen bonds stabilizing this form involve glycine C=O and NH groups only.

The $\beta_D \rightleftarrows \beta_L$ conformations of c-(PSG)$_2$ are the principal but not the only ones present. The peptide proton spectrum (Figure 1, inset) shows minor resonances appearing as *equal pairs* of glycine NH triplets and serine NH doublets. (Related resonances can be seen upon close inspection of the α-CH region, e.g. near 5.5 τ.) Variation of temperature and solvent does not alter the relative intensities of these resonances, but does alter somewhat their ratio to the principal spectrum. They correspond to an *asymmetric* conformation, designated A, with one proline residue now *cis*, and the other *trans*, in equilibrium with the major symmetric conformation. Their temperature dependence is consistent with one intramolecular glycine NH hydrogen bond, the other glycine and both serine NH protons being exposed to solvent. A proposed structure for this conformation is shown in Figure 4. Its observation is particularly

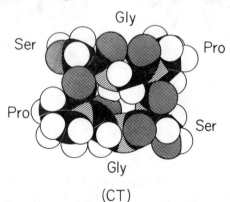

(CT)

Figure 4: The A conformation of *cyclo*(-Pro-Ser-Gly-Pro-Ser-Gly-). Angles: Gly(NH H-bonded): ω = 180°; all other ω = 0°.

significant because it tells us that structures which
appear to have C_2 (or higher) symmetry from the structural
formula alone need not actually have such symmetry.

The retroisomeric hexapeptide *cyclo*(-Ser-Pro-Gly-Ser-
Pro-Gly-), designated *c*-(SPG)$_2$, shows different conformational behavior from *c*-(PSG)$_2$. In H$_2$O, the NMR spectrum
indicates a conformation similar to the β_L structure of
the latter.[22] The β_D structure is excluded as its energy
is substantially higher. The serine and glycine residues
have now reversed roles, the serine NH being internally
hydrogen bonded, as shown by its small temperature dependence (Figure 5), whereas the glycine NH protons are

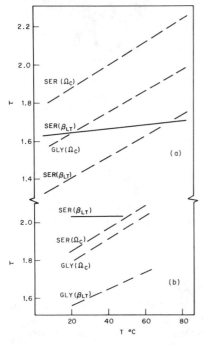

Figure 5: The temperature
dependence of the chemical
shifts of the NH resonances
of *cyclo*(-Ser-Pro-Gly-Ser-
Pro-Gly-) in (a) H$_2$O-CH$_3$
COOH, 98:2 by vol.; (b)
d$_6$-DMSO (ref. 22).

exposed to solvent. This conformer is designated β_{LT} and
is shown in Figure 6 (top). A minor fraction of another
symmetrical conformer can be detected in the spectrum. In
d$_6$-DMSO solution this becomes the major conformation.
Figure 7 shows the appearance of the NH spectrum as a
function of solvent composition. The NH temperature dependence demonstrates that this second form has no internal
hydrogen bonds; this observation and the $J_{N\alpha}$ values are
consistent with a conformation, Ω_C, in which both proline
residues are *cis* (Figure 6 bottom). It is strongly folded

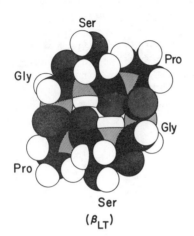 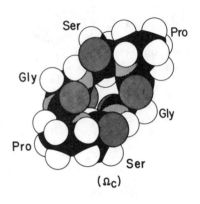

Figure 6: The β_{LT} and Ω_C conformations of *cyclo*(-Ser-Pro-Gly-Ser-Pro-Gly-). Angles for β_{LT}: all $\omega = 0°$; $(\phi, \psi)_{Ser}$ = 30°, 330°, Pro $\phi = 300°$; $(\phi, \psi)_{Gly}$ = 300°, 150°. Angles for Ω_C: $(\phi, \psi, \omega)_{Ser}$ = 25°, 325°, 180°; $(\phi, \psi)_{Pro}$ = 300°, 0°; $(\phi, \psi, \omega)_{Gly}$ = 240°, 90°, 0°.

rather than planar like β_{LT}. Energy calculations are at present unable to deal with such solvent interactions, and so even when dramatic changes such as those of c-(SPG)$_2$ occur, the most that can be concluded is that both the β_{LT} and Ω_C conformations represent low energy forms.

The even more striking conformational behavior of the related cyclic hexapeptide, *cyclo*(-Pro-Gly-)$_3$, will be described by Dr. Deber later in this meeting.

Of the many cyclic polypeptides of more complex structure which have been investigated by pmr, let us briefly mention here the mushroom toxin antidote antamanide,[9,10,23-26] an all-L decapeptide having no element of symmetry:

```
        8   9   10  1   2
        Pro-Phe-Phe-Val-Pro
         |               |
        Pro-Phe-Phe-Ala-Pro
        7   6   5   4   3
```

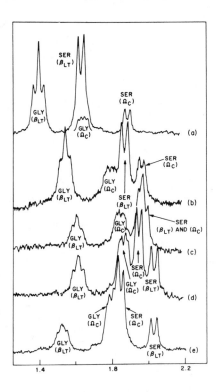

Figure 7: The 220 MHz NMR spectrum of the NH protons of
cyclo(-Ser-Pro-Gly-Ser-Pro-Gly-); (a) in H_2O-CH_3COOH,
98:2 by vol.; (b), (c), and (d) in d_6-DMSO:H_2O-CH_3COOH
(98:2); (e) in d_6DMSO. In (a) through (e) the mole
fraction of d_6DMSO is 0.0, 0.5, 0.7, 0.8, and 1.0. Each
spectrum is the result of 15-30 accumulated CAT scans
(ref. 22).

Figure 8 shows the 220 MHz pmr spectrum of antamanide in
CD_3CN at 34°. I will not discuss it in detail but show
it as illustrative of the spectra of the more complex
polypeptides which lack symmetry. The main parameters we
need are the values of $J_{N\alpha}$, each assigned if possible to
a specific residue in the sequence. An essential procedure
is the association of the NH resonances (at 2-3 τ) with
their corresponding α-protons (5-6 τ) and the association
of these with the side-chain β-protons at higher field,
and so on. This is done by *double resonance*. The irradiation of an α-CH multiplet causes the associated NH doublet
to collapse to a singlet; the assignment may be confirmed
by reversing the procedure and observing the resulting

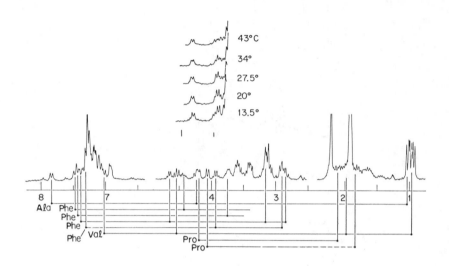

Figure 8: The 220 MHz proton spectrum of antamanide in
   CD₃CN at 34°. The lines below the spectrum connect NH,
   α- and β-protons shown by double resonance to be
   vicinally coupled. The inset spectra above show the
   temperature dependence of the NH portion of the spectrum,
   which in this solvent is relatively small.

perturbation of the α-region of the spectrum. This operation is then extended to the β, γ, etc. protons of the
side-chain. The lines in Figure 8 connect multiplets
corresponding to groups of spins demonstrated by this means
to be (vicinally) coupled. From the form of the α-CH and
sidechain multiplets the residues can be assigned by type
but not specific sequence position. All the $J_{N\alpha}$ had
relatively large values of 6.0 - 8.5 Hz. In CD₃CN and in
CD₃CO₂H temperature coefficients were small for the chemical
shifts of all six of the peptide NH doublets, two Phe
resonances showing somewhat larger slopes than the other
four residues. There are thus four strong intramolecular
hydrogen bonds and two weaker ones. (An earlier study of

antamanide in CDCl3[24] showed fast exchange rates for *all*
NH resonances.) In strong hydrogen bond accepting solvents
such as dioxane and N,N'-dimethylformamide, antamanide assumes a different conformation in which all peptide NH
protons are exposed to solvent.

The proton studies of antamanide are hampered by the
difficulty of having no entirely clear-cut test of whether
the proline residues are *cis* or *trans* or both, although
there does appear to be a correlation between the form of
the pro α-CH multiplet and the state of the X-Pro bond
(quartet for *trans* and "doublet" for *cis*). As we shall
shortly see, however, cmr can provide very useful and
important information on this point.

Antamanide is known to complex Na^+ quite strongly,[23,26]
probably by folding to form a cavity in which several
carbonyl groups can bond to the ion. It is found[10] that
on adding NaSCN to a CD_3CN solution of antamanide, the
peaks broaden as the mole ratio of Na^+ to polypeptide
approaches 0.5 and then narrow again while assuming new
positions corresponding to the Na^+ complex. There is no
actual "doubling" of the spectrum at any point. Analysis
of this behavior shows that the barrier in the process

$$\overset{*}{A}nt \cdot Na^+ + Ant \rightleftarrows \overset{*}{A}nt \cdot Na^+$$

is no greater than *ca.* 15 kcal. and that therefore no
significant alterations in X-Pro peptide bond conformations
occur on complexation. Again, cmr, which we now discuss,
provides important confirmatory information.

Carbon-13 NMR

As is well known, the inherent observing sensitivity of
C-13 is low owing to a natural isotopic abundance of 1.1%
and a gyromagnetic ratio only about one-fourth that of the
proton. This disadvantage has been very successfully overcome in the last 3-4 years by multiple scan techniques and,
still more recently, by Fourier transform spectroscopy.
Further increase in signal-to-noise ratio is accomplished
by abolishment of $^{13}C-^1H$ J couplings using a noise-modulated
proton decoupling field; this has the further advantage of
an accompanying nuclear Overhauser enhancement of as much
as 3-fold. As we have seen, one is repaid for all this
extra trouble by a range of chemical shifts *ca.* 30-fold
larger than for protons, with corresponding increase in
discrimination of structural and conformational features.

In a general way, the chemical shift positions of
carbon resonances in polypeptide spectra (commonly, as here,

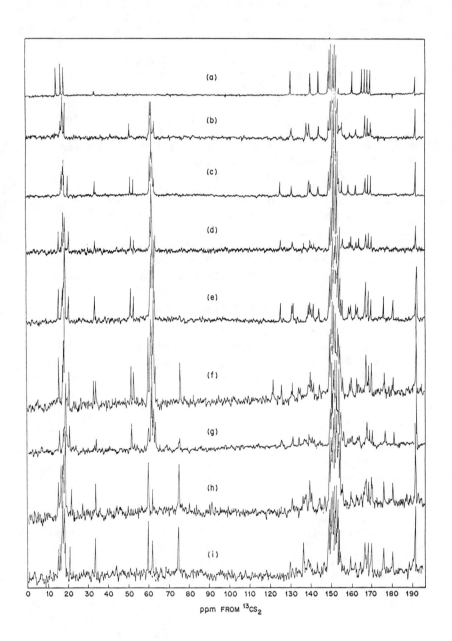

expressed with respect to $^{13}CS_2$ as zero) are fairly well
understood,[2-10] but there are marked perturbations arising
from secondary influences, probably chiefly conformational,
of which only a glimmer of understanding can be claimed at
present. It is fortunate that such perturbations occur,
for if a C-13 spectrum were only a summation of the resonances of the individual amino acid residues it would not
be very informative.

In Figure 9 are shown the C-13 spectra of the precursor
peptides of oxytocin[7] from Pro-Leu-Gly-NH$_2$ (a) through
the nonapeptide (g) and finally closing the ring (h);
[D-Pro$_7$]-oxytocin is shown as (i). I will comment only in
a general way on these spectra, my purpose in showing them
being to indicate the general appearance of C-13 spectra
of polypeptides. In the least shielded region, between
10 and 30 ppm, are the carbonyl resonances of the peptide
groups and the Asn, Gln, and Gly-NH$_2$ carboxamide groups.
Aromatic resonances of Tyr and the *S*-benzyl, *O*-benzyl, and
carbobenzoxy blocking groups (and the carbobenzoxy carbonyl)
appear between 35 and 74 ppm. The α-carbons appear between
130 and 150 ppm, Gly being the most shielded and Pro the
least. Benzylic methylene carbons of Tyr-*O*-benzyl, of
carbobenzoxy, and of Cys-*S*-benzyl appear just below and

Figure 9:   25 MHz C-13 spectra of the precursor peptides
of oxytocin; oxytocin; and [7-D-proline]-oxytocin.
(a)   Pro-Leu-Gly-NH$_2$
(b)   Cys(Bzl)-Pro-Leu-Gly-NH$_2$
(c)   Z-Asn-(S-Bzl)Cys-Pro-Leu-Gly-NH$_2$
(d)   Z-Gln-Asn-Cys(Bzl)-Pro-Leu-Gly-NH$_2$
(e)   Z-Ile-Gln-Asn-Cys(Bzl)-Pro-Leu-Gly-NH$_2$
(f)   Z-Tyr(Bzl)-Ile-Gln-Asn-Cys(Bzl)-Pro-Leu-Gly-NH$_2$
(g)   Z-Cys(Bzl)-Tyr-Ile-Gln-Asn-Cys(Bzl)-Pro-Leu-Gly-NH$_2$
(h)   oxytocin
(i)   [D-Pro$_7$]

just above this range of chemical shifts, respectively.
The large multiplet at *ca.* 150 ppm is that of the d_6-DMSO
solvent; the deuterons are not decoupled by the proton
decoupler, and give a 1:3:6:7:6:3:1 septet with *ca.* 20 Hz
spacing; in the oxytocins, this masks the peaks of Asnβ,
Leuβ, $Cys_1β$, and $Cys_6β$. The Pro δ-carbons appear at *ca.*
146 ppm in all spectra. Between 152 and *ca.* 167 ppm the
β-carbons appear and beyond this up to 182 ppm are the
Pro γ-carbons and those still further out on the Leu and
Ile sidechains. The latter can be unambiguously identified
by use of the empirically derived rules for carbon chemical
shifts in alkanes.[27-30]

Careful analysis of these spectra permits detailed
assignments of all the carbon peaks. These are shown in
schematic form in Figures 10, 11 and 12; dotted lines
connect peaks for corresponding carbons in different
spectra. Little deviation occurs in the open-chain polypeptide spectra beyond what can be readily attributed to
end-groups. But when the nonapeptide is closed to yield
oxytocin, there are alterations of many resonance positions,
particularly of the α-carbons. The assignment of the Cys
peaks in oxytocin is based on deuterium labelling, which
causes the labelled carbon resonance to broaden and virtually disappear. Some of these assignments await further
confirmation.

Extensive further study and interpretation will be
required before the power of cmr in such investigations is
fully realized. The most immediately useful result of our
present studies is the determination of the *cis-* or *trans*
state of the X-Pro bond. As we have seen, pmr is ambiguous
for this purpose; it clearly reveals both forms when present
but does not identify them. Cmr is more positive in this
regard, although not always entirely unambiguous. Figure
13 shows schematic cmr spectra of proline and a number of
simple proline derivatives and peptides, including poly-L-
proline I and II. The assignments are based on the observations of the polyprolines and of a number of simple
amides[31] and appear to be secure. The heights of the lines
indicate approximately the apparent proportions of *cis* and
trans conformers. One can see that whereas the α, δ and
(particularly) carbonyl resonances do not show a fixed
pattern of relative chemical shifts for *cis* and *trans*, the
pattern of the β and γ carbons is, with only minor perturbations, entirely consistent. These resonances show the
"back-to-back" intensity pattern to be expected if the β-
and γ-carbons experience the same relative environmental
influences when *syn* or *anti* to the carbonyl group. (The

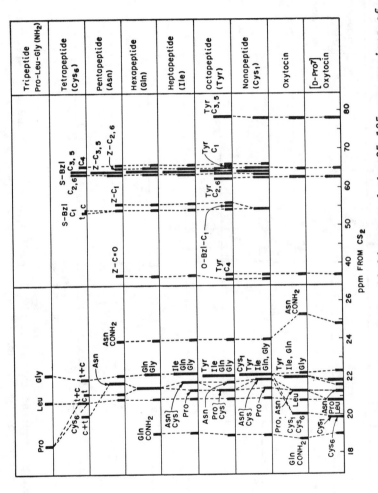

Figure 10: Schematic representation of the 18-26 ppm and the 35-125 ppm regions of the C-13 spectrum of precursor peptides of oxytocin; oxytocin; and [7-D-proline]-oxytocin (ref. 7).

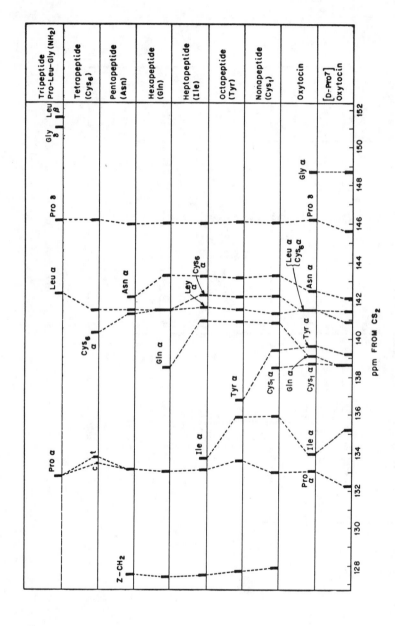

Figure 11: Schematic representation of the 127-155 ppm region of the C-13 spectrum of precursor peptides of oxytocin; oxytocin; and [7-D-proline]-oxytocin (ref. 7).

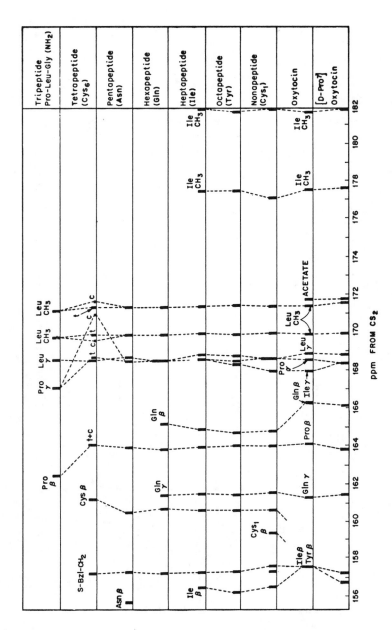

Figure 12: Schematic representation of the 154-182 ppm region of the C-13 spectrum of precursor peptides of oxytocin; oxytocin; and [7-D-proline]-oxytocin (ref. 7).

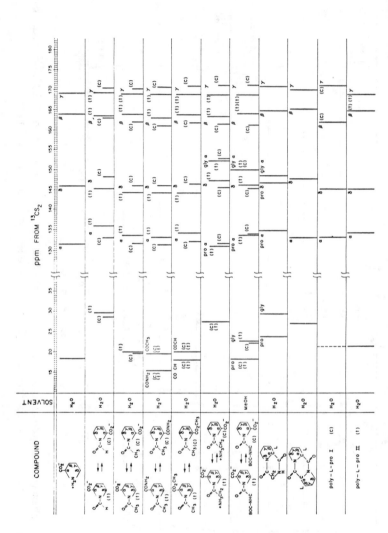

Figure 13: Schematic C-13 spectra of proline residues in simple proline derivatives and poly-L-proline, forms I and II (ref. 31).

origin of these chemical shift differences is not at present clear; electric field effects have been suggested as a possible cause.[32]) A similar pattern appears in the α and δ resonances for some compounds but for others it is reversed.

Application of these criteria to larger molecules will clearly be significant. Examples are illustrated in Figure 14. For example, the Pro β- and γ-carbon resonances in gramicidin A (data of Gibbons *et al.*[3]) are unmistakably in the *trans* positions, confirming the structure deduced for this molecule.[3,15,33] The eight line β- and γ-carbon pattern of antamanide[9] shows a grouping into two *cis* and two *trans* X-Pro bonds, resolving the ambiguity of the proton spectrum. At the same time, the close spacing of the lines in each group confirms the near two-fold symmetry of the proposed conformation in acetonitrile.[9] (There is an extra line, assigned to Val β, in the lowest field group.) Of equal significance is the fact that this pattern is retained with little change in the spectrum of the Na^+ complex. This is consistent with the conclusion from proton studies, discussed earlier, that no X-Pro bond conformations are altered when the complex is formed. The conformation deduced from pmr and cmr[9,10] for antamanide is shown in Figure 15, and for the Na^+ complex in Figure 16.

In Figure 14 are also shown the assigned resonances for Pro α, β, γ and δ carbons in the oxytocin precursor peptides, extracted from Figures 10 and 11. When Cys is added to Pro-Leu-Gly-NH$_2$, *cis* and *trans* X-Pro conformers become possible and are observed. But when Asn is added, the *cis* conformer is suppressed. The last finding agrees with the findings of Smith *et al.*, reported in this meeting. The Pro carbons maintain a consistent *trans* pattern in the open-chain peptides and upon closing the ring.

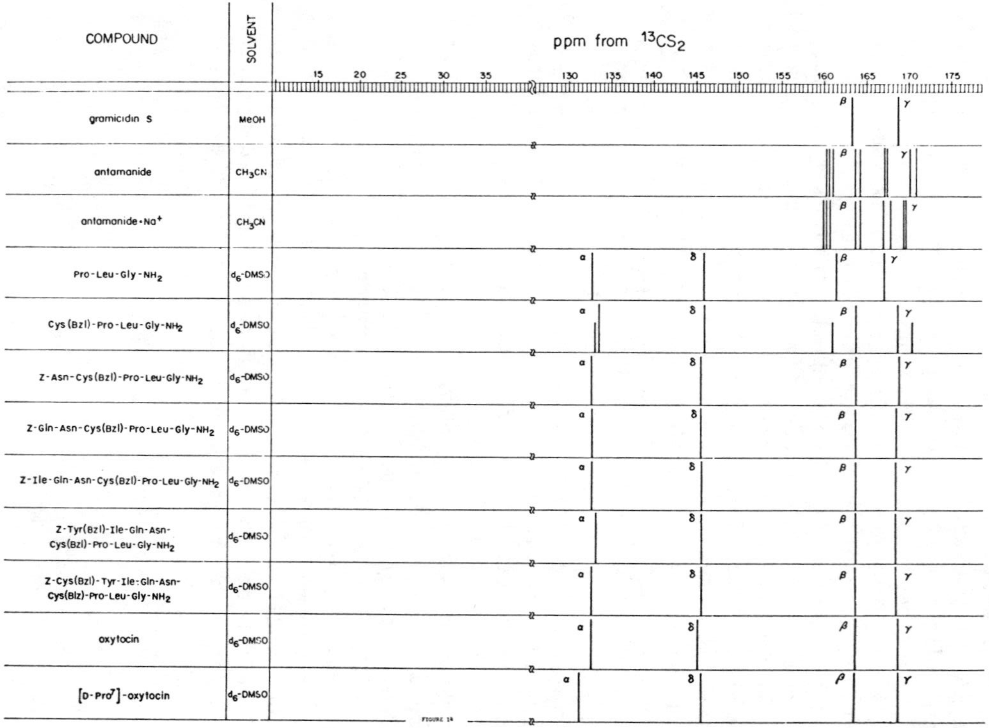

Figure 14: Schematic C-13 spectra of proline residues in gramicidin S, antamanide, antamanide sodium complex, oxytocin precursor peptides, oxytocin, and [7-D-proline]-oxytocin.

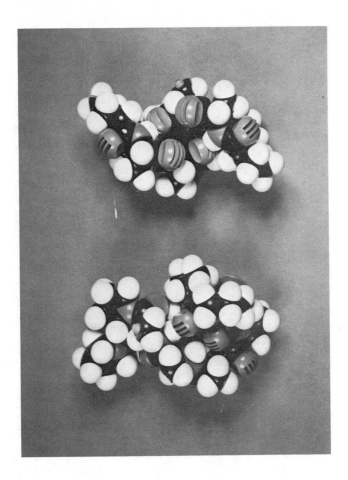

Figure 15: Two views of the "1,6-*cis*-I" conformation of antamanide in weakly hydrogen bond acceptor solvents (CD3COOH and CD3CN); all peptide NH are internally hydrogen bonded. (Phenyl groups are omitted for clarity.) Main chain rotational angles are:

	Val[1],Phe[6]	Pro[2],Pro[7]	Pro[3],Pro[8]	Ala[4],Phe[9]	Phe[5],Phe[10]
ϕ	60°	120°	120°	30°	60°
ψ	330°	330°	270°	120°	120°
ω	180°	0°	0°	0°	0°

(ref. 9).

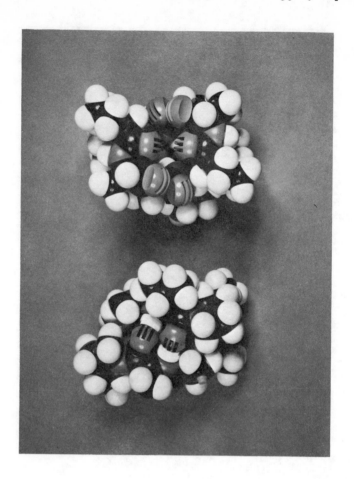

Figure 16: Two views of the Na⁺ complex of antamanide.
(Phenyl groups are omitted for clarity.) Main chain
rotational angles are:

	Val¹,Phe⁶	Pro²,Pro⁷	Pro³,Pro⁸	Ala⁴,Phe⁹	Phe⁵,Phe¹⁰
ϕ	30°	120°	120°	120°	90°
ψ	330°	360°	120°	120°	120°
ω	180°	0°	0°	0°	0°

(ref. 10).

References

1. Bovey, F. A., A. I. Brewster, D. J. Patel, A. E. Tonelli and D. A. Torchia. Acc. Chem. Res. $\underline{5}$, 193 (1972).
2. Horsley, W. J., H. Sternlicht, and J. S. Cohen. J. Amer. Chem. Soc. $\underline{92}$, 680 (1970).
3. Gibbons, W. A., J. A. Sogn, A. Stern, L. C. Craig and L. F. Johnson. Nature $\underline{227}$, 841 (1970).
4. Paolillo, L., T. Tancredi, P. A. Temussi, E. Trivellone, E. M. Bradbury, and C. Crane-Robinson. Chem. Comm. $\underline{1972}$, 335.
5. Ohnishi, M., M. C. Fedarko, J. D. Baldeschwieler, and L. F. Johnson. Biochem. Biophys. Res. Comm. $\underline{46}$, 312 (1972).
6. Allerhand, A., D. Doddrell, V. Glushko, D. W. Cochran, E. Wenkert, P. J. Lawson, and F. R. N. Gurd. J. Amer. Chem. Soc. $\underline{93}$, 544 (1971); V. Glushko, P. J. Lawson, and F. R. N. Gurd. J. Biol. Chem. $\underline{247}$, 3176 (1972).
7. Brewster, A. I., F. A. Bovey, and V. Hruby. Proc. Nat. Acad. Sci. (US), in press.
8. Patel, D. J., and A. E. Tonelli. Biochemistry, in press.
9. Patel, D. J. Biochemistry, in press.
10. Patel, D. J. Biochemistry, in press.
11. Lauterbur, P. C. J. Chem. Phys. $\underline{26}$, 217 (1957).
12. Holm, C. H. J. Chem. Phys. $\underline{26}$, 797 (1957).
13. Edsall, J. T., P. J. Flory, J. C. Kendrew, A. M. Liquori, G. Nemethy, G. Ramachandran and H. A. Scheraga. Biopolymers $\underline{4}$, 121 (1966); J. Biol. Chem. $\underline{241}$, 1004 (1966); J. Mol. Biol. $\underline{15}$, 399 (1966).
14. Kendrew, J. C., W. Klyne, S. Lifson, T. Miyazawa, G. Nemethy, D. C. Phillips, G. N. Ramachandran, and H. A. Scheraga. Biochemistry $\underline{9}$, 3471 (1970); J. Biol. Chem. $\underline{245}$, 489 (1970); J. Mol. Biol. $\underline{52}$, 1 (1970).
15. Stern, A., W. A. Gibbons, and L. C. Craig. Proc. Nat. Acad. Sci. (US) $\underline{61}$, 734 (1968).
16. Tonelli, A. E. J. Amer. Chem. Soc. $\underline{94}$, 346 (1972).
17. Torchia, D. A., A. Di Corato, S. C. K. Wong, C. M. Deber, and E. R. Blout. J. Amer. Chem. Soc. $\underline{94}$, 609 (1972).
18. Deber, C. M., F. A. Bovey, J. P. Carver and E. R. Blout. J. Amer. Chem. Soc. $\underline{92}$, 6191 (1970).
19. Deber, C. M., D. A. Torchia, and E. R. Blout. J. Amer. Chem. Soc. $\underline{93}$, 3893 (1971).

20. Torchia, D. A., and F. A. Bovey. Macromolecules 4, 246 (1971).
21. Torchia, D. A. Macromolecules 4, 440 (1971).
22. Torchia, D. A., S. C. K. Wong, C. M. Deber, and E. R. Blout. J. Amer. Chem. Soc. 94, 616 (1971).
23. Ivanov, V. T., A. I. Miroshnikov, N. D. Abdullaev, L. B. Senyavina, S. F. Arkhipova, N. N. Uvarova, K. Kh. Khalilulina, V. F. Bystrov, and Yu. A. Ovchinnikov. Biochem. Biophys. Research Comm. 42, 654 (1971).
24. Tonelli, A. E., D. J. Patel, M. Goodman, F. Naider, H. Faulstich, and T. Wieland. Biochem. 10, 3211 (1971).
25. Faulstich, H., W. Burgermeister, and Th. Wieland. Biochem. Biophys. Research Comm., in press.
26. Wieland, Th., H. Faulstich, W. Burgermeister, W. Otting, W. Möhle, M. M. Shemyakin, Yu. A. Ovchinnikov, V. T. Ivanov, and G. G. Malenkov. FEBS Letters 9, 89 (1970).
27. Speisecke, H., and W. G. Schneider. J. Chem. Phys. 35, 722 (1961).
28. Grant, D. M., and E. G. Paul. J. Amer. Chem. Soc. 86, 2984 (1964).
29. Lindeman, L. P. and J. Q. Adams. Anal. Chem. 43, 1245 (1971).
30. Carhart, R. H., D. E. Dorman and J. D. Roberts. Unpublished results.
31. Dorman, D. E., and F. A. Bovey. In preparation.
32. McFarlane, W. Chem. Comm. 1970, 418.
33. Ovchinnikov, Yu. A., Y. T. Ivanov, V. Bystrov, A. I. Miroshnikov, E. N. Shepel, N. D. Abdullaev, E. S. Efremov, and L. B. Senyavina. Biochem. Biophys. Res. Commun. 39, 217 (1970).

A CARBON-13 NUCLEAR MAGNETIC RESONANCE STUDY OF
NEUROHYPOPHYSEAL HORMONES AND RELATED OLIGOPEPTIDES

Ian C. P. Smith, Roxanne Deslauriers. Division of
Biological Sciences, National Research Council of
Canada, Ottawa, Canada K1A OR6

Roderich Walter. Department of Physiology, Mount
Sinai School of Medicine, City University of New
York, New York, New York 10029

PROTON MAGNETIC RESONANCE (PMR) has been useful for deriving
conformational information and constructing molecular models
for oxytocin[1-3] and lysine-vasopressin[4-8] in solution.
However, the twenty-fold greater range of chemical shifts
for carbon-13 implies that much better resolution of individual resonances could be obtained. The chemical shifts
of ^{13}C have been shown to be very sensitive to conformational
effects in small molecules.[9,10] Coupling constants between
carbon and hydrogen,[11] and carbon and phosphorus,[12,13] have
also been shown to be conformation-dependent. The relaxation
times, T_1, of individual carbon atoms are indicative of
mobilities of different regions of a molecule.[14] The value
of ^{13}C magnetic resonance (CMR) has not yet been conclusively
demonstrated in conformational studies on peptides other than
valinomycin.[15]

   We have obtained the CMR spectra of oxytocin (Figure 1),
lysine-vasopressin (LVP), 1-deamino lysine-vasopressin (DLVP)
(Figure 1), and arginine-vasopressin (AVP) in deuterated
dimethylsulfoxide (DMSO-d_6) and deuterium oxide (D$_2$0). As
expected, a high degree of resolution was obtained, with
separate resonances being observed for almost every carbon
atom in the molecule. However, due to the large number of
similar resonances, and to slight changes in chemical shifts
associated with neighbour effects, it was not possible to

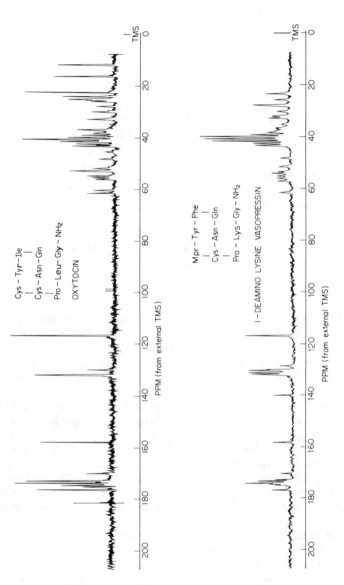

Figure 1: Fourier transformed CMR. (25.16 MHz, Varian XL-100-15) spectra of oxytocin (40 mg/0.2 ml, 84,000 accumulations) and 1-deamino lysine-vasopressin (40 mg/0.2 ml, 112,000 accumulations) in DMSO-d₆ at 37°C. Carbon-hydrogen spin couplings have been removed from the spectra by broad band (noise) decoupling of hydrogen. The sweep width is 5000 Hz, and data (4096 points) were accumulated with an aquisition time of 0.4 seconds. The sharp line at 182 ppm in the spectrum of oxytocin is an instrument artifact and should be disregarded. The regions at 40 ppm contain seven line multiplets from the CD₃ groups of DMSO-d₆. Mpr = β-mercaptoproprionic acid.

assign all the resonances by reference merely to data for the constituent amino acids or model dipeptides. We therefore obtained the CMR spectra of the constituent peptides of oxytocin, from the carboxamido-terminal tripeptide to the partially protected, uncyclized nonapeptide. Thus we were able to follow the influences of specific neighbouring groups, and of cyclizing the nonapeptide. Detailed assignments of the various spectra[16] cannot be presented here; they are available on request and will be published elsewhere. We discuss herein some of the observed effects and their significance.

Discussion

In the CMR spectra of the peptides, Z-Pro-Leu-Gly-OEt and Z-Pro-Leu-Gly-NH₂, two resonances were observed for the α, β, and δ carbons of Pro (the γ resonance is obscured partly by the δ of Leu, and doubling cannot be easily detected). We attribute this to the presence of both *cis* and *trans* isomers of the carbobenzoxylated Pro in these tripeptides. Elongation of the deprotected tripeptide to the protected tetrapeptide, Z-Cys(Bzl)-Pro-Leu-Gly-NH₂, results in loss of one set of Pro resonances, as is the case with all higher-membered peptide intermediates studied as well as with the hormones. This indicates a strong preference for the *trans* conformation of Pro. PMR (220 MHz) of the protected tetrapeptide confirms the *absence* of the *cis* conformer,[17] and the *presence* of the *cis* conformer in the tripeptide.[18]

In successive additions of protected N-terminal residues to the oligopeptides, the principal changes in ^{13}C chemical shift were observed in the penultimate residue. For example, on adding Z-Gln to Z-Asn-Cys(Bzl)-Pro-Leu-Gly-NH₂, the only observed changes in chemical shift are for Asn (α, -2.0; β, -0.3; carbonyl, +0.2 ppm). The one exception to this rule is Pro, whose α resonance is insensitive to *N*-peptide formation, possibly because it is a secondary amine to begin with. Further studies of proline-containing peptides are required in order to test the generality of this observation. The protecting groups (Z, Bzl, *etc.*) affect mainly the resonances of the protected amino acids.

The deprotected oligopeptides were studied in both D_2O and DMSO-d₆. This served as a measure of direct (non-conformational) solvent effects. In general the chemical shifts in DMSO-d₆ relative to those in D_2O were: α to higher field; β to higher or lower, depending on the amino acid; CH₃ to lower field; carbonyl 2.8 ppm to higher field.

This demonstrates that caution must be taken in interpreting data taken in non-aqueous solvents, and in comparing them with data taken in water.

Cyclization of the protected nonapeptide to form oxytocin gives rise to numerous changes in the CMR spectrum. The major changes are in the α and carbonyl carbon resonances of the amino acids of the cyclized portion. This in itself suggests that the effects have a conformational origin, as do temperature studies (*vide infra*). In general the α resonances move to lower field on cyclization. Formation of hydrogen bonds, as have been proposed[3] between the peptide NH of Asn and the carbonyl of Tyr, and between the side peptide NH of Gly and the carbonyl of Cys-6, should result in downfield shifts of the carbonyl resonances involved.[19,20] To estimate the magnitude of the downfield shift, we studied the system guanosine-cytidine in DMSO-d_6. PMR has shown that these compounds form dimers by hydrogen bonding.[21] Under optimal conditions for hydrogen bonding, we observed low field shifts in the hydrogen-bonded carbonyls of 0.27 ppm (cytidine) and 0.66 ppm (guanosine). Displacements of this magnitude, in addition to displacements caused by other effects of cyclization, makes unambiguous assignment of the carbonyl resonances difficult. In an attempt to resolve this problem, we studied a series of closely related hormones.

Figure 1 compares the CMR spectra of oxytocin and DLVP, which differ in three positions. Expected differences are the two Ile methyl resonances at 12.8 and 17.1 ppm in the spectrum of oxytocin, the five aromatic Phe resonances centered around 130 ppm in the spectrum of DLVP and the cysteine in position 1, in which the NH_2 has been replaced by H. There are of course greater differences between the spectra of oxytocin and the vasopressins, than there are between the spectra of the various vasopressins themselves. In particular, the α resonances of Asn and Cys-6 differ by +1.2 and −0.4 ppm, respectively, on comparing oxytocin and DLVP. Most of the other spectral differences can now be interpreted, although there remain uncertainties regarding our assignments in the carbonyl region.

Elevation of the temperature of oxytocin solutions in DMSO-d_6 results in changes in only a few resonances. In particular, the β resonance of Gln and the α resonance of Ile move towards the values found for the protected nonapeptide of oxytocin. This suggests that both the Gln β and Ile α resonances are conformationally sensitive.

Comparison of the CMR spectra of LVP and AVP reveals very few differences in the chemical shifts of the amino acids held in common.

Differences in the spectra of the various hormones in
D_2O or DMSO-d_6 parallel for the most part the solvent-induced
changes in the uncyclized peptides. We have therefore no
evidence for any dramatic conformational differences for the
hormones in these two solvents; however, this may be a
limitation in the sensitivity of ^{13}C chemical shifts to
small conformational changes in peptides.

Conclusion

The CMR spectra of the nonapeptide hormones manifest
the expected high resolution. By studying the constituent
oligopeptides almost all resonances can be assigned. For
some resonances a study of slightly modified hormones is
required for a complete assignment. Carbonyl and α-carbon
resonances appear to be conformationally sensitive, but
models for the observed behaviour cannot be constructed
until further information is available on conformational
influences on ^{13}C chemical shifts. Using the chemical
shifts arbitrarily, one can say that the conformations of
LVP, DLVP and AVP are very similar. It appears that the
conformations of the hormones are similar in DMSO-d_6 and
D_2O.
Although the conformational effects on ^{13}C chemical
shifts are relatively small (1-2 ppm), they should be useful
for determining the structures of such hormones in solution.
Considerable difficulty may be encountered with larger peptides, however, and enrichment of specific amino acids in
^{13}C may be necessary. The measurement of relaxation times
in such systems should also prove fruitful, especially in
those with enriched amino acids where the time required for
the experiment becomes reasonable. Couplings between carbon
and hydrogen also provide hope for future conformational
insight via CMR.

ACKNOWLEDGMENT

This work was supported by U.S.P.H.S. grant AM-13567
and by the National Research Council of Canada.

References

1. Johnson, L. F., I. L. Schwartz, and R. Walter. Proc.
   Nat. Acad. Sci. U. S. <u>64</u>, 1269 (1969).
2. Urry, D. W., M. Ohnishi, and R. Walter. Proc. Nat.
   Acad. Sci. U. S. <u>66</u>, 111 (1970).
3. Urry, D. W., and R. Walter. Proc. Nat. Acad. Sci.
   U. S. <u>68</u>, 956 (1971).

4. Deslauriers, R., and I. C. P. Smith. Biochem. Biophys. Res. Comm. <u>40</u>, 179 (1970).

5. Walter, R. In *Structure-Activity Relationships of Protein and Polypeptide Hormones,* part 1, ed. M. Margoulies and F. C. Greenwood (Amsterdam: Excerpta Medica, 1971) pp 181-193.

6. Von Dreele, P. H., A. I. Brewster, H. A. Scheraga, M. F. Ferger, and V. du Vigneaud. Proc. Nat. Acad. Sci. U. S. <u>68</u>, 1028 (1971).

7. Von Dreele, P. H., A. I. Brewster, F. A. Bovey, H. A. Scheraga, M. F. Ferger, and V. du Vigneaud. Proc. Nat. Acad. Sci. U. S. <u>68</u>, 3088 (1971).

8. Walter, R., J. D. Glickson, I. L. Schwartz, R. T. Havran, J. Meienhofer, and D. W. Urry. Proc. Nat. Acad. Sci. U. S. <u>69</u>, 1920 (1972).

9. Stothers, J. B. *Carbon-13 NMR Spectroscopy* (New York: Academic Press (in press)).

10. Pregosin, P. S. and E. W. Randall. In *Determination of Organic Structures by Physical Methods,* Vol. 4, ed. F. C. Nachod and J. J. Zuckerman (New York: Academic Press, 1971) pp 263-322.

11. Lemieux, R. U., T. L. Nagabhusan, and B. Paul. Can. J. Chem. <u>50</u>, 773 (1972).

12. Mantsch, H. H. and I. C. P. Smith. Biochem. Biophys. Res. Comm. <u>46</u>, 808 (1972).

13. Smith, I. C. P., H. H. Mantsch, R. D. Lapper, R. Deslauriers, and T. Schleich. In *Conformation of Biological Molecules and Polymers,* ed. B. Pullman and E. Bergmann (Israel Academy of Arts and Science, Jerusalem (in press)).

14. Allerhand, A., D. Doddrell, and R. Komoroski. J. Chem. Phys. <u>55</u>, 189 (1971).

15. Ohnishi, M., M. C. Pedarko, J. D. Baldeschwieler, and L. F. Johnson. Biochem. Biophys. Res. Comm. <u>46</u>, 312 (1972).

16. Deslauriers, R., Ph. D. Thesis, Department of Biochemistry, University of Ottawa, 1972.

17. Walter, R., unpublished results.

18. Hruby, V. J., A. I. Brewster, and J. A. Glasel. Proc. Nat. Acad. Sci. U. S. <u>68</u>, 450 (1971).

19. Maciel, G. E. and G. C. Ruben. J. Amer. Chem. Soc. <u>85</u>, 3903 (1963).

20. Maciel, G. E. and J. J. Natterstad. J. Chem. Phys. <u>42</u>, 2752 (1965).

21. Newmark, R. A. and C. R. Cantor. J. Amer. Chem. Soc. <u>90</u>, 5010 (1968).

ENERGY REFINEMENT OF THE NMR STRUCTURE OF OXYTOCIN AND ITS
CYCLIC MOIETY

David Kotelchuck. Department of Chemistry, Cornell
University, Ithaca, N.Y. 14850 and Department of
Physiology, Mt. Sinai School of Medicine of the City
University of New York, New York, N.Y. 10029.

Harold Scheraga. Department of Chemistry, Cornell
University, Ithaca, N.Y. 14850.

Roderich Walter. Department of Physiology, Mt. Sinai
School of Medicine of the City University of New York,
New York, N.Y. 10029.

IT HAS PREVIOUSLY BEEN SHOWN[1,2] that the combined use of
conformational energy calculations and NMR measurements
can resolve ambiguities present in both techniques, and
thereby provide structural information about polypeptides
in solution. The main ambiguity in the computational
method [which was encountered in earlier calculations on
oxytocin[3]] is the existence of many local minima in the
energy surface of the peptide; in the NMR method a major
ambiguity presently resides in the inability to obtain a
unique structure from the limited available experimental
data. However, if a possible structure is proposed for a
polypeptide on the basis of the NMR data, its energy and
that of its variants can be examined to assess the possible validity of the proposed structure. This was done
for the proposed NMR structure of oxytocin,[4] which consists
of two β-turns, involving the sequences Tyr-Ile-Gln-Asn in
the six-residue ring and Cys-Pro-Leu-Gly-NH$_2$ in the acyclic
tail (Figure 1). The resulting conformation is valid only
if the proposed structure is close to the correct one, and

35

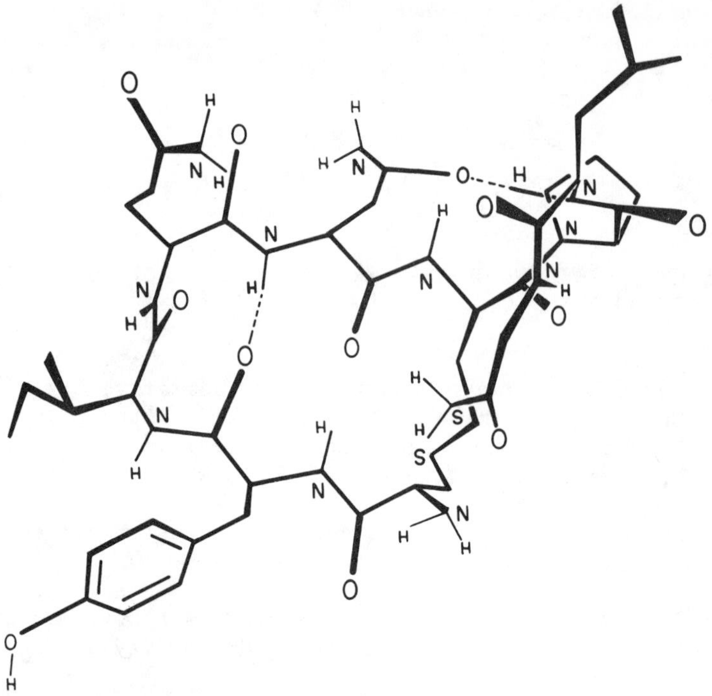

Figure 1: Proposed conformation of oxytocin in DMSO.[4]

if reliable geometry parameters and energy functions are
used in the computations.

The isolated six-residue ring of oxytocin, the cyclic
moiety, was examined first.   Initial dihedral angles were
measured from space-filling models based on the proposed
structure.[4]   Twenty related conformations were generated
by allowing the dihedral angles to vary randomly up to
$\pm 30°$ from these measured values.   After minimization, none
of the 20 conformations had two backbone hydrogen bonds in
the ring, and only four had a single hydrogen bond.   Furthermore, the average final energy of the four conformations
with a single hydrogen bond was the same as that for conformations without a hydrogen bond.   To assure that no
low-energy hydrogen bonded states existed that were in-
accessible from the above conformations because of the
existence of local minima, an exhaustive examination of
conformation space was conducted looking for structures
with hydrogen bonds.   Out of 1134 ring conformations

examined, only 16 had geometries with two hydrogen bonds, of which four were selected for further study (Table I).

Table I

Selected Conformations of the Tyr-Ile-Gln-Asn Sequence with Two Hydrogen Bonds

Confor-mation Number	Tyr		Ile		Gln		Asn	
	ϕ	ψ	ϕ	ψ	ϕ	ψ	ϕ	ψ
1	-140	170	-50	-40	-120	30	-140	140
2	-170	170	-50	120	120	-30	-170	140
3	-140	170	-70	-20	-90	0	-170	110
4	-140	170	-70	120	90	0	-140	110

Energy minimization of these four broke all initial hydrogen bonds, except for one weak hydrogen bond in the Conformation 3 of Table I. The final energies of the four conformations were the same or greater than the average of the 20 conformations examined above. When the acyclic tripeptide tail was added to each of the four conformations, forming a complete oxytocin molecule, all but one of the initial ring hydrogen bonds broke during energy minimization. We conclude that oxytocin conformations with two hydrogen bonds in the ring do not form, while conformations with a single hydrogen bond in the ring can form, but have the same energies as those without such bonds. This result is consistent with the one hydrogen bond proposed for the ring of oxytocin,[4] but in disagreement with the two hydrogen bonds proposed for the ring of deamino-oxytocin.[5] The discrepancy could be due to differences in solvation between DMSO and water—the measurements were made in DMSO and the calculations for water as a solvent. (Calculations ruled out any direct effect of the terminal amino group on hydrogen bonding in the ring.)

As regards the tail of oxytocin, in three of the four conformations considered the proposed hydrogen bond between the Gly peptide NH and the Cys-6 C=O remained intact after minimization. The hydrogen bond between the Leu peptide NH and the Asn side chain C=O, which was proposed to form when the tail is over the ring, did not form in any of these cases. However, a study of the dependence of the energy on the angle ψ of the Cys-6 residue shows an allowed conformation with the tail over the ring, as in the proposed structure,[4] as well as one with the tail away from the ring. The Tyr side chain also appears to have two allowed orientations--one folded over the ring, and the other stretched away from it.

In summary, the structures calculated for oxytocin are not inconsistent with the proposed NMR structure, making allowance for the use of water as a solvent in the calculations and DMSO in the NMR studies. However, the range of allowed conformations, all of roughly equal energy, suggests that there may be considerable flexibility in the oxytocin molecule. Fuller details are given in reference 6.

References

1. Silverman, D. N., and H. A. Scheraga. Biochemistry 10, 1340 (1971).
2. Gibbons, W. A., G. Némethy, A. Stern, and L. C. Craig. Proc. Nat. Acad. Sci. U. S. 67, 239 (1970).
3. Gibson, K. D., and H. A. Scheraga. Proc. Nat. Acad. Sci. U. S. 58, 1317 (1967).
4. Urry, D. W., and R. Walter. Proc. Nat. Acad. Sci. U. S. 68, 956 (1971).
5. Urry, D. W., M. Ohnishi, and R. Walter. Proc. Nat. Acad. Sci. U. S. 66, 111 (1970).
6. Kotelchuck, D., H. A. Scheraga, and R. Walter. Proc. Nat. Acad. Sci. U. S., in press (1972).

CHEMISTRY AND BIOLOGY OF PEPTIDES

© Copyright 1972 *Ann Arbor Science Publishers*

PROTON AND ^{13}C NMR STUDIES OF CONFORMATIONS OF *CYCLO* (-PRO-GLY-)$_3$

C. M. Deber, E. R. Blout. Department of Biological Chemistry, Harvard Medical School, Boston, Massachusetts 02115

D. A. Torchia, * *D. E. Dorman, F. A. Bovey.* Bell Laboratories, Murray Hill, New Jersey 07974

IN CONTRAST TO SEVERAL CYCLIC hexapeptides whose solution conformations have been studied,[1-4] the presence in *cyclo* (-Pro-Gly-Pro-Gly-Pro-Gly-) (abbreviated *c*-(PG)$_3$) of three L-prolyl residues in alternating positions precludes the formation of intramolecularly hydrogen-bonded antiparallel β-type structures. Instead, this peptide may have a conformation with a 3-fold symmetry and in this way resembles *cyclo*(tri-L-prolyl).[5] In *c*-(PG)$_3$, however, glycyl residues intervene between each prolyl residue, resulting in a situation where the three Pro-Gly peptide bonds must be *trans*, while the three Gly-Pro peptide bonds may be either *cis* or *trans*.[6,7]

220 MHz proton NMR spectra have shown[8] that *c*-(PG)$_3$ dissolved in methylene chloride-d$_2$ (CD$_2$Cl$_2$) exists predominantly in a C$_3$-symmetric conformation, as judged by the magnetic equivalency of the three Pro-Gly units. The Gly NH region of the spectrum, shown in Figure 1a, consists of a single resonance indicated as S, coupled to the Gly C$_\alpha$H$_2$ protons with J$_{N\alpha}$ of 2.5 and 4.0 Hz. This information, taken in conjunction with symmetry requirements and other indications from molecular models, suggests that the backbone

---

*Present address: Polymer A-209, National Bureau of Standards, Washington, D. C. 20234

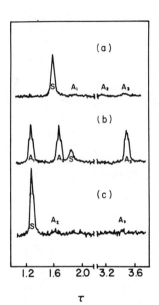

Figure 1: Peptide NH region of the proton NMR spectra of
  cyclo(-Pro-Gly-)$_3$ at 220 MHz. Solvents: (a) CD$_2$Cl$_2$;
  (b) DMSO-d$_6$; (c) DMSO-d$_6$ + NaSCN. In (c), the molar
  ratio NaSCN/*c*-(PG)$_3$ ≈ 7. Chemical shifts (τ scale)
  given in ppm downfield from TMS.

conformation of *c*-(PG)$_3$ in methylene chloride consists of
three *cis* Gly-Pro peptide bonds, three *cis*' Pro C$_\alpha$-C=O
bonds, and three sets of Gly (ϕ,ψ) angles of *ca.* (0°,0°).
    The striking sensitivity of *c*-(PG)$_3$ conformations to
solvent is seen in Figure 1b. Upon changing from CD$_2$Cl$_2$
(1a) to dimethylsulfoxide-d$_6$ (DMSO-d$_6$), the three Gly NH
protons now appear as three resonances of equal area,
demonstrating that the predominant structure is asymmetric
in DMSO-d$_6$ solution. Mixed CD$_2$Cl$_2$- DMSO-d$_6$ solvent studies
demonstrated[8] a dynamic equilibrium between the two *c*-(PG)$_3$
conformers, and minor populations of each conformer can be
seen in both pure solvents (Figures 1a and 1b).

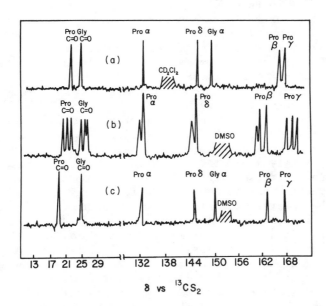

Figure 2: ^{13}C NMR spectra of *cyclo*(-Pro-Gly-)$_3$ at 25.2 MHz. Solvents: (a) CD_2Cl_2; (b) DMSO-d$_6$; (c) DMSO-d$_6$ + NaSCN. In (c), the molar ratio NaSCN/*c*-(PG)$_3$ ≈ 7. The Gly C$_\alpha$ resonances are obscured by solvent in (b). A resonance for the SCN$^-$ carbon atom appeared in (c) near 62 ppm. Chemical shifts given in ppm upfield from $^{13}CS_2$. ^{13}C spectra have been redrawn from the original spectra for clarity.

Addition of sodium thiocyanate to DMSO-d$_6$ solutions in increasing molar amounts converts the asymmetric structure to a *new C$_3$-symmetric conformer*, as shown in Figure 1c, where ~90% of *c*-(PG)$_3$ molecules are C3-symmetric when the molar ratio NaSCN/*c*-(PG)$_3$ ≈ 7. This transformation may be attributed to the formation of a stable cyclic peptide-sodium thiocyanate complex, and analysis of its complete NMR spectrum suggests[8] a backbone conformation containing three *trans'* Pro C$_\alpha$-C=O bonds and three *trans* Gly-Pro peptide bonds. Other alkali metal cations also convert *c*-(PG)$_3$ to C3-symmetric conformers in DMSO-d$_6$. Under

conditions similar to those used with NaSCN, the following
percentages of complexed cyclic peptide were observed:
K^+, 75%; Li^+ = Rb^+, 50%; and Cs^+, 25%. Models suggest that
either the three Gly carbonyl groups and/or the three Pro
carbonyl groups in the all *trans'/trans* c-(PG)$_3$ conformer
are well-oriented to bind metal cations, but the overall
stoichiometry of the complex is presently undetermined. A
2:1 peptide-cation "sandwich" complex, where the ion is
bound to six peptide carbonyl oxygens between two cyclic
peptide molecules, seems possible.

In addition to the proton spectra of c-(PG)$_3$, it was of
interest to examine the corresponding ^{13}C NMR spectra to
ascertain if these various conformational transitions could
be effectively monitored with this method; if further conformational information would become available; and to
determine if chemical shift differences between the spectra
could be related to conformational features with confidence.
Figure 2a, b, c shows the complete ^{13}C spectra of c-(PG)$_3$
under the conditions corresponding to Figure 1a, b, c.
Both the Gly and Pro peptide carbonyl carbon atoms may be
seen in 2a, whereas in proton NMR, proline residues give
no resonance in the peptide NH region. (Assignments of
carbonyl resonances were made by synthesizing c-(PG)$_3$ ~60%
enriched with ^{13}C in one Gly carbonyl of the linear hexapeptide precursor Gly-Pro-Gly-Pro-Gly-Pro-nitrophenyl ester
hydrochloride, which gives, after cyclization, c-(PG)$_3$ with
each Gly carbonyl enriched ~20%.)

The asymmetry induced by dissolving c-(PG)$_3$ in DMSO-d$_6$
is reflected clearly in the carbonyl region of ^{13}C spectrum
2b, where three carbonyl resonances are present for *both*
the three non-equivalent Gly carbonyl carbons and the non-
equivalent Pro carbonyls. Furthermore, when sodium thiocyanate is added to the DMSO-d$_6$ solution of c-(PG)$_3$ in about
7-molar excess, spectrum 2-c demonstrates the transformation
back to three-fold symmetry, as observed in proton NMR. The
downfield movement of the Pro carbonyl carbon while the Gly
remains constant (compare 2a versus 2c) suggests involvement
of the Pro carbonyls with salt. A similar effect has been
recently observed in valinomycin-K^+ complexation.[9]

The upfield portions of spectra 2a and c also indicate
the C$_3$-symmetry present in contrast to the asymmetry evident
in 2b. Assignments of resonances are made by comparison
with a number of proline ^{13}C spectra.[10] Chemical shifts
are generally in regions "normally" observed for most proline peptides,[10] with the notable exception of the Pro C$_\beta$-
carbon resonance in 2a, which falls about 4 ppm upfield
from the position expected for a proline residue involved

in a *cis* peptide bond. Since c-(PG)$_3$ dissolved in CD$_2$Cl$_2$ is a proline peptide proposed to contain *cis'* Pro C$_\alpha$-C=O bonds, the upfield shift may be diagnostic for this conformational feature which, to date, has not been directly observed experimentally.

ACKNOWLEDGMENT

The work at Harvard was supported, in part, by U.S. Public Health Service Grants AM-07300 and AM-10794.

References

1. Schwyzer, R., and U. Ludescher. Helv. Chim. Acta 52, 2033 (1969).
2. Kopple, K. D., M. Ohnishi, and A. Go. J. Amer. Chem. Soc. 91, 4264 (1969).
3. Torchia, D. A., A. Di Corato, S. C. K. Wong, C. M. Deber, and E. R. Blout. J. Amer. Chem. Soc. 94, 609 (1972).
4. Torchia, D. A., S. C. K. Wong, C. M. Deber, and E. R. Blout. J. Amer. Chem. Soc. 94, 616 (1972).
5. Deber, C. M., D. A. Torchia, and E. R. Blout. J. Amer. Chem. Soc. 93, 4893 (1971).
6. Madison, V., and J. A. Schellman. Biopolymers 9, 511 (1970).
7. Deber, C. M., F. A. Bovey, J. P. Carver, and E. R. Blout. J. Amer. Chem. Soc. 92, 6191 (1970).
8. Deber, C. M., D. A. Torchia, S. C. K. Wong, and E. R. Blout. Proc. Nat. Acad. Sci. U. S. 69, 1825 (1972).
9. Ohnishi, M., M. C. Fedarko, J. D. Baldeschwieler, and L. F. Johnson. Biochem. Biophys. Res. Comm. 46, 312 (1972).
10. Bovey, F. A. In *Chemistry and Biology of Peptides, Proceed. 3rd American Peptide Symposium,"* Meienhofer, J. ed. (Ann Arbor, Michigan: Ann Arbor Science Publishers, 1972), pp 3-28.

CARBON-13 NUCLEAR MAGNETIC RESONANCE OF SOME PENTAPEPTIDES

Frank R. N. Gurd, Philip Keim, V. G. Glushko, P. J. Lawson, * *R. C. Marshall, A. M. Nigen, and R. A. Vigna.*
Department of Chemistry, Indiana University, Bloomington, Indiana

THE FOLLOWING PROGRESS REPORT deals with ^{13}C NMR studies on a series of pentapeptides. The peptides were constructed as models for comparison with observations on proteins. A previous report has indicated that the central residue of a pentapeptide is fairly well isolated from terminal-residue effects, whereas the central residue of a tripeptide may not be so well isolated.[1] The peptides reported here all have glycyl residues in the first, second, fourth and fifth positions with some particular L-amino acid represented in the third position. The solid phase technique was used for synthesis, with *t*-Boc-diglycine employed to reduce the number of steps. Cleavage from resin and deblocking were done in trifluoracetic acid-HBr. Several standard variants were used and will be presented in full publications. Purity was checked by amino acid analysis, thin-layer chromatography and by the ^{13}C NMR itself.

Samples were prepared of the order of $1M$ solutions in water with an internal dioxane standard. The pH was adjusted with $5N$ NaOH or $6N$ HCl and measured at 25°. NMR observations of the ^{13}C nucleus at natural abundance were made at 15.1 MHz under ^{1}H decoupling conditions. The pulsed Fourier transform mode was used. Resolution was taken to about 0.12 ppm. Chemical shifts were expressed as parts per million upfield from CS_2,[1] with dioxane taken as 126.3 ppm on this scale. Measurements were made at 26 to 28°. The pulse-delay sequence (Inversion Recovery Method) for measurements of the spin-lattice relaxation time, T_1, has been described.[2-4]

*Present Address: Department of Chemistry, University of Queensland, St. Lucia, Qld. 4067, Australia

45

The Broader Objective

Before showing the types of results obtained, it is useful to have in mind the objectives in protein chemistry that this study is intended to serve. These are to establish chemical shift positions to be expected for the carbon nuclei in amino acid residues in the interior and at the ends of polypeptide sequences, to determine the effects of changes in protonation state at nearby sites, and to gain a picture of the T_1 values observed in a short, free segment of peptide chain. In the long term the most important applications to protein studies will probably come from enrichment in which the skill of the peptide synthetic chemist will be decisive. Carboxymethylation with enriched bromoacetate has recently shown that a single enriched locus can be observed by ^{13}C NMR with such proteins as myoglobin and ribonuclease A.[5]

Chemical Shifts

The main point to note about the chemical shift results is that a given carbon nucleus in a central residue of a pentapeptide does seem to fall in a position matching its attribution in a denatured protein spectrum.[4] Of nearly equal importance is the strong pH dependence of the shifts of nuclei close to sites of variable protonation. Some examples of magnitude and direction of changes in chemical shift with pH, and of the corresponding pK values, are shown in Table I. Such effects may prove useful in proteins for observing the influence of an ionizing group on a near neighbor, especially if the latter is enriched with respect to ^{13}C.

Relaxation Studies

In addition to a number of spectral measurements at various pH values to obtain the information outlined above, relaxation measurements have been made to obtain T_1 values, often at two pH values for each pentapeptide. The technique is like other perturbation techniques in which the time course of the reestablishment of equilibrium is followed. For molecules in the size range under consideration the $^{13}C-^{1}H$ dipolar relaxation mechanism is dominant for carbons attached directly to protons. The spin-lattice relaxation time, T_1, describes this exponential process.[3,6] For molecules in this size class it is also characteristic that

Table I

Chemical Shifts and pK Values for Some Central Amino Acids
in Gly–Gly–X–Gly–Gly Pentapeptides

All chemical shifts are expressed as ppm upfield of CS_2.
Titrations were performed on *ca.* 1*M* peptide solutions and
pK values were obtained by computer fitting to a simple
Henderson-Hasselbalch curve

| | Tyrosine | | | Histidine | | |
Carbon	$\delta Acid$	$\delta Base$	pK	$\delta Acid$	$\delta Base$	pK
C^γ	64.9	71.7	10.01	64.4	60.0	6.76
C^δ	62.4	62.4	--	75.4	75.4	--
C^ε	77.4	74.1	9.95	59.2	56.6	6.75
C^ζ	38.2	27.7	10.02			

those protonated carbon nuclei undergoing the most rapid
tumbling motion will generally show the longest values of T_1.

Figure 1 is a diagram showing T_1 values in milliseconds
for the pentapeptides containing the following central
residues: (A) alanine, (B) lysine and (C) tyrosine. The
T_1 values shown are the actually measured values multiplied
by the number of directly attached hydrogens for direct
comparison (*i.e.* NT_1). In each case the peptide backbone
is represented only by the α-carbons. All side chain car-
bon values are shown, including C^γ and C^ζ of tyrosine which
are not directly protonated and have much longer values of
T_1 that are not interpretable in the same terms.

For each of the three cases, and for the others that
we have studied as well, the α-carbon NT_1 values are greater
at the ends of the peptide chains than in the center, with
intermediate values for C^α of residues 2 and 4. The result
for C^β of the alanine peptide presumably reflects contribution from the spinning of the methyl group around the C^α–C^β
axis. The side chain carbon nuclei of the lysine residue
show a gradation in NT_1 that fits qualitatively with the
idea that rotational motion is freer the farther out the
chain that one considers. This idea parallels the argument
in explanation of the longer values for the C^α nuclei of

(A)

Gly-Gly-Ala-Gly-Gly

670-400-294-400-662 α

1350 β

(B)

Gly-Gly-Lys-Gly-Gly

508-302-180-302-524 α

232 β

432 γ

690 δ

914 ε

(C)

Gly-Gly-Tyr-Gly-Gly

662-258-180-258-598 α

232 β

1762 γ

242　　242 δ

242　　242 ε

1687 ζ

Figure 1: Schematic representation of NT_1 values for the pentapeptides in neutral solution: A, glycylglycyl-L-alanylglycylglycine; B, glycylglycyl-L-lysylglycylglycine; C, glycylglycyl-L-tyrosylglycylglycine. The values for α-carbons are shown horizontally.

the terminal glycine residues compared to the more central residues. These interpretations follow those of Allerhand and his coworkers in similar situations.[3,6] The observations on the tyrosine peptide are interesting in that NT_1 is comparable for all the hydrogen-bearing carbons of the side chain. The simplest explanation is that here motion is dominated by components in which the side chain revolves about its axis of attachment to the peptide chain, again a type of behavior observed in a somewhat similar system by Allerhand's group.[7] Information of this type, treated with suitable caution, should be of value in interpreting observations on proteins.[8]

ACKNOWLEDGMENT

Supported by Public Health Service Grants HL-05556 and HL-14680.

References

1. Gurd, F. R. N., P. J. Lawson, D. W. Cochran, and E. Wenkert. J. Biol. Chem. 246, 3725 (1971).
2. Allerhand, A., D. Doddrell, V. Glushko, D. W. Cochran, E. Wenkert, P. J. Lawson, and F. R. N. Gurd. J. Amer. Chem. Soc. 93, 544 (1971).
3. Allerhand, A., D. Doddrell, and R. Komoroski. J. Chem. Phys. 55, 189 (1971).
4. Glushko, V., P. J. Lawson, and F. R. N. Gurd. J. Biol. Chem. 247, 3176 (1972).
5. Nigen, A. M., P. Keim, R. C. Marshall, J. S. Morrow, and F. R. N. Gurd. J. Biol. Chem. 247, 4100 (1972).
6. Doddrell, D., V. Glushko, and A. Allerhand. J. Chem. Phys. 56, 3683 (1972).
7. Allerhand, A., and R. K. Hailstone. J. Chem. Phys. 56, 3718 (1972).
8. Gurd, F. R. N., and P. Keim. Methods Enzymol., in press (1972).

STUDIES ON THE CYCLIZATION TENDENCY OF PEPTIDES

M. Rothe, R. Theysohn, D. Mühlhausen, F. Eisenbeiß, and W. Schindler. Organisch-Chemisches Institut, Universität Mainz, Germany

CYCLO-PEPTIDES CAN BE synthesized according to two basic approaches: (1) by ring closure of activated linear peptides at high dilution, and (2) by insertion reactions into preformed rings, *e.g.* by aminoacyl incorporation into diketopiperazines.

As to the first approach, the phosphite method has proved to be particularly useful in our group.[1] In the meantime we succeeded in preparing biological active *cyclo*-peptides, *e.g.* gramicidin S and its analogues, fungisporin (M. Zamani) and valinomycin oligomers (W. Kreiss). Unprotected peptides react with a chlorophosphite or pyrophosphite to give a mixed anhydride which cyclizes spontaneously. The main advantages of this method are the experimental simplicity and the high yields up to 80%.

We also used this method as well as the active ester cyclization in our investigations on the formation tendency of *cyclo*-peptides. In particular, the influence of the chain length of the peptides as well as the peptide concentration, and the nature and configuration of the amino acids have been studied.

As is well known, the cyclization tendency is particularly high in the case of the 6-membered *cyclo*-dipeptides. On the contrary, 7-membered peptide rings containing one β-amino acid are not formed spontaneously. The product obtained by Sekiguchi[2] from β-alanyl-glycine using the azide method has been found not to be the 7-membered *cyclo*-dipeptide, as envisaged, but the 14-membered *cyclo*-tetrapeptide (mp<350°). Its structure was determined by

51

incorporation of two β-alanine residues into glycine diketopiperazine. Using the phosphite method, the much more soluble 7-membered ring (mp 171-172°) could be isolated by gel filtration. In more concentrated solutions (0.01 M), the 14- and 21-membered dimers and trimers are formed in yields of 28% and 4.5%, respectively (monomer: 28%).

If glycine is replaced by proline which contains a rigid N-C$_\alpha$ bond, the cyclization tendency will be increased considerably. Even at the relatively high concentration of 0.01 M, 41% of the highly strained *cyclo*(-Pro-βAla-) was obtained without any formation of higher ring homologues. Model studies show that both peptide bonds must deviate considerably from planarity (by about 30% each).

It is well known that 9-membered *cyclo*-tripeptides can only be obtained in exceptional cases from linear peptides containing tertiary peptide bonds. As the first example, we could synthesize *cyclo*-triprolyl[3] in yields of more than 80%. Later on, *cyclo*-trisarcosyl has been prepared by Dale.[4] In order to obtain a series of *cyclo*-tripeptides in which proline is successively replaced by sarcosine, we now have synthesized *cyclo*(-Sar-Pro-Pro-) and *cyclo*(-Sar-Sar-Pro-).

During the cyclization of the tripeptides containing sarcosine, a series of oligomeric ring peptides is formed, even at high dilution, in addition to the *cyclo*-tripeptides (Figure 1). In more concentrated solution, the corresponding 18-, 27-, and 36-membered cyclic hexa-, nona-, and dodecapeptides could readily be isolated by gel chromatography. The amount of the higher homologues increases with increasing flexibility of the peptide chain, *i.e.* with decreasing content of the rigid proline rings. Accordingly, *cyclo*-trisarcosyl has been formed only in small amounts in addition to cyclic hexa-, nona-, and dodecasarcosyl. On the contrary, the introduction of 1 or 2 proline residues leads to *cyclo*-tripeptides, even in considerably higher concentrations (0.01 M), along with the higher rings up to dodecapeptides.

As side products, the corresponding diketopiperazines are formed in each case by peptide cleavage, especially at higher temperatures. This is pronounced in the cyclization of trisarcosine with predominant formation of *cyclo*-disarcosyl. The prevailing formation of *cyclo*(-Pro-Sar-) from Pro-Sar-Sar active esters shows that the cleavage is effected essentially from the amino end (cf. [5]).

Table I shows that the formation of cyclic tripeptides increases very much with increasing number of proline residues, even in relatively concentrated solution. The high

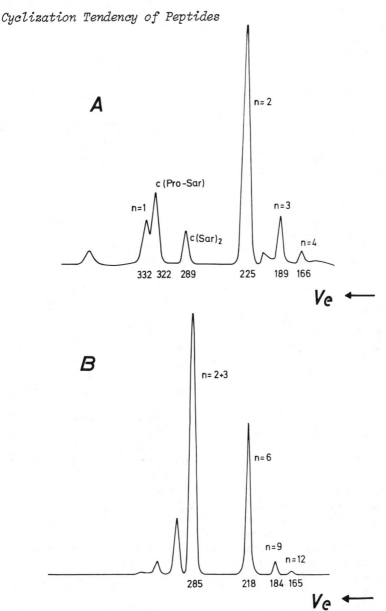

Figure 1: Gel chromatography of tripeptide cyclization
reactions. (A) Pro-Sar-Sar-OPcp (*c*, 0.01 *M* in pyridine)
giving *cyclo*(-Pro-Sar-Sar-)$_n$ (B) Sar-Sar-Sar-OPcp (*c*,
0.001 *M* in pyridine) giving *cyclo*(-Sar-)$_n$ V_e, elution
volume, Merckogel PGM 2000 column, 1.7 x 100 cm, water

Table I

Formation of Cyclic Tripeptides (and Hexapeptides)
from Linear Tripeptides*

| | $yield\ in\ \%$ $c^{(M)}$ | |
	<<0.01	0.01
Pro–Pro–Pro	83	60 (–)
Sar–Pro–Pro	28	(38)
Sar–Sar–Pro	11	3 (33)
Sar–Sar–Sar	16	– (38)

*Cyclization of the tripeptide pentachlorophenyl esters in
pyridine

yields found in the synthesis of *cyclo*-triprolyl can be
attributed to the easy occurrence of an all-*cis* conformation
of the linear tripeptide resembling the transition state of
the cyclization. According to model considerations as well
as NMR and CD studies it corresponds to one helix turn in
polyproline I.

The yields of the *cyclo*-hexapeptides formed are given
in parentheses. *Cyclo*-hexaprolyl is not formed at all because the chain ends of hexaproline are fixed in a rigid helix
with two turns and therefore cannot react with each other.

A similar behaviour is shown by Sar–Pro–Pro and Sar–
Sar–Pro although these peptides are less rigid and can
increasingly form *cyclo*-hexapeptides. It is quite interesting that the CD spectra of these *cyclo*-tripeptides
(Figure 2) resemble strongly those of the corresponding
polytripeptides. From these results and from mutarotation
studies we assume that *poly*(Sar–Pro–Pro) exists in two
solvent-dependent conformations similar to polyproline.
Hence, the peptide chromophores of the form I polymer must
be similarly arranged as in the rigid *cyclo*-tripeptide
which corresponds to a helix turn with the same number of
units and the pitch of zero.

Introduction of 1 or 2 β-alanine residues into linear
tripeptides should lead to medium-sized *cyclo*-peptides
with 10, and 11 ring atoms. The phosphite cyclization of
Gly–Gly–βAla and of Gly–βAla–βAla in 0.01 M solution, however, led to a predominant formation of the corresponding
20- and 22-membered *cyclo*-hexapeptides. In addition, the

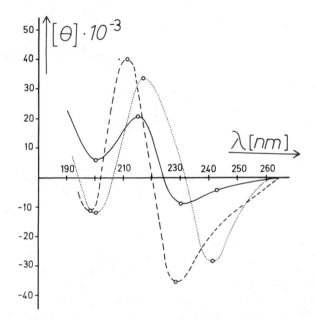

Figure 2: CD Spectra of cyclic tripeptides.
  cyclo(-Pro-Sar-Sar-) ------
  cyclo(-Pro-Pro-Sar-) oooooo
  cyclo(-Pro-Pro-Pro-) _____
    in trifluoroethanol

30- and 40-membered ring homologues could again be separated
by gel filtration. However, the less strained 11-membered
cyclo-tripeptide with two β-alanine residues is readily
formed, in contrast to the 10-membered *cyclo*-tripeptide of
Gly-Gly-βAla which could not be found at all.

   On the other hand, we succeeded in preparing this
cyclo-tripeptide as well as others, *e.g.* *cyclo*(-βAla-Gly-
Pro-), by aminoacyl incorporation from β-alanyl diketopiperazines with the intermediate formation of cyclols. The
synthesis of 9-membered *cyclo*-tripeptides, however, was
not possible in this way.

   Instead of the *cyclo*-tripeptides, bicyclic acylamidines
(I) or ketene aminals (II) have been obtained in this case
by elimination of water[6] (Equation 1). They have been
identified by spectroscopic methods in the case of the
reaction products of the *N*-glycyl derivatives of glycine

Equation 1   (1)

anhydride and glycyl-sarcosine anhydride (→I), and of the
N-prolyl derivatives of glycine anhydride, glycyl-sarcosine
anhydride, and glycyl-L-proline anhydride (→II). Because
of the close steric proximity of the peptide groups in the
medium-sized ring, they are formed in a transannular reaction via cyclol intermediates.

Cyclodimerizations may also occur with tetrapeptides
and pentapeptides. We have cyclized by the phosphite method
various linear pentapeptides containing the sequence of
gramicidin S and its analogs in which the D-Phe in position
4 is replaced by L-Phe or Gly[7] (Table II). Again, the

Table II

Gramicidin S and Analogs
Cyclization of Linear Pentapeptides **by** the Phosphite Method*

| *Pht (or Boc)* | *c* | *Yield* | *cyclic* | |
H-Val-Orn-Leu-X-Pro-OH (_M_)	(_M_)	(%)	*Pentapeptide*	*Decapeptide*
X = D-Phe	0.005	87	62	38
	0.01	92	57	43
L-Phe	0.005	60	40	60
	0.007	66	26	74
Gly	0.001	45	91	9
	0.005	67	80	20
	0.01	70	57	43
	0.04	41	54	46

*Diethyl phosphite as solvent; 30 min, 100°, N_2

doubling reaction was found to depend on the peptide concentration. Moreover, the nature and the configuration of the amino acids involved play an important role. Incorporation of a D-amino acid which shortens the distance between the chain ends yielded considerably more cyclic pentapeptide than decapeptide. At the same time, steric interaction caused by bulky side chains apparently exists as can be shown by the remarkable difference in yields between the Gly and the L-Phe analogues.

Finally, the nature of the solvent seems to be very important. During the cyclization of the pentapeptide active esters of gramicidin S in pyridine, Izumiya[8] obtained relatively more *cyclo*-decapeptide than pentapeptide (ratios 55:45, 0.0003 M, and 68:32, 0.003 M, resp.) even in ten-fold higher dilution as compared with our studies. This can be explained by a stronger association occurring in pyridine rather than in diethyl phosphite used in the phosphite method. Accordingly, in the latter solvent we found reversed yield ratios of both the ring peptides under comparable concentrations (38:62, 0.005 M).

References

1. Rothe, M., I. Rothe, H. Brünig, and K.-D. Schwenke. Angew. Chem. 71, 700 (1959).
2. Sekiguchi, H. Compt. Rend. 256, 4012 (1963).
3. Rothe, M., K.-D. Steffen, and I. Rothe. Angew. Chem. 77, 347 (1965).
4. Dale, J. and K. Titlestad. Chem. Commun. 1969, 656.
5. Cf. Meienhofer, J., Y. Sano, and R. P. Patel. In *Peptides: Chemistry and Biochemistry*, Weinstein, B. and S. Lande, eds. (New York, Marcel Dekker, 1970) pp 419-434.
6. Rothe, M., W. Schindler, R. Pudill, M. Kostrzewa, R. Theysohn, and R. Steinberger. Proc. 11th Europ. Peptide Sympos., Wien, 1971, in press.
7. Rothe, M., and F. Eisenbeiß. Z. Naturforsch. 21b, 814 (1966); Angew. Chem., internat. Edit. 7, 883 (1968).
8. Izumiya, N., T. Kato, H. Aoyagi, S. Makisumi, M. Waki, O. Abe, and N. Mitsuyasu. In *Peptides: Chemistry and Biochemistry*, Weinstein, B., and S. Lande, eds. (New York, Marcel Dekker, 1970) pp 45-53.

FORMATION OF CYCLIC DIPEPTIDES AND BI- AND TRICYCLIC
PRODUCTS FROM LINEAR TETRAPEPTIDES

Kirsten Titlestad. Kjemisk Institutt, Universitetet i
Oslo, Oslo 3, Norway

IN THE COURSE OF OUR SYNTHESES of cyclic tetrapeptides for
conformational studies,[1,2,3] we have noticed that not all
linear tetrapeptides cyclize to the corresponding cyclic
tetrapeptides. Cyclic dipeptides and bicyclic and tricyclic
products have all been encountered.

First observed was the formation of cyclic dipeptides
from several linear tetrapeptides of sarcosine combined
with either alanine or glycine.[4] Their 2,4,5-trichloro-
phenyl esters were heated in pyridine, and the results
(Table I) illustrate that the sequence of the linear
tetrapeptide is important both for the size and type of
cyclized products. Whenever cleavage of the peptide bond
occurs, it is always the central amide bond which breaks,
proving that the cyclic dipeptides are formed directly
from the linear peptide and not from an initially formed
cyclic tetrapeptide. It is also obvious that the second
amino acid plays an important role. If it is a *N*-methylated
amino acid (such as sarcosine) a *cis* configuration is
about as likely as a *trans* configuration. The free amino
end (I) can then easily move close to the second carbonyl
group and may react with it to form the unstable cyclol
(II).[5] This splits up to a cyclic dipeptide (III) and a
dipeptide active ester (IV), which subsequently cyclizes
to a second molecule of cyclic dipeptide. The presence of
a *N*-methylated amino acid at the fourth position will, however, mean that the active ester group can also be close to
the free amino end (V), and formation of the cyclic tetrapeptide (VI) will then be preferred to formation of the unstable

59

Table I

Cyclization of Tetrapeptides in Pyridine at 115°

Linear tetrapeptide 1 2 3 4	Cyclic tetrapeptide	yield	m.p.	Cyclic dipeptides, yields
H-Sar-L-Ala-L-Ala-Sar-OTcp	cyclo(-Sar$_2$-L-Ala$_2$-)	10%	290° subl.	cyclo(-Sar$_2$-) 30%, cyclo(-L-Ala$_2$-) 13%
H-Sar-Sar-L-Ala-L-Ala-OTcp	cyclo(-Sar$_2$-L-Ala$_2$-)	none		
H-Sar-Gly-Gly-Sar-OTcp	cyclo(-Sar$_2$-Gly$_2$-)	10%	310°	
H-Sar-Sar-Gly-Gly-OTcp	cyclo(-Sar$_2$-Gly$_2$-)	traces		cyclo(-Sar$_2$-) 25%, cyclo(-Gly$_2$-) 10%
H-Gly-Sar-Sar-Gly-OTcp*	cyclo(-Sar$_2$-Gly$_2$-)	none		cyclo(-Gly-Sar-) 55%
H-Sar-Sar-Sar-Sar-OTcp	cyclo(-Sar$_4$-)	43%	>350°	
H-L-Ala-Sar-Sar-Sar-OTcp	cyclo(-Sar$_3$-L-Ala-)	25%	315° subl.	
H-Gly-Sar-Sar-Sar-OTcp	cyclo(-Sar$_3$-Gly-)	25%	318°	
H-Sar-D-Ala-Sar-L-Ala-OTcp	cyclo(-Sar-D-Ala-Sar-L-Ala-)	30%	>350°	
H-Sar-Gly-Sar-Gly-OTcp	cyclo(-Sar-Gly-Sar-Gly-)	40%	>350°	

OTcp = 2,4,5-trichlorophenyl.
*Also cyclized in dimethylformamide-triethylamine at 25° with the same result, but lower yield of cyclic dipeptide.

cyclol. If, on the other hand, the second amino acid is alanine or glycine, the N-H amide bond prefers the *trans* configuration, cyclol formation no longer occurs and cyclic tetrapeptide is formed (Table I). NMR spectroscopy has proved to be a particularly useful technique for studying these cyclic tetrapeptides. The spectrum of *cyclo*(-Sar₂-L-Ala₂-) is shown (Figure 1) and clearly illustrates the favoured ring conformation with its *cis, trans, cis, trans* arrangement, even when, as in this case, one of the N-H amide bonds is forced into a *cis* configuration.

It was now of interest to try to prepare a cyclic tetrapeptide which does not adopt this conformation. The cyclic tetrapeptide of α-methylalanine was chosen since four trans amide bonds are needed to accommodate the four *gem*-dimethyl groups in their preferred corner positions.

Peptides of α-methylalanine[10] were activated by formation of the 2,4,5-trichlorophenyl ester. Cyclization of the dipeptide yielded the diketopiperazine with unexpected difficulty, while the larger peptides only formed polymers. The more reactive acid chloride was then prepared by treating the tetrapeptide with PCl₅ in acetyl chloride. A precipitate

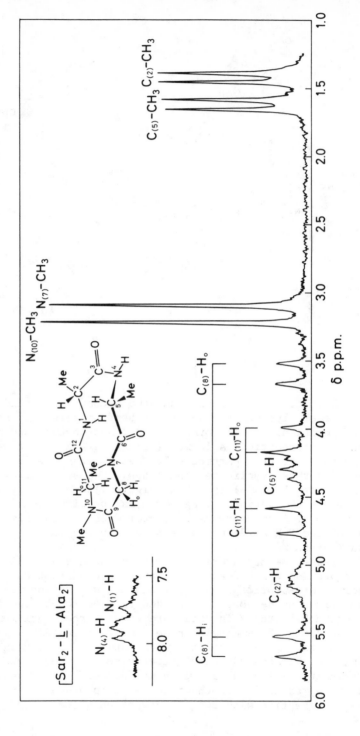

Figure 1:   The 100 MHz NMR spectrum in TFA solution of *cyclo*(-Sar₂-L-Ala₂-).

was deposited, and its infrared spectrum showed bands at 1790 (indicating an acid chloride), 1730 and 1590 cm^{-1}. Attempted cyclizations of this precipitate in pyridine, however, produced two crystalline compounds, a bicyclic imidazolone (XI) and a tricyclic bis-imidazolone (XII). Direct sublimation of the precipitate produced these products in a very low yield.

In marked contrast, treatment of the tripeptide with SOCl$_2$ formed the oxazolone (IX) characterized by an infrared band at 1820 cm^{-1}. Kenner *et al.*[6],[7] had shown that this oxazolone can be cyclized to the bicyclic imidazolone (XI), but this reaction is unique to the tripeptide, as the tetra- and higher peptides formed polymers.

To gain further insight into the mechanism, the methyl groups in α-methylalanine were fully deuterated, and the tri- and tetrapeptides synthesized with one deuterated residue in various positions (Table II). [14]C labelling experiments[7] had shown that the last amino acid in the tripeptide oxazolone ends up in the imidazolone ring, and

Table II

Results of the Deuteration Experiments

	bicyclic XI	*tricyclic XII*
Melting point	255°	253°
Yield from tetrapeptide	26%	10%
Double bond abs. in IR	1730, 1670, 1640 cm^{-1}	1720, 1630 cm^{-1}
NMR shifts (ppm) in CDCl$_3$	1.35, 1.68, 1.78	1.33, 1.89
Relative intensities:		
From tetrapeptide	1 : 1 : 1	2 : 2
From tetrapeptide d$_6$ in 1	1 : 1 : 1	1 : 2
From tetrapeptide d$_6$ in 4	1 : 1 : 0.2	2 : 1
From tripeptide d$_6$ in 3	0 : 1 : 1	– – –

this information allowed us to assign the line at 1.35 δ
(Table II) to the *gem*-dimethyl groups in this ring. By
analogy the lines at 1.33 and 1.89 δ in the tricyclic compound can be assigned to the methyl groups in the five- and
six-membered rings respectively. The reactions of the
deuterated tetrapeptide showed that the imidazolone (XI) is
formed with loss of the first amino acid residue, while in
the bis-imidazolone (XII) takes the specific sites indicated
on the diagram. This absence of scrambling indicates that
the reaction does not go via a mono cyclic tetrapeptide, and
leads us to propose the reaction mechanism VIII → X ⇄ $^{XI}_{XII}$.
α–Methylalanine peptides with their N–H amide bonds will
prefer to adopt *trans* configurations and the chain will be
folded at the α-carbon positions. In the tripeptide (VII)
with the amino end protected as the hydrochloride, the acid
chloride can only attack the 2-carbonyl oxygen leading to
the oxazolone (IX).

The alternative possibility, imidazolone[8,9] occurs in
the case of the tetrapeptide as this then brings the acid
chloride close to the amide group of the first peptide
linkage and cyclization to a fused piperazine system (X)
can occur. This can then either eliminate the side-chain
yielding the bicyclic imidazolone (XI) or cyclize to the
bis-imidazolone (XII) after liberation of the amino group.

In conclusion, it should be noted that all the cyclic tetrapeptides we have been able to prepare seem to adopt the same configuration with alternating *cis* and *trans* amide groups. It would therefore appear that this is an important factor in controlling the reactions.

References

1. Dale, J., and K. Titlestad. Chem. Comm. 656 (1969).
2. Dale, J., and K. Titlestad. Chem. Comm. 1403 (1970).
3. Dale, J., and K. Titlestad. Chem. Comm. 255 (1972).
4. Titlestad, K. Chem. Comm. 1527 (1971).
5. Lucente, G., A. Romeo. Chem. Comm. 1605 (1971).
6. Jones, D. S., G. W. Kenner, J. Preston, and R. C. Sheppard. J. Chem. Soc. 6227 (1965).
7. Jones, D. S., G. W. Kenner, J. Preston, and R. C. Sheppard. Tetrahedron 21, 3209 (1965).
8. Karrer, P., and C. Gränacher. Helv. Chim. Acta 7, 763 (1924).
9. Mohr, E. J. Prakt. Chem. 81, 49 (1910).
10. Ali, M. Y., J. Dale, and K. Titlestad. To be published in Acta Chem. Scand.

CHEMISTRY AND BIOLOGY OF PEPTIDES

CONFORMATIONAL STABILIZATION OF SIDE CHAINS IN AROMATIC DIPEPTIDES

Henry Joshua, Charles M. Deber.* Merck Sharp & Dohme Research Laboratories, Division of Merck & Co., Inc., Rahway, New Jersey 07065

Nuclear Magnetic Resonance Spectral Differences Between Diastereomeric Dipeptides

Differences in the NMR spectra of some aromatic diastereomeric dipeptide pairs have been noted for some time[1,2] and have been used to quantitatively determine racemization in coupling reactions.[3,4] In particular, these studies showed an upfield shift for the protons of an aliphatic side chain in an L-D (or D-L) dipeptide consisting of an aliphatic and an aromatic amino acid residue when compared to the L-L (or D-D) compounds.

Recently we have demonstrated[5] that this upfield shift is most pronounced for the γ-methylene protons of the aliphatic side chains in L-Phe-D-Abu[†] and L-Phe-D-Nva[†] (See Table I).

When the side chain of the aliphatic residue terminates in a polar group attached to the β position as in Asp and Asn the upfield shift of the β-methylene protons is unaffected. However, when the polar group terminating the aliphatic side chain is attached to the γ, δ or ε positions, the magnitude of the upfield shift increases and it is the γ-methylene protons which are most affected.

---

*Present Address: Department of Biological Chemistry, Harvard Medical School, Boston, Massachusetts 02115
[†]Abu, α-Aminobutyric acid; Nva, norvaline.

Table I

Chemical Shifts and Chemical Shift Differences for
C-Terminal Side Chain Protons of
Diastereomeric Dipeptides

| Compound | Protons | Chemical Shift | | Δppm |
		L–L	L–D	
Phe–Abu	β	2.03	1.86	0.17
	γ	1.18	0.91	0.27
Phe–Nva	β	2.01	1.80*	0.21
	γ	1.61	1.22*,†	0.39
	δ	1.21	1.09*	0.12
Phe–Asn	β	3.05	2.84	0.21
Phe–Asp	β	3.13	2.90	0.23
Phe–Gln	β	2.40	2.06	0.34
	γ	2.58	2.06	0.52
Phe–Glu	β	2.36	2.10	0.26
	γ	2.66	2.16	0.50
Phe–Arg#	β	2.00	1.82	0.18
	γ	2.00	1.31	0.69
	δ	3.52	3.28	0.24
Phe–Lys(Ac) #,√	β	2.04	1.71	0.33
	γ	1.77	1.20	0.57
	δ	1.77	1.71	0.06
Cha–Gln#,**	β		2.40	
	γ		2.70	
Ala–Gln	β	2.40		
	γ	2.69		

NMR spectra were recorded at 100 MHZ on samples of 15–20 mg
dipeptide/500 μl solvent at 32° in D_2O. Chemical shifts are
given in parts per million (ppm) downfield from hexamethyldisiloxane (HMS) as external standard. Values of pH were
adjusted to fall between 5.5 and 7.5 and each row of dipeptides was compared at constant pH ± 0.5 pH unit. For peptides
containing Asp and Glu, the pH was 3.5.

*The enantiomer D-Phe-L-Nva was used in this case.
†Due to complex spectra, values given are estimated ± 0.03 ppm.
#Proton assignments were confirmed by spin-decoupling
 experiments.
√Values for Δppm for ε-protons of Lys were less than 0.1 ppm.
**Cha, β-Cyclohexylalanine.

The chemical shifts of the aliphatic side chain protons for some diastereomeric dipeptide pairs are listed in Table I.

Lande[6] as well as Lemieux and Barton[7] have proposed conformations for diastereomeric dipeptides which require a trans peptide bond as well as a trans relationship of both α-hydrogen atoms in the plane of the peptide bond. We found[5] that the J_{N_α} vicinal coupling constants of some representative L–L and L–D dipeptides are 7.5 ± 0.5 Hz, which is consistent[8] with the two general regions of preferred backbone conformations proposed by Lande.[6]   In either of these two cases, if one considers the peptide bond as a virtual single bond between the two α-carbons, then the side chains would have a *trans* and *gauche* relationship in the L–L and L–D dipeptides, respectively.

The *gauche* relationship of the side chains allows shielding of the aliphatic side chain protons by the aromatic side chain in an L–D dipeptide, thus accounting for the upfield shifts. Models suggest[5] that the γ-methylene protons in particular can be positioned close to the face and center of the aromatic ring.

The increase in the upfield shift of the γ-methylene protons due to polar substituents on the aliphatic side chains indicates a closer proximity on a time averaged basis of these protons to the aromatic ring due to an *intramolecular attractive interaction* of the side chains. The nature of this interaction may be similar to that causing folding of aromatic side chains in diketopiperazines[9–14] and is probably due to an association of the polarizable π electrons of the aromatic ring with the positive end of the dipoles of the polar substituents.

This interpretation is supported by the following:

(1) This type of association has been invoked[15] to explain the forces responsible for the *intermolecular* collision complexes which display upfield shifts of the same *magnitude* as the increase in upfield shift due to the presence of polar groups in the aliphatic side chains of L–D dipeptides.

(2) The increase in upfield shift of the γ-methylene protons in going from L-Phe-D-Nva (0.39)* to, for instance, L-Phe-D-Gln (0.52) is eliminated[5] on moderate heating from 32° to 80°C indicating a weak interaction with *thermodynamic parameters* similar to those found[15] for intermolecular collision complexes.

(3) The increase in upfield shift of the side chain γ-methylene protons for L-Phe-D-Glu (0.50) at pH 3.5 over

---

*Chemical shift difference ppm. See Table I.

L-Phe-D-Nva (0.39) is absent at pH 11.5, *i.e.* the postulated
attractive interaction giving rise to the increase in upfield
shift is eliminated when the net partial *positive charge* of
the polar substituent is removed as in going from carboxyl
to carboxylate. The increase in upfield shift for L-Phe-D-
Gln is unaffected by pH changes.[5]

(4) Elimination of the *aromaticity* of the Phe residue
in L-Phe-D-Gln by hydrogenation to L-β-cyclohexylalanyl-D-
glutamine brings the chemical shift of the Gln side-chain
protons of the latter compound into equivalence with those
of L-Ala-L-Gln[5] (see Table I).

Parallel chemical shift differences have been observed
in diastereomeric dipeptide pairs of reversed sequence such
as Arg-Phe and Ala-Phe[1-3,5,16] (Table II). Replacement of

Table II

Chemical Shifts and Chemical Shift Differences for N-Terminal Side Chain Protons of Diastereomeric Dipeptides

Compound	Protons	Chemical Shift L-L	L-D	Δppm
Ala-Phe	β	1.84	1.58	0.26
Abu-Phe	β	2.24	2.02	0.22
	γ	1.29	0.95	0.34
Arg-Phe	β	2.22	1.94	0.28
	γ	1.94	1.34	0.60
	δ	3.50	3.26	0.24
L-Ala-(R) Amphetamine	β	1.99		} 0.26
L-Ala-(S) Amphetamine	β		1.73*	

NMR spectra were recorded at either 60 or 100 MHZ on samples
of 15-20 mg/500 μl D_2O at 32° pH ∼ 2. Chemical shifts are
given in part per million downfield from hexamethyldisiloxane
as external standard.

*The downfield doublet of two overlapping methyl doublet
pairs, the higher one centering at 1.69.

the C-terminal carboxyl group by methyl in L-L and L-D
Ala-Phe gives L-Ala-(R)-Amphetamine[17] and L-Ala-(S)-
Amphetamine respectively.   As listed in Table II, this
chemical transformation hardly affects the direction and
magnitude of the chemical shift differences between the
diastereomeric pairs and proves that the C-terminal carboxyl
group is not necessary for the conformational preferences
in these compounds to be observed.

Chromatographic Behavior of Diastereomeric Dipeptides

The conformational preferences of diastereomeric dipeptides which are proposed to explain the NMR patterns in
the preceding section are also consistent with their observed
relative chromatographic behavior on paper,[18] silica gel[2]
and the neutral polystyrene based Amberlite XAD-2.[19]
Chromatography on silica gel and paper showed consis-
tently[2],[18] a higher R_f value indicating a *lower polarity*
for the L-L isomer of a diastereomeric dipeptide pair.
On XAD-2, however, we found that L-L dipeptide diastereomers eluted first (Table III) indicating a *higher
polarity*.   This apparent contradiction can be resolved

Table III

Relative Chromatographic Behavior on XAD-2 Resin of
Diastereomeric Dipeptides and Some Standards

| *Compound* | *Peak Cut Number   (1.5 ml/cut)* | | |
		L-L	*L-D*
Acetic Acid	10		
L-Ala	9		
L-Gln	8		
L-Ile	12		
L-Phe	23		
Ala-Phe		33	68
Phe-Ala		28	71
Phe-Gln		20	44
Ile-Ala		13	26

The column used was 0.9 x 30 cm filled with sieved crushed
XAD-2 resin 200-325 mesh; upflow elution was employed with
distilled degassed water for 40 cuts (60 ml) followed by a
1:1 linear gradient of 50 ml H_2O and 50 ml 10% ethanol.
The charge consisted of 0.5 ml containing 10-20 mg dipeptide.
The elution was monitored by UV or by ninhydrin spotting of
cuts.

considering the adsorption mechanisms involved. Aromatic
and aliphatic areas of a molecule are adsorbed by hydrophobic or van der Waals-London forces, whereas ionic groups
repel the resin. It follows that an L-D dipeptide diastereomer with non-ionized side chains is more susceptible
to adsorption on XAD-2 since the side chains are on the
same side of the peptide bond and simultaneously accessible
to the resin.

In contrast, the mechanism of adsorption on paper and
silica gel involves ionic-polar interactions. There too
the L-D dipeptide diastereomer is more susceptible to
adsorption since it has both the N-terminal amino group
and the C-terminal carboxy group on the same side of the
peptide bond, and both polar groups can interact simultaneously with the polar adsorbent.

Acknowledgment

Charles M. Deber, Merck Sharp and Dohme Research Fellow
in Biophysics, 1970.

References

1. Bovey, F. A., and G. V. D. Tiers. J. Amer. Chem. Soc.
   81, 2870 (1959).
2. Wieland, Th. and H. Bende. Chem. Ber. 98, 504 (1965).
3. Weinstein, B. In *Peptides: Chemistry and Biochemistry*,
   Weinstein, B., and S. Lande, eds. (New York: Marcel
   Dekker, 1970) pp 371-387.
4. Dewey, R. S., E. F. Schoenewaldt, H. Joshua, W. J.
   Paleveda, Jr., H. Schwam, H. Barkemeyer, B. H. Arison,
   D. F. Veber, R. G. Denkewalter, and R. Hirschmann. J.
   Amer. Chem. Soc. 90, 3254 (1968).
5. Deber, C. M., and H. Joshua. Biopolymers (in press).
6. Lande, S. Biopolymers 7, 879 (1969).
7. Lemieux, R. U., and M. A. Barton. Can. J. Chem. 49,
   767 (1971).
8. Bystrov, V. F., S. L. Portnova, V. I. Tsetlin, V. T.
   Ivanov, and Yu. A. Ovchinnikov. Tetrahedron 25, 493
   (1969).
9. Hardy, P. M., G. W. Kenner, and R. C. Sheppard. In
   Peptides, Young, G. T., ed. (Oxford: Pergamon Press,
   1963) pp 245-251.
10. Gawne, G., G. W. Kenner, N. J. Rogers, R. C. Sheppard,
    and K. Titlestad. In *Peptides 1968*, Bricas, E., ed.
    (Amsterdam: North-Holland Publ., 1968) pp 28-38.
11. Webb, L. E., and Chi-Fan Lin. J. Amer. Chem. Soc. 93,
    3818 (1971).

12. Kopple, K. D., and M. Ohnishi.   J. Amer. Chem. Soc. 91, 962 (1969).
13. Ziauddin and K. D. Kopple.   J. Org. Chem. 35, 253 (1971).
14. Karle, I.   J. Amer. Chem. Soc. 94, 81 (1972).
15. Ledaal, T.   Tetrahedron Lett. 1683 (1968).
16. Dewey, R. S., E. F. Schoenewaldt, H. Joshua, W. J. Paleveda, Jr., H. Schwan, H. Barkemeyer, B. H. Arison, D. F. Veber, R. G. Strachan, J. Milkowski, R. G. Denkewalter, and R. Hirschmann.   J. Org. Chem. 36, 49 (1971).
17. Crabbe, P. and W. Klyne.   Tetrahedron 23, 3499 (1967).
18. Sokolowska, T., and J. F. Biernat.   J. Chromatogr. 13, 269 (1964).
19. The use of neutral polystyrene resin Porapak Q for rapid desalting and fractionation of non-polar amino acids and non-polar oligopeptides has been reported by Niederwieser, A.   J. Chromatogr. 61, 81 (1971).

CHEMISTRY AND BIOLOGY OF PEPTIDES
© Copyright 1972 *Ann Arbor Science Publishers*

DETERMINING SOLVENT EXPOSURE OF PEPTIDE PROTONS BY PROTON
MAGNETIC RESONANCE

Kenneth D. Kopple, Thomas J. Schamper. Department of
Chemistry, Illinois Institute of Technology, Chicago,
Illinois 60616

MODELS OF CYCLIC PEPTIDE CONFORMATION based on proton
magnetic resonance studies rely in part on assignment of
peptide (N-H) protons to internal and solvent-exposed
categories. Such assignments have been made as interpretations of temperature-induced chemical shift changes that
occur in hydrogen bond accepting solvents. They have also
been inferred from relative exchange rates measured in
solvent mixtures containing exchangeable deuterium. Cor-
respondence between slow exchange and small temperature
coefficient is not always observed.[1-3] This should not
be surprising, since neither phenomenon measures solely
solvent exposure of exchangeable protons. Any hydrogen
bonded proton can be expected to have a temperature-dependent
chemical shift;[4] and the exchange rate depends on the probability of peptide and solvent molecules reaching an
appropriate transition state, which probability is only
indirectly related to the equilibrium conformation of the
peptide.

   A more direct measure of solvent exposure would be the
effects of solvent variation on the N-H resonances, provided the resulting spectral changes can be shown not to
result from conformational transitions. Effects of transfer
of cyclic peptides to trifluoroacetic acid from more basic
solvents have been known for some time,[5,6] but their interpretation has been doubtful because of the high acidity of
trifluoroacetic acid and its known structure-breaking
effects on amino acid polymers.

75

Urry and Pitner[7] reported recently that the resonances of the phenylalanine and ornithine protons of the cyclic decapeptide gramicidin S, solvent exposed according to both temperature dependence and exchange studies, move strongly upfield on transfer from methanol to trifluoroethanol, probably because they are no longer hydrogen bonded to solvent in the latter. The resonances of the internal leucine and valine peptide protons of gramicidin S are relatively unaffected by this solvent change.

We have been using hexafluoro-2-propanol (HFiPA) in the same kind of studies, and have examined gramicidin S, the cyclic heptapeptide evolidine,[1] the cyclic hexapeptides *cyclo*(-Gly-L-Leu-Gly-)$_2$ and *cyclo*(-Gly-L-Tyr-Gly-)$_2$,[8] and the cyclic pentapeptide, *cyclo*(-L-Ala-L-Tyr-L-Asp-Gly-Gly-d_2-).* Assignment of peptide proton resonances of the HFiPA solutions to particular residues was carried out by following the resonances through continuous change of solvent composition from a solvent in which assignments were known from H-N-C$_\alpha$-H decoupling studies. Decoupling studies in HFiPA are not possible because the α-proton region is masked by intense solvent absorption. In the cases we studied, the change from dimethyl sulfoxide (DMSO) to HFiPA brings about only small changes in the H-N-C$_\alpha$-H coupling constants, making tenable the assumption of no major conformational change.

Given the conformational conclusions reached for these peptides on the bases of temperature dependences, H-N-C$_\alpha$-H coupling constants, and model building, the solvent-dependent changes in chemical shift are consistent with weakened hydrogen bonding of the solvent-exposed protons in the fluorinated alcohol. On going from DMSO to HFiPA, protons considered to be internal move from 0 to 0.5 ppm downfield, and protons considered to be freely exposed to solvent move upfield up to 2 ppm. *N*-Methylacetamide moves upfield 1.3 ppm.

We have measured the peptide proton temperature coefficients of *cyclo*(-Ala-Tyr-Asp-Gly-Gly-d_2-) in HFiPA; these run parallel with the coefficients determined for DMSO solutions of the same peptide, and range from 0.3 to 6.0 X 10^{-3} ppm/degree. The existence of so large a range, which

---

*In this peptide, on which a detailed report will be made elsewhere, there appears to be a β-turn of the L-L type formed by the residues -Gly-d_2-Ala-Tyr-, with the peptide proton of the aspartic acid residue is internal, transannularly hydrogen bonded, and diamagnetically shielded by the π cloud of the Ala-Tyr peptide bond.

is no less than what we observe for DMSO solutions, indicates that there must be a temperature dependent interaction of the peptide protons with HFiPA, possibly hydrogen bonding of the type

$$O=C \diagdown_{\overset{|}{N-H}....F-\overset{|}{\underset{|}{C}}-}^{F}$$

In this connection we should report that in mixtures of *N*-methylacetamide and HFiPA the chemical shift of the amide protons is relatively independent of the mole fraction amide (at low amide), while the HFiPA hydroxyl proton resonance shifts from 4.6 ppm in pure HFiPA to about 6.2 ppm in 25 mole per cent amide. Thus the alcohol acts as proton donor and the amide as acceptor in association of the two.

Figure 1 presents a correlation between the temperature coefficient of peptide proton chemical shift in dimethyl

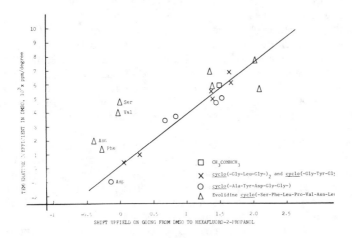

Figure 1: Correlation between shift upfield of peptide proton resonances on transfer from solution in dimethyl sulfoxide to solution in hexafluoro-2-propanol (HFiPA) at 30° and temperature coefficient of chemical shift of the same resonances in dimethyl sulfoxide solution. Shifts upfield are positive.

sulfoxide and the upfield shift on changing solvent from
dimethyl sulfoxide to HFiPA. The line is a least squares
fit excluding the evolidine points (triangles), and represents the correlation to be expected between the two kinds
of observation in those cases where weak internal hydrogen
bonds or specific solvation is not involved.

The effects of a change in solvent probably reflect the
average solvent exposure of peptide protons more closely
than do temperature coefficients or exchange rates, but
peptides will not always be conformationally stable to the
required change in environment. We recently reported[9] that
hydrogen bonded association of peptide protons with the
nitroxide free radical 3-oxyl-2,2,5,5-tetramethyloxazolidine
produces line broadening that can be used to identify solvent-
associated peptide protons at radical concentrations of under
three per cent; in such measurements the perturbation of
peptide environment is minor. In the gramicidin S test
case, extensive broadening of the lines of the external
phenylalanine and ornithine peptide protons, with relatively
little change in width of the internal leucine and valine
peptide proton lines, occurs at two per cent radical in
methanol.

Infrared studies ($\Delta\nu_{O-H}$) of the association of 3-oxyl-
2,2,5,5-tetramethyloxazolidine with phenol and with HFiPA
in dilute carbon tetrachloride solution indicate that,
according to published correlations,[10] the radical is a
hydrogen bond acceptor of basicity comparable to acetone
or an ether, but is less basic than dimethyl sulfoxide.
It exhibits most obvious effects on peptide proton resonances
in methanol or HFiPA solutions, although distinctions are
also possible in dimethyl sulfoxide. Figure 2 illustrates
the effects of added radical on the spectrum of evolidine
in methanol.

It is difficult simply to express quantitatively the
extent of line broadening by the radical, but we suggest
expressing this as a linewidth increment, the increase in
width at half height across both components of the doublet
for each per cent of radical added, in the range 0-3 per
cent radical. Our observations on the peptides listed
earlier are that those protons with DMSO - HFiPA shifts less
than zero (See Figure 1), which include the gramicidin S
valine and leucine protons not in the figure, have linewidth increments comparable to that of tetramethylsilane,
of the order of 1-2 Hz/% in methanol or HFiPA and 3 Hz/%
in DMSO. Protons to the right of the zero line of Figure 1
are increasingly sensitive to the presence of radical
(5-15 Hz/%), but, because of overlaps, there are few precisely enough known data to make more quantitative distinctions.

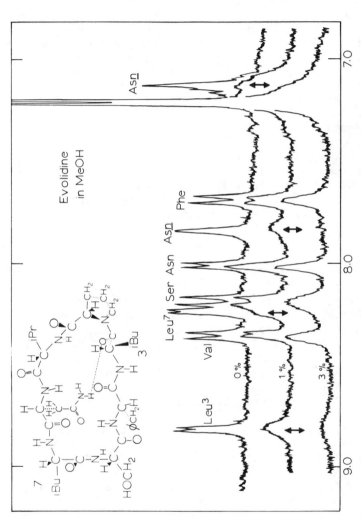

Figure 2: 250 MHz proton resonance spectrum (7–9 ppm below TMS reference) of evolidine in methanol, 30°, showing the effect of added 3-oxyl-2,2,5,5-tetramethyl-oxazolidine (0–3%) on the peptide proton resonances. The structure shown is the conformation proposed in reference 1.

It is clear from our observations that both the DMSO-HFiPA shift and the line broadening by radical measure solvent exposure of peptide protons. Therefore it is interesting to note that the temperature coefficients of the five most solvent-shielded protons in Figure 1 range from −1.0 to 4.8 x 10^{-3} ppm/degree. We suggest that these differences in temperature coefficient result from differences in stability of intramolecular hydrogen bonds, not from differences in solvent exposure. An additional source of the variation might, however, be temperature dependent changes in the average conformation of magnetically anisotropic side chain groups.

Acknowledgment

This work was supported by grants from the U.S. Public Health Service, National Institute of General Medical Sciences, GM 14069 and GM 47357. It was done in part at the NMR Facility for Biomedical Studies, supported by National Institutes of Health grant RR-00292.

References

1. Kopple, K. D. Biopolymers <u>10</u>, 1139 (1971).
2. Ivanov, V. T., A. I. Miroshnikov, N. D. Abdullaev, L. B. Senyavina, S. F. Arkhipova, N. N. Uvarova, K. Kh. Kalilulina, V. F. Bystrov, and Yu. A. Ovchinnikov. Biochem. Biophys. Res. Commun. <u>42</u>, 654 (1971).
3. Patel, D. J., A. E. Tonelli, P. Pfaender, H. Faulstich, and Th. Wieland. In press.
4. Muller, N., and R. C. Reiter. J. Chem. Phys. <u>42</u>, 3265 (1965).
5. Kopple, K. D., M. Ohnishi, and A. Go. J. Amer. Chem. Soc. <u>91</u>, 4264 (1969).
6. Schwyzer, R., and U. Ludescher. Helv. Chim. Acta. <u>52</u>, 2033 (1969).
7. Pitner, T. P., and D. Urry. J. Amer. Chem. Soc. <u>94</u>, 1399 (1972).
8. Kopple, K. D., A. Go., R. H. Logan, Jr., and J. Savrda. J. Amer. Chem. Soc. <u>94</u>, 973 (1972).
9. Kopple, K. D., and T. J. Schamper. J. Amer. Chem. Soc. <u>94</u>, 3644 (1972).
10. Purcell, K. F., J. A. Stikeleather, and S. D. Brunk. J. Amer. Chem. Soc. <u>91</u>, 4019 (1969).

SYMPOSIUM DISCUSSIONS

Summarized by Johannes Meienhofer

THE MAJORITY OF QUESTIONS and comments about the application
of proton and carbon-13 nuclear magnetic resonance spectroscopy to the study of the conformations of peptides
were concerned with the difficulties and problems of
assigning signals and peak positions and of interpreting
spectra in terms of conformational change. This is a
consequence of the present limitations, since a complete
interpretation of a given spectrum is usually not possible.
Additional complications arise from the fact that plots of
the upfield shifts of NH signals with temperature are not
always linear and occasionally even negative slopes have
been observed. This renders correlations with deuterium
exchange rates for the determination of external hydrogen
bonds to solvent very problematic, and it was recommended
to approach such studies with great care and caution. The
comment was made that room temperature is actually an
arbitrary point for taking NMR spectra. On cooling resonances might separate while on warming colaescense can be
observed. Therefore, taking both high temperature and low
temperature NMR spectra could provide a wider range of
information. Several questions and comments concerned
solvent effects and whether the observed spectral changes
might perhaps be caused by conformational changes. The
fashionable question which conformation might most closely
resemble that assumed during interaction with cellular
receptors remains at present open to everybody's guess.
Much consideration was given to conformational analyses of
oxytocin and related peptides (pp 16-24, 29-33, 35-38).
It was pointed out that differences in the NMR spectra of
unprotected and protected model peptides can be attributed
to the protecting groups present and that the effects of

81

protecting groups have to be calibrated for meaningful
interpretation of spectral changes. The NMR spectrum of
reduced oxytocin has not yet been studied, but it was agreed
that this should be very interesting. Energy calculations
were applied to oxytocin as a supplementary technique for
the refinement of the conformation derived from NMR studies
(pp 35 - 38).   Only the ϕ-ψ space in the region corresponding to the observed J_{N_α} was investigated, and it was
pointed out that *ab initio* energy calculations for oxytocin
or vasopressin have not been successful because of the large
number of possible conformations.

For some peptides (angiotensin, cyclic proline-containing
peptides) two forms have been separated by Sephadex chromatography or thin film dialysis, but unfortunately no difference in CD spectra could be observed, perhaps due to
interconversion. Whether these forms were conformational
isomers or due to aggregation phenomena remained unresolved.
It was pointed out that for *cyclo*(-Pro-Ser-Gly-)$_2$ (p 7) an
asymmetric form was observed, however, it does not become
the exclusive form and occurs to the extent of 10-15% of
the symmetric form. In the *retro* isomer, however, the
cis-cis form can become almost the exclusive form when
working with dimethylsulfoxide. Observations of asymmetric
conformers of symmetrical molecules are very remarkable and
somehow contrary to a, probably quite common, preference
for aesthetically pleasing symmetrical shapes.

The question was raised whether in cyclic peptides, as
cyclo(-Pro-Gly-)$_3$ (pp 39 to 43), which bind Na$^+$ or other
metal ions, these ions could catalyze conformational transitions, and this might indeed be the case. Whether Na$^+$
is or can be incorporated into the cavity of *cyclo* (-Pro-
Gly-)$_3$ in complexed form was discussed without a definite
conclusion at this time.

The point was also made that carbon-13 NMR spectra
will be a useful supplement to proton NMR spectra for observing *cis* and *trans* isomers of proline-containing peptides
especially since the proline side-chain carbon atoms also
give rise to separate resonances for the two forms.
Carbon-13 NMR studies (pp 45 to 49) of myoglobin and
ribonuclease, treated with [C^{13}]-enriched bromoacetate,
did not yet allow the location of the carboxymethyl groups,
but their signals have been very well identified.

The description of side product formation (diketopiperazines, bicyclic compounds) during cyclization of tetrapeptides (pp 59 to 65) lead to an inquiry about the driving
force of diketopiperazine formation. This was left open
during the discussion. In my opinion it might be explained

by the removal of the diketopiperazine (III) from an
equilibrium between the unstable cyclol (structure II on
p 61) and the linear tetrapeptide (I) as discussed for the
decomposition of D-Val-Pro-Sar to form *cyclo*(-D-Val-Pro-)
and sarcosine.[1] The suggestion was made that tetrapeptide
cyclization at lower temperature than that used (115°)
might minimize diketopiperazine formation. Apparently
that was not the case, even at room temperature; but the
overall cyclization yield was considerably smaller at
lower temperatures.

Reference

1. Meienhofer, J., Y. Sano, and R. P. Patel. In *Peptides:
    Chemistry and Biochemistry*, Weinstein, B., and S. Lande,
    eds. (New York: Marcel Dekker, 1970) pp 419-434.

SECTION II

FORUM DISCUSSION ON CYCLIC PEPTIDES

Session Chairmen

Yuri A. Ovchinnikov and Elkan R. Blout

Edited by

Roderich Walter

FORUM DISCUSSION ON CYCLIC PEPTIDES

Session Chairmen: *Yuri A. Ovchinnikov, Elkan R. Blout.*

Panel: *F. A. Bovey, C. M. Deber, M. Goodman, C. H. Hassall, N. Izumiya, I. Karle, K. D. Kopple, H. Lackner, M. Rothe, R. Schwyzer, P. Von Dreele, R. Walter, T. Wieland.*

Edited by: *Roderich Walter.*

OVCHINNIKOV: Ladies and Gentlemen: Dr. Blout, my co-Chairman, was very kind to give this microphone to me and to suggest that I say a few words. I would first of all like to express my deep gratitude for the flattering invitation to this Third American Peptide Symposium. I was really very surprised to see so many of my European colleagues in this hall.

It is not surprising and it is highly gratifying that the organizers of the Symposium should allocate a special forum discussion to cyclic peptides. Cyclic peptides are in a special position among the peptide substances, being actually the starting points of the modern study of their linear counterparts and of the more involved proteins. The study of these compounds is not only a school for the learning of various aspects of peptide and protein chemistry in which many scientists have received their training but, moreover, different concepts have been subjected to stringent tests. Cyclic peptides attract attention in their own right. Among them we encounter multifarious biologically active compounds: antibiotics and hormones, toxins and antitoxins, alkaloids and other types of substances strongly affecting the bodily functions. Cyclic peptides have become powerful tools in the hands of biochemists and have served as the means whereby great strides have been made in the study of a number of biochemical processes. One need only mention the role of actinomycin D in the study of protein biosynthesis and of valinomycin in the study of membranes. The cyclic peptides

87

are also unique from a chemical standpoint. Here we meet
with a wide range of ring sizes, with members containing
unusual, non-protein amino acids, including N-methylamino
acids and D-amino acids, and with the weak interactions
resulting from the ring constraint. It is apparently not
accidentally that nature has chosen a non-ribosomal path
for cyclic peptide synthesis, as has been shown for example
with gramicidin S. Surprises await us when studying the
spatial structure of the cyclic peptides; thus, we have the
pleated sheet conformation of gramicidin S, the tennis ball
seam of the closely related antamanide, the disc of enniatins
and the bracelet of valinomycin.

In the study of such structures, many of which are highly
medium-dependent, X-ray crystallography does not assume a
predominant role despite the power of this method; successfully competing with it are methods for studying the conformation of cyclic peptides in solution, and a proper
balance is being struck between them. I believe that the
same will be true of proteins and in the study of cyclic
peptides we are breaking the trail to these coveted
substances.

It is a pleasure to see here practically all the leading
figures in the area of cyclic peptide chemistry. Hopefully
each will avail himself of the opportunity to take the floor.

BLOUT: I will get to the more practical matters because
time is somewhat limited. If you will look at the forum
participants I am sure probably everybody realizes that each
one of the members could speak for the full two hours and
do it interestingly. However, since this forum is in the
nature of an experiment, we're going to try to limit any
presentation, including any formal presentation, to a maximum
of five minutes and hope that the audience will participate
in the discussion. Professor Ovchinnikov and I decided to
divide the morning into two parts, interrelated but somewhat
different, namely, the problems of synthesis of cyclic peptides and the problems of conformational determinations.
So, with no further ado we should like to call on the man
who has volunteered to start off the morning, Professor
Rothe.

ROTHE: I wish to discuss briefly a cyclization reaction
which occurs during solid-phase synthesis and which was
mentioned by Drs. Khosla and Brunfeldt (see Section III).

POSSIBLE SIDE-REACTIONS DURING SOLID-PHASE PEPTIDE SYNTHESIS.
II.[1] REACTION BETWEEN NEIGHBORING CHAINS. FORMATION OF
HYDROXY GROUPS ON THE RESIN AND THEIR CONSEQUENCES

Manfred Rothe, Jan Mazánek. Organisch-Chemisches
Institut, Universität, Mainz, Germany

WE HAVE OBSERVED TWO HITHERTO unknown possibilities of
formation of undesired side products during solid-phase
peptide synthesis: (1) the esterification of the hydroxymethyl groups formed on the polymer support during
cyclization of the peptide chain with the Boc-amino acid
used for chain lengthening, (2) an aminolytic cleavage of
the benzyl ester bond of the resin-bound amino acid by a
neighboring peptide chain leading to peptides with extended
sequences which can couple in subsequent steps. As we
could show these reactions may be important during the
first two steps of the synthesis; in particular, cyclization at the dipeptide stage may cause considerable difficulties during the synthesis of higher peptides. In
principle, both reactions can lead to cyclic dipeptides
(diketopiperazines) and consequently to hydroxymethyl
groups (a) by cyclization of a resin-bound dipeptide

$$aa_2-aa_1-resin \rightarrow \boxed{aa_2-aa_1} + HO-resin \quad (aa, \text{ amino acid}) \quad (a)$$

(b) by formation of a new peptide in side reaction (2) which
may cyclize subsequently, e.g. during the first step of the
synthesis

$$aa_1-resin \rightarrow aa_1-aa_1-resin \rightarrow \boxed{aa_1-aa_1} + HO-resin \quad (b)$$
$$\underset{aa_1}{|} \qquad\qquad \underset{OH}{|} \qquad\qquad\qquad \underset{OH}{|}$$

Reaction (b) has been investigated with resins esterified
with glycine, sarcosine, and proline (capacity 0.82, 0.77,

and 0.98 meq/g, respectively). In the combined filtrates
of the neutralization reaction (triethylamine in CH_2Cl_2,
10 min) after deprotection and of the subsequent washing
operations (4 x CH_2Cl_2, 5 min each), we could show the
presence of the corresponding cyclic dipeptides cleaved
from the resin. Their amount was about 0.1% of the
original substitution with the three resins examined. It
increased after 20 hr shaking in CH_2Cl_2 to about 0.6% and
after a total of 100 hr to about 1%. The dipeptide-resin
which is not cyclized can react further in the following
coupling steps, *e.g.* with the formation of chains of the
type resin--aa_1-aa_1-aa_2-... As the extent of this side
reaction is small with low amino acid substitution,
it appears to be not significant under the usual conditions of solid-phase synthesis. However, it should be
considered that it can occur in each step increasing the
amounts of side products which can no longer be neglected.

From these results it may be concluded that the formation of cyclic peptides and consequently of hydroxy groups
can occur especially at the dipeptide stage [equation (a)].
Indeed, in certain cases we could show the formation of
considerable amounts of diketopiperazines.

In order to examine the reaction (a), all possible
dipeptide-resins of glycine, sarcosine, and proline have
been investigated. The deprotection and neutralization
(total 30 min) reactions were carried out as usual. From
these filtrates and after shaking the resin in methylene
chloride (60 min) we could isolate the corresponding cyclic
dipeptides in chromatographically pure state (Table I).

Forum Discussion Table I

Formation of Cyclic Dipeptides During
Solid-Phase Peptide Synthesis

Dipeptide-resin	*Percent Cyclic Dipeptide (Diketopiperazine) Formation*	
	30 min	*90 min*
Gly-Gly	5	6
Pro-Gly	3.5	4
Sar-Gly	4.5	6
Gly-Pro	15	17
Pro-Pro	20	46
Sar-Pro	22	24
Gly-Sar	29	48
Pro-Sar	42	62
Sar-Sar	61	73

Their amounts correspond to the hydroxymethyl groups formed.
Table I shows that the cyclization tendency depends on the
nature and sequence of the amino acids. As expected, it
has been found to be particularly high with *N*-alkylated
amino acids and amounts up to 90% with Sar-Sar and Pro-Pro
after 24 hours. The formation of diketopiperazines does
not only decrease the substitution of the resins. The
hydroxymethyl groups formed on the polymer support can
react in each coupling step with the Boc-amino acid and
carbodiimide with the formation of esters and thus become
starting points for new peptide chains. Although this
reaction occurs only in low yields,[1] even small amounts of
amino acids bound in this way can lead to peptide mixtures
which are no longer separable. Just this was observed by
R. Pudill in our group during the synthesis of an antamanide
sequence starting with a Pro-Pro-resin. Beginning with the
tripeptide stage small amounts of the amino acid used for
chain lengthening as well as further products formed from
these new starting points could be detected as impurities
during each of the following steps of the synthesis.

Reference

1. Rothe, M., and J. Mazánek. Angew. Chem. **84**, 290 (1972);
    Angew. Chem., Int. Ed. **11**, 293 (1972).

GOODMAN: In the course of a general procedure using the
resin-bound amino acid dipeptide, what would you say the
competition would be from cyclization? I see here you allow
them to sit 30 or 90 min without adding another amino acid.

ROTHE: The usual time for neutralization and washing the
substituted resin requires 30 min. I think we have to
shorten this time in the Merrifield synthesis.

BODANSZKY: For which other amino acids did you check the
percentage of cyclodipeptide formation?

ROTHE: We found it also when tyrosine was coupled with
resin-bound serine. *Zahn* found it with Gly-Leu and Gly-
Val, I believe, but it's far more pronounced with Pro-Pro.

BODANSZKY: The formation of diketopiperazines from dipeptide esters is not unexpected. They will form more readily

when one amino acid belongs to the L and the other to the
D configuration. This was the case in the study reported
by Merrifield.

FACTORS FAVORING CYCLIZATION OF PEPTIDE CHAINS

C. M. Deber. Department of Biological Chemistry,
Harvard Medical School, Boston Massachusetts 02115

SINCE THE CYCLIZATION STEP in the synthesis of cyclic peptides frequently proceeds in poor yield, it is of importance
to examine the features which play a dominant role in
determining the propensity of a given peptide chain to
undergo cyclization. For this purpose, the qualitative
mechanistic outline of a cyclization reaction (in this
instance of the linear tripeptide active ester I) shown
in Figure 1 is helpful. The synthesis of the cyclic peptide is viewed here in terms of a series of most-probable

Forum Discussion Figure 1: Proposed reaction pathways of
linear tripeptide active ester.

93

events in a reaction sequence of alternative pathways.
The starting tripeptide I may "cyclize" directly to give
cyclic tripeptide II, it may "double" linearly to give
hexapeptide active ester III, which may subsequently
"cyclodimerize" to cyclic hexapeptide IV, or continue to
"oligomerize" linearly to the series of extended chain
peptides V. Which pathway will predominate, and what
factors will determine this?

A survey[1] of the approximately 100 homodetic cyclic
hexapeptides whose syntheses have been reported to date
revealed that in all cases (where products were well
characterized) the peptide chain contained either (a) a
D residue, (b) a Gly residue, and/or (c) an imino acid
residue such as proline or sarcosine. Also, only those
chains consisting entirely of imino acids (Pro-Pro-Pro,[2]
Pro-Pro-Hyp,[3] Pro-Pro-Sar,[4] Pro-Sar-Sar,[4] and Sar-Sar-Sar[5])
gave cyclic tripeptides. Thus, it appears that an LD sequence in the chain (or an L-Gly sequence) provides an
elbow-like bend in the peptide backbone which aids in
accomplishing the objective of bringing reactive amino and
activated carboxyl termini into reaction proximity. Then,
stabilized by intramolecular hydrogen bonds in transition
states such as suggested by Schwyzer, cyclization of the
chain occurs.[6] However, numerous peptide chains containing
imino acids (lacking amide protons in positions necessary
for intramolecular H-bonding) nevertheless give good yields
of cyclic products, suggesting that a further conformational
influence at or near the transition state of the cyclization
may be the presence of *cis*-peptide bonds. Such bonds are
"allowed" when the sequence X-imino acid is present, and
models indicate that their effectiveness in bringing chain
ends together may be greater than that of the LD sequence.
Only when the two existing peptide bonds of a precursor
tripeptide I are *cis*-allowed (*i.e.* in Pro-Pro-Pro) will
cyclic tripeptide formation be an important pathway.

Note that III is the same material which is used as
the starting material when cyclic hexapeptides are synthesized directly from linear hexapeptides (which is essential,
of course, in the absence of sequential C_2-symmetry). In
fact, about two-thirds of the known synthetic cyclic hexapeptides were prepared in this manner. It seems likely
that this linear hexapeptide active ester III is a discrete
intermediate in reactions leading to cyclic hexapeptide *via*
cyclodimerization of tripeptide active esters, as distinguished from a concerted ring closure involving simultaneous
formation of two new peptide bonds. Experimental support
for this suggestion is presently lacking, however. If one

considers a hypothetical case in which cyclization leads
initially to a pair of "kinetic" cyclic peptide conformations, *i.e.* conformers differing by *cis-trans* peptide bond
isomers formed in varying relative ratios depending, perhaps,
upon the point of ring closure of the incipient cyclic
peptide, it may become possible to define the role of III.
In our laboratory *cyclo*(-L-Pro-L-Ser-Gly-)$_2$ was synthesized[7]
by three different routes: from cyclodimerization of Pro-
Ser-Gly-ONp; from cyclodimerization of Ser-Gly-Pro-OPcp;
and from cyclization of Pro-Ser-Gly-Pro-Ser-Gly-ONp. However, when the solution conformations of these three preparations were compared by 220 MHz NMR spectroscopy, identical
populations of major conformer (85%, all *trans* peptide
bonds) and minor (15%, asymmetric conformer, one *cis* Gly-
Pro peptide bond) in two polar solvents, H$_2$O and DMSO, were
observed. These results imply that the establishment of
"thermodynamic" ratios of conformers had already occurred,
and perhaps cyclization and NMR studies in less polar media
will be preferable.

Temperature may also profoundly influence the pathways
of a given peptide cyclization reaction. Preliminary experiments in our laboratory (and results of Dale and
Titlestad[5,8]) suggest that formation of cyclic tetrapeptides (from linear tetrapeptide active esters) in good
yield may require elevated temperature (*i.e.* refluxing
pyridine). Attempts to cyclize the same tetrapeptide chain
at room temperature might produce only small yields of the
cyclodimer cyclic octapeptide (and perhaps a small yield
of cyclic tetrapeptide). These results reflect the difference in reaction rates as a function of temperature for the
cyclization process, due to different activation parameters
for each pathway, and suggest that the outcome of the overall reaction could be directed to favor a desired product
by adjusting the temperature.

This discussion has presented in speculative terms how
the working hypothesis shown in Figure 1 can be used
empirically to make an informed estimate as to which conditions, and which configurational *vs.* conformational
features will guide the cyclization reaction. Of course,
the choice of the method of activation of the linear chain
is also of primary importance. A more thorough understanding
of the diverse factors controlling peptide cyclization
mechanisms would clearly be desirable.

References

1. Deber, C. M., S. C. K. Wong, and E. R. Blout. Unpublished.
2. Rothe, M., K. D. Steffen, and I. Rothe. Angew. Chem. Int. Ed. 4, 365 (1965).
3. Deber, C. M., D. A. Torchia, and E. R. Blout. J. Amer. Chem. Soc. 93, 4893 (1971).
4. See: Rothe, M., R. Theysohn, D. Mühlhausen, F. Eisenbeiss, and W. Schindler. This volume, pp 51–57.
5. Dale, J., and K. Titlestad. Chem. Commun. 1969, 665.
6. Schwyzer, R., and U. Ludescher. Helv. Chim. Acta 52, 2033 (1969), and references therein.
7. Torchia, D. A., A. di Corato, S. C. K. Wong, C. M. Deber, and E. R. Blout. J. Amer. Chem. Soc. 94, 609 (1972).
8. Dale, J., and K. Titlestad. Chem. Commun. 1970, 1403; 1972, 255.

PROBLEMS OF PEPTIDE CYCLIZATION ON SOLID SUPPORTS

T. Wieland. Max-Planck-Institut für medizinische
Forschung, Abteilung Chemie, Heidelberg, Germany

I WOULD LIKE TO ASK SOME QUESTIONS of the audience. Everybody who works on cyclic peptides tries to get as high a
yield as possible in the cyclization step, and one of the
methods which seemed very promising was to synthesize the
linear peptide precursor of desired length on a solid support, and then bring about the cyclization by activating
the carboxyl group of the peptide on the resin. Five years
ago or so, I reported with C. Birr[1] that it is possible to
have the activated carboxyl group of an amino acid or peptide attached to a phenol resin which possesses a nitro
group in the *o*-position or a sulfonyl group in the *p*-
position. Patchornik independently reported at the same
time that he had succeeded in making cyclic peptides in
very good yields by the same principle. Therefore, we
continued our attempts to make cyclic peptides by this
method, and we proceeded in the following manner: T. Lewalter
in my laboratory esterfied *p*-hydrazinobenzoic acid onto a
polystyrene resin, and then the first amino acid in the
solid-phase synthesis could be coupled to the hydrazino
group. Proceeding stepwise, he finally got a linear peptide with the antamanide sequence. On oxidation of the
substituted hydrazine with *N*-bromosuccinimide, the *N*-bound
carboxyl was activated. The activation of this linkage
was proved by its great reactivity with benzylamine or
other amino compounds. Although aminolysis was very good,
the internal cyclization reaction which occurred on depro-
tonation of the protonated amino end yielded only traces
of the cyclic peptide.

The other approach was to introduce a handle which consists essentially of a glycol grouping:

$$HO - CH_2 - \underset{\underset{OH}{|}}{C} - \text{Resin}$$

This was shown by P. Fleckenstein, who esterified the first amino acid in a Merrifield synthesis with the primary hydroxyl of the resin-bound glycol and proceeded stepwise to get a peptide chain of 10 or more residues. The activation of the carbonyl group occurred by elimination of water by treatment with trifluoroacetic acid, which yielded an enol-ester. This enol-ester is an activated ester as was shown by virtue of its reactivity towards amines. However, also in this case no cyclization, or almost no cyclization, occurred. So, I put the question to the audience: has anyone of you ever tried to make cyclic peptides on a solid support by an approach similar to the one just described?

MITCHELL: I have been working (initially with R. W. Roeske and now with R. B. Merrifield) on the development of a polystyrene support that will allow (1) a stepwise synthesis of a linear peptide and (2) a *non-racemic* removal of the peptide from the support via an intramolecular aminolysis by the N-terminal amino group of the peptide resin. Wieland's work[2] in which the *p*-hydrazinobenzoylated resin failed to give significant cyclization complements our data which indicate that the polystyrene backbone *may* be sterically hindering the desired intramolecular aminolysis of peptide resin. We are investigating the use of hydrocarbon bridges to increase the separation of peptide and leaving group (*e.g.* *p*-nitrophenol, *p*-hydrazinobenzoic acid, catechol) from the polystyrene backbone. This approach will allow us to determine the effect of polystyrene proximity on the efficiency of cyclic peptide formation via intramolecular aminolysis of peptide resin.

PATCHORNIK: There are polymers which are very sterically hindered and you are not even successful in forming a diketopiperazine. For example, if you remove the *N*-carbobenzoxy group--or for that matter any blocking group-- from a dipeptide attached to a nitrated styrene resin (Novolak), you obtain after neutralization no diketopiperazine. On treating this dipeptide active ester with a free peptide or an amino acid ester, you get peptides in nice yields. However, there are instances where steric hindrance will interfere even in these reactions. Several years ago we reported that we were able to obtain tetraalanine, and glycyltrialanine.[3] We are checking now cyclization of longer linear peptides on "new polystyrene" derivatives hoping to achieve a good cyclization reaction. Similar experiments are being carried out in our laboratory on the formation of cyclic ketones on a polymeric support.

GOODMAN: I should like to make a comment to that. In passing, I think, Dr. Patchornik may have indicated a way of avoiding the side reaction that was alluded to by Dr. Rothe. Namely, that if one uses a Merrifield resin with sufficient steric hindrance, then one might avoid the diketopiperazine side-reaction that can arise in the dimer stage.

OVCHINNIKOV: I would like to mention that Dr. Patchornik sent his substances to our laboratory for mass spectrometry, and it proved to be a real cyclic tetrapeptide, at least in the case of *cyclo*(-Ala-Ala-Ala-Ala-).

References

1. Wieland, T., and C. Birr. Angew. Chem. <u>78</u>, 303 (1966); Angew. Chem., Int. Ed. <u>5</u>, 310 (1966).
2. Wieland, T., J. Lewalter, and C. Birr. Justus Liebigs Ann. Chem. <u>740</u>, 31 (1970).
3. Fridkin, M., A. Patchornik, and E. Katchalski. J. Amer. Chem. Soc. <u>87</u>, 4646 (1965).

PREPARATION OF GRAMICIDIN S ANALOGS AND EFFECTS OF RING
SIZE ON ANTIBACTERIAL ACTIVITY

Nobuo Izumiya. Laboratory of Biochemistry, Faculty of
Science, Kyushu University, Fukuoka, Japan.

WE HAVE BEEN CARRYING OUT studies on the structure-activity
relationship of gramicidin S (GS) and tyrocidines through
syntheses of their analogs. Here I would like to talk on
synthetic methods used to prepare GS and its analogs and
then on the effect of ring-size of GS on its antibacterial
activity.

For many years we have been synthesizing GS and its
analogs by cyclization of the linear peptide *p*-nitrophenyl
esters in pyridine as described by Schwyzer and Sieber.[1]
Yields are usually 40% to 50%. Recently, we found that the
N-hydroxysuccinimide ester (Figure 2) or azide methods give
better yields, usually 60% to 70%.

$$Boc-(Val-\overset{Z}{Orn}-Leu-D-Phe-Pro)_2-OH \quad (a)$$

$$\Big\downarrow HONSu + DCC$$

$$Boc-(Val-\overset{Z}{Orn}-Leu-D-Phe-Pro)_2-ONSu$$

$$\Big\downarrow TFA, 5 min (0°C)$$

$$TFA·H-(Val-\overset{Z}{Orn}-Leu-D-Phe-Pro)_2-ONSu$$

$$\Big\downarrow pyridine, 2 hr (25°C)$$

$$cyclo(-Val-\overset{Z}{Orn}-Leu-D-Phe-Pro-)_2 \quad (78\% \text{ from a})$$

$$\Big\downarrow H_2-Pd \text{ in } HCl-CH_3OH$$

$$GS·2HCl \quad (90\%)$$

Forum Discussion Figure 2: Synthesis of gramicidin S (GS) *via*
N-hydroxysuccinimide ester-mediated cyclization

101

Some years ago we synthesized *semi*GS. Recently, we prepared *sesqui-* and *di*GS as the macro-ring analogs. As the control peptides, several linear peptides were also synthesized (Table II).

The ORD of several cyclic and linear peptides was measured in ethanol and 8 *M* urea (Figure 3). The "troughs"

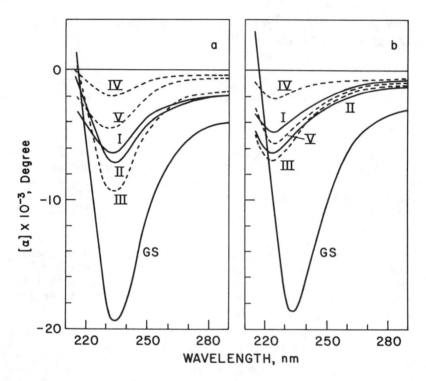

Forum Discussion Figure 3: ORD of cyclic and linear analogs of gramicidin S (GS). Solvent: a, ethanol; b, 8 *M* urea. I, *sesqui*GS; II, *di*GS; III, decapeptide; IV, pentadecapeptide; V, eicosapeptide.

of GS remained constant in both solvents. On the contrary, the position of the troughs of *sesqui*GS, *di*GS, linear deca-, linear pentadeca-, and linear eicosapeptide changed in 8 *M* urea. These results suggest that the conformation of the macro-ring analogs is similar to that of the linear analogs; whereas the conformation of GS is very stable even in 8 *M* urea as suggested already by Hodgkin, Schwyzer, Craig, and other investigators.

Forum Discussion Table II

Inhibitory Activity of Cyclic and Linear Analogs of Gramicidin S on Microorganisms

Compound	Minimum Inhibitory Concentration*	
	Staphylococcus aureus	Bacillus subtilis
↑Val-Orn-Leu-D-Phe-Pro┐ *semi*GS	>100	>100
↑Val-Orn-Leu-D-Phe-Pro┐ └Pro←D-Phe←Leu←Orn←Val↵ GS	5	5
↑Val-Orn-Leu-D-Phe-Pro-Val-Orn-Leu-D-Phe-Pro┐ └Pro←D-Phe←Leu←Orn←Val↵ *sesqui*GS	50	50
↑Val-Orn-Leu-D-Phe-Pro-Val-Orn-Leu-D-Phe-Pro┐ └Pro←D-Phe←Leu←Orn←Val←Pro←D-Phe←Leu←Orn←Val↵ *di*GS	20	10
Linear Analogs: H-(Val-Orn-Leu-D-Phe-Pro)$_n$-OH		
Pentapeptide (n = 1)	>100	>100
Decapeptide (n = 2)	50	50
Pentadecapeptide (n = 3)	50	50
Eicosapeptide (n = 4)	20	20
GS + *sesqui*GS†	5	5
GS + *di*GS†	5	2
GS + linear decapeptide†	5	5
GS + linear pentadecapeptide†	5	5
GS + eicosapeptide†	5	5

*MIC, µg/ml.

†Each mixture is composed of 1:1 (by weight).

The antibacterial properties of the macro-ring analogs were similar to those of the corresponding linear analogs (Table II). For example, compare *di*GS with the linear eicosapeptide. Furthermore, the activity increased with increasing molecular size. These results afford an additional support for the contention that the decapeptide sequence in GS is important to form the rigid structure, which is required for the high and specific activity of the cyclic decapeptide.

Reference

1. Schwyzer, R., and P. Sieber. Helv. Chim. Acta <u>40</u>, 624 (1957).

SCHWYZER: I think the hydroxysuccinimide method will prove to be very valuable. There is just one question: how good are the yields in the preparation of the hydroxysuccinimide decapeptide? With the nitrophenyl esters prepared by the DCCI method you have low yields, and one has to use the di-*p*-nitrophenyl sulfite to get high yields.

IZUMIYA: I believe about 70 or 80%.

CHEMISTRY AND BIOLOGY OF PEPTIDES
© Copyright 1972 *Ann Arbor Science Publishers*

SYNTHESIS OF A HYDROPHOBIC ALKALI ION BINDING CYCLIC PEPTIDE

B. F. Gisin. Department of Physiology, Duke University
Medical Center, Durham, N.C. 27710.

R. B. Merrifield. Rockefeller University, New York,
N.Y. 10021.

THE SO-CALLED "ION CARRIERS" are able to complex with alkali
ions and render them soluble in non-polar media. This
quality is of great interest in the study of the ionic
permeability of membranes, and it also provides a tool to
investigate the relationship between primary structure and
properties of a given molecule. Here we report the synthesis of a neutral cyclic peptide designed to solubilize
alkali salts in an organic phase through complexation with
the cation.[1]
    Clues necessary to devise a peptide sequence which would
contain all the information to display such a property are
found in the well established three-dimensional structure
of the potassium complex of valinomycin, *cyclo*(-L-valyl-D-
α-hydroxyisovaleryl-D-valyl-L-lactyl-)3.[2,3] It exhibits
the two main features of molecular architecture that are
common to all of the ion carriers: there is a polar interior and a non-polar exterior. All six ester carbonyl
oxygens point towards the center where they encompass the
potassium ion. The backbone of this *cyclo*dodecadepsipep-
tide encircles the cation in three complete waves, thus
forming six loops. Each peptide carbonyl is engaged in a
hydrogen bond with the closest peptide NH to form a bridge
across each of the loops. These hydrogen bonds are considered essential in stabilizing this symmetrical folding
of the backbone. All of the side chains point outwards
and thus shield the potassium ion and the hydrogen bonds
from the solvent.

Space-filling models show that the choice of changes in
the nature or chirality of the hydroxy and amino acid residues that would not distort the symmetrical arrangement of
the coordinating atoms is very limited. Every other residue
should provide an NH which can stabilize the "active conformation" through hydrogen bonding. The orientation of
the side chains (which is dictated by the optical configuration of the individual hydroxy or amino acid residues)
should be so chosen that they could not prevent the formation
of these hydrogen bonds. The models indicate the overall
geometry of the molecule is not significantly altered if
the ester groups are substituted by amide groups. Nevertheless, to exclude additional NH groups that might stabilize
other conformations through hydrogen bonding or decrease
the lipophilic character of the compound, these amide bonds
should be part of an imino acid, such as proline.

Based on these considerations we chose to synthesize
cyclo(-L-Val-D-Pro-D-Val-L-Pro-)$_3$. The solid phase method[4]
was used with the aid of an automatic Beckman Peptide Synthesizer. The tendency of H-D-Val-L-Pro sequence to cyclize
to give D-Val-L-Pro diketopiperazine[5] was particularly high
for H-D-Val-L-Pro-resin in the presence of a carboxylic
acid.[6] Therefore, DCC was added prior to the Boc-amino acid
in the coupling step. In the "regular" DCC coupling the
loss of dipeptide was *ca.* 70% and with this "reversed" procedure *ca.* 10-20%. The loss of peptide chains during the
synthesis was monitored by the picric acid method.[7] It indicated an amine content of the resin of 75% of its original
value at the tripeptide and 65% at the pentapeptide stage,
and it remained at that level throughout the synthesis. The
linear dodecapeptide was purified by gel chromatography and
cyclized with Woodward's Reagent K. The crystalline *cyclo*-
dodecapeptide gave the expected elemental analysis and showed
the calculated molecular weight of 1176 by mass spectrometry.
Amino acid analysis according to Manning and Moore[8] indicated
equimolar amounts of L-Val, D-Pro, D-Val and L-Pro.

The macrocycle was demonstrated to bind alkali ions as
follows. A known amount of the peptide was dissolved in
CH_2Cl_2 and solid potassium picrate (which is insoluble in
this solvent) was added. The yellow picrate went immediately
into solution and its spectrophotometric determination indicated a 1:1 complex with the peptide. Upon evaporation
of the solvent the compound was obtained in large crystals.
In CH_2Cl_2-H_2O the peptide was able to extract Li, Na, K, Rb,
and Cs picrate into the organic phase. In contrast,
valinomycin does not extract Li or Na picrate under these
conditions.

The IR spectra of the peptide and its potassium complex are shown in Figure 4. The two stretch bands of the uncomplexed peptide indicate the presence of both free (3397 cm^{-1}) and hydrogen-bonded (3317 cm^{-1}) amide hydrogens. Upon complexation these bands merge and undergo a bathochromic shift to form a single sharp band at a frequency of 3280 cm^{-1}.

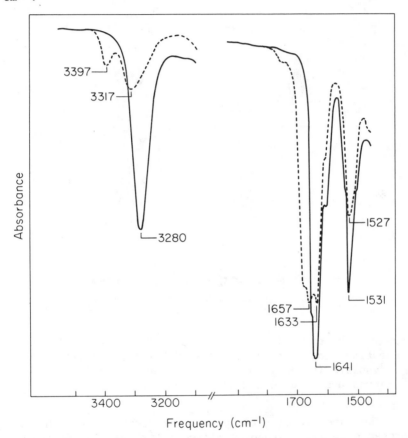

Forum Discussion Figure 4: Infrared spectrum of the free peptide (broken line) and of its potassium picrate complex (solid line) in CH_2Cl_2 (*c*, 2.5×10^{-3} *M*). The potassium thiocyanate complex gave the same spectrum.

The NMR spectrum (Figure 5) shows the presence of several non-equivalent amide hydrogens (6.5 to 8.2 ppm) in the free peptide while there is only one NH signal in the complex (8.12 ppm).

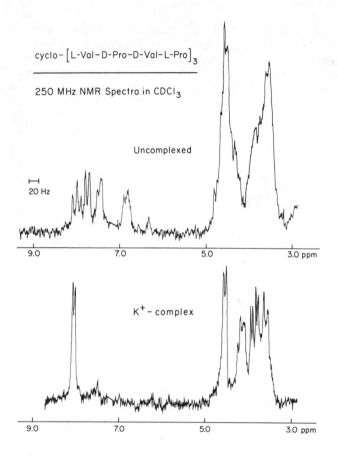

cyclo−[L-Val−D-Pro−D-Val−L-Pro]₃

250 MHz NMR Spectra in CDCl₃

Uncomplexed

20 Hz

9.0   7.0   5.0   3.0 ppm

K⁺- complex

9.0   7.0   5.0   3.0 ppm

Forum Discussion Figure 5:   250 MHz NMR spectrum of the free
   peptide (top) and its complex with potassium trinitro-
   cresolate (bottom) in CDCl₃. The sodium complex also
   showed a single NH signal (δ = 7.53 ppm).

   The data for the free peptide are consistent with the
presence of a mixture of conformers with varying degrees
of intramolecular hydrogen bonding. Complexation, on the
other hand, appears to eliminate all non-hydrogen-bonded
amide hydrogens in favor of one single type of hydrogen-
bonded NH as would be required for a symmetrical compact
structure analogous to the valinomycin-potassium complex.

Acknowledgments

   We thank Dr. D. C. Tosteson for inspiring discussions,
Dr. D. G. Davis for the NMR spectra, and Mr. Arunkumar

Dhundale for skillful technical assistance. Supported by
NIH Grant HE 12157, US Public Health Service Grant AM 1260,
and the Hoffmann-La Roche Foundation.

References

1. Gisin, B. F., and R. B. Merrifield. J. Amer. Chem. Soc.
   94, 6165 (1972).
2. Pinkerton, M., L. K. Steinrauf, and P. Dawkins. Biochem.
   Biophys. Res. Commun. 35, 512 (1969).
3. Ivanov, V. T., I. A. Laine, N. D. Abdulaev, L. B.
   Senyavina, E. M. Popov, Yu. A. Ovchinnikov, and M. M.
   Shemyakin. Biochem. Biophys. Res. Commun. 34, 803 (1969).
4. Merrifield, R. B. Advan. Enzymol. 32, 221 (1969); J.
   Amer. Chem. Soc. 85, 2149 (1963).
5. Meienhofer, J., Y. Sano, and R. P. Patel. In *Peptides:
   Chemistry and Biochemistry*, Weinstein, B., and S. Lande,
   eds. (New York: Marcel Dekker, 1970) pp 419-433.
6. Gisin, B. F., and R. B. Merrifield. J. Amer. Chem. Soc.
   94, 3102 (1972).
7. Gisin, B. F. Anal. Chim. Acta 58, 248 (1972).
8. Manning, J. M., and S. Moore. J. Biol. Chem. 243, 5591
   (1968).

PATEL: I should like to ask you whether you intend replacing
the proline by other amino acids? The choice of proline was
very good because the conformational maps of an L-proline
and an L-ester residue are similar.

GISIN: Yes, we are planning to synthesize analogs where the
prolines are replaced by other residues.

WIELAND: What was the yield on the cyclization reaction,
please?

GISIN: Cyclization was with Woodward's Reagent K according
to a procedure by Rudinger. The yield of pure crystalline
peptide was 16%.

WIELAND: Do you know something about the binding constant
of the peptide with the potassium ion, as compared with
valinomycin?

GISIN: In the two-phase system I have mentioned it is
comparable to or higher than for valinomycin.

OVCHINNIKOV: I would just like to mention that if you
replace in the valinomycin molecule any ester bonds for

N-methylamide, just replacing lactic acid by *N*-methylalanine, you can get complexation two orders of magnitude greater than with valinomycin itself. We have synthesized two such analogs.

GISIN: Did you replace all of the hydroxy acids?

OVCHINNIKOV: Both or either one of them.

GOODMAN: We can expand on that if we take something which is a valine analog in the ester field, this is hexahydro-mandelic acid. It has the steric restriction of a kind of isopropyl side chain plus the ester rigidity; and then one gets what Dr. Patel was alluding to, a very specific con-formational region allowed. When we get a chance, I'd like to report on our synthesis where we do the same thing with the idea of restricting the orientation of the carbonyl groups so that complexation can take place. This is alternating L-valine and D-hexahydromandelic acid cyclic hexamer.

EDITOR: Here we insert the manuscript of Yu. A. Ovchinnikov (Chairman) who courteously restrained from giving his talk to allow others ample speaking time.

THE CONFORMATION OF ANTAMANIDE IN NON-POLAR SOLVENTS

*Yu. A. Ovchinnikov, V. T. Ivanov, V. F. Bystrov,
A. I. Miroshnikov.* Shemyakin Institute for Chemistry
of Natural Products, USSR Academy of Sciences, Moscow,
USSR

STUDIES OF ANTAMANIDE (Figure 6) are of interest not only
for the purpose of ascertaining the mode of its antitoxic
action, but, of course, also as a compound selectively
stimulating Na^+ permeability in biological and artificial
membrances. It is a truism to say that the biological

Forum Discussion Figure 6:   Antamanide.

111

activity of peptides is highly dependent upon their conformational properties and antamanide is certainly no exception
to this rule, so that a necessary condition for comprehending
the peculiarities of its biological behavior is knowledge
of its conformational states under various conditions.

On the basis of IR, NMR spectra and ORD studies we
postulated[1] the Na⁺ complex of antamanide to have the rigid
conformation shown in Figure 7, stabilized by four intramolecular hydrogen bonds (IaMHB) and possessing a pseudo
axis of symmetry. The sodium ion enclosed in this structure
interacts more strongly with two of the carbonyls than with
the other four, a fact unequivocally confirmed in a study
of the ^{13}C NMR spectra of [Val[6], Ala[9]]-antamanide and its
Na⁺ complex.[2]

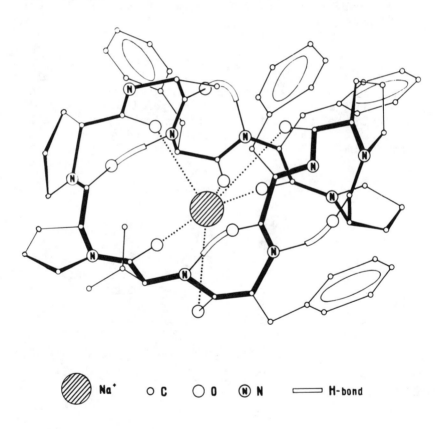

Na⁺   O C   ◯ O   Ⓝ N   ▭ H-bond

Forum Discussion Figure 7:   Conformation of the Na⁺ complex
   of antamanide (side view).

Free antamanide, on the other hand, as evidenced by spectral data, exists in the form of a highly involved conformational equilibrium whose position depends upon the composition of the medium. In non-polar media, what we have called form A, containing six IaMHB is predominant whereas on gradually passing over to alcoholic media several new forms successively appear in which the IaMHB system is more and more broken down. When passing from chloroform to dimethylsulfoxide a new form appears with *cis* amide bonds. The content of this form in the mixture can attain a value of 60%.

The structure of form A was established by systematic examination of all possible antamanide conformations possessing a twofold symmetry axis and six IaMHB. Of these the conformations were picked whose NH–CH protons were oriented in compliance with the requirements of the $^3J_{NH-CH}$ values determined from the NMR spectra, taking into account the possibility of both *cis* and *trans* amide bonds. The analysis showed that only two structures were in accord with the above requirements (all *trans* amide bonds). One of these structures was considered of little probability because of its high energy level according to the conformational energy maps and the poor agreement between the experimental and calculated dipole moments (5.2 - 5.8 D *vs.* 2.4 D). The second structure has the following parameters:

	Val^1, Phe^6	Pro^2, Pro^7	Pro^3, Pro^8	Ala^4, Phe^9	Phe^5, Phe^{10}
ϕ	−80	−60	−55	−100	60
ψ	165	−40	−40	10	−70

and the dipole moment (4.5 D) is in good agreement with the experimental values. As can be seen from Figure 8 this structure has two IaMHB of the type $3 \rightarrow 1$ and four IaMHB of the type $4 \rightarrow 1$; the latter securing the proline residues in a conformation corresponding to a 3_{10} helix. Figure 9 shows that the conformations of the peptide chain of form A and of the Na-complex of antamanide are much the same and as a first approximation one may represent transition of the complex conformation into the form A conformation as a twist of the planes of the secondary amide groups formed by Pro^3 and Pro^8. Thereby the CO^3 and CO^8 groups are shifted from the middle of the molecule to its periphery, while the NH^4 and NH^9 groups approach CO^1 and CO^6 to form with them IaMHB. Concurrently, there is a certain shift of the CO^1, CO^6, CO^5 and CO^{10} groups from the center of the internal cavity. Hence, whereas in the complex the six carbonyls are spatially nearer to one another and point

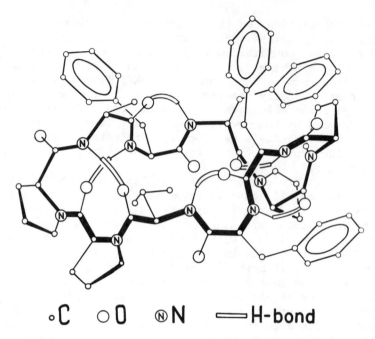

°C   ○O   ⊗N   ═══H-bond

Forum Discussion Figure 8:   Conformation of antamanide in
non-polar solvents (side view).

within the molecule, in the free antamanide molecule these
groups are much more remote from each other, excluding any
significant dipole-dipole repulsion.

It is noteworthy that Tonelli *et al.*,[3] on the basis of
CD and NMR data similar to ours, have arrived at quite
different conclusions regarding the antamanide structure.
The authors believe that antamanide has the same conformation in all the media they investigated (dioxane,
chloroform, methanol, *etc.*), characterized by the absence
of IaMHB and by location of all carbonyls on the same side
of the average plane of the ring. However, the dipole
moment of this structure, calculated on the basis of the
ϕ and ψ values they presented, is 16.6 D, while localization of the amide A bands in the 3350 – 3300 cm^{-1} region
of the IR spectrum leaves no doubt as to the existence of
IaMHB. Thus this conformation cannot be the prevailing
one in non-polar solvents, although it cannot be excluded
for polar media.

Prof. Th. Wieland in a private communication has informed us that he and his coworkers[4] have now accepted the
conformation proposed by us, which has been reported at the

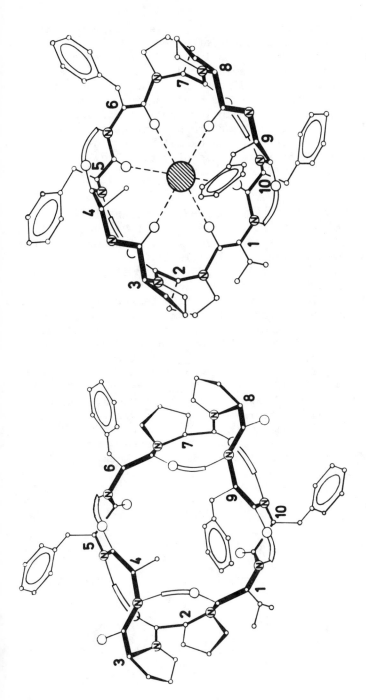

$\bigodot$Na$^+$ $\bullet$C $\circ$O $\otimes$N ━━H-bond

Forum Discussion Figure 9: Conformation of antamanide in non-polar solvents and in the Na$^+$-complex (view along the pseudo symmetry axis).

Granada Symposium (May, 1971).[5] At the same time considering the intensity of the Cotton effects at 223 nm on the
antamanide CD curves in polar solvents Wieland and co-
workers suggested that the amide bonds in the Pro^3-Ala^4
and Pro^8-Phe^9 fragments were nonplanar. We believe, however, that this requires further confirmation, since the
deviation from planarity should lead not only to augmented
intensity but to a red shift of the n→π* transition (as
observed by Deber *et al.*[6] on the example of cyclo-triproline).
There is no such shift in the case of antamanide.

References

1. Ivanov, V. T., A. I. Miroshnikov, N. D. Abdullaev,
    L. B. Senyavina, S. F. Arkhipova, N. N. Uvarova, K. Kh.
    Khalilulina, V. F. Bystrov, Yu. A. Ovchinnikov. Biochem.
    Biophys. Res. Commun. <u>42</u>, 654 (1971).
2. Bystrov, V. F., V. T. Ivanov, S. A. Koz'min, I. I.
    Mikhaleva, K. Kh. Khalilulina, Yu. A. Ovchinnikov,
    E. I. Fedin, P. V. Petrovski. FEBS Letters <u>21</u>, 34 (1972).
3. Tonelli, A., D. J. Patel, M. Goodman, F. Naider, H.
    Faulstich, T. Wieland. Biochemistry <u>10</u>, 3211 (1971).
4. Faulstich, H., W. Burgermeister, and Th. Wieland.
    Biochem. Biophys. Res. Commun. <u>47</u>, 975 (1972).
5. Ovchinnikov, Yu. A., V. T. Ivanov, and A. M. Shkrob.
    In *Proc. Int. Symp. Molecular Mechanisms of Antibiotic
    Action on Protein Biosynthesis and Membranes*, Granada,
    Spain, May 1971.(Amsterdam: Elsevier, 1972).
6. Deber, C. M., A. Scatturin, V. M. Vaidya, and E. R.
    Blout. In *Peptides: Chemistry and Biochemistry*,
    Weinstein, B., and S. Lande, eds. (New York: Marcel
    Dekker, 1970) pp 163-173.

CONFORMATION AND CRYSTAL STRUCTURE OF CYCLIC PEPTIDES

Isabella L. Karle. Laboratory for the Structure of
Matter, Naval Research Laboratory, Washington, D.C.
20390

I HAVE JUST A FEW MINUTES to present a lot of information,
so all I can really do is to bring out some highlights from
the various crystal structure determinations. In some sense
the great advantage of doing structure determination in the
solid state is that all the atoms are in fixed positions
so that the conformations and geometries can be determined
very exactly. On the other hand, this is a rather artificial
state for the molecule, and it is possible that its conformation is different when it is actually reacting. Figure
10 is an historic one: the structure determination was
made 10 years ago. However, it is very instructive, and
that is why I bring it up now. The material is cyclic
hexaglycyl and in the same cell there are four different
conformations for the molecule. Altogether there are 8
molecules in the unit cell, four of conformation a, two of
b, and one each of c and d. Since these conformers co-
crystallize, obviously, they must have fairly equivalent
energies. The conformers b, c and d do not have any internal hydrogen bonding, whereas conformer a has two internal hydrogen bonds. You are now quite familiar with
this particular type of internal hydrogen bonding. It
must be a fairly basic characteristic of polypeptides. It
has been found since in the crystalline state of several
other hexapeptides. In the valinomycin molecule, a cyclic
decapeptide, it occurs four times (there are also two
hydrogen bonds of another type). It occurs in an analog
of viomycin, which is an approximate cyclic pentapeptide.
It even occurs in linear peptides. It is found in the

117

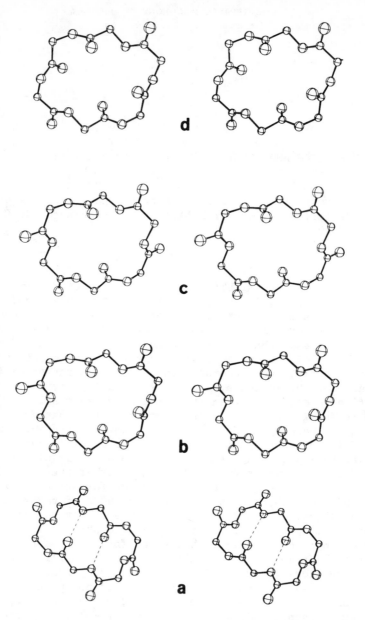

Forum Discussion Figure 10:   A stereodiagram of four different conformers which exist in one unit cell of *cyclo-*
   hexaglycyl·1/2 H$_2$O.   The different size spheres depict
   C, N and O atoms in order of size.[1,2]

tetrapeptide from the carbon terminal of oxytocin, and since
the conformation in the region of the internal hydrogen bond
is the same as in conformer *a*, the linear peptide is not
really linear but curled up into a *C*.

Now, if we take a look at the individual peptide groups
shown in Figure 10, each one is in a *trans* conformation.
What represents the difference between the conformers is
the arrangement of the residues. If we examine the portion
from nitrogen to nitrogen in conformer *a*, for instance, we
will see a *cis* arrangement, another *cis*, then *trans*, *cis*,
cis, *trans*. When we look at conformers *b*, *c* and *d*, we see
not only a *trans* and a *cis* type of conformation, but also
what I have called skew, that is, about 60° away from the
trans. Each one of the conformers *b*, *c* and *d*, has a number
of skew residues. As I have already mentioned, all the
peptide groups are *trans* and are planar. Under what conditions or where do we find *cis* peptide groups? Figure 11
shows the actinomycin molecule. The structure analysis was
performed by Sobell.[3] Actinomycin has two pentapeptide rings
which are the same. Let us examine the bottom one. I have
indicated the α-carbon atoms by the black coloration. Here
you see that there is a *cis* conformation for the peptide
group containing the prolyl moiety and also a *cis* conformation in the peptide group adjacent to the prolyl group.
Viomycin also has a 16-membered ring similar to the one in
actinomycin, but in viomycin the conformations for all the
peptide groups are *trans*. Figure 12 shows a cyclic tetradepsipeptide. The α-carbon atoms are at the corners of
the cyclic part of the molecule. The two ester groups (top
and bottom) are in the *trans* conformation whereas the two
peptide groups (on either side) are in the *cis* conformation.

Figure 13 shows a cyclic dipeptide which, of course,
contains a diketopiperazine ring. Here the peptide groups
are constrained to be *cis* and they are planar, or essentially
planar, as in all the studies of diketopiperazine rings of
which I am aware. On the other hand, the diketopiperazine
ring takes on various conformations. Sometimes it is planar
and sometimes it is a twist-boat. Here the diketopiperazine
ring is very definitely in the boat conformation. There
is a fold along the line joining the α-carbon atoms with a
dihedral angle of about 140° between the planes of the two
peptide groups. There are two ways in which the side groups
can be attached, either in the equatorial position or, if
the diketopiperazine ring is folded in the opposite direction, in the axial position. From the small amount of
information we have so far in the solid state, it seems
that the attachment is in the equatorial position when the

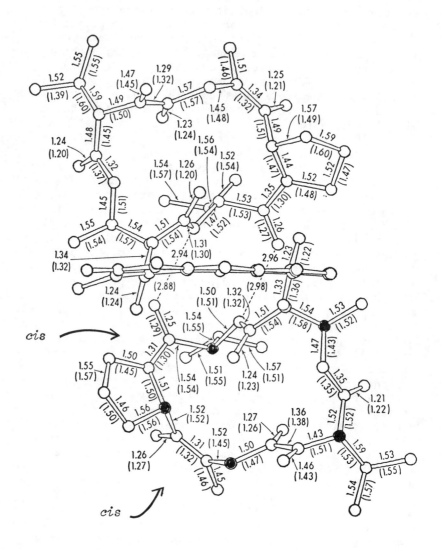

Forum Discussion Figure 11: The conformation of the
actinomycin molecule. In the lower peptide ring the
shaded atoms are the α-carbon atoms.[3]

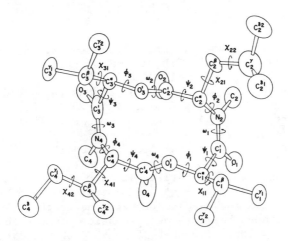

Forum Discussion Figure 12: The structure of the cyclic
tetradepsipeptide D–HyIv–L–MeIle–D–HyIv–L–MeLeu.[4]

Forum Discussion Figure 13: A stereodiagram of cyclic
L–Pro–L–Leu.[5]

side group is aliphatic, as in this case. Hence the two
hydrogen atoms on the α–carbons are in the axial positions
and quite close together. However, if one of the side
groups is aromatic, the axial position for attachment
appears to be preferred and the side chain with the
aromatic ring is folded over the diketopiperazine ring.

I know that this presentation is a bird's eye view,
but limited time prevents a more thorough discussion.

References

1. Karle, I. L., and J. Karle. Acta Cryst. 16, 969 (1963).
2. Karle, I. L., J. W. Gibson, and J. Karle. J. Amer.
   Chem. Soc. 92, 3755 (1970).
3. Sobell, H. M., S. C. Jain, T. D. Sakore, and C. E.
   Nordman. Nature New Biology 231, 200 (1971).
4. Konnert, J., and I. L. Karle. J. Amer. Chem. Soc. 91,
   4888 (1969).
5. Karle, I. L. J. Amer. Chem. Soc. 94, 81 (1972).

BLOUT: Thank you, Dr. Karle. It is remarkable to me that
with the power of the X-ray method there has been so little
discussion of X-ray results of peptides at this meeting,
although we have had lots of discussion of solution conformation results. I wonder if somebody would like to
comment on or to contrast the solution results in specific
cases with solid-state X-ray diffraction results?

KOTELCHUCK: I was surprised, when we began doing energy
calculations on oxytocin, to discover the degree of flexibility of its proposed structure--that in fact, we couldn't
find a unique low energy conformation. Today Dr. Karle
has reported on the many structures found in X-ray crystals,
and previously there were reports of loose structures in
NMR. Thus, it seems to me that we ought to consider the
peptides, not as having a unique structure, but rather a
range of related structures, as found with oxytocin. I
think we are seeing evidence at this symposium, from different points of view, that this is the case.

PATEL: I will try to answer your question on comparison
between solution, theoretical calculations and X-ray methods.
For metal complexes like valinomycin-K and enniatin-Na, the
solution conformations derived from spectroscopic studies
coupled with theoretical calculations are in excellent
agreement with the crystallographic results. The solution
conformations of uncomplexed peptides and depsipeptides
are highly solvent dependent. For instance, studies from
Professor Ovchinnikov's and our laboratory suggest that
valinomycin exists in different conformations in hydrocarbon, polar and aqueous media. These conformations which
exhibit three-fold symmetry have been defined in terms of
their backbone rotation angles. The crystal structure for
valinomycin has just been reported and does not exhibit
three-fold symmetry. The backbone rotation angles are
considerably different from those predicted for the solvent

dependent conformations in solution. In this particular
case, solution and crystallographic analysis are in disagreement. A comparison can be made between a conformation
for actinomycin-D calculated by DeSantis[1] from conformational
analysis and a crystal structure analysis by Sobell[2] on
actinomycin-D deoxyguanosine (1:2) complex. There is excellent agreement between the two studies for the structure
of the antibiotic. In the crystal, the two pentapeptide
lactones are related by a dyad symmetry axis while the solution data and conformational calculations suggest a small
nonequivalence between the conformations of the two rings.

References:

1. Conti, F., and P. DeSantis. Nature <u>227</u>, 1239 (1970).
2. Sobell, H. M., S. C. Jain, T. D. Sakore, and C. E.
   Nordman. Nature New Biology <u>231</u>, 200 (1971).

BOVEY: I just wanted to make a very brief contribution to
the subject that's been alluded to: the comparison between
the solution structure as observed by NMR and the crystal
structure. There is one other example (perhaps still others
that I don't know of) of a simpler compound other than what
has been mentioned so far, namely, *cyclo*(-D-Ala-D-Ala-Gly-
Gly-Gly-Gly-). Here the crystal structure was determined
by Dr. Karle and her associates a couple of years ago, and
somewhat later the NMR and energy calculations were reported
from our laboratory by Tonelli and Brewster. There is an
obvious discrepancy between them which is perfectly all
right, but just illustrates the point that the solution
structure may be very different. According to NMR and
according to energy calculations, the solution conformation
of this molecule in dimethyl sulfoxide probably has about
24 or 25 oscillating structures between which it is
equilibrating quite rapidly. None of these correspond at
all to the crystalline structure, which was actually very
similar to the hexa-Gly shown as *a* on your first slide
(see Figure 10), if I'm not mistaken. The reason for the
difference is that there are three molecules of water of
crystallization in the crystals and, as Dr. Karle's work
has shown, a very efficient system of hydrogen bonds: 9
hydrogen bonds to the water of crystallization, one intramolecular hydrogen bond, as shown in one of the hexa-Gly
structures, and also one intermolecular hydrogen bond
between peptide molecules, for a total of 12 hydrogen
bonds. Multiplied by a nominal value for the hydrogen
bond energy this clearly enables this structure to exist

in the crystal, although according to energy calculations it
is about 16 kilocalories above the solution conformation.
So here is one clearcut case of a difference between the
crystal and the solution structure.

RUDINGER: I should like to propose a very primitive consideration which suggests why the solution conformation of
smaller peptides might differ more from the crystal conformation that we have got used to seeing with proteins.
If we assume, very crudely, that the conformation-holding
forces, which operate in the "interior" of the molecule,
are proportional to the volume of the molecule, and the
intermolecular forces, acting superficially either in the
crystal or in solution, are proportional to the surface
area, then we would expect that the larger the molecule,
and hence the higher the volume-to-surface ratio, the more
the conformation-holding forces will predominate over the
intermolecular forces and the less sensitive the conformation will be to the molecular environment (solution or
crystal).

GURD: On the other hand, I think from the point of view
of the protein chemist you must be very careful to leave
plenty of room for differences around your surface where
you're making your point. I think this has to be remembered. The argument you have here pleases me because I
think what you have been putting forth fits in with what
we have to contend with in the protein field all the time,
i.e. discrepancies between crystal and solution structure
and detail. You can't possibly have a structure sitting
prefectly still at the active site of an enzyme while it's
working, and that kind of thing.

GLICKSON: I would like to cite some experimental evidence
in support of Dr. Rudinger's contention that proteins are
generally more rigid than peptides, and that consequently
with proteins crystal structures are better approximations
to the solution structures. In the course of NMR studies
of the denaturation of proteins by heat, acid, and chemical
denaturants, Drs. McDonald, Phillips, and I observed that
unfolding of proteins was usually slow on the NMR time
scale. As a result, one generally observes in the transition range an NMR spectrum which is the weighted super-
position of spectra of the native and denatured protein.[1]
By contrast, most of the conformational changes of peptides
that have been studied by NMR spectroscopy are fast on the
NMR time scale, and throughout the transition a single

spectrum is observed which continuously changes its characteristics from the spectrum of the initial state in the
transition to the spectrum of the final state. There are,
of course, some exceptions to this rule; but, for the most
part, peptides unfold more rapidly than proteins, which
implies that more free energy must be expended to overcome
the greater cohesive forces which stabilize protein conformations. This evidence together with various comparisons
of protein structure in the crystalline state and in solution[2] explains why x-ray studies are generally more relevant
to solution studies of proteins than of peptides. The
cyclic peptide-cation complexes mentioned by Dr. Gisin
are notable exceptions to this generalization.

Because of evidence that proteins such as hen egg white
lysozyme, whose crystal structure has been characterized
by x-ray diffraction, for the most part retain this structure in solution, it is possible to use these proteins to
assess the extent to which the temperature dependence of
NH resonances reflects the exposure to the solvent of NH
hydrogens. Inspection of the three dimensional model of
hen egg white lysozyme shows that the indole NH protons
of Trp-28, Trp-108, and Trp-111 are internally hydrogen
bonded, whereas those of Trp-62, Trp-63, and Trp-123 are
for the most part exposed to the solvent. Recently, Drs.
Phillips, Rupley, and I assigned the five resolved
tryptophan indole NH resonances of hen egg white lysozyme
to their specific tryptophan residues.[3] In Figure 14, we
display the temperature dependence of the chemical shifts
of these resonances. For comparison, we have also included
the indole NH resonances of the amino acid tryptophan and
the six fold degenerate indole NH resonance associated
with thermally denatured hen egg white lysozyme. The
indole NH hydrogens associated with the free amino acid
and denatured protein are all exposed to the solvent. It
is apparent from this figure that there is no obvious correlation between intra- and intermolecular hydrogen bonds
and temperature dependence. Thus, the two resonances
associated with intramolecularly hydrogen bonded indole
NH protons (Trp-28 or 111 and Trp-108) yield similar
slopes to the resonances associated with the exposed
indole NH protons of Trp-62, Trp-63, tryptophan, and the
tryptophan residues of denatured hen egg white lysozyme.
This serves to illustrate that, at least in aqueous
solution, temperature dependence of NH resonances is not
a reliable criterion for distinguishing between intra- and
intermolecular hydrogen bonds. Whereas these studies were
confined to indole NH resonances, peptide NH resonances
are expected to behave similarly.

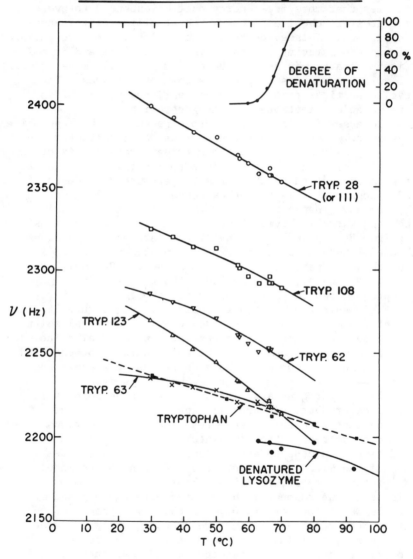

INDOLE N<u>H</u> RESONANCES OF 10% HEW LYSOZYME / H₂O, pH 3.3

Forum Discussion Figure 14: The temperature dependence of the chemical shifts of indole N<u>H</u> resonances (at 220 MHz) in H₂0.[4]

References

1. McDonald, C. C., W. D. Phillips, and J. G. Glickson.
    J. Amer. Chem. Soc. <u>93</u>, 235 (1971).
2. Rupley, J. A. In *Biological Marcomolecules*, Vol. 2,
    Timasheff, S. N., and G. D. Fasman, eds. (New York:
    Marcel Dekker, 1969), pp 291-352.
3. Glickson, J. D., W. D. Phillips, and J. A. Rupley.
    J. Amer. Chem. Soc. <u>93</u>, 4031 (1971).
4. Glickson, J. D., and W. D. Phillips. Unpublished.

WALTER: I would like to take this opportunity to discuss
two conformational assignments in the neurohypophyseal
hormone field which left me unsatisfied. The first concerns
one of the conformational assignments of the β-turn in the
tail portion of lysine-vasopressin (LVP) in dimethylsulfoxide (DMSO) considered by Von Dreele *et al.*,[1] and the
second the suggestion by Hruby *et al.*[2] that in the tetrapeptide, H-L-Cys(Bzl)-L-Pro-L-Leu-Gly-NH$_2$, the lone electron
pairs on the sulfur and the carboxamide nitrogen might
interact.

Figure 2 in Ref. 3 shows the proton NMR assignment of
LVP, which is similar to that reported by Von Dreele *et al.*[4]
In passing, I would like to mention that both groups have
used a different approach for the assignment of LVP. Von
Dreele *et al.*[1] followed the route we introduced some years
ago with oxytocin.[5] That is, in their study resonances
were assigned progressively from spectra of the C-terminal
dipeptide, tripeptide, etc., through the acyclic nonapeptide of LVP, and then, when possible, this information was
applied to the assignment of the LVP spectrum. We abandoned
this approach for the assignment of the proton NMR spectra
of neurohypophyseal peptides, because it suffers from some
intrinsic difficulties. In retrospect, we can say that
considerable changes occur in the chemical shifts of
resonances of certain residues upon ring closure of the
nonapeptide precursor of oxytocin to the hormone, and that
similar effects were observed with other neurohypophyseal
peptides. Therefore, we felt a more unequivocal way to
assign the proton NMR spectrum of a previously unassigned
neurohypophyseal hormone analog would be *via* a comparison
of its spectrum with the assigned spectrum of oxytocin.
The positions of resonances (side chain resonances and, by
subsequent spin decoupling experiments, the corresponding
peptide NH resonances) associated with residues present in
the analogs but not in oxytocin were estimated on the basis
of the study by McDonald and Phillips.[6] This approach was

found to be very satisfactory not only with LVP, but also with other neurohypophyseal peptides investigated subsequently.

Returning to the conformational analysis of the tail portion of LVP, it will be noticed that the α-CH-NH coupling constants of Lys in LVP and Leu in oxytocin are similar, as are their peptide NH temperature coefficients. The same holds for the NH peptide resonances of the Gly residues in both hormones.

This data was part of the evidence which suggested to us that in DMSO the peptide NH of Gly is hydrogen-bonded to the C=O of Cys-6 to form in LVP a β-turn comprised of the sequence -Cys-Pro-Lys-Gly-; the residues Pro and Lys occupy the corners of the β-turn. This conformation of the tail portion of LVP seems to be less preferred and much more sensitive to small changes in solvent composition than the β-turn in oxytocin, comprised of the sequence -Cys-Pro-Leu-Gly-. The conformational assignment of the latter seems to be well supported in view of the presentation by Dr. Kotelchuck, pp 35-38, in which he confirmed that conformational energy calculations starting with the proposed solution conformation of oxytocin are in agreement with the formation of such a β-turn. In addition, Rudko *et al.*[7] find an identical conformation of the crystal structure of (S-Bzl)-Cys-Pro-Leu-Gly-NH$_2$.

I wonder why Dr. Von Dreele or Dr. Bovey did not consider such a conformation for LVP but instead prefer as a possible structure for LVP in DMSO at room temperature one which would contain a hydrogen bond between the *trans* NH of Gly-NH$_2$ and the C=O of proline. On the basis of our experiences with H-Pro-Leu-Gly-NH$_2$, where we believe that such a β-turn does exist (*vide infra*), we would expect a large difference between the position of resonances of the *cis-* and *trans*-protons of the Gly-NH$_2$ in the LVP spectrum. However, the differences in chemical shift between these carboxamide protons are small and just about the same for oxytocin and LVP.

Hruby *et al.*[2] noted a large decrease in the differences of chemical shift between the nonequivalent Gly-NH$_2$ protons when they compared the proton NMR spectra of H-Pro-Leu-Gly-NH$_2$ and (S-Bzl)-Cys-Pro-Leu-Gly-NH$_2$. As mentioned above, it was suggested that this effect is due to interactions between the carboxamide -NH$_2$ and the sulfur atom. However, in the very same publication Hruby *et al.* show that Z-Pro-Leu-Gly-NH$_2$, which does not contain the sulfur, likewise exhibits a small difference in chemical shift in the resonances of the *cis* and *trans* Gly-NH$_2$ protons. Similar data were obtained when the proton NMR spectrum of (1,6-

aminosuberic acid, 2-alanine)-oxytocin, an analog in which
the disulfide group is replaced by an ethylene bridge
(Sakakibara *et al.*, unpublished) was recorded in DMSO.
Also, in the carbon-13 NMR study of Deslauriers *et al.*[8]
no changes in chemical shifts other than those associated
with the Pro residue were seen in comparing Z-Pro-Leu-Gly-
NH_2 with Z-Cys(S-Bzl)-Pro-Leu-Gly-NH_2. My question is
whether Dr. Hruby still retains his original proposal?

References:

1. Von Dreele, P. H., A. I. Brewster, F. A. Bovey, H. A.
   Scheraga, M. F. Ferger, and V. du Vigneaud. Proc. Nat.
   Acad. Sci. U.S. 68, 3088 (1971).
2. Hruby, V. J., A. I. Brewster, and J. A. Glasel. Proc.
   Nat. Acad. Sci. U.S. 68, 450 (1971).
3. Walter, R., J. D. Glickson, I. L. Schwartz, R. T.
   Havran, J. Meienhofer, and D. W. Urry. Proc. Nat.
   Acad. Sci. U.S. 69, 1920 (1972).
4. Von Dreele, P. H., A. I. Brewster, H. A. Scheraga,
   M. F. Ferger, and V. du Vigneaud. Proc. Nat. Acad.
   Sci. U.S. 68, 1028 (1971).
5. Johnson, L. F., I. L. Schwartz, and R. Walter. Proc.
   Nat. Acad. Sci. U.S. 64, 1269 (1969).
6. McDonald, C. C., and W. D. Phillips. J. Amer. Chem.
   Soc. 91, 1513 (1969).
7. Rudko, A. D., F. M. Lovell, and B. W. Low. Nature New
   Biology 232, 18 (1971).
8. Deslauriers, R., R. Walter, and I. C. P. Smith.
   Biochem. Biophys. Res. Commun. 48, 854 (1972).

HRUBY: No, not necessarily, this was only a suggestion.
Clearly, one can suggest a number of other functional groups
that could cause the quite small chemical shift changes
(0.1 to 0.3 ppm) resulting in the nearly identical chemical
shifts observed for the glycinamide carboxamide hydrogens.
The important observation is that the effect is apparently
caused by the presence of a group or groups attached to the
proline nitrogen, and this can be taken to imply that the
tripeptide and tetrapeptide you mentioned spend at least
some of their time in DMSO in a conformation which enables
the amino and carboxyl terminal ends of these peptides to
interact. The problem, of course, as you point out is to
unambiguously identify these interactions. The major point
of our paper was that *cis,trans* isomerism obtains about the
X-proline bond in the two peptides, and this seems firmly
established.

KOPPLE: I'm not as familiar as I should be with all the
ins and outs of the NMR spectra of these hormones, but some
of the conclusions do seem to be dependent on the temperature dependence of the chemical shift of amide protons.
I would like to ask if anyone has reasons to say that a
zero chemical shift dependence necessarily means a transannular hydrogen bond? I don't think it does so necessarily.
It may mean no hydrogen bond at all.

WALTER: I would agree with Dr. Kopple that a zero temperature shift does not absolutely have to be equated with the
presence of an intramolecular hydrogen bond, but the
assignment--I think you are referring to the intramolecular
hydrogen bond assignment in oxytocin--is not based solely
on the temperature plot but also on proton-deuterium exchange
studies.[1]

Reference

1. Walter, R., R. T. Havran, I. L. Schwartz, and L. F.
    Johnson. In *Peptides 1969*, Scoffone, E., ed.
    (Amsterdam: North-Holland Publ. Co., 1971) pp 255-265.

GOODMAN: I think that we have a case where this is actually
proved, that there isn't a hydrogen bond if the temperature
dependence is essentially zero.

KOPPLE: That seems reasonable. If there is nothing around
the hydrogen, its environment will not change with temperature.

BLOUT: In what solvent, Murray?

GOODMAN: In various solvents, the one I quote, I think, is
carbon tetrachloride. It also is the case in cyclohexane.

HAS THE MSH-RELEASE-INHIBITING HORMONE A PREFERRED
CONFORMATION?

Roderich Walter. Department of Physiology, Mount
Sinai School of Medicine, City University of New York,
New York, and Brookhaven National Laboratory, Upton,
New York.

Ivan Bernal. Brookhaven National Laboratory, Upton,
New York.

*LeRoy F. Johnson.** Varian Associates, Palo Alto,
California.

HAVING PRESENTED EVIDENCE that the C-terminal tripeptide
of oxytocin is the natural factor inhibiting the release
of melanocyte-stimulating hormone[1]--a finding substantiated
by the isolation of an active principle with this structure
from bovine hypothalami[2]--we turned to the question of
whether this small peptide, H-Pro-Leu-Gly-NH_2(I) possesses
a detectable preferred conformation.

The particular synthetic sample of I used in this investigation crystallized as a *hemihydrate* from water and
exhibited the properties described by Zaoral and Rudinger.[3]
The proton nuclear magnetic resonance (PMR) spectrum of I
taken in deuterated dimethylsulfoxide (DMSO-d_6) at 300 MHz
is shown in Figure 15. The assignments, determined by spin-
decoupling experiments, agree with those previously reported[4] except for the chemical shifts of the α-CH proton
resonance of proline (Figure 16). Splitting due to coupling
between Pro β-CH and NH cannot be seen; this may indicate a
J of zero, or, more likely because of the broad signal
around 3.2 ppm, may result from rapid exchange of the Pro
NH proton with H_2O.

---

*Present Address: Transform Technology, Inc., Palo Alto,
California.

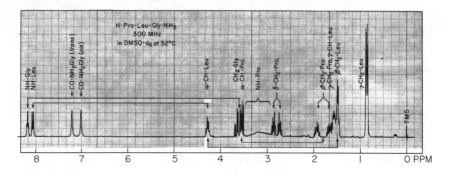

Forum Discussion Figure 15:   300 MHz PMR spectrum of MSH-release-inhibiting hormone.

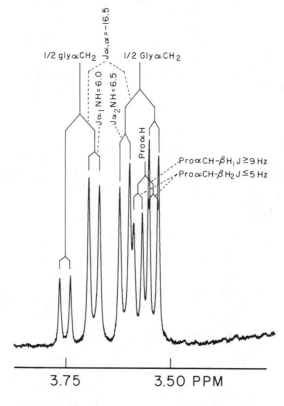

Forum Discussion Figure 16:   Expansion of Figure 15 in the 3.75 to 3.50 ppm region, which shows the chemical shifts and splitting patterns of the CH$_2$ of the glycine residue and the α-CH of the proline residue.

Note the conspicuously large difference in chemical shift between the resonances of the *trans* and *cis* carboxamide protons of the glycinamide moiety; in oxytocin this difference amounts to only 0.05 to 0.07 ppm.[5] We presently believe this large chemical shift difference suggests the presence of a hydrogen bond between the *trans* carboxamide proton and the C=O of proline, to form a 10-membered β-turn. In systems of at least four amino acid residues the Type I or II structures of Venkatachalam[6]—which are likewise 10-membered β-turns—seem to be generally preferred; such conformations have been found for the sequence –Cys–Pro–Leu–Gly–NH$_2$ in oxytocin[7] as well as for the isolated tetrapeptide[8] and for –Cys–Pro–Lys–Gly–NH$_2$ in LVP.[9] However if there are only three amino acid residues involved and the chemical prerequisites exist for the formation of an intramolecular hydrogen bond, then a conformation as suggested for I may be the peptide's choice. Hydrogen bonding between carboxamide protons and carbonyl oxygens has been reported in the literature, *e.g.* for succinamide.[10] The fact that the NH of proline does not enter into hydrogen-bond formation with either a carbonyl oxygen or the oxygen of a water molecule is in line with the low electrophilicity of the proton on the secondary nitrogen.

There is also preliminary evidence for the proposed conformation of I in the crystalline state. Two monomers would be held together by one H$_2$O molecule as shown in Figure 17, in which *both* protons of water interact with the carbonyl oxygens of the glycinamide residues of two molecules of I. An analogous situation has been described for the structure of glycyl-L-tryptophan · 2H$_2$O.[11] Moreover, the model is consistent with the crystallographic data listed in Table III. Given the crystal space

Forum Discussion Table III
Crystallographic Parameters of H–Pro–Leu–Gly–NH$_2$·1/2 H$_2$O

Space Group:	P6$_1$22 or P6$_5$22	Density (measured) :	1.20(2)
a = b :	10.594(1) Å	Density (calculated) :	1.19
c :	50.355(8) Å	No. molecules/cell :	12 tripeptides, 6H$_2$O
Chemical Composition:	C$_{13}$H$_{23}$O$_4$N$_4$·1/2 H$_2$O; thus there is a crystallographic requirement that the H$_2$O molecules be located at two-fold axes of the space group.		

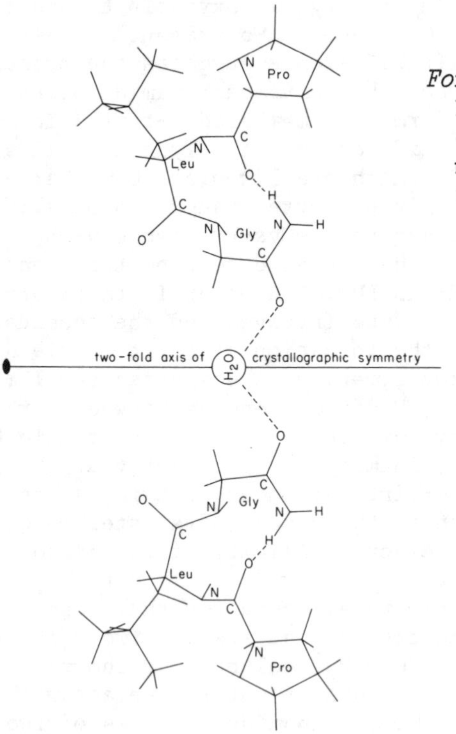

Forum Discussion Figure 17:
Preferred conformation
of H–Pro–Leu–Gly–NH$_2$
monomer and its packing
in the crystalline state.

group, cell constants, and density, the dimeric model has
to be placed on a two-fold axis for either choice of these
space groups, all of which is consistent with conformation
of the dimer shown in Figure 17. A detailed X-ray crystallographic investigation of I is in progress.

Acknowledgment

This work was in part supported by USPHS grant AM–13567
and by the U.S. Atomic Energy Commission.

References

1. Celis, M. E., S. Taleisnik, and R. Walter. Proc. Nat.
   Acad. Sci. U.S. <u>68</u>, 7 (1971).
2. Nair, R. M. G., A. J. Kastin, and A. V. Schally.
   Biochem. Biophys. Res. Commun. <u>43</u>, 1376 (1971).

3. Zaoral, M., and J. Rudinger. Coll. Czech. Chem. Commun. 20, 1183 (1955).
4. Hruby, V. J., A. I. Brewster, and J. A. Glasel. Proc. Nat. Acad. Sci. U.S. 68, 450 (1971).
5. Johnson, L. F., I. L. Schwartz, and R. Walter. Proc. Nat. Acad. Sci. U.S. 64, 1269 (1969).
6. Venkatachalam, C. M. Biopolymers 6, 1425 (1968).
7. Urry, D. W., and R. Walter. Proc. Nat. Acad. Sci. U.S. 68, 956 (1971).
8. Rudko, A. D., F. M. Lovell, and B. W. Low. Nature New Biology 232, 18 (1971).
9. Walter, R., J. D. Glickson, I. L. Schwartz, R. T. Havran, J. Meienhofer, and D. W. Urry. Proc. Nat. Acad. Sci. U.S. 69, 1920 (1972).
10. Davis, D. R., and R. A. Pasternak. Acta Cryst. 9, 334 (1956).
11. Pasternak, R. A. Acta Cryst. 9, 341 (1956).

VON DREELE: Let me begin by saying that it would have been impossible to obtain the results which I am going to explain today without the cooperation of a very talented engineer by the name of Dr. Joseph Dadok. Much of the structural information which is contained in a proton NMR spectrum of an oligopeptide is obtained from the peaks arising from the peptide protons in the form of coupling constants, temperature dependence on the chemical shift, and the HD exchange data which are related to dihedral angles, hydrogen bonding, or conformational interchange. In order to use this information to obtain the 3-dimensional structure of the oligopeptide, we must be able to assign each NH peak to a particular proton in the molecule. This assignment is generally made by establishing the spin decoupling relationships between each NH peak and a C-α-H peak, and then between that C-α-H peak and the C-β-H peaks and then using the chemical shifts and the splitting patterns of the C-β-H protons to assign this set of the coupling-related peaks to a specific amino acid in the molecule. When one attempts to perform this experiment in water, one generally fails, since the C-α-H protons are located under the large H_2O peak and when you attempt to irradiate them you experience experimental difficulties with the instrument. Therefore, most of the work that has been done so far on oligopeptides has been done in dimethylsulfoxide. This has led to a number of discussions which we have heard earlier this week as to whether a structure is necessarily the same in dimethylsulfoxide as it is in water, and whether

the peptides that you recover from dimethylsulfoxide have
been chemically modified by the solvent (see p 580).
I would like to describe a series of decoupling experiments
which have enabled us to avoid these problems by simply
working directly in H_2O. The main problem in a decoupling
experiment in any aqueous solution is the strong H_2O signal
entering the NMR spectrometer each time the second radio-
frequency field irradiates a line close to the water peak.
The strong signal will usually saturate one or more amplifier
stages and the lock or signal channels or both channels. A
saturated amplifier stage will prevent the proper function
of the appropriate channel and we lose the internal lock
or obtain a distorted spectrum or both. The most probable
stages to be overloaded are the last audiofrequency stages
and the audiofrequency synchronistic detectors. A successful decoupling experiment can be performed if we prevent
the saturation at any stage in the spectrometer by properly
adjusting the amplification along the path of the signals.
In the course of this work we have noted that under certain
circumstances the success of this decoupling experiment is
much more difficult to achieve than under others. It is
difficult if the C-α-H proton is located under the H_2O
peak, but not at the H_2O resonance frequency. It is easy
if the C-α-H proton being irradiated is not located under
the H_2O peak or is located at the resonance frequency of
H_2O. I have brought along with me today two slides which
will show successful experiments decoupling under the
water peak, and illustrate both an easy case and a hard
case. Figure 18 is that of oxidized glutathione. The NH
protons are located in the region near 8.1 ppm. The C-α-H
protons are located near 4.0 ppm under the very large water
peak which has been truncated, and the C-β-H protons are
located from 1.4 to 3.0 ppm. We have expanded the region
of the NH protons and it is shown in the insert. There is
a doublet from the cystine residue and a triplet from the
glycine residue. The frequency 4000 Hz is an off-resonance
frequency. The easy case corresponds to irradiating the
glycine residue where the C-α-H protons appear out from
under the water peak, and we see that if we irradiate it
at 895 hertz, we obtain the collapse of the characteristic
glycine triplet to a singlet. This then is an easy case
where the C-α-H peak is out from under the water. Another
easy case occurs when the C-α-H peak is directly at the
resonance frequency of the water, since irradiation at that
frequency saturates the water signal as well. An illustration of this is the cystine residue which happens to occur
at a frequency of 1095 Hz which is exactly the resonance of

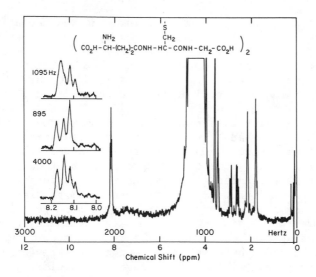

Forum Discussion Figure 18: PMR of oxidized glutathione in H$_2$O with selective NH, C-α-H decoupling.

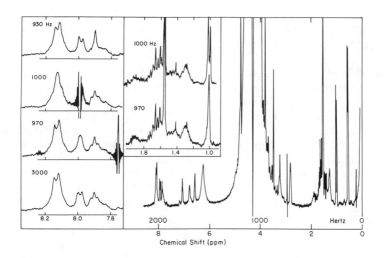

Forum Discussion Figure 19: PMR of an Ala, Arg, Gly-containing peptide in H$_2$O with selective NH, C-α-H decoupling.

the water frequency in this spectrum. We see the collapse
of the characteristic doublet of the cystine to a singlet.
Figure 19 shows a more difficult case, *i.e.*, the C-α-H
protons occur at intermediate frequencies. This NH region
has again been expanded in the insert and 3000 Hz is an
offset frequency of irradiation. We notice that if we
irradiate first at 930 hertz we collapse the characteristic
triplet of glycine to a singlet peak. As we move then in
towards the water and irradiate at 970 hertz, we collapse
a doublet to a singlet in the NH region, and if we observe
the C-β-H region we see a three-proton doublet has been
collapsed to a singlet. This is characteristic of the
behavior of alanine, and permits us to assign this particular
NH proton by a direct irradiation to the alanine resonance.
If we move further in towards the water peak and irradiate
at 1000 megacycles, we see a collapse of another doublet to
a singlet peak. There are also changes in a region which
is characteristic of the C-β-H_2 region of arginine. I have
shown you today the first proton spectrum in which we were
able to obtain selective NH, C-α-H decoupling of C-α-H
protons located under the H_2O peak in an aqueous solution
of the oligopeptide. It is no longer necessary to work in
dimethylsulfoxide. We can now establish the NH to C-α-H
to C-β-H spin decoupling relationships and do the concomitant
peak assignments directly in the biologically interesting
solvent, water.

COMMENT: Would you mind saying again what the instrumental
modification was that ....

VON DREELE: There isn't any modification, it's just a very
nice instrument. The success of the experiment depends in
part upon the dynamic ranges of the amplifiers in the instrument. You have to be able to hold in your amplifier
both the very small NH signal and the very large H_2O signal
which you obtain when irradiating at frequencies where the
H_2O signal has a substantial amplitude. To do this without
saturating the amplifier (which will introduce beats and
otherwise distort the spectrum) depends upon your ability
to control the settings on each amplifier that is in the
instrument. Some of the commercial instruments may have
preset values on the control knobs on the amplifiers or
amplifiers which do not have a large enough dynamic range.
This is something which you can ascertain by going over
your instrument with an oscilloscope and checking that
each amplifier is not saturated. When these conditions
are satisfied, you should be able to do this experiment,

just as I do this experiment, and therefore working directly in water rather than in dimethylsulfoxide.

WALTER: I would like to show you the assignment of the proton magnetic resonances of lysine vasopressin (Figure 20).

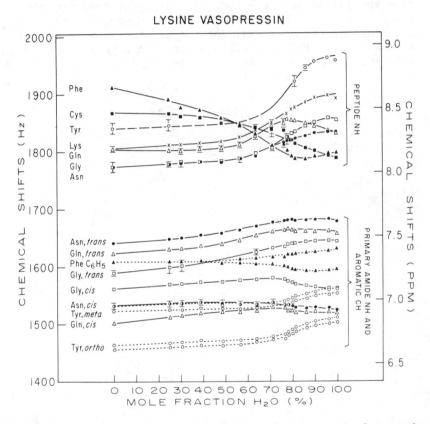

Forum Discussion Figure 20: Effect of progressively varying the solvent from pure DMSO to water, on the chemical shifts of NH (——) and aromatic CH (---) resonances of lysine vasopressin. The concentration of each hormone was main-tained at 3% w/v throughout the titration and the spectra were recorded at 24°C.

This work was carried out in collaboration with Drs. Glickson and Urry. Starting with the assignments of the hormone in DMSO, the chemical shifts of individual resonances are followed through the stepwise transition of the solvent from DMSO to water. Advantages of this titrimetric method are: a) its ability to detect conformational changes accompanying the solvent transition, and b) its applicability to the assignment of uncoupled resonances, *e.g.* carboxamide protons. Happily, several methods--each with some special advantages--seem to emerge for studying peptides in water. Besides our titrimetric method, the lowering amplification gain of the spectrometer below the saturation level as described by Dr. Von Dreele is promising and also the INDOR method as applied to peptides by Gibbons *et al.*[1] looks hopeful.

Reference

1. Gibbons, W. A., H. Alms, R. S. Bockman, and H. R. Wyssbrod. Biochemistry <u>11</u>, 1721 (1972).

SIDE CHAIN INTERACTIONS AND METAL ION COMPLEXES IN CYCLIC PEPTIDES

R. Schwyzer. Institut für Molekularbiologie und Biophysik, Eidg. Technische Hochschule Zürich-Hönggerberg, Zürich, Switzerland

I WOULD LIKE TO SAY A WORD about work on cyclic peptides that is going on in my group in Zürich. We're using cyclic polypeptides as models for research in two directions. One is to make compounds with rather stable conformations in order to study side chain interactions. The other is to prepare cyclic peptides which can complex alkali metal ions and at the same time hopefully provide a basis for the model study of chemically driven, active ion transport through membranes.

If you build a gramicidin S molecule with the ornithines replaced by cysteines, you observe a facile, rapid formation of a cystine disulfide bond across the homodetic ring. In this [2,7-cystine]-gramicidin S, the decapeptide backbone is locked in the same secondary structure as has been observed for gramicidin S (NMR studies). The disulfide bridge appears to be stabilized in its P (positive, right-handed) helical configuration shown in Figure 21, because NMR indicates shielding by sulfur of the two valine NH protons (0.2 ppm) and no effect on the leucine NH. Model building suggests a <u>large</u> dihedral disulfide angle ($\phi_{SS} \simeq 120°$). The inherent optical activity of the disulfide chromophore expresses itself in a CD couplet with (in EtOH) $\lambda_{min} = 271.5$ nm (rotational strength, $R \simeq -12.3 \times 10^{-40}$ erg·cm^3) and $\lambda_{max} = 230$ nm ($R \simeq +58.6 \times 10^{-40}$). <u>Another</u> bridged compound we studied was *cyclo*-L-cystine, ⌐Cys–Cys⌐. ^{1}H and ^{13}C NMR indicated a boat conformation of the diketopiperazine ring and a *chiral* arrangement of the cystine disulfide bond.

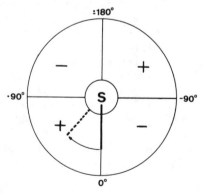

Forum Discussion Figure 21: Proposed structure of [2,7-cystine]-gramicidin S. The side chains of the individual amino acid residues are indicated by their abbreviations. The side chain of D-phenylalanine (phe) on the right is meant to be pointing upwards of the ring plane.

UV spectra and model building suggested $\phi_{SS} \approx 90°$. Yet, despite chirality, no *Cotton* effect in the long wave-length region indicating disulfide inherent optical activity could be detected. These observations support the view developed on the basis of MO-calculations by Linderberg and Michl,[1] that the optical activity of the chiral disulfide bond obeys a quadrant rule, Figure 22. This implies

Forum Discussion Figure 22: Quadrant rule for disulfides (Newman projection along the S-S bond).

that for the prediction of disulfide chirality from CD and
ORD data the dihedral angles must be approximately known.
Our work has recently been described in some detail;[2] we
are following it up in order to learn more about the
stabilization of ring conformations by bridges.

As to the other problem, we have demonstrated that
bis-cyclo peptides of the S,S'-*bis-cyclo*-glycyl-hemicystyl-
glycyl-glycyl-prolyl type can complex sodium and potassium
ions. As a working hypothesis, we have assumed that the
ion is "sandwiched" between the two cyclic peptide rings
that are held in positions adjacent to one another by the
disulfide link. If this is actually the case, then one
could possibly devise a trans-membrane carrier system
driven, for example, by a redox potential.[3] In order to
be able to construct efficient carrier peptides, we are
presently studying the conformation of such *bis-cyclo*
peptides and the constituent cyclic peptides by NMR.

If we dissolve *cyclo*(-glycyl-L-alanyl-glycyl-glycyl-L-
prolyl-) in deuterated dimethylsulfoxide at room temperature, we observe the ^{1}H and ^{13}C spectra of two different
molecular conformations ("conformation" = "ensemble average
seen by NMR technique") *M* and *m* with relative concentrations
of about 2:1. Figure 23 shows that, for example, each one
of the two amide protons gives rise to two signals, one

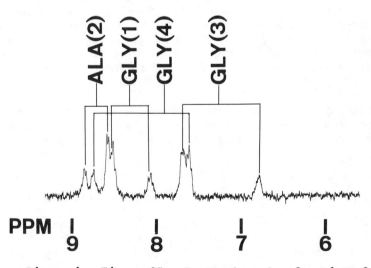

Forum Discussion Figure 23: Spectral region from 6 to 9
ppm in the proton NMR spectrum of *cyclo*(-Gly-Ala-Gly-Gly-
Pro-) in DMSO-d$_6$ at 22°C.

with double the intensity of the other. In the ^{13}C spectra, two signals for each carbon atom can also be observed (Figure 24). Positions and temperature dependence of the amide protons indicate two solvent shielded (intramolecularly

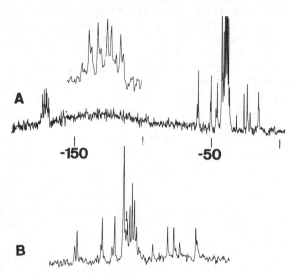

Forum Discussion Figure 24: Proton-decoupled ^{13}C Fourier-transform NMR spectrum at 25.1 MHz of *cyclo*(-Gly-Ala-Gly-Gly-Pro-) in DMSO-d₆ (the septet at 39.8 ppm corresponds to the solvent). A: 0-200 ppm region, carbonyl resonances between 165 and 175 ppm are also shown in the expanded insert. B: expansion of the 10 to 70 ppm region.

H-bonded) protons in *M* and only one in *m*. Double resonance experiments, coupling constants, and model building led to the proposition of the following conformation types:

M: Gly-Pro = *trans,trans'*
[see also ref. 4]

m: Gly-Pro = *cis,trans'*

The main difference between the two resides in the *cis-trans* isomerism of the Gly^3-Pro peptide bond. The relatively high activation energy of this transition offers a plausible explanation for the slow rate of $M \rightleftarrows m$ interchange. From the occurrence of separate NMR lines for the two species (calculated spectra see Figure 25) and from the observation in INDOR experiments of double resonance effects mediated between corresponding resonances of the two species by the interchange process, we conclude that the lifetime of the major species, M, must lie between 2×10^{-2} and 3×10^{-1} sec.

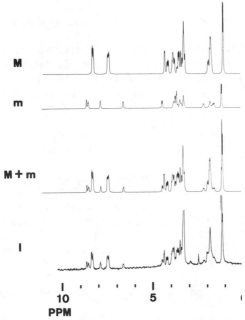

Forum Discussion Figure 25: Comparison of the spectra calculated for the M, m, and $M+m$ conformers of *cyclo* (-Gly-Ala-Gly-Gly-Pro-) with the observed (I) proton NMR spectrum.

A comparison of the ^{13}C spectra of M and m in Figure 24 reveals particularly large chemical shift changes of the proline C_β and C_γ resonances in the range of 22 to 32 ppm. In the *trans* prolyl situation of M they are rather close to one another (24.3 and 26.7 ppm), in *cis*, m, they are further separated (22.2 and 32.1 ppm). Such differences are to be found in analogous cyclic peptides containing proline, of which we have investigated four others. We therefore believe that these shifts will prove to be of great diagnostic value in the future.

It appears that in S,S'-*bis-cyclo*(-glycyl-hemicystyl-glycyl-glycyl-prolyl-) the two rings are also present as conformers *M* and *m* in DMSO, and they undergo considerable changes on complexation with potassium ion. For details of this work see ref. 5, 6.

References

1. Linderberg, J., and J. Michl. J. Amer. Chem. Soc. <u>92</u>, 2619 (1970).
2. Ludescher, U., and R. Schwyzer. Helv. Chim. Acta <u>54</u>, 1637 (1971); Donzel, B., B. Kamber, K. Wüthrich, and R. Schwyzer. Helv. Chim. Acta <u>55</u>, 947 (1972).
3. Schwyzer, R., Aung Tun-Kyi, M. Caviezel, and P. Moser. Helv. Chim. Acta <u>53</u>, 15 (1970); Schwyzer, R. Experientia (Basel) <u>26</u>, 577 (1970).
4. Carver, J. P., and E. R. Blout. In *Collagen*, Ramachandran, G. N., ed. (New York: Academic Press, 1967) p 452.
5. Meraldi, J. P., R. Schwyzer, A. Tun-Kyi, and K. Wüthrich. Helv. Chim. Acta <u>55</u>, 1962 (1972).
6. Wüthrich, K., J. P. Meraldi, A. Tun-Kyi, and R. Schwyzer. In *First European Biophysics Congress*, Broda, E., A. Locker, and H. Springer-Lederer, Eds. (Vienna: Verlag der Wiener Medizinischen Akademie, 1971) E I/20, p 93.

OVCHINNIKOV: What about selectivity of complexation?

SCHWYZER: Dr. Simon has shown that the sequence is: $K^+ > Na^+ > Li^+$. Using synthetic bilayer membranes, we believe that Rb^+ is better yet.

COMMENT: Does sulfur contribute to complexation?

SCHWYZER: We're not considering this in our model. It might be, although I don't know.

BOVEY: What does the shielding anisotropy of a disulfide bond look like? Is this known from other studies?

SCHWYZER: No, it's not known from other studies. We get a shielding of the valine NH by 0.2 ppm.

CHEMISTRY AND BIOLOGY OF PEPTIDES

ON CONFORMATIONS AND INTERANNULAR RELATIONSHIPS OF FREE AND ONE-SIDEDLY FIXED PENTAPEPTIDE LACTONES

Helmut Lackner. Institute of Organic Chemistry, University of Göttingen, Göttingen, Germany

DUE TO THEIR UNIQUE STRUCTURE containing two closely adjacent, one-sidedly fixed peptide groups, actinomycins (Figure 26, IIa) and their precusors are excellent objects

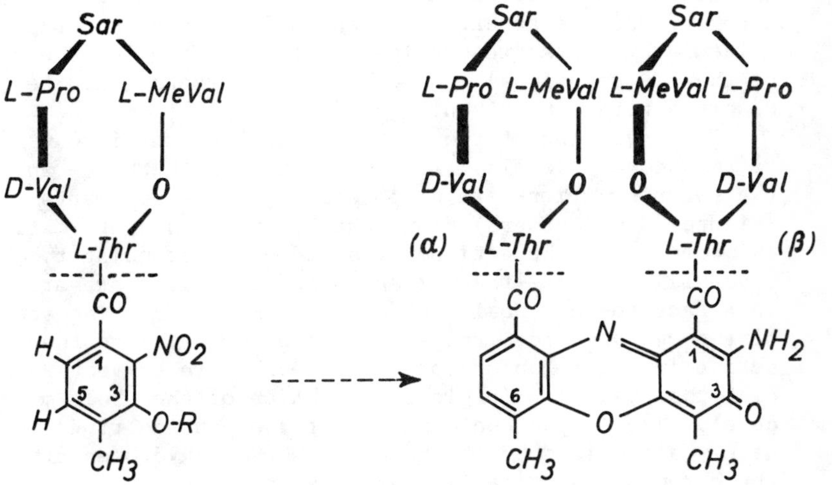

Forum Discussion Figure 26: The structures of actinomycin D and synthetic precursors.

Ia: R = Bzl IIa: Actinomycin D

b: R = CH_3 b: Lactone bonds opened (actinomycin D acid)

c: D-Thr-L-Val-D-Pro-Sar-D-MeVal-O in α or β

(na, *enantio*-actinomycins D) - (na = native)

for NMR investigations on conformation and peptide/peptide
relationships. The indispensable α,β-specific assignments
of the NMR spectra were reliably achieved by the aid of
various selectively labelled deuterio compounds[1] (Figure
27).

If not linked to actinocinyl the linear intermediates
of the pentapeptide lactone groups show well analysable
spectra (CDCl$_3$)[1] and a remarkable continuity of characteristic data, when the chain is lengthened. This suggests a
preservation of partial conformations up to the pentapeptides. Subsequent cyclisation changes the spectrum
considerably.

The *free pentapeptide lactones* (I) dimerize in dry
benzene by face-to-back association;[2] a redoubling of NMR
signals and molecular weights is observed. The very sharp
and characteristic spectra of I (and of related cyclic
peptides) in chloroform[1] and acetone (Figure 27) differ
from each other to such an unusual degree (δ- and J-values,
temperature dependence, hydrogen exchange), that two distinct conformations of the peptide ring must be assumed:
the *C* (chloroform)- and the *A* (acetone)-form.* In mixtures
of the two solvents or in acetone (or methanol)-water both
types of spectra appear simultaneously, but only normal
molecular weights are obtained in these cases (osmometry
of the NMR solutions). No comparable effects could be
observed with *linear* pentapeptides.

On conversion of the pentapeptide lactones into
actinomycines (Ia→IIa) (methanol-water) both the C- and
the A-conformations are disposable. NMR measurements
(Figure 28 and others) show, that only the *A*-conformation
is accepted by the actinomycins and--corresponding to the
crystalline form[3]--the α- and β-peptide rings are arranged
in a face-to-face position ("axial symmetry"). The spectra
of the *cyclo*peptide groups now linked by pairs to the
actinocinyl chromophore prove to be no more solvent dependent than usual[1,4] (rigid conformation of the whole molecule). The A-type conformation of the peptide rings is
mainly stabilized by interannular NH(Val)-hydrogen bridges,
the chromophore primarily serves as a clamp.

According to NMR results and chemical properties[1] this
mutual stabilization is considerably minimized, but not
cancelled, if *one* of the peptide rings is opened [actinomycin
acid (=IIb) monolactones or their esters]. The *cyclo*peptide

---

*Other solvents from benzene to tetrahydrofuran, dimethyl-
  sulfoxide and water do not cause a third type of spectrum.

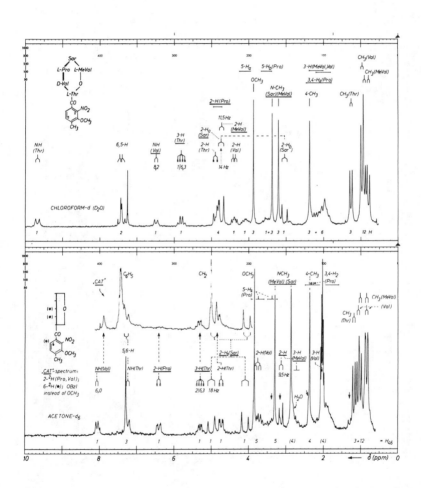

Forum Discussion Figure 27: 100 MHz NMR spectra of synthetic pentapeptide lactones (I) in chloroform (C-conformation) and in acetone (A-conformation; ↓ traces of the C-type). CAT: Control spectrum of a deuterio compound.

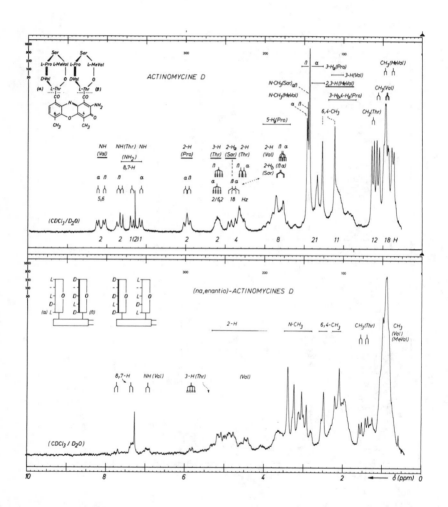

Forum Discussion Figure 28: 100 MHz NMR spectra of
actinomycin D (IIa; A-conformation of the peptide rings)
and of the α,β-isomeric na,*enantio*-actinomycins D (IIc;
ratio 1:1).

group still exerts a preformative influence on the adjacent linear peptide (yield of cyclisation, NMR analysis of the α- and β-deuterio compounds[1]). Opening of *both* rings however (IIb) causes a strong conformational inhomogeneity as indicated by very broad-lined and uncharacteristic spectra.

In the well crystallized, but bacteriostatically inactive (*na,enantio*)-*actinomycins* D (IIc) containing two enantiomer peptide lactones a stabilizing interaction between the rings is lacking. The broad-lined and very complex NMR spectrum (Figure 28; *N*-methyl region (!); no association effects) differs strongly from that of IIa and also contains characteristic signals of the *C*-conformation of the free peptide lactone I. This—contrary to IIa—nonuniform behavior of the two peptide groups is a good proof, that a mere clamping of the *cyclo*peptides is not sufficient for a conformational stabilization; this needs a marked interannular interaction, for instance by hydrogen bridges.

Some interesting chemical and optical properties of the IIc-molecules and the problem of a slow rotation of the peptide groups are under investigation.

References

1. Lackner, H. Chem. Ber. <u>104</u>, 3653 (1971).
2. Lackner, H. Tetrahedron Lett. <u>1970</u>, 2807; 3189.
3. Sobell, H. M., S. C. Jain, T. D. Sakore, and C. E. Nordman. Nature New Biology <u>231</u>, 200 (1971).
4. Arison, B. H., and K. Hoogsteen. Biochemistry <u>9</u>, 3976 (1970).

SOME STUDIES RELATING TO THE SYNTHESIS OF CYLINDRICAL
PEPTIDES

C. H. Hassall. Roche Products Ltd., Welwyn Garden City,
Herts, United Kingdom

I WOULD LIKE TO SAY SOMETHING, briefly, about some work
which is still very much in progress. It relates to an
attempt to synthesize particular cylindrical peptides.
These compounds are members of a family in which cyclic
peptides are linked in such a fashion that hydrogen bonding
between the individual rings contributes towards producing
a cylindrical form. Figure 29 illustrates this concept
for 14-membered oligopeptide rings and compares the dimensions for this case with those of a protein α-helix. The
reason for the choice of the 14-membered ring system is
further clarified by the consideration of Figure 30. The
design was based on the conformation assigned by Dale to
cyclotetradecane. If this representation in Figure 30 is
considered in detail, it will be observed that the NH-
function in one ring is attached to a carbon atom on the
14-membered ring below. As a result, a conformation which
is highly favored for inter-annular hydrogen bonding is
obtained.

In order to obtain some experimental evidence of the
conformations of such 14-membered cyclic peptides, a detailed investigation of the natural product serratamolide
and related synthetic cyclodepsipeptides has been undertaken.[1] In summary, the proton NMR evidence indicates a
remarkably rigid conformation--there is no change over a
temperature range of approximately 250°--but there are
some differences from compound to compound depending on
the chirality at the different centers of asymmetry. As
a general rule it appears that a pseudo-equatorial arrangement of side chains is favored, and the conformation of the
ring is influenced by this.

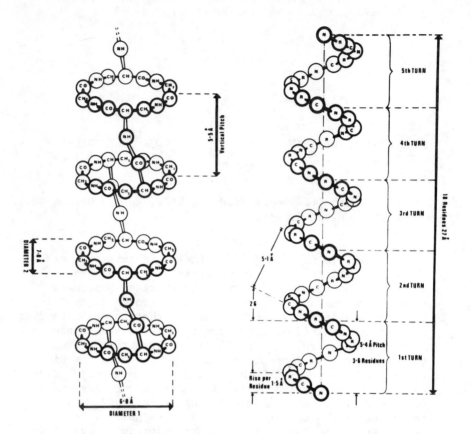

Forum Discussion Figure 29: Dimensions of cylindrical
peptides consisting of 14-membered cyclic peptide units
linked to one another by peptide bonds in comparison
with dimensions of the α-helix.

The synthesis of typical 14-membered cyclic oligopeptides[2] is illustrated in Schemes I and II. The synthesis
of the linear tetrapeptides utilizes conventional stages.
The ring closure step proceeds in quite good yields
(approximately 60%) through the use of *o*-phenylene
chlorophosphite reagent.
    The compound represented on the right-hand side of
Figure 31 has been prepared in 40% yield[3] by azide coupling
of the individual cyclic peptides.   The representations
of conformations follows from proton NMR studies

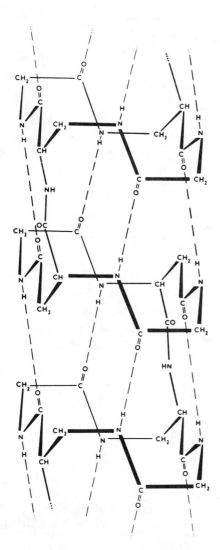

Forum Discussion Figure 30:
Projected hydrogen bond
formation in cylindrical
peptides consisting of
14-membered ring units

of serratamolide and analogues. Unfortunately, this product is not crystalline. The molecular structure cannot
therefore, be investigated by X-ray diffraction methods.
However, other combinations of such cyclic peptides with
different side chains are in preparation. It is intended
to pursue further studies to provide direct evidence of
the molecular structures of these compounds. Moreover,
particular compounds are designed to allow the investigation

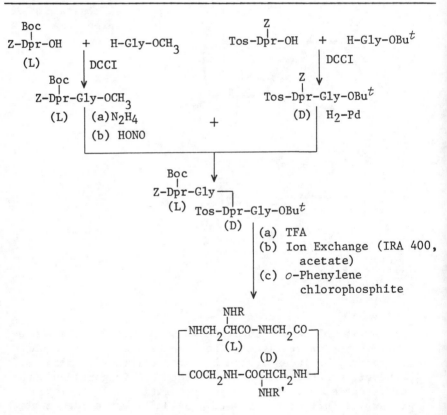

Scheme I

Scheme II

Dpr, α,β-diaminopropionic acid
R, Z; R', Tos   R=R', H
R, H; R', Tos   R=R', Z

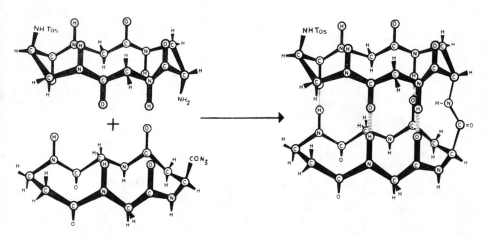

Forum Discussion Figure 31: Synthetic cylindrical peptide.
The conformation shown follows from NMR studies on
serratamolide models.

of neighboring-group interactions resulting from the
propinquity of the functions attached to separate 14-membered
rings.

References

1. Hassall, C. H., M. C. Moschidis, and W. A. Thomas,
   J. Chem. Soc. (B), 1757 (1971).
2. Hassall, C. H., D. S. Sanger, and B. K. Handa. J.
   Chem. Soc. (C), 2814 (1971) and unpublished data.
3. Chexal, K. K., B. K. Handa, and C. H. Hassall, unpublished data.

CONFORMATIONAL ANALYSIS OF A CYCLIC HEXADEPSIPEPTIDE,
CYCLO–TRI[D–HEXAHYDROMANDELYL–L–VALYL]

Alvin Steinfeld, Ugo Lepore, Murray Goodman. Department
of Chemistry, University of California, San Diego,
La Jolla, California 92037.

Alan Tonelli. Bell Telephone Laboratories, Murray
Hill, New Jersey

WE WOULD LIKE VERY BRIEFLY to describe our work on a cyclic
hexadepsipeptide composed of alternating D–hexahydromandelic
acid and L–valine residues (Figure 32), whose synthesis[1,2]
was carried out in our laboratories. NMR spectra for the
compound were measured on a 220 MHz apparatus in the following solvents: cyclohexane-d_{12}, benzene-d_6, CCl$_4$, CDCl$_3$,
p-dioxane-d_8: CD$_3$OD (2:1), trifluoroacetic acid (TFA), and
TFA-acetonitrile-d_3.

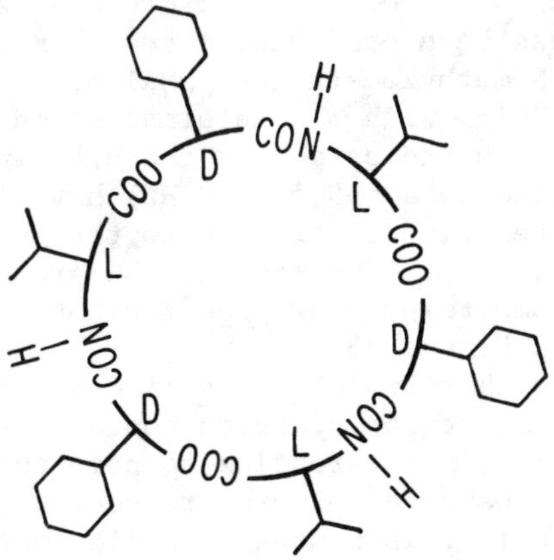

Forum Discussion Figure 32:
Cyclo-tri-(D-hexahydro-
mandelyl-L-valyl) a
cyclic hexadepsipeptide

In TFA a single NH peak is observed for all the valine residues. Likewise, all the α-CH protons from hexahydro-mandelic acid occur as a singlet, separate from the singlet observed for the α-CH protons from the valine residues. These results led us to conclude that the cyclic hexadep-sipeptide assumes a symmetrical conformation in TFA. Essentially identical results were obtained using acetonitrile as a solvent.

For less polar solvents, the NH and α-CH protons of each residue show absorptions at different chemical shifts. The NH protons appear as two widely separated doublets with a broad singlet between them. For the α-CH regions, the hexahydromandelic acid residues can be seen as two or three separate singlets while the valine residues exhibit two triplets and a singlet.

In all such cases, the coupling constants for the NH doublets vary between 9.5 and 10.0 Hz, while the signal appearing as a broad singlet has a coupling constant of less than 2.0 Hz. We were able to make an excellent cor-relation between the respective NH and α-CH peaks of the valine residues in CDCl3 (at -36°C) because of the clear and accurate coupling constants we could measure. One α-CH has a coupling constant of less than 2 Hz and is associated with the midfield NH showing the same coupling constant. A perfectly symmetrical triplet for another of the valine α-CH residues has a coupling constant of 10.0 Hz while links this absorption to the highfield valine N-H showing the same coupling constant. The remaining α-CH triplet can be correlated in the same manner with the downfield NH absorption.

The change in chemical shift of the NH protons as a function of temperature has been observed in cyclohexane, CDCl3: CD3OD (2:1) (with N-methylacetamide (NMA) as an internal standard), and CDCl3 (with NMA internal standard). The spectra in these solvents indicate that the high and low field NH protons (doublets, J = 9.5-10.0 Hz) have low temperature coefficients relative to NMA and to the third midfield NH proton (J < 2.0 Hz). The midfield NH proton and NMA show nearly the same temperature coefficient. Our results can be seen in Figure 33.

These results indicate that two of the valine NH pro-tons are involved in intramolecular hydrogen bonds or are hidden from the solvent. It is interesting to note that the high field NH shows a positive temperature coefficient with decreasing temperature. This unusual behavior may result from the placement of this NH in an environment where it is hidden from the solvent but not hydrogen bonded.

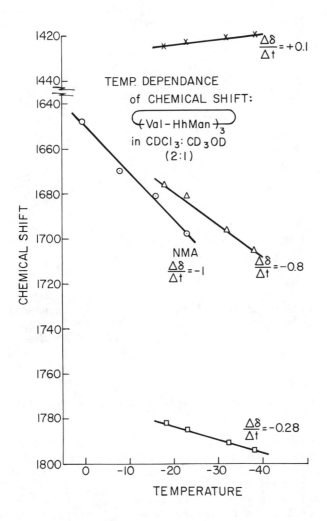

Forum Discussion Figure 33: Temperature dependence of chemical shift for (Val–HhMan)$_3$ in CDC1$_3$: CD$_3$OD (2:1). The curves are shown for the NH protons as referred to N-methylacetamide (NMA).

Any structure which we deduce must show that one of the valine N-H groups is exposed to the solvent. In this way, its behavior is expected to parallel the N-H group temperature dependence for N-methylacetamide.

Conformational energy $V(\phi,\psi,\chi)$ maps were calculated for the following *trans* peptide and ester bond fragments:

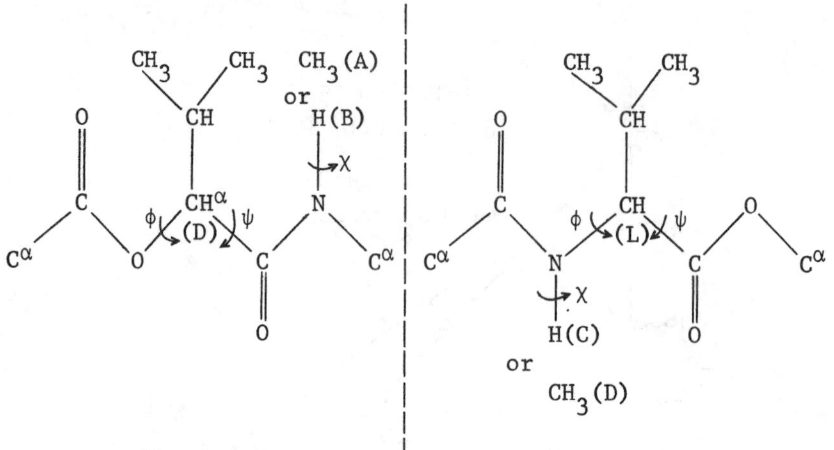

Map A + B   Map C + D

The potential functions, bond lengths, valence angles, 6-12 nonbonded potential constants, torsional barrier heights and partial atomic charges employed in these energy calculations were taken from Brant, Miller and Flory[3] and Brant, Tonelli and Flory.[4] The 6-12 potential constant $C_{N,O}$ and $A_{N,O}$ $\left(-\dfrac{C_{N,O}}{r_{N,O}^6} + \dfrac{A_{N,O}}{r_{N,O}^{12}} \right)$ appropriate to the nonbonded van der Waals interactions between the amide nitrogen and the ester oxygen atoms were evaluated in the usual manner.[3,4] Backbone rotations ϕ and ψ were varied in 10 degree increments, while rotation χ about the C^α - C^β bond was restricted to the three staggered conformations.

The resulting conformational energy maps agree closely with those calculated by Ovchinnikov *et al.*[5] in their studies of the solution conformations of enniatin B, valinomycin and their complexes with metal ions.

In the search for low energy cyclic conformations of the cyclic depsipeptide (Figure 32) only those conformations $(\phi,\psi)_{\text{amide(L)}}$ and $(\phi,\psi)_{\text{ester(D)}}$ that lie within the 5.0 kcal/mol energy contour of Maps B + C are considered.

Several attempts were made in this search, and in the final one each of the amide bonds was kept planar and *trans* while the ester bonds were allowed to adopt both the *cis* and the *trans* conformations.

Of the cyclic conformations generated in this search, one was found to be most consistent with the NMR data. This conformation $[(\omega,\phi,\psi)_{A1} = 0°, 240°, 240°; (\omega,\phi,\psi)_{E2} = 180°, 250°, 0-30°; (\omega,\phi,\psi)_{A3} = 0°, 120°, 0°; (\omega,\phi,\psi)_{E4} = 0°, 280°, 120°; (\omega,\phi,\psi)_{A5} = 0°, 240°, 240°;$ and $(\omega,\phi,\psi)_{E6} = 180°, 260°, 30°]$ possesses two intramolecularly hydrogen-bonded amide protons, $(N-H)_{A1}$ and $_{A5}$. Both amide protons belong to peptide residues with large $J_{N_\alpha}(\phi = 240°, \phi = 0°)$ and both are hydrogen-bonded to the C=O group of the amide 3 residue. One of these hydrogen bonds $[(N-HO_{A5}---(O=C)_{A3}]$ is a seven-membered hydrogen bond of the type discussed by Bystrov *et al.*[6] As noted previously,[7,8] this kind of hydrogen bond should be rather weak because of its marked nonplanar nature. However, as can be seen in the photograph of the proposed cyclic depsipeptide conformation, N-H$_{A5}$ is partially internally buried (Figure 34). Thus, even if its hydrogen bond to (C=O)$_{A3}$ is weak, one might still expect[9-14] its chemical shift to be nearly temperature independent. Figure 35 shows the other side of the molecule. As we predicted from the temperature dependence studies, one of the valyn N-H groups is completely exposed to the solvent.

The A1 and A5 residues both possess $\phi = 240°$ or $\phi' = 0°$ $[(J_{N\alpha} = 8.0$ Hz (calc.), 9.5 Hz (exp.)] in the proposed solution conformation of the depsipeptide, *i.e.*, the amide and α-protons in both residues are *cis* to each other.

The NMR spectra of the cyclic hexadepsipeptide reveal large coupling constants $J_{\alpha-\beta} = 6.5-9.5$ Hz between the α and β protons in each residue. If we assume only the staggered conformations of the side chain are allowed, this observation leads to the conclusion[15] that the three rotational states $\chi = 60, 180$ and $300°$ about the $C^\alpha -C^\beta$ bond are appreciably populated.

Moreover, we find no side chain-side chain or side chain-backbone steric interactions which would prevent the side chain in any of the six residues from adopting any of the three staggered conformations about the $C^\alpha -C^\beta$ bonds.

Acknowledgment

We gratefully acknowledge the support of research grants from the National Institutes of Health (GM 18694) and the National Science Foundation (GB 28467). Alvin Steinfeld was Postdoctoral Fellow at the Polytechnic Institute of Brooklyn, 1969-1971. Ugo Lepore was Postdoctoral Fellow at the University of California, San Diego, 1971-1972.

Forum Discussion Figure 34: Model of the molecule showing
the N-H$_{A5}$ partially internally buried.

Forum Discussion Figure 35: Photograph of the opposite
side of the molecule showing the N-H group completely
exposed to the solvent.

References

1. Shemyakin, M. M., E. I. Vinogradova, M. Yu. Feigina,
   N. A. Aldanova, V. A. Oladkina, and L. A. Shchukina.
   Dokl. Akad. Nauk. SSSR 140, 387 (1961); Shemyakin,
   M. M. Angew. Chem. 71, 741 (1949); Shemyakin, M. M.
   Angew. Chem. 72, 342 (1960).
2. Shemyakin, M. M., Yu. A. Ovchinnikov, V. T. Ivanov,
   and A. A. Kityushkin. Tetrahedron 19, 581 (1963).
3. Brant, D. A., W. G. Miller, and P. J. Flory. J. Mol.
   Biol. 23, 47 (1967).
4. Brant, D. A., A. E. Tonelli, and P. J. Flory.
   Macromolecules 2, 228 (1969).
5. Ovchinnikov, Yu. A., V. T. Ivanov, A. V. Evstratov,
   V. F. Bystrov, N. D. Abdullaev, E. M. Popov, G. M.
   Lipkind, S. F. Arkhipova, E. S. Efremov, and M. M.
   Shemyakin. Biochem. Biophys. Res. Commun. 37, 668
   (1969); Ivanov, V. T., I. A. Layne, N. D. Abdullaev,
   V. Z. Pletnev, G. M. Lipkind, S. F. Arkhopova, L. B.
   Senyavina, E. N. Mescherakova, E. M. Popov, V. F.
   Bystrov, and Yu. A. Ovchinnikov. Khim. Prir. Soedin.
   3, 221 (1971).
6. Bystrov, V. F., S. L. Portnova, V. I. Tsetlin, V. T.
   Ivanov, and Yu. A. Ovchinnikov. Tetrahedron 25, 493
   (1969).
7. Tonelli, A. E. J. Amer. Chem. Soc. 94, 346 (1972).
8. Tonelli, A. E. Macromolecules 4, 618 (1971).
9. Stern, A., W. Gibbons, and L. C. Craig. Proc. Nat.
   Acad. Sci. U.S. 61, 734 (1968).
10. Ohnishi, M., and D. W. Urry. Biochem. Biophys. Res.
    Commun. 36, 194 (1969).
11. Kopple, K. D., M. Ohnishi, and A. Go. J. Amer. Chem.
    Soc. 91, 4087, 4264 (1969).
12. Llinas, M., M. C. Klein, and J. B. Neilands. J. Mol.
    Biol. 52, 399 (1970).
13. Torchia, D. A., A. di Corato, S. C. K. Wong, C. M.
    Deber, and E. R. Blout. J. Amer. Chem. Soc. 94, 609
    (1972).
14. Torchia, D. A., S. C. K. Wong, C. M. Deber, and E. R.
    Blout. J. Amer. Chem. Soc. 94, 616 (1972).
15. Pachler, K. G. R. Spectrochimica Acta 20, 581 (1964).

SECTION III

SOLID-PHASE PEPTIDE SYNTHESIS

Session Chairmen

Jerker Porath and R. Bruce Merrifield

PELLICULAR SILICONE RESINS AS SOLID SUPPORTS FOR PEPTIDE
SYNTHESIS

Wolfgang Parr, Karel Grohmann. Chemistry Department,
University of Houston, Houston, Texas 77004

IT HAS BEEN SHOWN THAT silicone polymers can be successfully
bound to the surface of silicaceous materials through
chemically stable Si-O-Si-C bonds.[1-3] Because of their
reactivity modified halosilanes were chosen as monomers in
these investigations. The growing peptide chain could be
attached to the silicone polymer through formation of a
benzyl ester linkage because its behavior during the course
of synthesis is well established. The reactive monomers
used for the preparation of chemically bonded layers were
generally of the type:

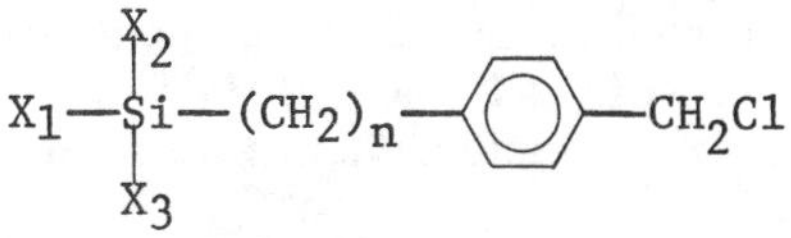

$$X_1-\underset{\underset{X_3}{|}}{\overset{\overset{X_2}{|}}{Si}}-(CH_2)_n-\!\!\!\bigcirc\!\!\!-CH_2Cl$$

n = 0,2,3,4

X_1 = Cl

X_2,X_3 = Cl or Methyl

There are two general methods for the preparation of the
desired monomers. The first one consists of a reaction of a suitably substituted Grignard reagent with
tetrachlorosilane. We have used this reaction for the
preparation of *p*-tolylmagnesium bromide with silicon
tetrachloride.[4] The obtained *p*-tolyltrichlorosilane was
then brominated by *N*-bromosuccinimide in CCl_4 solution at
reflux. Excess succinimide and solvents were removed and
the crude product was distilled twice *in vacuo*. The structure and purity of *p*-bromomethylphenyltrichlorosilane [I]
was determined by NMR spectroscopy. Higher homologs were
prepared by the addition of suitably substituted olefins

169

to trichlorosilane.[5] This reaction is catalyzed by
hexachloroplatinic acid. The suitable starting olefin
was prepared by the sequence of following reactions:[6]

$$Br-\!\!\langle O \rangle\!\!-Br \quad \xrightarrow[\text{2. Allylbromide}]{\text{1. Mg-Et}_2\text{O}} \quad Br-\!\!\langle O \rangle\!\!-CH_2-CH=CH_2$$

$$H_2C=CH-CH_2-\!\!\langle O \rangle\!\!-Br \quad \xrightarrow[\text{2. CH}_2\text{O (g)}]{\text{1. Mg-Et}_2\text{O}} \quad CH_2=CH-CH_2-\!\!\langle O \rangle\!\!-CH_2Cl$$

$$\text{3. HCl-Na}_2\text{SO}_4\text{ anhyd}$$

$$H_2C=CH-CH_2-\!\!\langle O \rangle\!\!-CH_2Cl + H-Si-Cl_3 \quad \xrightarrow{\text{H}_2[\text{PtCl}_6]\ \text{cat.}}$$

$$Cl-CH_2-\!\!\langle O \rangle\!\!-(CH_2)_3SiCl_3$$

[II]

The NMR data confirmed the presence of one isomer only,
formed by terminal addition of $SiCl_3$ group.
    These silicone monomers were bonded to porous silica
beads (Porasil E, average pore diameter 800–1500 Å, surface
area 10–20 m^2/g, manufactured by Waters Associates) or
porous glass beads (Bio-Glass, pore size 1000, 1500, or
2500 Å, surface area undisclosed, manufactured by Bio-Rad
Laboratories). The beads were first refluxed overnight
with concentrated hydrochloric acid in order to remove
absorbed cations, then washed successively with distilled
water and methanol, dried under vacuum and then heated
overnight to 105°C.
    Individual batches of beads were then stirred or shaken
overnight with 1% solution of monomers in benzene. Excess
monomers were removed by repetitive washing with dry benzene.
Hydrolysis of remaining Si-Cl groups was achieved by washing
the beads either with water-ethanol-benzene (5:45:50) or
with ethanol-water (1:1). After drying, the beads were
polymerized at 105°C for 24 hours.
    The silanized beads were washed with ethanol and benzene
in order to remove absorbed impurities. The capacity of
the beads was in the range of 0.03–0.1 mequiv Cl or Br/g
as determined by modified Volhard analysis. These values
correspond very well to the capacity values calculated for
monomolecular layers. The absorption of dyes have shown,
however, that the beads are coated quite unevenly. Also, the

presence of very fine particles after each manipulation
step shows that particles break down easily, which leads
to the losses of support.

Synthesis of Model Peptides

The first protected amino acids were esterified to the
benzylic group by refluxing the glass beads with triethylammonium salt of Boc-amino acid in dioxane for 24-38 hr.[7]
The beads were washed with dioxane, ethanol, benzene, and
petroleum ether, then dried *in vacuo* and, in aliquots of
beads, the Boc-group was cleaved. Chloride was titrated.
The yields of attachment were quite variable (30-50%) and
the extent of side reactions also varied to a considerable
degree.

Two peptides had been synthesized by standard batch
procedure.[8] In both cases, DCC coupling was used exclusively. The tetrapeptide H-Pro-Gly-Phe-Ala-OH was synthesized on the glass beads with the aryl group attached
directly to the silicone atom [I]; 120 mg of crude product
was obtained. Only 50% of the peptide was cleaved from
the support with HBr in TFA, as determined by total
hydrolysis of the remaining glass beads with $6N$ HCl and
amino acid analysis. The second peptide H-(Phe-Ala)$_6$-OH
was synthesized on glass beads silanized with monomer
[II], in which the aryl group was separated from the silicon
atom by three methylene groups. HBr cleavage was 90% complete in this case and 249 mg of crude product was obtained.
This discrepancy is easily explained by the strong electron
withdrawing effect of the SiO_3 group which destabilizes the
cationic reaction intermediate.

The tetrapeptide was checked for purity by amino acid
analysis (molar ratio Pro 0.95, Gly 1.00, Ala 1.00, Phe
1.05) and by thin-layer chromatography (tlc). None of the
techniques revealed the presence of lower peptides which
would be the result of incomplete reaction steps. The
dodecapeptide was partially hydrolyzed by $12N$ HCl, the
obtained amino acids and lower peptides were esterified,
trifluoroacetylated and analyzed by GC-MS combination
according to Bayer *et al.*[9] No dipeptides of the type
Ala-Ala or Phe-Phe were detected, but this result can be
considered only as semi-quantitative due to formation of
not only dipeptides, but also free amino acids and higher
peptides, where failure sequences could be hidden.

In the last experiment we have tested the suitability
of silicone supports for column procedures. All steps
starting from attachment of the second amino acid were

carried out by pumping reagents and solvents through a
column packed with silanized glass beads.

Bio-Glass 1500 silanized with [II] (capacity 0.108
mequiv Cl/g) was reacted with excess of triethylammonium
salt of Boc-Tyr(Bzl) as described previously. A chromatographic column (65 x 0.9 cm) was filled with 30 g of
aminoacyl glass beads. A second column served as the
solvent reservoir. All operations, but coupling, were
performed by pushing solvents or reactants with nitrogen
through the column. The effluent was collected and
tested for completeness of washing steps. *P*-Nitrophenyl-
ester coupling was used exclusively. Solutions of Boc-amino
acid-ONp were recycled through the system with a Beckman
Accu-Flow pump. The peptide synthesized was H-Gln-Gln-Gly-
Gly-Tyr(Bzl)-NH$_2$. The coupling of both Boc-Gly-ONp esters
was performed in chloroform and the speed of reaction was
followed by measuring adsorbance at 315 nm. The measurements showed that the reaction was practically complete
after 6 hours. No significant difference in coupling rates
of the second and third glycine was noted. The obtained
UV values were checked by the reaction of the unreacted
NH$_2$ groups with Pyridine-HCl and subsequent titration of
chloride as described by Dorman.[10]

The peptide was cleaved from the resin by transesterification with methanol-triethylamine for 24 hours; the
cleavage was monitored by measuring UV adsorbance at 280
mm; 530 mg of the crude peptide was obtained. The collected
peptide was purified by gel permeation chromatography on
Sephadex LH-20. Methanol was used as the eluant. The
purification step on Sephadex LH-20 revealed the presence
of small amounts of lower peptides. This finding was confirmed by amino acid analysis and tlc of the crude peptide.
The purified pentapeptide was amidated by ammonolysis in
saturated solution of ammonia in the mixture DMF-methanol
(2:1) for three days at room temperature. The Boc-group
was then cleaved by 1 N HCl in acetic acid for 30 minutes.
Deprotection of Tyr(Bzl) was carried out by catalytic
hydrogenation over 10% Pd-charcoal H$_2$ pressure 55 psi,
reaction time 12 hr.

The deprotected peptide was dissolved in methanol-water
(1:2) and placed into a refrigerator. The pentapeptide
precipitated out of solution in the form of a gel. Upon
drying the gel *in vacuo* a white powder formed. The peptide
was purified by gel permeation chromatography on Sephadex
G-15 column (43 x 2.5 cm) with methanol-water (1:2) used
as an eluant. The peptide was eluted as a single peak.
Tlc on silica gel in two solvent systems [solvent I,

methanol; solvent II, *n*-butanol-water-pyridine-acetic acid
(15:12:10:3)] also showed single spots.

The presented results show that peptides can be successfully synthesized on this kind of support. More precise and
sensitive analytical techniques will have to be developed,
however, before the advantages and disadvantages of these
supports can be clarified.

References

1. Abel, C. W., F. H. Pollard, P. C. Uelen, and G. Nickless.
   J. Chromatogr. 22, 23 (1966).
2. Aue, W. A., and C. R. Hastings. J. Chromatogr. 42,
   319 (1969).
3. Kirkland, J. J., and J. J. DeStefano. *Advances in
   Chromatography 1970*, Zlatkis, A., ed. (Houston, Texas)
   p 397.
4. Parr, W., and K. Grohmann. Tetrahedron Lett. 2633
   (1971).
5. Noll, W. *Chemistry and Technology of Silicones* (New
   York: Academic Press, 1968) p 49.
6. Parr, W., and K. Grohmann. Angew. Chem. 84, 266 (1972).
7. Merrifield, R. B. J. Amer. Chem. Soc. 85, 2149 (1963);
   86, 304 (1964).
8. Stewart, J. M., and J. D. Young. *Solid Phase Peptide
   Synthesis* (San Francisco: W. H. Freeman and Co.,
   1969).
9. Bayer, E., H. Eckstein, K. Hägele, W. A. König, W.
   Brünig, H. Hagenmaier, and W. Parr. J. Amer. Chem.
   Soc. 92, 1735 (1970).
10. Dorman, L. C. Tetrahedron Lett. 2319 (1969).

GRAFT COPOLYMERS AS INSOLUBLE SUPPORTS IN PEPTIDE SYNTHESIS

Geoffrey W. Tregear. Endocrine Unit, Massachusetts
General Hospital, and Department of Medicine, Harvard
Medical School, Boston, Massachusetts

GRAFT COPOLYMERIZATION[1] affords an opportunity of obtaining
a 0% crosslinked, physically stable polystyrene resin with
improved properties for use in solid-phase peptide synthesis.
The polystyrene chains can be insolubilized without crosslinking by anchoring them at one end to an inert, insoluble
core resin such as Teflon or Kel F (see Figure 1). This
may be accomplished experimentally by irradiating the core
polymer with ionizing radiation such as γ or x-rays, in the
absence of air and exposing the irradiated polymer to styrene
monomer. The irradiation produces free radical sites
throughout the bulk of the core polymer but with resins

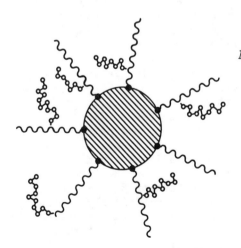

Figure 1: Diagrammatic
representation of a graft
copolymer resin for use
in solid-phase peptide
synthesis. The polysty-
rene chains which provide
the point of attachment
for the growing peptide
chain are insolubilized
by grafting to an inert
core resin.

175

such as Teflon or Kel F which are not swollen by solvents or monomer the styrene only "sees" the free radicals at the surface and polymerization of the grafted side chains occurs only at these points. By the appropriate selection of experimental conditions it is possible to achieve surface grafting with minimal crosslinking. The number of grafted side chains is determined by the number of free radical sites which is a function of the total dose of radiation. The length of the grafted chains is a function of the polymerization rate and is controlled by the dose-rate and experimental conditions such as temperature of irradiation and presence of solvent or chain terminating reagents.

A unique advantage of this system is that the backbone polymer can be prefabricated as film, discs, pellets or small diameter beads and this physical form is retained throughout the grafting procedure and the subsequent chloromethylation step.

A particularly useful graft copolymer for use in peptide synthesis consists of a core of small diameter beads of Kel F with a 10% by weight surface graft of polystyrene. We have been using this type of resin routinely in our laboratory and have been particularly impressed with its physical characteristics. Being 90% Kel F the resin displays essentially the properties of the core polymer. It does not swell appreciably, it is very dense and sinks rapidly in all solvents and does not stick to glassware.

To compare the performance of the grafted resin with the conventional 1% crosslinked polystyrene the two resins were mixed together in the same reaction vessel. Both resins were chosen to have the same chloromethyl substitution of the polystyrene chains (17%). At any point during the synthesis the two resins could be separated by the addition of methylene chloride. The crosslinked polystyrene resin floats while the graft copolymer resin being very dense, sinks. The separation is rapid and complete.

As a test synthesis, the amino terminal dodecapeptide sequence H-Ala-Val-Ser-Glu-Ile-Gln-Phe-Met-His-Asn-Leu-Gly-OH of bovine parathyroid hormone[2] was prepared in the Beckman-990 Peptide Synthesizer. The progress of the synthesis was monitored by the qualitative ninhydrin test of Kaiser *et al.*[3] Quantitative data on the extent of reaction was obtained using the Beckman-890C Sequencer[4] (see Figure 2). Both resins were surprisingly similar in their reaction characteristics. In general the carbodiimide couplings were complete on both resins after 5 to 10 min reaction time and the active ester couplings after 2 to 4

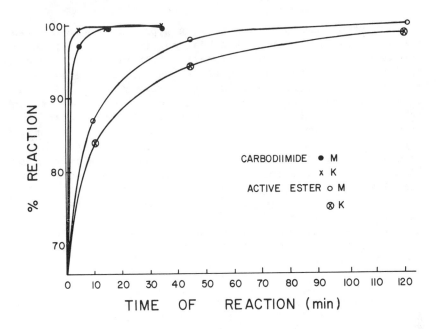

Figure 2: Rate of coupling reaction for the graft copolymer
resin (K) and the 1% crosslinked polystyrene resin (M)
during the synthesis of sequence 1-12 of bovine parathyroid hormone. The carbodiimide coupling step was Phe
to Met and the active ester step Asn to Leu. The extent
of reaction was determined from Automated Edman Sequence
Data.[4]

hr. Inefficient couplings requiring a repeat reaction or
change in solvent occurred at the same sequence positions
with both resins (Ile to Gln and Glu to Ile).

The yield of dodecapeptide after preliminary purification by gel filtration and based on the original amount of
chloromethyl groups available on the resin was 30.5% for
the 1% crosslinked resin and 25.8% for the graft resin.
The purity of the peptides obtained from each resin was
assessed by amino acid analysis, sequence determination
and thin layer chromatography. In all these systems the
products were indistinguishable.

The similarity in reaction characteristics and performance combined with the ease of separation of the two

resins has opened up an interesting application whereby two different peptides may be synthesized at the one time under the same experimental conditions. For example in the study of structure-activity relations at the carboxyl terminus of a peptide it is possible to use the graft resin esterified to one particular amino acid and the crosslinked polystyrene resin esterfied to a different amino acid. The remainder of the peptide chain can then be assembled under the same experimental conditions. The biological importance of a carboxyl terminal amide in a peptide may also be investigated by using one of the resins as the benzhydrylamine derivative and the other resin in the chloromethyl form.

Although the current experiments have demonstrated that the graft copolymer resins are comparable to but not significantly better than the conventional 1% crosslinked resins it should be pointed out that the graft resins offer a considerably flexibility in design which has not yet been fully exploited and which should lead to greatly improved resin properties. The optimal physical characteristics of the core polymer and the effects of altering the number and length of the grafted polystyrene chains have not yet been fully evaluated. An improved spatial arrangement of the side chains should also be possible by block copolymerizing the grafted styrene with other monomers to obtain a more controlled distribution of styrene units and to render the grafted side chains comparable in polarity with the growing peptide chain.

Acknowledgment

The graft copolymer resin described in this report was synthesized by Dr. H. Battaerd, Mr. G. Lang and Mr. M. Scandrett at the Central Research Laboratories, Imperial Chemical Industries (Australia) Ltd. The chloromethylated 1% crosslinked polystyrene resin was kindly provided by Dr. B. Gisin of the Rockefeller University, New York.

References

1. Battaerd, H. A. J., and G. W. Tregear. *Graft Copolymers*. (New York: John Wiley & Sons, Interscience, 1967).
2. Potts, J. T., Jr., G. W. Tregear, H. T. Keutmann, H. D. Niall, R. Sauer, L. J. Deftos, B. F. Dawson, M. L. Hogan, and G. D. Aurbach. Proc. Nat. Acad. Sci. U.S. 68, 63 (1971).
3. Kaiser, E., R. L. Colescott, D. D. Bossinger, and P. I. Cook. Anal. Biochem. 34, 595 (1970).
4. Niall, H. D., G. W. Tregear, and J. Jacobs. This volume, p 695.

SOLID-PHASE SYNTHESIS OF PROTECTED PEPTIDE FRAGMENTS

Su-Sun Wang. Chemical Research Department, Hoffmann-
La Roche Inc., Nutley, New Jersey, 07110

THE PROBLEM OF DECREASING purity in the chain elongation
process using the solid-phase technique[1] may be approxi-
mated by the binomial distribution law. The relationships
between the expected purity of the product, the chain
length of target peptide and the average coupling efficiency
can be illustrated in Figure 1. It is apparent that near
100% efficiency in each step and every cycle is required
when a large polypeptide is to be synthesized. The preparation of protected fragments by the solid phase technique
followed by fragment condensation might serve as an alternative. In the following, the preparation and application
of two resins suitable for the synthesis of protected
fragments is described.

Preparation of the Resins

Merrifield resin (chloromethylated copolystyrene-2%
divinylbenzene) was reacted with $CH_3OCO-C_6H_4-OH$ and $NaOCH_3$
to form $CH_3OCO-C_6H_4-OCH_2-C_6H_4-Resin$. Reduction with $LiAlH_4$
gave p-alkoxybenzyl alcohol resin (I) ($HOCH_2-C_6H_4-OCH_2-C_6H_4-$
Resin). The same resin can also be prepared from Merrifield
resin and $HOCH_2-C_6H_4-OH$ plus $NaOCH_3$. Acylation of I with
phenyl chloroformate gave the phenyl carbonate resin which
on hydrazinolysis yielded p-alkoxybenzyloxycarbonylhydrazide
resin (II) ($H_2NNH-CO-OCH_2-C_6H_4-OCH_2-C_6H_4-Resin$).

179

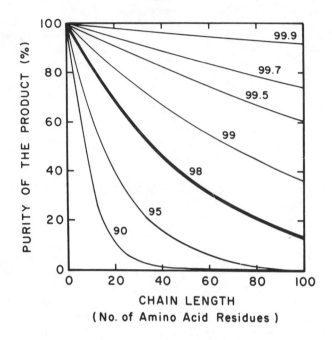

Figure 1: Relationship between average coupling efficiency and purity of products prepared by the solid-phase method.

Attachment of Amino Acids to the Resins

2-(*p*-Biphenylyl)isopropyloxycarbonylamino acids (Bpoc-amino acids) were attached to I using dicyclohexylcarbodiimide (DCC) (1 equiv. of pyridine as catalyst) or by the *p*-nitrophenyl ester method (10 equiv. of imidazole as catalyst). The unreacted hydroxyl group on the Bpoc-amino acyl resin was then blocked by benzoylation (benzoyl chloride and pyridine). No racemization was detected during the esterification process. Bpoc-amino acids were attached to resin II efficiently with the DCC method. Substitutions on I and II were normally 0.3-0.6 mmol/g.

Application of the Resins

Bpoc-Phe-Resin I was deprotected with 0.5% trifluoroacetic acid (TFA), neutralized and coupled with Bpoc-L-Val, Bpoc-L-Leu and Z-L-Leu to give Z-Leu-Leu-Val-Phe-Resin I by a procedure similar to that described before.[2] Treatment with 50% TFA in CH_2Cl_2 (30 min) liberated the peptide from the resin. Pure crystalline Z-Leu-Leu-Val-Phe (mp 216-219°) was isolated in an overall yield of 68% calculated from the phenylalanine content of Bpoc-Phe-Resin I. During the synthesis, a four-fold excess of amino acid derivatives and coupling agent (DCC) was used in each cycle. The product was converted to its methyl ester (Z-Leu-Leu-Val-Phe-OCH$_3$; mp 204-206°) by dilute methanolic HCl.

Bpoc-Gly-ONp was reacted with I in the presence of imidazole to give Bpoc-Gly-Resin I. Solid-phase synthesis with Bpoc-L-Phe, Bpoc-L-Phe and Z-L-Lys(Z) gave Z-Lys(Z)-Phe-Phe-Gly-Resin I. Cleavage with 50% TFA (30 min) afforded crystalline Z-Lys(Z)-Phe-Phe-Gly (mp 220-222°) in 60% overall yield. This fragment was condensed with a dipeptide methyl ester (Leu-Met-OCH$_3$·HCl) by the DCC method. The product, hexapeptide ester Z-Lys(Z)-Phe-Phe-Gly-Leu-Met-OCH$_3$ (mp 180-184°) upon ammonolysis yielded the crystalline hexapeptide amide Z-Lys(Z)-Phe-Phe-Gly-Leu-Met-NH$_2$ (mp 238-242°). The carbobenzoxy groups were removed by hot TFA (80°, 3 hr) in the presence of anisole and mercaptoethanol. The crude peptide was purified by counter-current distribution followed by gel filtration. The desired compound Lys-Phe-Phe-Gly-Leu-Met-NH$_2$ was obtained as an amorphous white solid, which was homogeneous on electrophoresis and thin layer chromatography. Acid hydrolysis and amino acid analysis gave the expected values.

Angiotensin II was also prepared using resin I. The protected octapeptide intermediate Z-Asp(OBzl)-Arg(NO$_2$)-Val-Tyr(Bzl)-Val-His(Tos)-Pro-Phe was obtained in a similar manner and the protecting groups were removed by treatment in liquid HF.[3] Purification by counter-current distribution and gel filtration gave an analytically pure produce (15% overall yield), which was homogeneous on electrophoresis and thin layer chromatography.

Resin II was utilized to prepare Z-Phe-Val-Ala-Leu-HNNH$_2$ by a procedure similar to that described previously.[2] The desired protected peptide hydrazide was obtained in 42% yield (mp 252-254°). It was found to be identical with a product prepared previously[2] by an alternate route.

Acknowledgments

The author thanks Dr. R. B. Merrifield for helpful discussions and Drs. G. Saucy and A. M. Felix for valuable suggestions and help with the manuscript.

References

1. Merrifield, R. B.   J. Amer. Chem. Soc. <u>85</u>, 2149 (1963);
   Advan. Enzymol. <u>32</u>, 221 (1969).
2. Wang, S. S. and R. B. Merrifield.   J. Amer. Chem. Soc.
   <u>91</u>, 6488 (1969); Int. J. Protein Res. <u>1</u>, 235 (1969).
3. Sakakibara, S., Y. Shimonishi, M. Okada and Y. Kishida.
   In *Peptides*, Beyerman, H. C., A. van de Linde and W.
   Maassen van den Brink, eds. (Amsterdam: North-Holland
   Publ., 1967) pp 44-49.

AUTOMATIC MONITORING OF SOLID-PHASE PEPTIDE SYNTHESIS BY PERCHLORIC ACID TITRATION

K. Brunfeldt, D. Bucher, T. Christensen, P. Roepstorff, O. Schou, P. Villemoes. The Danish Institute of Protein Chemistry, affiliated to the Danish Academy of Technical Sciences, 33, Finsensvej, DK-2000, Copenhagen F, Denmark

I. Rubin. Department of Biochemistry A, University of Copenhagen. Copenhagen, Denmark

STEPWISE YIELDS NEAR 100% are required for the synthesis of long peptide chains.[1,2] A monitoring system for solid-phase synthesis must therefore possess a high degree of accuracy. A determination of the amount of liberated α-amino groups is a direct measure of the yield of protecting group cleavage. A direct determination of the coupling yield seems very difficult to achieve. Instead a measurement of the amount of residual α-amino groups after coupling may be used.

Titration of free α-amino groups with perchloric acid has been investigated in our laboratory.[3,4] The method is easily automated and sufficiently rapid (approx. 2 hr) to be used in automated peptide synthesis. The method is in principle non destructive and might be carried out without sampling. Our experience is, however, not extensive enough to ensure that damage in all cases (*e.g.* with Try- or Cys-containing peptides) can be avoided.

The titration is carried out in a mixture of methylene chloride and acetic acid 1:1 (v/v) on the whole batch in the reactor. A circulation system ensures a continuous washing of the walls of the reactor. A glass electrode

---

*Present Address: Department of Biochemistry A, University of Copenhagen.

and a calomel electrode with secondary salt bridge (10% sat aq LiCl in HOAc) are used. Approximately 0.05 N HClO$_4$ in HOAc is used as titrant.

The high concn of HOAc shifts the equilibrium R-NH$_2$ + HOAc $\rightleftharpoons$ R-NH$_3^+$ + OA$_c^-$ to the right, the so-called levelling effect.[5] Therefore, the acetate ions are titrated by the HClO$_4$; and we used the same endpoint irrespective of the N-terminal amino acid. The method is not specific for α-amino groups and other groups may be titrated.

Loss of peptide and blocking of amino groups will both result in decreases of titration values, and other analytical methods must be used in addition for a correct interpretation.

All operations are coded on punched tape and automatically carried out by the combined synthesizer titration equipment.[6] Treatments before and stirring during titration are controlled by the synthesizer's control unit. When the code is read for titration, an interface based on sequential logic controls the titration equipment, consisting of a titrator, autoburette, stripprinter, and recorder. The titration is controlled by the titrator. The variations of the potential against time are recorded and serve as a control for proper electrode function as well as for demonstration of the course of the titration. When a titration is finished the total volume of added titrant is printed out and the interface signals to the synthesizer's control unit to read the next and all fol following codes.

The accuracy of titration is *ca.* $\pm$ 0,01 mequiv (single expt at 1 mequiv amine) and *ca.* $\pm$ 0,004 mequiv (at 0.1 mequiv amine). An increase in accuracy by optimizing the procedure seems possible. The technique is useful. We have thus demonstrated the presence of amino-group-blocking impurities in methylene chloride by repeated titration of an alanyl-resin.[7] In the repetitive solid-phase procedure even minute artefacts in each step may accumulate and thus significantly reduce the final yield. Therefore, purity of solvents and reagents is of a similar importance as in automated Edman degradation. Our automatic monitoring system may be used for checking solvents and reagents for solid-phase synthesis.

During an attempt to synthesize antamanide considerable amounts of dipeptide were cleaved from a H-Pro-Pro-O resin as diketopiperazine[6],* during the Et$_3$N treatment and subsequent washings. The titration values showed that the degree of diketopiperazine formation was prohibitive for

---

*Identified by mass spectrometry.

obtaining a reasonable yield of final product. Therefore
the antamanid sequence H-Phe-Pro-Pro-Phe-Phe-Val-Pro-Pro-
Ala-Phe-O-resin[8] was chosen as target. During this synthesis, the titrations after Boc-group cleavage strongly
indicated irreversible blocking[6] since repetition of
deblocking at the tetra- and pentapeptide stage did not
change the titration value (see Figure 1). The difference

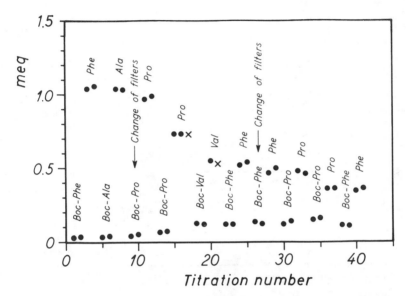

Figure 1: Titration values during the synthesis of a
sequence of antamanid H-Phe-Pro-Pro-Phe-Phe-Val-Pro-Pro-
Ala-Phe-O-resin. X indicates that the procedure for
cleavage of the Boc-group was repeated before the
titration. (By the courtesy of FEBS Letters.)

between the titration values with the Boc-group intact and
after cleavage was used as a measure of amino acid incorporation at the respective step of the synthesis. The
reasonable agreement between the titration values and amino
acid analysis data, Table I, supported that a blocking had
occurred, because a loss of peptide would have resulted in
a pronounced discrepancy between the two tests. A loss of
peptide would lead to a lower total yield but it would not
influence the relative amino acid content of the product.

Table I

Amino Acid Content of Synthetic Decapeptide Determined by
Titration and by Amino Acid Analysis

Boc-Phe-O-resin	Amino acid	Theoretical	Titration	Amino acid analysis		
	Phe	4	4.00	4.00*	4.00[†]	4.00[#]
2.2 g	Ala	1	1.97	1.74	1.88	1.92
1.00 mequiv	Pro	4	4.23	4.52[††]	4.55[††]	4.73[††]
	Val	1	0.82	0.76	0.77	0.82
	Phe	4	4.00	4.00*	4.00[#]	4.00**
2.9 g	Ala	1	1.20	1.13	1.09	1.01
1.32 mequiv	Pro	4	4.02	3.92[††]	4.16[††]	3.97[††]
	Val	1	0.98	0.97	1.06	1.01
	Phe	4	4.00	4.00*	4.00[#]	4.00**
6.0 g	Ala	1	1.27	1.08	1.05	1.00
2.74 mequiv	Pro	4	4.04	3.57[††]	3.59[††]	3.96[††]
	Val	1	0.98	0.96	0.92	0.98

*Resin-bound product
[†]Cleaved crude product
[#]Ether precipitated product
**Cyclicized peptide (antamanid)
[††]Proline (amino acid analysis) corrected for concentration
   dependency of calibration factor

The amino acid analyses of the resin bound and cleaved
product differed only slightly. Therefore, the hydrolysis
of the resin bound peptide seems to have been reliable in
this case. It was performed at 110° for 96 hr in a 1:1
mixture of 6 N HCl-glacial acetic acid in evacuated, sealed
ampoules, followed by evaporation at reduced pressure and
low temperature. The hydrolysis of the cleaved product
was performed in 6 N HCl at 110° for 24 hr.

For the synthesis 2.2 g of Boc-Phe-O-resin with a
substitution degree of 0.456 mequiv/g was used. For removal of the Boc-group N HCl-HOAc was used. The coupling
reagent was DCC. The yield of the resin bound decapeptide

was determined by the titration to 20%. The crude cleaved product weighed 296 mg.

The titration results indicated the presence of tri-, tetra-, octa-, and decapeptide. The cleaved product was deuteroacetylated and permethylated.[9] Mass spectrometry then showed that the blocking was due to acetylation as the expected tri-, tetra-, and octapeptide were acetylated and only the decapeptide was deuteroacetylated. Furthermore, some acetylated heptapeptide was found. Minor amounts of failure sequences missing a proline or phenylalanine in positions where Pro-Pro or Phe-Phe was expected, were also detected. The acetylation was mainly due to acetic acid leaching out from all the teflon parts of the reaction system[6] into the coupling mixture.

The synthesis was twice repeated in an all glass system using 2.9 and 6.0 g of Boc-Phe-O-resin. The overall titration yields were 64% and 62% giving 1.06 and 2.20 g of crude products, respectively. Practically identical titration results were obtained in both experiments (Figure 2). No pronounced decrease was observed at any stage. The second titration values of proline at tetra- and nonapeptide stages were found to be lower than the first. A third titration was identical with the second. As in the first synthesis, Figure 1, an increase of the titration value after Boc-valine incorporation was observed. Repeated coupling did not improve this result, neither did succeeding couplings result in lower values. We are, at present, unable to explain these two observations. The latter may be due to steric hindrance prohibiting coupling but not protonization.

In the synthesis using 6 g Boc-Phe-O-resin the coupling with Boc-phenylalanine following valine led to a high titration value suggesting an incomplete coupling. A repeated coupling after change of the solutions lead to a value not different from the ones obtained before and after this coupling. From the crude cyclisized product a cycloundecapeptide containing 5 phenylalanine was isolated in addition to the expected cyclodecapeptide (antamanid). The unexpected high titration value thus must have been due to an undesirable partial deblocking of the amino group at the hexapeptide stage and not to an incomplete coupling with Boc-phenylalanine. The unexpected high titration value thus certainly revealed a human error.

The amino acid analysis of both resin bound final products and the amino acid compositions calculated from the titration values again agreed (Table I). Probably, the linear decrease in the titration values is partly due to blocking. It was not possible to verify this by mass spectrometry

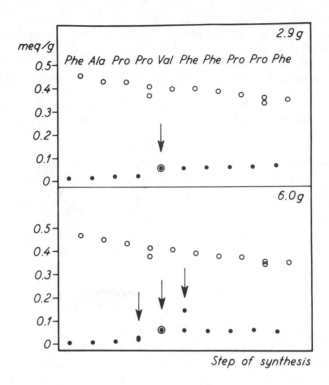

Figure 2: Titration values during synthesis of the sequence
shown on the figure: 2,9 or 6,0 g of Boc-Phe-0-resin was
used. Values are means of double determinations except
the upper values of the Pro titrations in step 4 and 9,
which are single determinations. meq, At the ordinate,
refers to mequiv titrated per g of Boc-Phe-0-resin.
•, Before Boc-group cleavage. 0, After Boc-group cleavage.
↓, Repeated coupling. ⊖, Repeated double determination
resulting in identical values.

because of the small amount of each component. Some
acetylation seems likely to occur as acetic acid is difficult
to wash out of the resin completely, as shown by utilizing
[14]C-acetic acid. Furthermore, loss of peptide certainly
did occur during these syntheses.[10,11]
    The titration is not specific for α-amino groups.
During the synthesis of renin substrate tetradecapeptide
(1.8 g Boc-Ser(Bzl)-0-resin, 0.440 mequiv g) the titration
values increased due to imidazole-N after N^α-Boc-N^{im}-Dnp-
histidine incorporated. The titration value after Boc-group

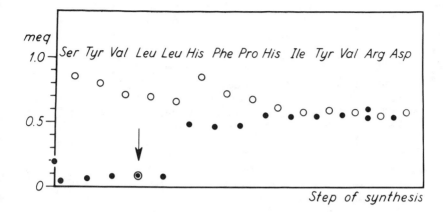

Figure 3: Titration values during synthesis of renin substrate tetradecapeptide. Values are means of double determinations except the first titration on Boc-Ser(Bz1)-0-resin and the upper one of Boc-Arg(NO2)-peptide, which was the first of three determinations. Signatures as in Figure 2.

cleavage is the double of that obtained before cleavage (Figure 3). The titration value of the preceeding Boc-Leu was subtracted. The decrease from the first to the second titration of Boc-Ser(Bz1)-0-resin was due to removal of ionically bound Boc-Ser(Bz1)-OH.

Extensive acetylation nullified the yield after incorporation of the second histidine in the teflon containing reactor system. Mass spectrometry confirmed the presence of acetylated C-terminal penta-, hexa-, hepta-, and octa-peptides. The decrease in titration values at the di-, and tripeptide stage was mainly due to formation of Tyr(Bz1)-Ser(Bz1) diketopiperazine which was identified by mass spectrometry and use of ^{14}C-tyrosine. The diketopiperazine formation begins during the triethyl amine treatment and lasts throughout the end of the following Boc-valine coupling. C-Terminal diketopiperazine formation has also been observed in other laboratories.[12-14]

A modification of the titration procedure in which the present titration medium is replaced by halogenated alkanes containing a quarternary ammonium salt as carrier electrolyte, is under investigation.

References

1. Baas, J. M. A., H. C. Beyerman, B. van de Graaf, and E. W. B. de Leer. In *Peptides 1969*, Scoffone, E., ed. (Amsterdam: North-Holland, 1971) pp 173-176.
2. Bayer, E., H. Eckstein, K. Hägele, W. A. König, W. Brünig, H. Hagenmaier, and W. Parr. J. Amer. Chem. Soc. 92, 1735 (1970).
3. Brunfeldt, K., P. Roepstorff and J. Thomsen. Acta Chem. Scand. 23, 2906 (1969).
4. Brunfeldt, K., P. Roepstorff, and J. Thomsen. In *Peptides 1969*, Scoffone, E., ed. (Amsterdam: North-Holland, 1971) pp 148-153.
5. Hantzsch, A. Z. Physik. Chem. A 134, 406 (1928).
6. Brunfeldt, K., T. Christensen, and P. Villemoes. FEBS Letters 22, 238 (1972).
7. Brunfeldt, K., and T. Christensen. FEBS Letters 19, 345 (1972).
8. Wieland, Th., C. Birr, and F. Flor. Justus Liebigs Ann. Chem. 727, 130 (1969).
9. Brunfeldt, K., T. Christensen, and P. Roepstorff. FEBS Letters, in press.
10. Merrifield, R. B. J. Amer. Chem. Soc. 91, 2501 (1969).
11. Norris, K., J. Halstrøm, and K. Brunfeldt. Acta Chem. Scand. 25, 945 (1971).
12. Lunkenheimer, W., and H. Zahn. Justus Liebigs Ann. Chem. 740, 1 (1970).
13. Rothe, M., and J. Mazánek. Angew. Chem. 84, 290 (1972).
14. Gisin, B. F., and R. B. Merrifield. J. Amer. Chem. Soc. 94, 3102 (1972).

IMPROVED PROTECTING GROUPS FOR SOLID-PHASE PEPTIDE SYNTHESIS

Bruce W. Erickson, R. B. Merrifield. The Rockefeller
University, New York, New York 10021

THE SUCCESSFUL SYNTHESIS of pure peptides by the solid-phase
method requires a careful choice of protecting groups. When
the Boc group is used for temporary protection of the α-amino
groups, side chain protecting groups must be stable during
acidolysis of the Boc groups and yet be removed by a stronger
acid at the end of the synthesis.
We have established the relative stability of the benzylic
protecting groups toward acid by determining the apparent
first-order rate constants for the deprotection of seven
side chain-protected amino acids with 50% (v/v) trifluoroacetic
acid in dichloromethane [50% TFA] at 20° (Table I). The

Table I

Apparent First-order Loss of Benzylic Side Chain
Protecting Groups in 50% TFA -- CH_2Cl_2 at 20°

Protected amino acid	k_1 $(10^{-4} hr^{-1})$	*Relative rate*
Thr(Bzl)	2.5	[1.0]
Glu(OBzl)	3	1
Asp(OBzl)	3	1
Ser(Bzl)	3.9	1.5
Cys(4-CH_3OBzl)	29	12
Lys(Z)	142	57
Tyr(Bzl)	181	73

191

relative amounts of the free and the protected amino acid
were measured on a 13-cm column of sulfonated polystyrene
eluted with pH 7 buffer at a temperature between 56 and 96°.

Thr(Bzl) is sufficiently stable to survive extensive
treatment with 50% TFA during solid-phase synthesis, since
exposure to 50% TFA for 40 hr caused only 1% loss of the
benzyl group. The stability of Glu(OBzl), Asp(OBzl), and
Ser(Bzl) toward 50% TFA is essentially the same as that
of Thr(Bzl). Cys(4-CH$_3$OBzl) was deprotected 12 times
faster than Thr(Bzl), which suggests that the 4-methoxybenzyl
group may not provide sufficient protection for cysteine
residues during the synthesis of large peptides.

Protection of tyrosine as the benzyl ether is not
recommended for two reasons. Not only was Tyr(Bzl) deprotected 73 times faster than Thr(Bzl), but this reaction
furnished a mixture of 63% tyrosine and 37% 3-benzyltyrosine.[1]
The ratio of these products was constant for 100 hr, during
which 85% of the Tyr(Bzl) reacted. Since essentially the
same results were obtained when 100 equivalents of anisole
were present per equivalent of Tyr(Bzl), the 3-benzyltyrosine
is probably formed by an intramolecular rearrangement.[2-4]

In contrast, O-(2,6-dichlorobenzyl)tyrosine, which was
stable to 50% TFA for at least 350 hr at 20°, is deprotected
at least 4000 times slower than Tyr(Bzl) and at least 60
times slower than Thr(Bzl). Both Tyr(Bzl) and Tyr(2,6-Cl$_2$Bzl)
were completely deprotected on treatment with HF for 10 min
at 0° to provide a mixture of 60% tyrosine and about 40% of
the respective 3-benzylated derivative. After exposure to
50% HF--anisole for 10 min at 0°, 13% of the Tyr(Bzl) was
converted into 3-benzyltyrosine but only 4% of Tyr(2,6-Cl$_2$Bzl)
was isomerized to the ring-alkylated derivative. Protection
of the phenolic hydroxyl group of tyrosine as the 2,6-
dichlorobenzyl ether is an improvement over use of the benzyl
ether, but it does not solve the problem of intramolecular
benzylation of the phenolic ring.

Loss of the N^ε-protecting group from a lysine residue
during solid-phase synthesis generates a free ε-amino group
that can couple subsequently with an activated amino acid
to form a branched peptide. In order to examine the distribution of branched peptides, we have synthesized decalysyl-
valine using Z to protect the ε-amino group of lysine.
Boc-valyl-oxymethyl-polystyrene resin was deprotected for
1 hr with 50% TFA, neutralized with diisopropylethylamine,
coupled for 1 hour with three equivalents each of Boc-Lys(Z)
and DCC, and again neutralized and coupled. These excessive
deprotection conditions were purposely used to accentuate
the formation of branched peptides. Repetition of this

cycle nine times and treatment of the resin with HF and
anisole for 1 hr at 25° furnished a crude peptide mixture
that was analyzed[5-7] on a 50-cm column of carboxymethylcellulose using a sodium chloride gradient. After elution
of decalysyl-valine, a series of peaks due to branched
peptides containing 11-19 lysine residues was observed.

To a first approximation, each of the branched peptides
was formed as a result of the loss of one Z group somewhere
along the polylysyl-valine chain during a certain deprotection step. The peptide containing 18 lysine residues,
for example, arose by loss of either Z group from a fully
protected lysyl-lysyl-valine chain. Thus the mole percent
found for each branched peptide provides an estimate of
the rate at which the Z groups are lost during removal of
the Boc groups with 50% TFA. The average loss of Z groups
from the protected peptide resin was 0.8% per hour per
Lys(Z) residue. This result agrees well with the loss of
1.4% per hour observed for the deprotection of Lys(Z) in
solution. This agreement suggests that the rates of deprotection of side chain-protected amino acids in solution
are close to the rates of deprotection of amino-acid
residues in a resin-bound peptide chain.

Lys(Z), which was deprotected 57 times faster than
Thr(Bzl) in 50% TFA, is too unstable for use when the
α-amino group is protected by Boc. Since the addition of
electron-withdrawing substituents to Z is known to increase
its stability toward acid,[8-13] the stability of six
chlorinated derivatives of Lys(Z) toward 50% TFA was
examined (Table II). The 4-chloro derivative, which was

Table II
Apparent First-order Loss of Chlorinated Benzyloxycarbonyl(Z)
Groups from Lysine in 50% TFA--CH_2Cl_2 at 20°

Protected amino acid	k_1 $(10^{-4}\,hr^{-1})$	Relative rate
Lys(Z)	142	57
Lys(4-ClZ)	50	20
Thr(Bzl)*	2.5	[1.0]
Lys(2-ClZ)[†]	2.3	0.9
Lys(2,4-Cl$_2$Z)[†]	1.8	0.7
Lys(3,4-Cl$_2$Z)[†]	0.86	0.3
Lys(3-ClZ)	0.18	0.07
Lys(2,6-Cl$_2$Z)	0.14	0.06

*Inserted as a standard for comparison.
[†]Recommended as a suitable side chain protecting group
in solid-phase synthesis.

deprotected 20 times faster than Thr(Bzl), is too acid-
labile. In contrast, the 3-chloro and the 2,6-dichloro
derivatives were deprotected about 15 times slower than
Thr(Bzl). Since the 3-ClZ group was not completely removed
on treatment with HF for 1 hr at 0°, these derivatives are
probably too stable for general use in solid-phase synthesis. The 2-chloro, the 2,4-dichloro, and the 3,4-
dichloro derivatives, however, were deprotected at essentially the same rate as Thr(Bzl) in 50% TFA and were
completely deprotected on treatment with HF for 1 hr at
0°. Thus each of the latter derivatives exhibited a
stability toward acid that is suitable for the solid-phase
synthesis of large peptides when Boc is used for α-amino
protection.

Finally, decalysyl-valine was prepared as before but
using the 2,4-Cl₂Z group to protect the ε-amino group of
lysine. Chromatography of the crude product revealed the
complete absence of any peptides containing more than 10
lysine residues. This result indicates that the formation
of peptides that branch at the ε-amino groups of lysine
residues can be avoided and confirms the prediction that
protecting groups whose acid stability is close to that of
Thr(Bzl) are suitable for the solid-phase synthesis of large
peptides.

References

1. Iselin, B. Helv. Chim. Acta 45, 1510 (1962).
2. Tarbell, D. S., and J. C. Petropoulos. J. Amer. Chem.
   Soc. 54, 244 (1952).
3. Cullinane, N. M., R. A. Woolhouse, and G. B. Carter.
   J. Chem. Soc. 2995 (1962).
4. Spanninger, P. A., and J. L. von Rosenberg. J. Amer.
   Chem. Soc. 94, 1973 (1972).
5. Schlossman, S. F., A. Yaron, S. Ben-Efraim, and H. A.
   Sober. Biochemistry 4, 1638 (1965).
6. Yaron, A., and S. F. Schlossman. Biochemistry 7, 2673
   (1968).
7. Grahl-Nielsen, O., and G. L. Tritsch. Biochemistry 8,
   187 (1969).
8. Bláha, K., and J. Rudinger. Coll. Czech. Chem.
   Commun. 30, 585, 599 (1965).
9. Meienhofer, J. In *Peptides*, Zervas, L., ed. (Oxford:
   Pergamon Press, 1966) pp 55-59.
10. Homer, H. B., R. B. Moodie, and H. N. Rydon. J. Chem.
    Soc. 4403 (1965).
11. Noda, K., S. Terada, and N. Izumiya. Bull. Chem. Soc.
    Jap. 43, 1883 (1970).

12. Sakakibara, S., T. Fukuda, Y. Kishida, and I. Honda. Bull. Chem. Soc. Jap. <u>43</u>, 3320 (1970).
13. Schnabel, E., H. Klostermeyer, and H. Berndt. Justus Liebigs Ann. Chem. <u>749</u>, 90 (1971).

CHEMISTRY AND BIOLOGY OF PEPTIDES

NEW PROTECTING GROUPS FOR AMINO ACIDS IN SOLID-PHASE PEPTIDE SYNTHESIS

Donald Yamashiro, Richard L. Noble, Choh Hao Li.
The Hormone Research Laboratory, University of
California, San Francisco, California 94122

THE STABILITY OF SIDE-CHAIN protecting groups in solid-phase peptide synthesis becomes increasingly important as the length of the target sequence increases. If N^α-Boc protection is employed along with final removal of the side-chain protecting groups by hydrogen fluoride, four necessary conditions are required: (1) the side-chain protecting groups must survive the repeated use of deprotecting agents for removal of the Boc groups, (2) these protecting groups must be removed efficiently by hydrogen fluoride, (3) the protecting groups should not give rise to side-products at any stage of synthesis and (4) the derivatives must couple efficiently to the peptide-resin. With these considerations in mind we have undertaken a study of protecting groups for those amino acids possessing side-chain functions. Since 50% trifluoroacetic acid in dichloromethane[1] is frequently used for removal of N^α-Boc groups, we have tested the stabilities of various N^α-acetylamino acid amide derivatives toward this reagent (Table I).

It has been reported that γ-benzyl protection of glutamic acid is not entirely stable,[2] and in some preliminary tests we found that γ-benzyl glutamate lost 5% of the benzyl protection in 5.5 hours. When the test was carried out with the N^α-acetyl amide derivative (Table I) where the influence of the amino and α-carboxyl groups is removed, the protection was found to be more stable. It is in fact the most stable derivative listed in Table I. It was for

197

Table I

N^α-Acetylamino Acid Amide Derivatives and
Their Stabilities in Trifluoracetic Acid

Acetylated amide*	mp	Loss of side chain protection after 23 hr in 50% TFA in CH_2Cl_2
Benzyl-N^α-acetyl-isoasparaginate	129–130°	4%[†]
N^α-Acetyl-O-benzyl-serinamide	170–172°	3%[†]
N^α-Acetyl-S-(p-methoxybenzyl)-cysteinamide	150–152°	20%[†] 27%[#]
Benzyl-N^α-acetyl-isoglutaminate	148–150°	2%[†]
N^α-Acetyl-O-benzyl-threoninamide	219–220°	5%[†]
N^α-Acetyl-O-benzyl-tyrosinamide	191–195°	50%[†] 62%[√]

*Prepared by conversion of the Boc-amino acid to the amide
by the mixed anhydride procedure followed by treatment
with TFA and acetylation with acetic anhydride in pyridine.
Each compound showed a single spot in tlc on silica gel in
chloroform-methanol (1:1). Correct elemental analyses
were obtained for each compound.

[†]Estimated on tlc in n-butanol-acetic acid-water (4:1:1)
against measured amounts of HF-treated samples.

[#]Estimated by the Ellman reagent.[12]

[√]Estimated by absorption at 295 nm in $1N$ NaOH-DMF (1:1).

this reason that the tests were extended to the acetylated
amides of other amino acid derivatives. The difficulty of
extrapolating results of experiments performed in solution
to those on solid phase is recognized, but we have proceeded
on the assumption that the reactions are several times
slower on solid phase than in solution. All the side-chain
protecting groups shown in Table I are completely removed
in hydrogen fluoride in 10 to 15 minutes at 0° and we refer
to these as standard test conditions.

Table II

Derivatives of Lysine, Cysteine and Tyrosine

*Derivative**	*mp*	$R_f(tlc)^+$	$[\alpha]^{24}_D$
Lys (p-BrZ)	247-249°	0.65 (BAW)	+11.2° (c 1,80% HOAc)
Boc-Lys (p-BrZ)	102-104°	0.62 (CM)	+ 6.0° (c 4.6, CHCl₃)
Lys (o-BrZ)	220-223°	0.50 (BAW)	+ 9.6° (c 1.1,80% HOAc)
Boc-Lys (o-BrZ) DCHA salt	106-108°	----	+ 6.3° (c 2.4, CHCl₃)
Ac-Cys(3,4-Me₂Bzl)-NH₂	147-147.5°	0.77 (CM)	----
Cys(3,4-Me₂Bzl)	195-197°	0.57 (BAW)	-15.7° (c 2,80% HOAC)
Boc-Cys(3,4-Me₂Bzl) DCHA salt	122-124°	----	-20.0° (c 2.3,80% HOAc)
Tyr(m-BrBzl)	218-220°	0.60 (BAW)	- 6.5° (c 1,80% HOAc)
Boc-Tyr(m-BrBzl)	96- 97°	0.62 (CM)	+23.4° (c 2.2, EtOH)
Tyr(2,6-Cl₂Bzl)	200-203°	0.65 (BAW)	-10.6° (c 2,80% HOAc)
Boc-Tyr(2,6-Cl₂Bzl)	108-110°	0.75 (CM)	+21 ° (c 2, EtOH)

*Correct elemental analyses were obtained for all compounds.

†Solvents used: BAW, n-butanol-acetic acid-water (4:1:1); CM, chloroform-methanol (1:1).

Of the basic amino acids it is known that N^ϵ-benzyloxy-carbonyl (Z) protection of lysine is not entirely stable.[3] We have prepared N^ϵ protected Lys(p-BrZ) (Table II) and have found it to be four times more stable to 50% trifluoro-acetic acid in dichloromethane than the parent Z compound in both solution (Table III) and on solid phase.[4] Both protecting groups were about four times more stable on solid phase than in solution. More recently we have prepared[5] the o-bromo isomer (see Table II) and found it to be 60-fold more stable than the Z protection (Table III).

Table III

Stabilities of New Derivatives in Trifluoracetic Acid

Derivative	*Time of Treatment in 50% TFA in CH$_2$Cl$_2$ (hr)*	*Loss of Protection*
Lys(Z)	20	42%*
Lys(p-BrZ)	20	12%*
Lys(o-BrZ)	20	0.7%*
Ac-Cys(p-MeOBzl)-NH$_2$	23	27%[†]
Ac-Cys(3,4-Me$_2$Bzl)-NH$_2$	23	0.2%[†]
Tyr(Bzl)	21	55%[#]
Tyr(m-BrBzl)	21	1.6%[#]
Tyr(2,6-Cl$_2$Bzl)	21	1.4%[#]

*Determined by quantitative amino acid analysis.

[†]Determined by the Ellman reagent.[12]

[#]Estimated by absorption at 295 nm in 1N NaOH.

Both new protecting groups are removed completely in HF under standard test conditions. It is of interest to note that these conditions were not sufficient to completely remove the m-BrZ protection.

For histidine and arginine we have recently employed N^{im}-Boc and N^G-tosyl protection, respectively, in the

synthesis of the heptapeptide[6] Ala-His-Arg-Leu-His-Gln-Leu
(I) which occurs in human growth hormone[7,8] (HGH). Use of
these protecting groups in the synthesis[9] of the nonadeca-
peptide α^{1-19}-adrenocorticotropic hormone has further
demonstrated the usefulness of this combination. In the
case of the heptapeptide I nitro protection of arginine
led to substantial amounts of side-products containing
ornithine whereas tosyl protection gave no such difficulties.

For protection of the cysteine thiol group the advantages of *p*-methoxybenzyl protection over that of benzyl
protection are known.[10] The results in Table I indicate,
however, that a problem of stability of the former to TFA
could arise for very large peptides. In an effort to find
a protecting group of intermediate stability between the
p-methoxybenzyl and benzyl protections we prepared[11] N^{α}-
acetyl-*S*-(3,4-dimethylbenzyl) cysteinamide (see Table II).
The derivative lost only 0.2% of its protection in TFA
(Table III). Furthermore, the protecting group was completely removed in HF under standard test conditions. The
suitability of this protecting group for cysteine was then
demonstrated by synthesis of the C-terminal cyclic dodecapeptide[11] of HGH Val-Gln-Cys-Arg-Ser-Val-Glu-Gly-Ser-Cys-
Gly-Phe.

The serious instability of benzyl protection of the
tyrosine hydroxyl group suggested by the data in Table I
led to an examination of the *m*-bromobenzyl and 2,6-
dichlorobenzyl protecting groups (see Table II). Both
were found to be about 50-fold more stable than benzyl
protection (Table III), and both were removed completely
in HF. We synthesized[4] the octapeptide Phe-Lys-Gln-Thr-
Tyr-Ser-Lys-Phe (II) occurring in HGH with *m*-bromobenzyl
protection of tyrosine in addition to *p*-BrZ protection of
lysine. Peptide II was isolated in good yield and with
slightly less by-product formation than in a synthesis
where benzyl protection of tyrosine was employed. The
by-products, as judged by amino acid analyses, are presumably octapeptide II containing an altered tyrosine
residue.[13] Although a benzyl type of protection for
tyrosine can be stabilized to TFA treatment and used with
success, by-product formation in HF is exceedingly difficult to eliminate entirely.

Acknowledgment

We thank Mr. W. F. Hain and Mr. Kenway Hoey for
technical assistance. This work was supported in part by
the American Cancer Society, the Geffen Foundation and the
Allen Foundation.

References

1. Gutte, B., and R. B. Merrifield. J. Amer. Chem. Soc. 91, 501 (1969).
2. Gutte, B., and R. B. Merrifield. J. Biol. Chem. 246, 1922 (1971).
3. Yaron, A., and S. F. Schlossman. Biochemistry 7, 2673 (1968).
4. Yamashiro, D., and C. H. Li. Int. J. Protein Res., in press.
5. Yamashiro, D., and C. H. Li. Unpublished observations.
6. Yamashiro, D., J. Blake, and C. H. Li. J. Amer. Chem. Soc. 94, 2855 (1972).
7. Li, C. H., J. S. Dixon, and W.-K. Liu. Arch. Biochem. Biophys. 133, 70 (1969).
8. Li, C. H., and J. S. Dixon. Arch. Biochem. Biophys. 146, 233 (1971).
9. Blake, J., K.-T. Wang, and C. H. Li. Biochemistry 11, 438 (1972).
10. Sakakibara, S., Y. Shimonishi, M. Okada, and Y. Kishida. In *Peptides*, Beyerman, H. C., A. van de Linde, and W. Maasen van den Brink, eds. (Amsterdam: North Holland, 1967) pp 44-49.
11. Yamashiro, D., R. L. Noble, and C. H. Li. Manuscript in preparation.
12. Ellman, G. L. Arch. Biochem. Biophys. 82, 70 (1959).
13. Compare the preceding paper: Erickson, B. W., and R. B. Merrifield. In *Chemistry and Biology of Peptides, Proceedings of the Third American Peptide Symposium,* Meienhofer, J., ed. (Ann Arbor, Michigan: Ann Arbor Science Publ. 1972) pp 191-195.

EXPERIMENTS WITH ACTIVE ESTERS IN SOLID–PHASE PEPTIDE
SYNTHESIS

*M. Bodanszky, R. J. Bath, A. Chang, M. L. Fink, K. W.
Funk, S. M. Greenwald, Y. S. Klausner.* Department of
Chemistry, Case Western Reserve University, Cleveland
Ohio 44106

THE APPLICATION OF ACTIVE ESTERS in solid-phase peptide
synthesis[1] (SPPS) was proposed earlier.[2,3] Subsequently,
the use of some active esters led to successful syntheses
in some cases,[4] while in other instances difficulties[5] and
even failure[6] were reported. Nevertheless, there are good
reasons to persist with the search for new active esters,
specially designed for SPPS. The most general method of
acylation through activation of acylamino acids with
dicyclohexylcarbodiimide[1] (DCC) requires "global protection"[7] while the use of active esters for the same purpose[2]
allows considerable freedom in this repect, *e.g.,* side
chain hydroxyl and carboxyl groups can be left unprotected.
This detail becomes quite important if ammonolysis[2] or
alcoholysis[3,8] is planned for the removal of the completed
chain. An additional problem related to activation with
DCC was recognized with the surprising observation[9] that
the addition of DCC to solutions of Boc-α-amino acids in
dichloromethane or dimethylformamide, *etc.* results in the
formation of ninhydrin positive byproducts, one of which
is the free amino acid. All common Boc- and Aoc-α- amino
acid derivatives were tested and all showed decomposition,
while Boc-β-alanine did not. In retrospect this side reaction should not be surprising: acidolytic removal of
protecting groups is based on the presence of groupings
ready to form carbonium ions, while activation rests on an
electron-withdrawing substituent. An interaction between

two oppositely polarized centers within the same molecule
is to be expected: higher activation or higher sensitivity

to acids of the protecting group should render the lability
of the protected and activated derivatives of amino acids
a serious problem. At higher temperatures, benzyloxycarbonyl
amino acid chlorides yield benzylchloride and N-carboxyan-
hydrides (NCA's). The decomposition of biphenylylisopropyl
p-nitrophenyl carbonate[10] may provide a second analogy for
the intramolecular displacement on O-Boc-aminoacylisoureas.
Our investigations so far have not furnished convincing
evidence of an NCA intermediate, but the decomposition itself cautions against uncritical activation with DCC in
SPPS. Moderately active esters may not present similar
problems and therefore their application in SPPS could be
well justified.

The low rates of acylation with hindered p-nitrophenyl
esters[11,12] prompted experiments with different active
esters. Hindrance by the resin-matrix and by the growing
peptide chain[13,14] can be compensated by appropriately
selected active esters, as shown in Table I. The values
in Table I were obtained in a solvent chosen for reasons
of convenient measurements: ethyl acetate. The rate of
aminolysis of p-nitrophenyl esters is about an order of
magnitude lower in ethyl acetate than in dimethylformamide.[12]
The solvent dependence of aminolysis rates is much less
pronounced in the case of o-nitrophenyl ester (Table II).
This favorable property together with their higher reactivity, reflected also in ir frequency of their carbonyl
bands (Table III), were considered auspicious for further
experimentation. The experiments in progress aim at
the development of optimal conditions for the preparation
of o-nitrophenyl esters of Boc-α-amino acids and for their
application in SPPS.

Table I

Rates of Reaction of Different Active Esters
with the Growing Peptide Chain*

Nucleophile	*Boc-L-Leu*			
	ONo	*ONp*	*OPcp*	*OTcp*
Gly–R	150	840	1100	
Val–R	600	2400	1800	
Gly–Leu–Val–R	50	120	900	60
Val–Leu–Val–R	275	900	3500	300
Gly–Leu–Gln–Gly–Leu–Val–R	220	600	2600	
Val–Leu–Gln–Gly–Leu–Val–R	700	>10,000	>10,000	

*The numbers represent half reaction times (in min) of
active esters of t-butyloxycarbonyl-L-leucine with the
different nucleophiles. The reactions were carried out
in ethyl acetate at room temperature with 0.02 M concentration of the reactants. Rates were measured by the uv
absorption of the liberated phenols. *Abbreviations:* ONo:
o-nitrophenyl ester; ONp: p-nitrophenyl ester; OPcp:
pentachlorophenyl ester; OTcp: 2,4,5-trichlorophenyl
ester; R: resin.

Table II

Solvent Dependence of Aminolysis Rates of Z-Leu
Para and Ortho Nitrophenyl Esters*

Solvent	*Z-Leu-ONp*	*Z-Leu-ONo*
Ethyl acetate	10	1.5
Methylene chloride	>110	3
Dimethylformamide	1.5	0.65

*The numbers represent half reaction times (in min) between
active esters of benzyloxycarbonyl-L-leucine and benzylamine,
at room temperature, with 0.02 M concentration of the
reactants. Rates were measured by the uv absorption of
the liberated phenols and of the consumed active esters.

Table III

Physical Properties of Some *o*-Nitrophenyl Esters

Active Ester	*Mp*	$[\alpha]_D^{25}$	*ir (C=O stretch)*
Z-Asn-ONo	156.5-157.5°	-42° (*c* 2; DMF)	5.63µ
Boc-Gly-ONo	97-98.5°	-----	5.63µ
Boc-Leu-ONo	55-57°	-68° (*c* 1; DMF)	5.63µ
Z-Phe-ONo	109-110°	-63° (*c* 1.07; DMF)	5.62µ
Boc-Phe-ONo	146-146.5°	-65° (*c* 1; DMF)	5.62µ

References

1. Merrifield, R. B. J. Amer. Chem. Soc. <u>85</u>, 2139 (1963).
2. Bodanszky, M., and J. T. Sheehan. Chem. & Ind., 1423 (1964).
3. Bodanszky, M., and J. T. Sheehan. Chem. & Ind. 1597 (1966).
4. Hörnle, S. Z. Physiol. Chem. <u>348</u>, 1355 (1967); Inukai, N., K. Nakano, and M. Murakami. Bull. Chem. Soc. Japan <u>41</u>, 182 (1968).
5. Beyerman, H. C., C. A. M. Boers-Boonekamp, and H. Maassen Van Den Brink-Zimmermannova. Rec. Trav. Chim. <u>87</u>, 257 (1968).
6. Klostermeyer, H. Chem. Ber. <u>101</u>, 2823 (1968).
7. Wünsch, E. Angew. Chem. Internat. Edit. <u>10</u>, 786 (1971).
8. Beyerman, H. C., H. Hindriks and E. W. B. De Leer. Chem. Commun., 1668 (1968).
9. Bodanszky, M., and J. T. Sheehan, unpublished, 1965.
10. Sieber, P., and B. Iselin. Helv. Chim. Acta <u>52</u>, 1525 (1969).
11. Lübke, K. In *Peptides 1969, Proc. 10th Europ. Peptide Symp.*, Scoffone, E. Ed. (Amsterdam: North Holland Publishing Co., 1971) p 154; *cf.* also Rudinger, J., and V. Gut. In *Peptides, Proc. 8th Europ. Peptide Symp.*, Beyerman, H. C., A. Van De Linde, and W. Maassen Van Den Brink, Eds. (Amsterdam: North Holland Publishing Co., 1967) p 89.

12. Bodanszky, M., and R. J. Bath.   Chem. Commun. 1259
    (1969).
13. Bath, R. J.   Ph.D. Dissertation, Case Western Reserve
    University, Cleveland, Ohio, 1970.
14. Hagenmaier, H.   Tetrahedron Letters, 283 (1970).

PEPTIDE SYNTHESIS BY OXIDATION-REDUCTION CONDENSATION

Teruaki Mukaiyama, Masaaki Ueki, Rei Matsueda.[†]*
Laboratory of Organic Chemistry, Tokyo Institute of
Technology, Ookayama, Meguro-ku, Tokyo, Japan

IT WAS ESTABLISHED THAT various peptides and amino acid
active esters are prepared without any accompanying side
reactions by the oxication-reduction condensation method
with use of triphenylphosphine (TPP) and 2,2'-dithiodi-
pyridine (DTP) as shown in the following scheme.[1]

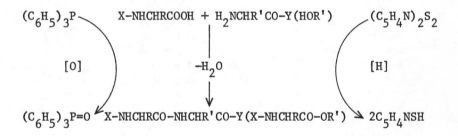

This new method is characterized by the generation of
an active dehydrating agent, a phosphorane, in the reaction
vessel from two components which are individually stable
and safely stored. It is also a favorable feature for use
in solid-phase synthesis that the two co-products produced

*Present Address: Department of Chemistry, Science Univer-
sity of Tokyo, Kagurazaka, Shinjuku-ku, Tokyo, Japan.
†Present Address: Product Development Laboratories,
Sankyo Co., Ltd., Hiromachi, Shinagawa-ku, Tokyo, Japan.

along with peptides are very soluble in the usual organic
solvents including methylene chloride. A most important
merit is the attachment of the first amino acid to the
resin support by the same procedures using the same reagents
as for subsequent chain elongations.

All Boc-amino acids are incorporated to the hydroxymethyl
resin by esterification to the extent of 0.11-0.44 mmol/g
simply by shaking for 24 hr at ambient temperature with
TPP-DTP in methylene chloride or N,N-dimethylformamide
(DMF), Table I. Boc-dipeptides were also attached to resin
under the same conditions[2] (Table I).

Table I

Attachment of BOC-Amino Acids and Peptides to Hydroxymethyl
Resin by Oxidation-Reduction Condensation*

Boc-Amino Acid	Solvent	Content (mmol/g)	Boc-Amino Acid and Boc-Peptide	Solvent	Content (mmol/g)
Boc-Asn	DMF	0.19	Boc-Gly	CH_2Cl_2	0.44
Boc-Gln	"	0.21	Boc-Ala		0.25
Boc-Arg(NO$_2$)	"	0.12	Boc-Phe	"	0.38
Boc-Try	"	0.12	Boc-Pro	"	0.16
Boc-Cys(Bzl)	CH_2Cl_2	0.18	Boc-Ileu	"	0.11
Boc-Met	"	0.15	Boc-Ala-Ala	"	0.11
Boc-His(Bzl)	DMF	0.44	Boc-Lys(Z)-Phe	"	0.26

* 3 Equiv   triphenylphosphine-2,2'-dithiopyridine + Boc-amino
acid shaking with hydroxymethyl resin for 24 hr at room
temperature.

A solid-phase synthesis of [Phe[2]]-lysine-vasopressin by
oxidation-reduction condensation was tried starting from
Boc-Gly-resin obtained as described above. Couplings were
performed by a 2+(3+4) fragment condensation on the tetra-
peptide-resin Aoc-Cys(MeOBzl)-Pro-Lys(Z)-Gly-resin, prepared
stepwise. The first fragment (Boc-Phe-Gln-Asn, 3 equiv) was
introduced by shaking with 3 equiv of TPP-DTP and 6 equiv
of 2(1H)-pyridinethione in DMF for 8 hr at -15° and 2 hr
at ambient temperature. After cleavage of Boc by TFA-
CH_2Cl_2(1:1,v/v) and neutralization, the fragment Boc-Cys(Bzl)-

Phe (3 equiv) was coupled by shaking with 3 equiv of TPP-DTP in CH_2Cl_2 for 8 hr at ambient temperature. Protected nonapeptide amide was obtained in 62% yield based on Gly by ammonolytic cleavage from protected peptide resin and purification by Sephadex LH-20 chromatography (DMF). Amino acid ratios: Cys 0.96, Phe 2.02, Glu 1, Asp 0.98, Pro 0.97, Lys 1.02, Gly 1.04, NH_3 3.31. One hundred mg of this protected nonapeptide amide showed 2840 units of pressor activity after deprotection by HF at room temp for 1.5 hr and oxidative cyclization by aeration at pH 6.7 for 3 hr.

In similar fashion, the nonapeptide amide corresponding to the sequence (2-10) of LH-RH[3] was also successfully synthesized. Fragment condensations with Boc-Ser(Bzl)-Tyr(Bzl)-Gly (1.5 equiv) and Boc-His(Tos)-Trp (3 equiv), were mediated for 24 hr with TPP-DTP in methylene chloride to give Boc-His(Tos)-Trp-Ser(Bzl)-Tyr(Bzl)-Gly-Leu-Arg(NO_2)-Pro-Gly-resin. Ammonolytic cleavage and purification by Sephadex LH-20 (elution by DMF) and Avicel (BuOH-AcOH-H_2O, 4:1:1) afforded protected nonapeptide amide monohydrate in 48% yield based on Gly. Amino acid ratios: His 1.01, Ser 0.94, Trp 0.82, Tyr 0.96, Gly 2.04, Leu 1, Arg 0.88, Pro 0.94, NH_3 1.38. All protecting groups were removed by HF in the presence of anisole and 2(1H)-pyridinethione. After neutralization with Amberlite IRA-400 in methanol, the decapeptide (LH-RH) was obtained by reaction with pyroglutamic acid pentachlorophenyl ester in DMF. The crude synthetic product obtained was converted to the acetate form. Purification was carried out by Sephadex G-25 (1 M acetic acid), CM-Sephadex C-25 (0.15 M ammonium acetate) and finally by Sephadex G-10 (1 M acetic acid). LH-RH was isolated as diacetate trihydrate. Amino acid ratios: Glu 1.02, His 0.98, Trp 0.84, Ser 1.01, Tyr 0.99, Gly 2.06, Leu 1, Arg 0.92, Pro 0.96, NH_3 1.35. Synthetic product showed higher activity than natural LH-RH (AVS 77-33#215-269).

References

1. Mukaiyama, T., R. Matsueda, and M. Suzuki. Tetra. Lett. 1901 (1970).
2. Matsueda, R., E. Kitazawa, H. Maruyama, H. Takahagi, and T. Mukaiyama. Chem. Lett. 379 (1972).
3. Matsuo, H., Y. Baba, R. M. G. Nair, A. Arimura, and A. V. Schally. Biochem. Biophys. Res. Commun. 43, 1334 (1971); Baba, Y., H. Matsuo, and A. V. Schally. Biochem. Biophys. Res. Commun. 44, 459 (1971).

PEPTIDE SYNTHESIS THROUGH 4-NITROSO-5-AMINO-PYRAZOLE INSOLUBLE ACTIVE ESTERS

M. Guarneri, R. Ferroni, P. Giori, C. A. Benassi.
Istituto di Chimica Farmaceutica, Università di
Ferrara, Ferrara, Italy

THE USE OF INSOLUBLE ACTIVE ESTERS derived from crosslinked polymers and *N*-protected amino acids in the synthesis of peptides has been previously described.[1-6] Very recently Fridkin *et al.*[7] investigated the preparation and the use of *N-t*-Boc-amino acid esters of a poly (ethylene-*co-N*-hydroxymaleimide) in order to overcome the difficulties encountered in peptide synthesis by using other polymeric reagents.

We had already reported[8] the use of 1-phenyl-3-methyl-4-nitroso-5-aminopyrazoline esters (OPmp-esters) in peptide liquid phase synthesis. We describe now the preparation of insoluble active esters derived from *N*-protected amino acids and a styrene-2% divinylbenzene polymer into which a similar pyrazole derivated is inserted. After unsuccessful attempts to introduce the pyrazole handle to a aminomethyl polymer we attained the object by reacting a hydroxymethyl styrene-2% divinylbenzene-polymer[9] with a pyrazolo-oxadiazinone II, synthetized from 4-nitroso-5-aminopyrazoles I and phosgene:

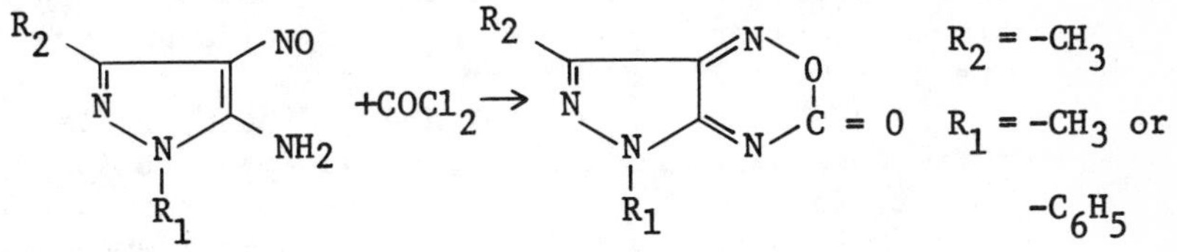

I   II

$R_2 = -CH_3$

$R_1 = -CH_3$ or

$-C_6H_5$

Scheme I

Compound II easily reacts with the hydroxymethyl function of the polymer to give a green colored derivative III which forms active esters with Z, Boc and Nps protected amino acids when suspended in methylene chloride or dioxane in the presence of DCC as a coupling agent. The best results are obtained when $R_1=R_2=CH_3$ giving orange colored *N*-5(polystyryl-4-methyloxycarbonyl)-imino-4-oximino-1,3-dimethyl-2-pyrazoline esters (OPsp esters, IV) of *N*-protected amino acids.

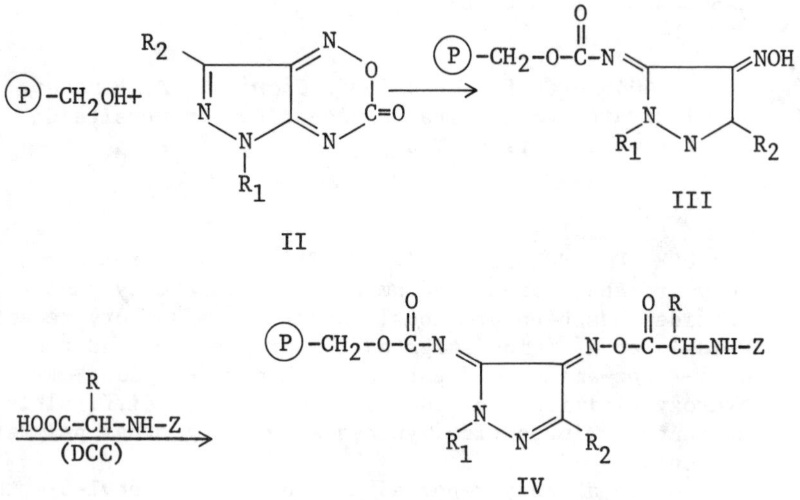

Scheme II

As it appears from Table I, difficulties have been encountered to introduce sterically hindered amino acids as isoleucine for which a 48 hr reaction time has been required.

The polymeric orange colored active esters III react very rapidly with nucleophiles in a molar ratio of 1:1 at room temperature. The reaction is completed with benzylamine in few seconds and with amino acids and peptide esters between 15 and 60 min.

The synthesis of several model dipeptides as well as of the N-terminal hen egg-white lysozyme tetrapeptide Z-Lys(Z)-Val-Phe-Gly-OEt was carried out with a high yield (80% overall yield for the lysozyme tetrapeptide). Moreover, these OPsp esters present the advantage that coupling reaction can be followed from the discoloration of the polymer that reverts to green colored at the end of the reaction.

Table I

N-Protected Amino Acid Esters of
N-5(polystyryl-4-methyloxycarbonyl)-imino-4-oximino-1,
3-dimethyl-2-pyrazoline (OPsp esters)

Amino acid derivative	Acylation reaction time (hr)	nmol of amino acid bound/g of polyester
Z-Gly-OH	4	0.85
Z-Phe-OH	3	0.99
Z-Val-OH	3	1.04
Z-Lys(Z)-OH	4	0.97
Boc-Gly-OH	5	0.90
Boc-Val-OH	24	0.54
Boc-Leu-OH	24	1.00
Boc-Ileu-OH	48	0.24
Nps-Phe-OH	5	0.58

The acylation reaction is carried out in methylene chloride
using an equivalent of polymer with 1-3 equivalents of
N-protected amino acid and dicyclohexylcarbodiimide as
coupling agent.

References

1. Fridkin, M., A. Patchornik, and E. Katchalski. J. Amer. Chem. Soc. 88, 3164 (1966).
2. Fridkin, M., A. Patchornik, and E. Katchalski. J. Amer. Chem. Soc. 90, 2953 (1968).
3. Sklyarov, L. Y., V. I. Gorbunov, and L. A. Shchukina. Zh. Obshch. Khim. 36, 2220 (1966).
4. Wieland, T., and C. Birr. Angew. Chem. 78, 303 (1966).
5. Wieland, T., and C. Birr. Chimia 21, 581 (1967).
6. Laufer, D. A., T. M. Chapman, D. I. Marlborough, V. M. Vaidya, and E. R. Blout. J. Amer. Chem. Soc. 90, 2696 (1968).
7. Fridkin, M., A. Patchornik, and E. Katchalski. Biochemistry 11, 466 (1972).

8. Guarneri, M., P. Giori, and C. A. Benassi. Tetrahedron Lett. <u>8</u>, 665 (1971).

9. Merrifield, R. B. Biochemistry <u>3</u>, 1385 (1964).

10. Mizoguchi, T. Unpublished; In Stewart, J. M., and Young, J. D. *Solid Phase Peptide Synthesis* (San Francisco: W. H. Freeman and Co., 1969) pp 27.

A SUITABLE REGENERABLE, SOLID-PHASE COUPLING REAGENT FOR
AUTOMATED PEPTIDE SYNTHESIS*

J. Brown, D. R. Lauren, R. E. Williams. Division of
Biological Sciences, National Research Council of
Canada, Ottawa, Ontario, Canada K1A OR6

THE REAGENT EEDQ (*N*-ethyloxycarbonyl-2-ethyloxy-1,2-
dihydroquinoline), first described by Belleau and Malek,[1]
is finding increasing application in the field of peptide
synthesis. When this reagent is used to synthesize a
peptide ethanol, carbon dioxide and quinoline are formed
as by-products. Removal of the quinoline from the reaction
product can sometimes constitute a problem. However, if
the reagent were to be incorporated into an insoluble
polymer this problem could be overcome. Filtration of the
reaction mixture would remove the insoluble quinoline
polymer produced during the reaction and the protected
peptide could then be isolated by simply evaporating the
solvent. In addition, the quinoline polymer could be
reactivated to recover the coupling reagent.

Polymerisation of 6-isopropenylquinoline, styrene and
divinylbenzene (200:300:8 w/w/w/) gave us[2] the desired
insoluble quinoline polymer I, Figure 1. Nitrogen analysis
of the product indicated approximately 1.33 mmol nitrogen
as (N_2) per gram. Activation of the polymer by reaction
with ethyl chloroformate, ethanol and triethylamine in
methylene chloride solution[3] afforded the solid-phase
coupling reagent II, Figure 1.

The number of "active sites" per gram was assayed by
reacting a known weight of resin with an excess of Dnp-Leu
and Gly-OEt. Isolation of the product by thin-layer

---

*Issued as NRCC No. 12864.

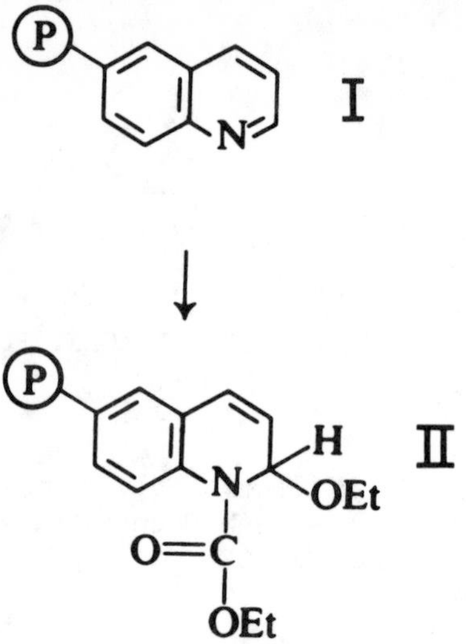

Figure 1: Solid-phase coupling reagent. I, Quinolyl polymer; II, N-ethyloxycarbonyl-2-ethyloxy-1,2-dihydroquinolyl polymer (EEDQ polymer); P, crosslinked polystyrene

chromatography and colorimetric estimation of the amount formed allowed calculation of the number of "active sites" per gram. Several batches of resin have been activated and assayed in this manner and found to contain between 0.2 and 0.5 mmol "active sites" per gram of resin.

The preparative utility of the insoluble reagent was assessed by coupling Z-Phe and Gly-OEt on a 0.5 mmol scale. The coupling reaction was carried out in methylene chloride solution (20 ml) using a 1.1 molar excess of the reagent. Work-up of the reaction involved only filtration and gave a crude yield of 84%. One crystallisation from ethyl acetate-petrol (30-60°) gave pure product in 71% yield.[2] Similarly, Z-Phe and Leu-OBzl were coupled (1 mmol scale) and gave a crude yield of 91%. Crystallisation from ethyl acetate-petrol (30-60°) gave pure product in 74% yield (m.p. 96-98°C; $[\alpha]_D$ -16.0 ± 0.3° (c 2.0, acetic acid); C,71.75; H,6.91; N,5.75. $C_{30}H_{34}N_2O_5$ requires C,71.69; H,6.82; N,5.58.). Repetition of this latter reaction using the monomeric reagent EEDQ in solution showed only a slight increase in the overall yield of pure product (86%).

Racemisation caused by each form of the "quinoline" coupling reagent was assayed under standardized reaction conditions by using the method of Izumiya et al.[4] This entailed preparation of the tripeptide, Gly-Ala-Leu, and separation of the diasteriomeric tripeptides by ion-exchange chromatography. Both forms of the reagent produced 6.0% of the undesired Gly-D-Ala-L-Leu.

These results, along with the previously demonstrated regenerability of the polymer,[2] suggest that this solid-phase coupling reagent could be profitably applied to the automation of peptide syntheses.

References

1. Belleau, B., and G. Malek. J. Amer. Chem. Soc. 90, 1651 (1968).
2. Some aspects of this work have been described in a preliminary communication, Brown, J. and R. E. Williams. Can. J. Chem. 49, 3765 (1971).
3. Fieser, M., and L. F. Fieser. *Reagents for Organic Synthesis*, Vol. 2 (New York: Wiley-Interscience) p 191.
4. Izumiya, M., M. Muraoko, and H. Aoyagi. Bull. Chem. Soc. Jap. 44, 3391 (1971).

SOME PROBLEMS IN SOLID-PHASE PEPTIDE SYNTHESIS

John Morrow Stewart, Gary R. Matsueda. Department
of Biochemistry, University of Colorado School of
Medicine, Denver, Colorado 80220

WHILE NEW TYPES OF RESIN SUPPORTS will probably eventually
give significant improvements in solid-phase peptide
synthesis, good results can usually be obtained with the
currently used polystyrenedivinyl benzene resin supports
if certain factors are taken into consideration. Especially
important are crosslinking of the resin and the use of
solvents and reagents which swell the resin maximally.

The encephalitogenic peptide of the myelin A protein,
Thr-Thr-His-Tyr-Gly-Ser-Leu-Pro-Gln-Lys-Gly, has been reported to be impossible to synthesize by the solid-phase
method, because of failure of deprotection at two places.[1]
Synthesis of this peptide on a 1%-crosslinked resin, using
TFA-chloroform (1:3) for deprotection, and chloroform for
DCC coupling (2.5 x reagents), afforded the desired peptide
in good yield, in a perfectly routine synthesis. Monitoring
of coupling reactions by the ninhydrin reaction[2] showed that
all went to completion within the usually allotted times.
All but the first three were complete within 10 min. When
the synthesis was repeated on a 2% crosslinked resin, using
the same reagents, the coupling of glutamine (nitrophenyl
ester) and proline were incomplete in four and two hours
respectively, and uncoupled chains were blocked by acetylation with acetylimidazole. There was no evidence of significant failure of deprotection, and the desired product
was still obtained, although in lower yield.

Further evidence of the importance of resin crosslinking
was obtained in the synthesis of the peptide described by
Folkers[3] as having luteotropin releasing activity,

<Glu-Tyr-Arg-Trp-NH$_2$. When the synthesis of this peptide
was attempted on a 2% crosslinked chloromethylated resin,
no coupling reaction was complete in 2 hr, and acetylation
was used at every step. In contrast, the synthesis on a
1% resin went smoothly, with all coupling reactions being
complete within 30 min. In each case, ammonolysis of the
peptide-resin gave the desired tetrapeptide amide.

The peptide Pro-Pro-Thr(Bzl)-Ile-Val-Val-His(Tos)-Gly
is proposed as a very difficult sequence for use as a test
peptide to study effectiveness of resins, solvents and
reagents. It contains the amino acids found to be most
difficult to couple in solid-phase synthesis.[4] Histidine
is included as a marker for detection of terminated
sequences. Synthesis of this peptide on aliquots of a 1%
crosslinked Gly-resin (0.36 mmol/g) was used to compare
chloroform and dichloromethane as solvents for synthesis.
Deprotection was by 30 min treatment with 25% TFA in the
test solvent. Indole (1 mg/ml) was routinely included as
a scavenger in this reagent. Although the test peptide
does not contain tryptophan, any aldehydes or anhydride
present in the reagent could cause undesirable termination
of any peptide. A pre-wash with the reagent was used to
avoid excessive dilution of the reagent by solvent in the
resin. Neutralization was by 5 min treatment with 10%
triethylamine in the test solvent, following a pre-wash.
For coupling reactions, 2.5 equivalents of Boc-amino acid
and DCC were used.

Although chloroform has been used in this laboratory
recently for many syntheses, dichloromethane appears to be
better for the synthesis of this peptide. When chloroform
was used throughout, no coupling reaction went fully to
completion in 2 hr, as shown by persistence of weak ninhydrin
color on the resin. Acetylation with acetic anhydride-
triethylamine or acetylimidazole was used at each step. As
previously reported,[5] acetylimidazole appears to be more
effective as a terminating agent. When dichloromethane
was used throughout, or when deprotection was in the chloroform reagent and coupling in dichloromethane, only coupling
of threonine and the first proline were incomplete in 2 hr.
Examination of the crude products by paper electrophoresis
showed that the desired peptide was obtained as the major
product from the dichloromethane and chloroform-dichloro-
methane runs, while from the run using chloroform only,
about half of the product consisted of acetylated peptides
(ninhydrin negative, Pauly positive). The observed better
result with dichloromethane was surprising, since chloroform
swells the resin slightly better than dichloromethane, and
has a lower dielectric constant.

Under optimum conditions of resin crosslinking and solvent, most solid-phase coupling reactions go very rapidly, and are complete within ten minutes. Monitoring of coupling can lead to significant acceleration of overall synthesis rates. When deprotection is monitored, this should be done on the amine salt of the resin, so that the peptide-resin is not left in the free amine form during the monitoring process.

With increasing importance of peptide amides, the benzhydrylamine (BHA) resin proposed by Marshall[6] has received much attention. Crosslinking is even more important in this resin, and hindered Boc-amino acids are difficult to attach to the 2% crosslinked BHA resin. These C-terminal residues are also difficult to remove from the resin by acid hydrolysis. HCl-dioxane gives very poor removal, HCl-propionic acid[7] gives better removal, but for quantitative removal it is necessary to cleave the amino acid-resin with HF.

Several general principles should be followed to minimize peptide-resin steric interactions. 1) The lowest degree of crosslinking of the resin consistent with resin stability should be used. 2) The chloromethylation should be conducted very carefully to avoid additional crosslinking.[8] The degree of chloromethylation should be kept low, and the temperature and time of the reaction should be carefully controlled. 3) The first amino acid should be esterified to the resin in a solvent (*e.g.*, ethanol) which does not swell the resin extensively, thereby limiting substitution to the most accessible sites on the resin. 4) All of the reactions of the synthesis should be done in solvents which swell the resin maximally. Pure trifluoroacetic acid does not swell the resin appreciably, and is not an effective deprotection reagent for solid-phase synthesis. The 25% trifluoroacetic acid reagent is an excellent swelling solvent and a very effective reagent.

Acknowledgments

This work was aided by grants HE-12325 from the USPHS and M71.81c from the Population Council. Katherine Hill, Martha Knight, Roberta Tudor and William Howland contributed valuable technical assistance.

References

1. Chow, F. C., R. K. Chawla, R. F. Kibler, and R. Shapira. J. Amer. Chem. Soc. **93**, 267 (1971).

2. Kaiser, E., R. L. Colescott, C. D. Bossinger, and P. I. Cook. Anal. Biochem. 34, 595 (1970).
3. Folkers, K., F. Enzmann, J. Boler, C. Y. Bowers, and A. V. Schally. Biochem. Biophys. Res. Commun. 37, 123 (1969).
4. Ragnarsson, U., S. Karlsson, and B. Sandberg. Acta Chem. Scand. 25, 1487 (1971).
5. Markley, L. D., and L. C. Dorman. Tetrahedron Lett. 1787 (1970).
6. Pietta, P. G., and G. R. Marshall. J. Chem. Soc. D., London 650 (1970).
7. Scotchler, J., R. Lozier, and A. B. Robinson. J. Org. Chem. 35, 3151 (1970).
8. Green, B., and L. R. Garson. J. Chem. Soc. D., London 401 (1969).

SYNTHESIS OF ENCEPHALITOGENIC PEPTIDES ON 1% CROSSLINKED
POLYSTYRENE RESIN

Raymond Shapira, Frank Chuen-Heh Chou, Robert F. Kibler.
Department of Biochemistry and Medicine, Emory University,
Atlanta, Georgia 30322

WE HAVE BEEN SUCCESSFUL in using the 1% crosslinked polystyrene resin beads from Bio-Rad* to synthesize the
encephalitogenic decapeptide Thr-Thr-His-Tyr-Gly-Ser-Leu-
Pro-Gln-Lys (I) and analogs of I. Previously described
difficulties in deblocking[1] and coupling using 2% crosslinked resin were virtually eliminated by using the 1%
crosslinked resin.

   Chloromethylated polystyrene-1% divinylbenzene resin
was esterified with the first Boc-amino acid. Double
deblocking at 0°C in 5.6 N dry HCl in dioxane and double
coupling (using DCCI) at room temperature usually gave
excellent yields as determined by the ninhydrin test[2] and
by amino acid analysis. Cleavage of the peptides from the
resin with liquid hydrogen fluoride in purified anisole
under a stream of dry nitrogen was essentially quantitative.
The resulting peptides showed a high degree of purity based
on chromatography, using a 16 feet Sephadex G-25 column,
and on amino acid analysis. The following modifications
have been successfully made:
   1. Replacement of Lys with Arg.
   2. Elongation of the C-terminal with Gly, and Ala-Gln-Gly.
   3. Elongation of the N-terminal with Arg and Ala-Arg.
   4. Substitution of Bzl-His and Dnp-His for His.
   5. Substitution of other aromatic residues for Tyr.
   6. Combinations of these modifications

---

*Bio·Rad Laboratories, Richmond, Ca 94804.

The synthetic peptides are being tested for encephalitogenic activity in rabbit, rat, guinea pig, and monkey. The
nature of the encephalitogenic peptide receptor site in
these species will be studied and the ability of these
peptides to produce and block the disease. We have reported
previously[3] that decapeptide I possesses moderate encepha-
lytogenic activity. This peptide occupies an active site
(positions 66–75) of bovine basic myelin protein (170
residues) which produces experimental allergic encephalo-
myelitis in animals, characterized clinically by onset of
weakness, ataxia, and other neurological signs 2 to 3 weeks
after challenge and pathologically by multiple perivenular
areas of demyelination and cellular infiltration.[4]

Acknowledgment

This work was supported by the National Multiple
Sclerosis Society Grant No. 464-D-8 and the National
Institutes of Health Grant No. NS 08278.

References

1. Chou, F. C.-H., R. J. Chawla, R. F. Kibler, and R.
    Shapira. J. Amer. Chem. Soc. <u>93</u>, 267 (1971).
2. Kaiser, E., R. L. Colescott, C. D. Bossinger, and
    P. L. Cook. Anal. Biochem. <u>34</u>, 595 (1970).
3. Shapira, R., F. C.-H. Chou, S. McKneally, E. Urban,
    and R. F. Kibler. Science <u>173</u>, 736 (1971).
4. Kibler, R. F., and R. Shapira. J. Biol. Chem. <u>243</u>,
    281 (1968).

DIFFICULTIES IN SOLID-PHASE PEPTIDE SYNTHESIS OF SOME
ANALOGS OF ANGIOTENSIN II

M. C. Khosla, R. R. Smeby, F. M. Bumpus. Research
Division, Cleveland Clinic Foundation, Cleveland, Ohio
44106

DURING SYNTHESES OF ANGIOTENSIN II analogs by the solid-
phase procedure of Merrifield some problems were observed.
Our experience with these problems and the efforts made to
overcome these difficulties are reported here.

*Formation of Failure Sequence
Due to Intramolecular Aminolysis*

Solid-phase synthesis of the octapeptide, Asp-Arg-Val-
Tyr-Ile-His-Pro-MePhe ([8-*N*-methylphenylalanine]-angiotensin
II) did not yield the desired peptide. Repetition of this
synthesis and analysis at each step revealed that almost
85% of Pro-MePhe was cleaved from the polymer during
neutralization of HCl.H-Pro-MePhe-polymer with 10% Et$_3$N in
DMF. Filtrates yielded *cyclo*(-Pro-MePhe-) while further
coupling resulted in direct esterification of Boc-His(Bzl)-
OH on the hydroxymethyl polymer so formed. The compound
formed in the original synthesis was therefore identified
as the N-terminal hexapeptide of angiotensin II. Cleavage
of Ala-MePhe (80%), Pro-Pro (60%) and Ala-Pro (40%) also
occurred from respective dipeptide polymers. Gisin and
Merrifield[1] observed similar cleavage with H-D-Val-Pro-
polymer and reported that the cleavage of dipeptide was
catalyzed by carboxylic acid and took place during the
next coupling step. These conclusions are surprising since
the formation of diketopiperazine is supposedly due to the
stabilization of the *cis* conformation which is possible in

peptide units containing proline[2] (or other secondary amino acids); and ring closure is further expected to be accelerated if the α-amino group is not protonated. However, our conditions of neutralization step are different than those reported by Gisin and Merrifield[1] in that these authors used 5% N, N-diisopropylethylamine in CH_2Cl_2 for neutralization of the HCl-salt.

The failure sequence was avoided by not neutralizing the HCl-salt and carrying out condensation with Boc-His (Bzl)-OH either through N-ethyl-N'-(3-dimethylaminopropyl) carbodiimide hydrochloride salt (EDAC) in $CHCl_3$ at -10 to 0° or through the corresponding pentachlorophenyl ester in DMF. However, racemization of histidine residue was higher with pentachlorophenyl ester than with EDAC.

Racemization of Histidine Residue During the Coupling Step

Earlier studies in our laboratories and by Windridge and Jorgensen[3] indicated that almost 35-40% racemization of histidine residue occurred during solid-phase synthesis of angiotensin II analogs. We have now been able to isolate[4] the D-histidine containing octapeptides from the L-histidine analogs of angiotensin II by ion-exchange chromatography with buffers of varying pH followed by partition chromatography on Sephadex G-25. Steric homogeneity of these octapeptides was determined by incubating the acid hydrolysates with L-amino acid oxidase (*Agkistrodon p. piscivorus*). Pressor as well as antagonistic properties of the D-histidine containing analogs of angiotensin II were found to be very low as compared to the L-histidine analogs. For example, [Ile^5,D-His^6]-angiotensin II possessed 4% of the pressor response of the parent hormone, Hypertensin Ciba.

Studies to avoid racemization revealed that extensive racemization of the histidine residue occurred (30-40%) when DMF was used as a solvent for coupling Boc-His(Bzl)-OPcp or Boc-His(Bzl)-OH with H-Pro-Leu-polymer through DCC or EDAC. Coupling of Boc-His(Bzl)-OH by "reversed" DCC procedure in CH_2Cl_2 gave similar results. Racemization was suppressed (15-20%) when Boc-His(Bzl)-OH was condensed in CH_2Cl_2 or $CHCl_3$ at -10° or when the imidazole moiety in histidine was protected with the tosyl group. Use of N-hydroxysuccinimide in CH_2Cl_2 along with DCC reduced the racemization considerably but did not abolish it as reported;[3] besides, the coupling reaction was very slow and the yields of the desired peptides were lowered due to the formation of β-alanyl peptides as a side product.[5] Azide coupling[6] with Z-Ile-His(Bzl)-N$_3$ did not give racemization but the condensation reaction could not be completed even after shaking for 3 days at 5°.

Formation of Side Products During Cleavage of Peptides with HBr

During the synthesis of Tyr-Ile-His-Pro-Leu, peptides were removed at every intermediate stage with HBr to (a) judge the progress of synthesis and to (b) obtain intermediate peptides. In spite of the correct amino acid ratios in the peptide polymer as well as in each cleaved peptide, tlc of the tripeptide and the tetrapeptide revealed the presence of a number of products (Figure 1). However, the

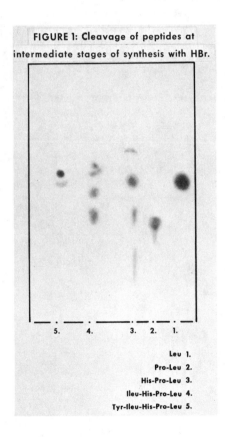

FIGURE 1: Cleavage of peptides at intermediate stages of synthesis with HBr.

5.   4.   3.   2.   1.

Leu 1.
Pro-Leu 2.
His-Pro-Leu 3.
Ileu-His-Pro-Leu 4.
Tyr-Ileu-His-Pro-Leu 5.

Figure 1: Cleavage of peptides at intermediate stages of synthesis with HBr.

pentapeptide showed a single homogeneous product and cleavage by transesterification of the intermediate peptides gave single components. This indicates that the formation of side products is due to cleavage with HBr and is presumably due to the peculiarity of this susceptible sequence.

References

1. Gisin, B. F., and R. B. Merrifield. J. Amer. Chem. Soc. <u>94</u>, 3102 (1972).
2. Ramachandran, G. N., and V. Sasisekharan. *Advances in Protein Chemistry*, Anfinsen, C. B., J. T. Edsall, M. L. Anson, and F. M. Richards, Eds., Vol. 23 (New York and London: Academic Press, 1968) pp 342.
3. Windridge, G. C., and E. C. Jorgensen.   Intra-Sci. Chem. Rep. <u>5</u>, 375 (1971).
4. Khosla, M. C., R. A. Leese, L. W. Maloy, A. T. Ferreira, R. R. Smeby, and F. M. Bumpus. J. Med. Chem. <u>15</u>, 792 (1972).
5. Weygand, F., W. Steglich, and N. Chytil. Z. Naturforsch. <u>23</u>(B), 1391 (1968) German.
6. Visser, S., and K. E. T. Kerling. Rec. Trav. Chim. Pays-Bas <u>89</u>, 880 (1970).

CHEMISTRY AND BIOLOGY OF PEPTIDES

SOLID PHASE SYNTHESIS OF PROTEIN WITH HIGH SPECIFIC
LYSOZYME ACTIVITY

*Leon E. Barstow, Dennis A. Cornelius, Victor J. Hruby,
Tadahisa Shimoda, and John A. Rupley*. Department of
Chemistry, The University of Arizona, Tucson, Arizona
85721

John J. Sharp, Arthur B. Robinson, and Martin D. Kamen.
Department of Chemistry, The University of California
at San Diego, La Jolla, California 92038

THE PRINCIPAL RESULTS from four solid phase (Merrifield)
syntheses of lysozyme[1-3] are summarized in Table I. Each
synthesis gave crude protein which had 0.5% to 1.3% the
specific enzymic activity of the native enzyme even though
reaction conditions were varied substantially from one
synthesis to another.

The specific enzymic activity of native lysozyme is
13% after treatment with liquid HF under the condition used
for cleavage of the protein resin ester in the experiments
of Table I (20°C, 90 min, added Trp, Met, Gln and anisole).
The activity of the material decreased further to 6% when
the native protein was reduced and reoxidized. This suggests that there is considerably more active protein on
the resin than is being obtained. Thus far, changes in
the reaction time (1 hr to 24 hr), the temperature (from
20°C to -20°C) and the kind and quantity of additives have
failed to give a significant improvement in the yield of
active enzyme. We have noted great variation in quality
between batches of HF.

The synthetic proteins have been characterized by
several measures of enzymic activity (cell lysis, hexa-*N*-
acetylglucosamine hydrolysis and glycosyl transfer) and by
chromatography, UV spectra, and amino acid analysis.

231

Table I

Solid-Phase Syntheses Conditions and Yields

Synthesis	1*	2*	3†	4†
Yield of resin ester/gram of starting resin	1.8 g	3.7 g	2.2 g	2.4 g
Yield of crude protein/gram resin ester	60 mg	160 mg	195 mg	220 mg
Specific enzymic activity of crude protein	1.3%	0.8%	0.5%	0.8%
Equivalent of lysozyme/gram of resin ester	0.8 mg	1.3 mg	1.0 mg	1.8 mg
Leu Substitution	0.41 mm/g	0.40 mm/g	0.30 mm/g	0.36 mm/g
Deprotection Reagent	50% TFA-CH_2Cl_2 mercaptoethanol	4N HCl-Dioxane ethanedithiol	50% TFA-CH_2Cl_2 ethanedithiol	40% TFA-CH_2Cl_2 ethanedithiol
Deprotection Time	30 min	30 min	35 → 17 min	2 x 6 min
DCC Coupling Times Solvent Number of Couplings	4 hr CH_2Cl_2 1	1 to 2 hr CH_2Cl_2,DMF,DMF urea 2 to 6	90 min, 60 min CH_2Cl_2,DMF 2	15 min CH_2Cl_2 2
Nitrophenyl Ester Coupling Times	4 hr	12 to 24 hr	10 to 12 hr	6 hr DIEA 2 hr
Total Synthesis Time	8 weeks	24 weeks	7 weeks	3.5 weeks

*Synthesis carried out at Dept. of Chemistry, University of California at San Diego; ref. 3.
†Synthesis carried out at Dept. of Chemistry, University of Arizona.

At this time, substantial work has been carried out on small portions from the fourth synthesis. About 10% of the protein has the same elution volume as the native on Sephadex G-75. All of the active protein is in this fraction. A combination of Bio-Rex 70 and affinity chromatography gave protein with more than 70% the specific lytic activity of the native enzyme.

Brown[4] has found that Residue 103 is Asn instead of Asp as had been reported.[5,6] The first two syntheses of lysozyme were prepared with the published sequence. This substitution apparently does not significantly lower the enzymic activity.

Substitutions in the active site of Trp for Tyr at position 53 and Tyr for Thr at position 51 give protein material which is totally inactive. This experiment demonstrates that only molecules with a tertiary structure like that of lysozyme have the lysozyme enzymic activity.

We are continuing to work on improvements in the synthesis of native lysozyme as well as the synthesis and characterization of other lysozyme analogs.

Acknowledgments

Research supported by NSF, USPHS, ACS (JAR); USPHS (VJH); NSF, USPHS (ABR, MDK); NIH Special Postdoctoral (LEB); NJH Predoctoral (JJS).

References

1. Sharp, J. J., A. B. Robinson, and M. D. Kamen. Federation Proceedings 30, 1287 (1971).
2. Barstow, L. E., V. J. Hruby, A. B. Robinson, J. A. Rupley, J. J. Sharp, and T. Shimoda. Federation Proceedings 30, 1292 (1971).
3. Sharp, J. J. Dissertation, University of California at San Diego, June 1971.
4. Brown, J. R. Personal Communication.
5. Jauregui-Adell, J., J. Jolles, and P. Jolles. Biochim. Biophys. Acta 107, 97 (1965).
6. Canfield, R. E., and A. K. Liu. J. Biol. Chem. 240, 1997 (1965).

AN IMPROVED SOLID PHASE SYNTHESIS OF LH-RH/FSH-RH: A PROGRESS REPORT

Maurice Manning, Esther Coy. Medical College of Ohio, Department of Biochemistry, Toledo, Ohio 43614.

Andrew Schally. Endocrine and Polypeptide Laboratories, Veterans Administration Hospital, Department of Medicine, Tulane University, School of Medicine, New Orleans, Louisiana.

SCHALLY AND CO-WORKERS RECENTLY reported[1] the isolation from porcine hypothalami of the peptide hormone which controls the secretion of both luteinizing hormone (LH) and follicle-stimulating hormone (FSH) from the pituitary gland. The amino acid sequence of this hypothalamic hormone, designated LH-releasing hormone/FSH-releasing hormone, I (LH-RH/FSH-RH), was shown[2] to be:

I: <Glu-His-Trp-Ser-Tyr-Gly-Leu-Arg-Pro-Gly-NH$_2$
    1   2   3   4   5   6   7   8   9   10

Subsequently, seven separate reports on the chemical synthesis of this decapeptide amide have thus far appeared in the literature.[3-9] In chronological order of submittal for publication, these are originating from the groups of (a) Arimura *et al.*,[3] (b) Monahan *et al.*,[4] (c) Sievertsson *et al.*,[5] (d) Geiger *et al.*,[6] (e) Matsuo *et al.*,[7] (f) Rivaille *et al.*,[8] (g) Sievertsson *et al.*[9] Reports (a-d) give no weight or percentage yield of I. Synthesis (e) resulted in an overall yield of 8% of I. Synthesis (f) afforded 10 mg of I but the overall yield appears to be low: from the data given it is not possible to calculate it but the final purification step alone resulted in only a 25% yield. Two different synthetic approaches are presented in (g) but

neither weights or percentage yields of the final purified
I are given.

It is thus clear that all of these synthetic approaches
possess serious shortcomings with regard to either a) synthetic strategy or b) complicated purification procedures;
resulting in both instances in most unsatisfactory yields
of I.

The present report deals with yet another approach
towards achieving a satisfactory synthesis of LH-RH/FSH-RH
and to thereby provide a readily reproducible method for
the synthesis of analogs of this hormone.

The approach followed is, with minor modifications,
essentially that used for the solid phase synthesis of
oxytocin[10] as utilized for the synthesis of [4-threonine]-
oxytocin.[11]

In a preliminary experiment, the protected decapeptide
resin: <Glu-His(Bzl)-Trp-Ser(Bzl)-Tyr(Bzl)-Gly-Leu-Arg(Tos)-
Pro-Gly-Resin (II) was synthesized by the Merrifield Method.[12]
Starting with 3.0 g Boc-Gly-Resin (purchased from Schwartz/
Mann Inc. and containing 1.2 mmole glycine) the appropriate
Boc-amino acids were added in stepwise fashion following the
general procedure outlined before.[11] Mercaptoethanol (1%)
was utilized in the deprotection and acid washing steps subsequent to the incorporation of the trypthophan residue.[13]
All coupling reactions were mediated by dicyclohexylcarbodiimide with methylene chloride being used as the coupling
solvent for all residues except those involving Boc-Arg(Tos),
Boc-Trp, Boc-His(Bzl) and pyroglutamic acid; for which
redistilled dimethylformamide was employed. The chloride
values (estimated by Volhard titration) following the
neutralization presented a very curious and anomalous pattern: starting with Gly and ending with the value for
ion-exchange bound chloride following the incorporation of
the pyroglutamic acid residue, these values were (in mmole)
i 1.56; ii 1.53; iii 1.91; iv 1.80; v 1.90; vi 1.65; vii
1.65; viii 1.65; ix 2.87; x 1.95. Thus the chloride value
exhibited a big increase following the neutralization of
the Arg(Tos) (iii) with the final ion-exchange bound value
(x) being higher than the starting value for Gly (i). We
have no explanation for these results. The protected decapeptide resin (II) weighed 4.58 g. The weight gain
represents a 93% incorporation on the resin. II was
ammonolyzed as described previously,[10,11] except that the
reaction time was extended to three days. The cleaved
resin was extracted with warm (50°) ethanol and the extract
evaporated to dryness to give 1.4 g of crude product. This
was recrystallized twice from methanol-ether to give the

protected decapeptide: <Glu-His(Bzl)-Trp-Ser(Bzl)-Tyr(Bzl)-
Gly-Leu-Arg(Tos)-Pro-Gly-NH$_2$ (III) as an off-white amorphous
powder (1.0 g). This represents a yield of 60% based on the
glycine content of the starting resin.

   III gave a satisfactory amino acid analysis and showed
only one spot (R$_f$ 0.45) with iodine and with the chlorine
reagent when an aliquot was examined by tlc in the system
butanol-acetic acid-water (4:1:5). M.P. 155-159° (softens
to a brown globule; melts finally >200°). The poor melting
point is not satisfactory and indicates the presence of an
impurity. C.H.N. values have thus not yet been obtained.
However, this material upon deprotection and subsequent
purification gave the desired pure I in excellent yield.
An aliquot (100 mg) of III was deblocked by the sodium
liquid ammonia procedure.[11,14] The final medium blue color
was discharged after only 10 sec. by the addition of a few
drops of dry acetic acid. Upon evapoaration of the ammonia,
the crude deblocked product was purified on Sephadex G-15
by a slight modification of the two step procedure used
for the purification of oxytocin and analogs.[11,15] In the
first step, the sample was applied to a column (2.7 x 110
cm) in 50% HOAc (1.5 ml) and washed-in with another 1.5 ml
of 50% HOAc. The column was eluted with 50% acetic acid
at a flow rate of 6.8 ml/hr and the elute monitored by uv
absorption at 280 nm. Ninety-five fractions were collected
with 4.1 ml in each of the tubes 1-52 and 2.0 ml in each of
the remaining tubes. The uv absorption spectrum indicated
the presence of a single large peak of peptide material in
tubes 56-86 with very minor amounts of absorbing material
in the tubes immediately preceding and following the peak.
Thin layer chromatographic examination on precoated silica
gel G plates of the material in tubes 56-86 in the system
butanol-acetic acid-water (4:1:5) showed the presence of
only one component, which was both Pauly positive and
chlorine reagent positive, in tubes 67-78 (FI). This
material gave the same R$_f$ (0.22) as standard LH-RH/FSH-RH.
Tubes 79-83 (FII) were found to contain a faint trace of a
faster moving Pauly positive component (R$_f$ 0.26) in addition
to I. Upon lyophilization, FI yielded 44.1 mg and FII gave
20.0 mg of white fluffy powders. Both of these fractions
were then purified separately by the second step[15] which
employs a different size column (1.5 x 110 cm) and 0.2 *N*
acetic acid as eluting solvent. Both FI and FII were each
eluted independently at a flow rate of 12 ml/hr at a fraction
size of 1.6 ml. Each gave a symmetrical peak of 280 nm
absorbing material. Tlc monitoring of the FII elute indicated that the two substances present in FII had been largely

resolved. Upon lyophilization, FI and FII gave 34.0 mg and 11.0 mg respectively of I for a total of 45 mg, $[\alpha]_D^{26}$ -48.0° (*c* 1, 1% acetic acid). This represents a yield of 60% from the deblocking step and an overall yield of 30%.

The side-product from the 0.2 M HOAc Sephadex G-15 purification of FII gave 3.0 mg. Amino acid analysis of this material indicated that it possessed the sequence of the N-terminal heptapeptide amide of I *i.e.* it lacked one residue each of Arg, Pro and Gly.

The synthetic product, I, was shown to be chemically and biologically indistinguishable from the natural hormone. This indicates that no appreciable racemisation had taken place during the incorporation of the Boc-His(Bzl) residue, or if it had, that the D-diastereoisomer was removed during the purification of the protected decapeptide. Bioassays of synthetic I were carried out as described previously.[16] *In vivo* LH-releasing activity of I was 120% (57-305%) of the potency of purified natural LH-RH.

Conclusion

The approach outlined here must still be regarded as work in progress rather than as a desired completed project. There are obviously some strange things happening especially in regard to the synthesis on the resin. Nevertheless our results to date are encouraging and, even at this stage, the final yield of I is, to our knowledge, the most satisfactory so far reported. A fuller report will be presented elsewhere upon completion of a truly satisfactory improved synthesis of FSH-RH/LH-RH.

Acknowledgments

This work was supported in part by a General Research Support Grant 5 S01 RR05700-03 and by the Medical College of Ohio. We thank Drs. T. C. Wuu and L. Balaspiri for helpful advice and criticism. We are grateful to Mr. Weldon H. Carter for LH-RH bioassays and statistical evaluation of results.

References

1. Schally, A. V., A. Arimura, Y. Baba, R. M. G. Nair, H. Matsuo, T. W. Redding, L. Debeljuk, and W. F. White. Biochem. Biophys. Res. Commun. **43**, 393 (1971).
2. Baba, Y., H. Matsuo, and A. V. Schally. Biochem. Biophys. Res. Commun. **44**, (1971).

3. Arimura, A., H. Matsuo, Y. Baba, L. Debeljuk, J. Sandow, and A. V. Schally. Endocrinology 90, 163 (1972).
4. Monahan, M., J. Rivier, R. Burgus, M. Amoss, R. Blackwell, W. Vale and R. Guillemin. C. R. Acad. Sci (Paris) 273, 508 (1971).
5. Sievertsson, H., J. K. Chang, C. Bogentoft, B. L. Currie, and K. Folkers. Biochem. Biophys. Res. Commun. 44, 1566 (1971).
6. Geiger, R., W. König, H. Wissmann, K. Geisen, and F. Enzmann. Biochem. Biophys. Res. Commun. 45, 767 (1971).
7. Matsuo, H., A. Arimura, R. M. G. Nair, and A. V. Schally. Biochem. Biophys. Res. Commun. 45, 822 (1971).
8. Rivaille, P., A. Robinson, M. Kamen, and G. Milhaud. Helv. Chim. Acta 54, 2772 (1971).
9. Sievertsson, H., J. K. Chang, A. Klaudy, C. Bogentoft, B. Currie, and K. Folkers. J. Med. Chem. 15, 222 (1972).
10. Manning, M. J. Amer. Chem. Soc. 90, 1348 (1968).
11. Manning, M., E. Coy and W. H. Sawyer. Biochemistry 9, 3925 (1970).
12. Merrifield, R. B. J. Amer. Chem. Soc. 85, 2149 (1963); Science 150, 178 (1969).
13. Marshall, G. R. In *Pharmacology of Hormonal Polypeptides and Proteins*, Back. N., L. Martini and R. Paoletti, eds. (New York: Plenum Press, 1968) pp 48–52.
14. Sifferd, R. H., and V. du Vigneaud. J. Biol. Chem. 108, 753 (1935).
15. Manning, M., T. C. Wuu and J. W. M. Baxter. J. Chromatogr. 38, 396 (1968).
16. Schally, A. V., T. L. Redding, M. Matsuo and A. Arimura. Endocrinology 90 (1972).

SOLID–PHASE SYNTHESIS OF THE PENTADECAPEPTIDES VALINE–GRAMICIDIN B AND C

Kosaku Noda, Erhard Gross. Section on Molecular Structure, Laboratory of Biomedical Sciences, National Institute of Child Health and Human Development, National Institutes of Health, Bethesda, Maryland 20014

THE AMINO ACID SEQUENCE (Figure 1) of the pentadecapeptide antibiotics valine-gramicidin B and valine-gramicidin C were determined in 1964.[1] Syntheses by conventional techniques of valine-gramicidin A and isoleucine-gramicidin A (Figure 1) were reported in 1965.[2]

```
                         1     2     3     4     5     6     7
                         L           L     D     L     D     L
Val-gramicidin A : HCO – Val – Gly – Ala – Leu – Ala – Val – Val –
Ile-gramicidin A : HCO – Ile – ............................
Val-gramicidin B : HCO – Val – ............................
Val-gramicidin C : HCO – Val – ............................


         8     9    10    11    12    13    14    15
         D     L     D     L     D     L     D     L
Val A : Val – Trp – Leu – Trp – Leu – Trp – Leu – Trp – NHCH2CH2OH
Ile A : ............ – Trp – ................. – NHCH2CH2OH
Val B : ............ – Phe – ................. – NHCH2CH2OH
Val C : ............ – Tyr – ................. – NHCH2CH2OH
```

Figure 1: The amino acid sequences of the valine-gramicidins and of isoleucine-gramicidin A.

Recently, we completed the synthesis of valine-gramicidin A by the solid-Phase method.[3] We have now synthesized valine-gramicidin B and valine gramicidin C by the solid-phase method.

It was the objective of our studies to provide a rapid and simplified synthetic route for the antibiotics and their analogs. Moreover, the nature of the amino acid residues constituting the peptides are of interest from the point of view of solid-phase synthesis. The peptides contain: (i) three tryptophan residues, which, as is well known, undergo frequently undesirable side reactions under acidic conditions,[4] (ii) the -Val-Val-Val- sequence, which is also considered a possible source of difficulty, since it has been reported that yields of coupling are usually poor when this sterically hindered amino acid is involved,[5] and (iii) an *N*-formyl group and the ethanolamide moiety at the COOH-terminus.

Figure 2 shows the outline for the syntheses of the desired peptides. The solid-phase synthesis followed

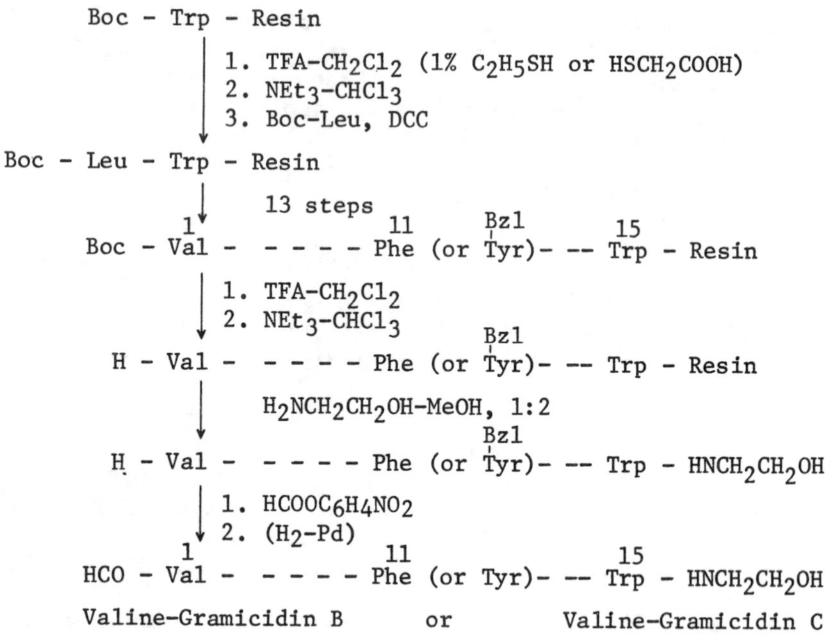

Boc - Trp - Resin

     1. TFA-CH_2Cl_2 (1% C_2H_5SH or $HSCH_2COOH$)
     2. NEt_3-$CHCl_3$
     3. Boc-Leu, DCC

Boc - Leu - Trp - Resin

     13 steps

       1   11   Bzl   15
Boc - Val -   - - - - Phe (or Tyr)- -- Trp - Resin

     1. TFA-CH_2Cl_2
     2. NEt_3-$CHCl_3$
               Bzl
H - Val -   - - - - Phe (or Tyr)- -- Trp - Resin

     $H_2NCH_2CH_2OH$-MeOH, 1:2
               Bzl
H - Val -   - - - - Phe (or Tyr)- -- Trp - $HNCH_2CH_2OH$

     1. $HCOOC_6H_4NO_2$
     2. (H_2-Pd)
       1   11   15
HCO - Val -   - - - - Phe (or Tyr)- -- Trp - $HNCH_2CH_2OH$

    Valine-Gramicidin B   or   Valine-Gramicidin C

Figure 2: Outline of the solid-phase synthesis of the valine-gramicidins B and C.

essentially the general procedure of Merrifield.[6] Boc-Amino
acids were used throughout, the coupling step being carried
out in methylene chloride with dicyclohexylcarbodiimide.
The *N*-protecting group was removed by treatment with 25%
trifluoroacetic acid in methylene chloride containing 1%
ethanethiol (synthesis of gramicidin B) or with 50% trifluoroacetic acid containing 1% of thioglycolic acid
(synthesis of gramicidin C). In the case of the gramicidin
B synthesis, the progress of amino acid coupling at each
cycle was determined by the ninhydrin method of Kaiser
et al.[7] It is noteworthy that all peptide bond forming
reactions were completed within 1 to 2 hours (*cf.* Table I)
even in the case of coupling of valine in the -Val-Val-Val-
sequence. Throughout the gramicidin C synthesis the coupling
time was fixed at four hours.

After removal of the Boc-group from the H_2N-terminal
valine, the *N*-formylated pentadecapeptide ethanolamide was
obtained by cleavage from the resin with ethanolamine in
methanol (1:2) and subsequent *N*-formylation with *p*-nitro-
phenylformate. Gramicidin C was obtained by removal of the
O-benzyl group from tyrosine by catalytic hydrogenation of
the formylated pentadecapeptide ethanolamide.

The crude peptides so obtained were then purified by
countercurrent distribution in a solvent system consisting
of chloroform:benzene:methanol:water (1:1:1.5:0.5 by volume).
Figures 3-a and 4-a show the distribution patterns of crude
synthetic valine-gramicidin B and C, respectively, for 200
transfers. Each main fraction corresponding to the natural
product was subjected to further distribution (Figures 3-b
and 4-b). The main fractions were isolated and lyophilized
from glacial acetic acid. Yields of purified material were
5% for gramicidin B and 6% for gramicidin C calculated on
the basis of Boc-Trp-resin.

The products of final purification, alone or in an
equimolar mixture with the natural compounds, showed single
symmetrical peaks on redistribution (Figures 3-c and d and
4-c and d). The synthetic peptides showed single spots of
the same R_f-value as the natural compounds on silica gel
plates in thin layer chromatography in several solvent
systems. Amino acid analyses are in good agreement with
the amino acid compositions of the natural products (Table
II).

The infrared spectra (KBr) are superimposable with
those of the natural compounds (Figure 5 and 6).

The synthetic products were tested for their antibiotic
activities. Against several microorganisms they were found

Table I

Determination of the Progress of Amino Acid Coupling at
each Cycle of the Synthesis of Valine-Gramicidin B*

Cycle Number	Amino Acid	Time (min)	Ninhydrin Color[†]	Total Coupling Time (min)
1	D-Leu	30	−	60
2	L-Trp	60		120
		90	±	
		120	−	
3	D-Leu	30	±	60
4	L-Phe	30	++	120
		60	++	
		90	+	
		120	±	
5	D-Leu	30	−	60
6	L-Trp	30	+	120
		45	±	
7	D-Val	10	+	75
		30	±	
		60	±	
8	L-Val	30	+	75
		60	±	
9	D-Val	30	+	75
		60	±	
10	L-Ala	30	−	45
11	D-Leu	30	−	45
12	L-Ala	30	−	45
13	Gly	30	−	45
14	L-Val	60	±	75

*Ninhydrin Method of Kaiser *et al.*[7]

[†]++: blue; +: weakly blue; ±: slightly blue; −: negative

Figure 3   Figure 4

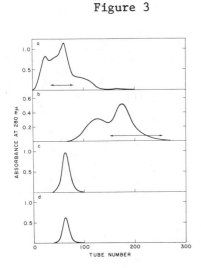

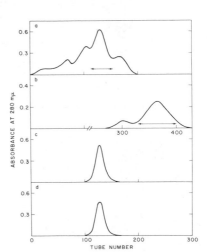

Figure 3:   Countercurrent distribution of synthetic valine-
    gramicidin B in the solvent system chloroform:benzene:
    methanol:water (1:1:1.5:0.5 by volume); a) distribution
    of crude synthetic peptide, 200 transfers; b) continued
    distribution of the main fraction (tube 36-80) up to 580
    transfers; c) redistribution for a total of 200 transfers
    of purified synthetic valine-gramicidin B (tubes 151-255
    of pattern b); d) distribution for a total of 200 trans-
    fers of a 1:1 mixture of synthetic and natural valine-
    gramicidin B.

Figure 4:   Countercurrent distribution of synthetic valine-
    gramicidin C in the solvent system chloroform:benzene:
    methanol:water (1:1:1.5:0.5 by volume); a) distribution
    of crude synthetic peptide, 200 transfers; b) continued
    distribution of the main fraction (tube 113-155) up to
    560 transfers; c) redistribution for a total of 200
    transfers of purified synthetic valine-gramicidin C
    (tubes 330-400 of pattern b); d) distribution for a total
    of 200 transfers of a 1:1 mixture of synthetic and natural
    valine-gramicidin C.

Table II

Amino Acid Analyses of Synthetic Gramicidins*

Amino Acid	Gramicidin B		Gramicidin C	
	Theory	Synthetic	Theory	Synthetic
Gly	1	0.95	1	1.09
Ala	2	2.00	2	1.97
Val	4	4.03	4	3.80
Leu	4	4.00	4	4.00
Tyr	–	–	1	0.86
Phe	1	1.09	–	–
Trp	3	2.80	3	2.64
Ethanolamine	1	1.06	1	0.95

*Hydrolyses were performed in 6N HCl containing 1% thiogly-colic acid at 110°C in tubes flushed with nitrogen and sealed under vacuum.

Table III

Antibiotic Activity of Natural and Synthetic
Valine-Gramicidins*

Organism	Minimum Inhibitory Concentration mg/ml			
	Gramicidin B		Gramicidin C	
	Natural	Synthetic	Natural	Synthetic
Bacillus subtilis	8.0	8.0	0.3	0.3
Sarcina lutea	0.3	0.3	0.3	0.3
Staphylococcus aureus	8.0	3.0	1.0	0.3
Streptococcus pyogenes	0.3	0.3	<0.005	<0.005
Escherichia coli	30.0	30.0	30.0	30.0
Pseudomonas aeruginosa	30.0	30.0	30.0	30.0

*Minimum inhibitory concentrations of the natural and syn-thetic peptides were determined by the tube dilution method. The medium used was yeast beef broth, pH 6.8.

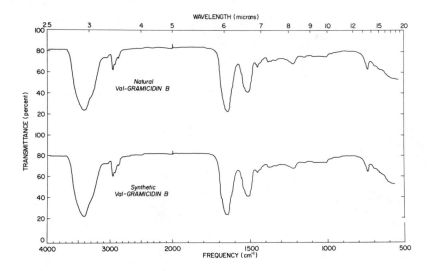

Figure 5: Infrared spectra (KBr) of natural and synthetic valine-gramicidin B.

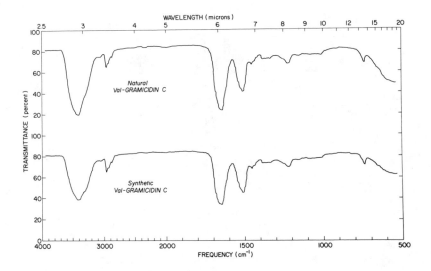

Figure 6: Infrared spectra (KBr) of natural and synthetic valine-gramicidin C.

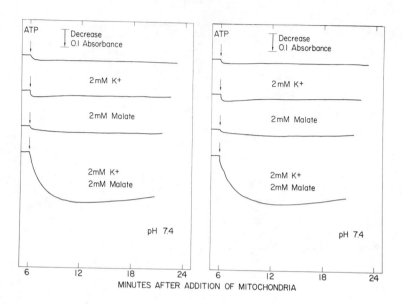

Figure 7: Valine-gramicidin B induced swelling of mitochondria; left: synthetic peptide; right: natural product

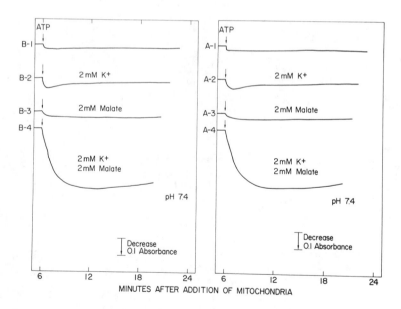

Figure 8: Valine-gramicidin C induced swelling of mitochondria; left: synthetic peptide; right: natural product

to possess the full complement of activity of the corresponding natural gramicidins (Table III). Finally, synthetic valine-gramicidin B as well as valine-gramicidin C displayed the same potential in causing mitochondrial swelling[8] as do the natural compounds, Figure 7 and 8.

We have demonstrated that the solid-phase technique, in combination with countercurrent distribution, makes readily available synthetic peptides of a high degree of purity of the gramicidin A, B, and C series. Access by synthesis to these peptides of unique physical, chemical, and biological properties occurred at a most opportune time, now that D. W. Urry and his associates[9] have demonstrated the singular type of helix which these peptides are capable of forming and that the head-to-head dimer of these gramicidins spans lipid bilayers and acts as transmembrane channel affecting ion transport.

It shall be most interesting to provide analogs of the gramicidins A, B, and C and to study their effect on ion transport across biomembranes and the ensuing biological consequences.

Acknowledgment

We gratefully record the use of a Peptide Synthesizer at Beckman Instruments, Inc. of Palo Alto, California, and the collaboration of Mr. Jon F. Harbaugh throughout the synthesis of gramicidin B. Dr. Marvin J. Weinstein and his associates of the Schering Corporation of Bloomfield, New Jersey, performed the antibiotic tests. Dr. Herbert I. Hadler determined the effect of the gramicidins on mitochondria. The IR-spectra were taken by Mr. Noel F. Wittaker.

References

1. Sarges, R., and B. Witkop. J. Amer. Chem. Soc. **86**, 1861, 1862 (1964); **87**, 2011, 2027 (1965); Biochemistry **4**, 2491 (1965).
2. Sarges, R., and B. Witkop. J. Amer. Chem. Soc. **87**, 2020 (1965).
3. Fontana, A., and E. Gross. Manuscript in preparation.
4. Alakov, Y. B., A. A. Kiryushkin, V. M. Lipkin, and G. W. Milne. Chem. Commun. 406 (1970).
5. Esko, K., S. Karlsson, and J. Porath. Acta Chem. Scand. **22**, 3342 (1968); Weygand, F., and R. Obermeier, Z. Naturforsch. **23b**, 1390 (1968).

6. Merrifield, R. B.   J. Amer. Chem. Soc. <u>85</u>, 2149 (1963);
   Merrifield, R. B., J. M. Stewart, and N. Jernberg.
   Anal. Chem. <u>38</u>, 1905 (1966).
7. Kaiser, E., R. L. Colescott, C. D. Bossinger, and P. I.
   Cook. Anal. Biochem. <u>34</u>, 595 (1970).
8. Falcone, A. B., and H. I. Hadler. Arch. Biochem.
   Biophys. <u>124</u>, 91, 115 (1968); Hadler, H. I. and A. B.
   Falcone. Arch. Biochem. Biophys. <u>124</u>, 110 (1968).
9. Urry, D. W. Proc. Natl. Acad. Sci. U.S. <u>68</u>, 672
   (1971); *cf.* also Urry, W. D. Biochim. Biophys. Acta
   <u>265</u>, 115 (1972).

GLOBIN DECAPEPTIDE SYNTHESIS AND LABELING FOR PARENTERAL USE

Julie Brown, Henry Brown. Sears Surgical Laboratory, Harvard Surgical Service, Boston City Hospital; Department of Surgery, Harvard Medical School.

Arnold Trzeciak. Children's Cancer Research Foundation, Boston, Massachusetts 02115.

THE INTRIGUING POSSIBILITY THAT amino acids in peptide linkage may be used in forming new hemoglobin was investigated. Previous animal experiments were done measuring effects of amino acids in peptide linkage from a peptic globin hydrolysate on hemoglobin regeneration.[1] Hemoglobin regeneration was better with amino acids in peptide linkage than with equivalent amounts of free amino acids. Using globin containing ^{14}C labeled amino acids, a decapeptide with higher specific activity than others was isolated from the peptic hydrolysate.[2] The amino acid sequence was found to be: Ser–Glu–Asp–Leu–Gly–Ala–Ser–Val–Ser–Leu.[3] It is
$$\underset{1}{}\ \underset{2}{}\ \underset{3}{}\ \underset{4}{}\ \underset{5}{}\ \underset{6}{}\ \underset{7}{}\ \underset{8}{}\ \underset{9}{}\ \underset{10}{}$$
probably from the α–chain, analogous to positions 129–138 in human hemoglobin.[4]

A peptic hydrolysate of rat globin containing ^{14}C labeled amino acids was administered intraperitoneally to rats made anemic by bleeding. This peptide, isolated from newly formed hemoglobin, again had higher specific activity than 26 other acidic peptides studied. The findings suggested incorporation of at least some amino acids from this sequence while remaining in peptide linkage.[2]

To test this hypothesis the decapeptide was synthesized. The Merrifield solid–phase method[5] was used with modifications as described by Manning *et al.*[6] in which all major steps of deprotection, neutralization and coupling are repeated. Two

percent cross-linked polystyrene was used to prepare the
starting Boc-leucine resin. Dicyclohexylcarbodiimide
mediated couplings using 3 equivalents of Boc-amino acids
were tested for completeness by the Kaiser ninhydrin procedure.[7] Benzyl group side-chain protection was employed
for Boc-Ser(Bzl), Boc-Asp(OBzl) and Boc-Glu(OBzl). Nα-t-
Butyloxycarbonyl protecting groups were cleaved by successive
treatments with 30 percent trifluoroacetic acid in methylene
chloride and 1N HCl in acetic acid. Cleavage was tested
for completeness by Volhard chloride titration.

The final decapeptide was cleaved from the resin by
anhydrous liquid HF. The peptide was purified by repeated
Sephadex G-15 gel filtrations in 0.1N and 0.2N acetic acid,
respectively. The synthetic peptide had the same N-terminal
serine as measured by the Dansyl and DNP techniques and
amino acid analysis as the natural peptide. The synthetic
and natural peptides had the same R_f value on filter paper
chromatography in n-butanol-water-acetic acid (4:1:5).
They also had the same electrophoretic mobility on filter
paper at 1700 volts D.C. both in 0.2N acetic acid, pH 2.7
and in formic acid-dioxan-water, pH 2.1. The yield of
peptide was 80% of theoretic.

A second synthesis was done using Boc-(1-^{14}C)-L-leucine.
This radioactive amino acid was incorporated into amino acid
position 4 but not at the C-terminal leucine. Specific
activity of the peptide was 0.21 µCi/mg.

For hemoglobin regeneration experiments 50.1 mg of
[1-^{14}C]-L-leucine labeled globin (10 µCi) was dissolved in
5 ml of lactated Ringer's solution. pH was brought to 7.4
with dilute sodium hydroxide. A 420 gram male rat was bled
by cardiac puncture 2 ml per day for 4 days. The dissolved
globin peptide was then injected intraperitoneally in
divided doses 15 minutes apart. Twenty-three hours later
the animal was exsanguinated by cardiac puncture. Hemoglobin was prepared from red cells and globin was separated
from heme by acid acetone by previously described methods.[1,2]
Specific activity of this globin was $1.3 \cdot 10^{-4}$ µCi/mg.
Studies are in progress to determine specific activities
of the two leucine residues in positions 4 and 10 of the
isolated decapeptide.

Acknowledgment

This work was supported in part by Research Grants
AM-0861 from the National Institute of Arthritis and
Metabolic Diseases, C-6516 from the National Cancer

Institute and FR-05526 from the Division of Research
Facilities and Resources, National Institutes of Health.

References

1. Brown, H., and J. Brown. Metabolism 8, 286 (1959).
2. Brown, H., and J. Brown. Metabolism 9, 587 (1960).
3. Brown, H. *Hepatic Failure.* (Springfield: Thomas, 1970)
   p 49.
4. Brown, H., and J. Brown. *Protein Nutrition.* (Springfield:
   Thomas, in press).
5. Merrifield, R. B. Advan. Enzymol. 32, 221 (1969).
6. Manning, M., E. Coy, and W. H. Sawyer. Biochemistry 9,
   3925 (1970).
7. Kaiser, E., R. L. Colescott, C. D. Bossinger, and P. I.
   Cook. Anal. Biochem. 34, 595 (1970).

STUDIES WITH SEMISYNTHETIC NONCOVALENT PROTEIN COMPLEXES

Irwin M. Chaiken. Laboratory of Chemical Biology,
National Institute of Arthritis and Metabolic Diseases,
National Institutes of Health, Bethesda, Maryland 20014

CURRENT ADVANCES IN CHEMICAL SYNTHESIS of large peptides[1,2]
have provided the potential for studying protein conformation and biological activity, at the level of amino acid
function, by the preparation of synthetic analogues. To
be sure, preparation of chemical mutants for proteins of
at least several thousand molecular weight is limited by,
among other things, the insolubility of large intermediates
for synthesis in solution and the difficulty of obtaining
pure products by solid-phase methods. Nonetheless, for the
biologically active noncovalent complexes staphylococcal
nuclease-T'[3] and bovine pancreatic ribonuclease-S',[4] a wide
variety of structural analogues have been synthesized,[5-11]
in each instance for a relatively small protein fragment
which binds to a complementary native fragment. In such
cases, the synthetic problem is simplified while at the
same time allowing protein systems of considerable size and
complexity to be examined. For both nuclease-T' and ribonuclease-S', it has now become possible to adapt solid-phase
synthesis for the preparation of semisynthetic complexes of
purity sufficient to allow detailed conformational and
functional characterization.
 One of the systems studied, nuclease-T', is the enzymically
active complex obtained from intact nuclease (foggi strain)
by limited trypsin treatment and composed of the two non-
covalently bound fragments containing, respectively, residues
6 through 48 [nuclease-T-(6-48)] and 49 through 149 [nuclease-
T-(49-149)].[3] The fragment corresponding to residues 6
through 47 has been synthesized by the Merrifield solid-phase

procedure[12],[13] as described previously.[14] As reported,
the resulting crude synthetic-(6-47) peptide (obtained after
cleavage from resin and deblocking) is only partially effective (due to impurities) in replacing nuclease-T-(6-48) in
forming an active complex with nuclease-T-(49-149). However,
methods have since been developed[15],[16] to obtain a highly
active complex, containing synthetic-(6-47) and nuclease-
T-(49-149), by a procedure involving trypsin treatment of
a mixture of the synthetic and native fragments followed by
phosphocellulose chromatography. The resulting semisynthetic nuclease-T' is quite similar to native nuclease-T'
in the enzymic properties of specific activity and binding
of substrates and inhibitors (Table I), as well as in the
structural properties of fluorescence emission and stability
to temperature and trypsin treatment.[16]

The ability to obtain purified semisynthetic nuclease-T'
provides the basis for the preparation and detailed characterization of important semisynthetic analogues. In one
case, the analogue [Asp^{43}]-semisynthetic nuclease-T', with
aspartic acid replacing glutamic acid at position 43, was
obtained.[17] This material was shown to be enzymically
inactive. On the other hand, the analogue complex is conformationally similar to normal nuclease-T' and still able
to bind both deoxythymidine-3',5'-diphosphate, an inhibitor
of nuclease and nuclease-T', and calcium, an atom directly
required for nuclease-T' activity and also important for
conformational stabilization. Taken together with the fact
that Glu^{43} is in the active site region in the crystal
structure of nuclease[18] and probably nuclease-T' also,[19]
the results indicate that Glu^{43} is critical for the organization of the active site of nuclease-T', either directly
in hydrolysis or perhaps in the orientation of a directly
participating group (such as the essential calcium ion).

By procedures similar to those used above, semisynthetic
ribonuclease-S', the complex of synthetic-(1-15)(the solid-
phase synthesized fragment corresponding to residues 1
through 15 of ribonuclease A[20],[21] and ribonuclease-S-(21-
124)(the native fragment containing residues 21 through 124
of ribonuclease), has been prepared[22],[23] in a form, after
sulfoethyl-Sephadex fractionation, about equally as active
as native ribonuclease-S' (Table II). For this semisynthetic
complex, analogues have been prepared which are labeled with
either enriched carbon-13 or fluorine-19 at specific residues
in the 1-15 region (Table II). Using the appropriate
analogue, ^{13}C and ^{19}F nuclear magnetic resonance spectra
have been obtained, in both cases allowing specific resolved
resonances to be characterized and assigned to individual

Table I

Functional Characteristics of Semisynthetic Nuclease-T'

| Protein species | Enzymic Activity | | | | $pdTp$[††] Dissociation Constant K_I^{app} |
	DNA $\Delta A_{260}/min/mg$ *	RNA $\Delta A_{260}/min/mg$ †	PNP-$pdTp$[#] V_{max} ($\Delta A_{330}/min/mg$)**	K_m ($M \cdot 10^4$)**	($M \cdot 10^6$)[##]
Nuclease-T'	171	10.1	0.051	2.1	2.1
Semisynthetic Nuclease-T'					
prep. a	153	9.4	---	---	2.6
prep. b	147	9.1	0.044	2.2	
Nuclease	1664	223	1.30	0.12	1.1

*Specific activity at room temperature with 50 µg/ml DNA (salmon sperm, heat denatured) in 0.05 M Tris-Hcl, pH 8.8, containing 0.01 M CaCl2.

†Specific activity at room temperature with 80 µg/ml yeast RNA in 0.05 M Tris-HCl, pH 8.8, containing 0.01 M CaCl2.[16]

#5'-p-Nitrophenylphosphoryl-deoxythymidine-3'-phosphate.

**Determined at 24.3° in 0.05 M Tris-HCl, pH 8.8, containing 0.01 M CaCl2.[16]

††Deoxythymidine-3',5'-diphosphate.

##Determined by equilibrium dialysis, using [14C-methyl]-pdTp, at 7° in 0.05 M Tris-HCl, pH 8.8, containing 0.01 M CaCl2.[17]

Table II

Enzymic Activity of Semisynthetic Ribonuclease-S' Analogues
Against 2',3'-Cyclic-Cytidine Monophosphate

Protein Species	Specific Activity* (ΔA_{286}/min/ mole)	K_m† (mM)
Ribonuclease-S'	38.8	1.1
Semisynthetic Ribonuclease-S'	41.8	1.3
[^{13}C-Phe^8]-Semisynthetic Ribonuclease-S'#	40.2	1.0
[p-fluoro-Phe^8]-Semisynthetic Ribonuclease-S'**	39.0	0.9
Ribonuclease A	40.8	1.0

*Determined at 25° with 0.25 mmol C-CMP in 0.05 M Tris-HCl, pH 7.13, containing 0.8 gm NaCl/250 ml.

†Determined at 25° in 0.05 M Tris-HCl, pH 7.13, containing 0.8 gm NaCl/250 ml.

#Position 8 contains L-Phe uniformly enriched at 13% in ^{13}C (the enriched amino acid was generously provided by Dr. Jack S. Cohen[24]).

**Position 8 contains p-fluoro-L-Phe, the latter a gift of Dr. N. L. Benoiton.[25]

atoms.[23] By such procedures, it is hoped that conformational and enzymic features of ribonuclease-S' can be defined, especially in relation to the participation of specific amino acid residues.

The semisynthetic noncovalent protein complexes described, when obtained in sufficient amounts, should be appropriate for characterization to a degree expected for native proteins and protein derivatives. Inasmuch as X-ray crystallographic studies have been, or are being, carried out for ribonuclease-S'[26] and nuclease-T',[19] it is expected that specific important semisynthetic analogues for these two complexes can be examined crystallographically, analogous to recent experiments with native abnormal hemoglobin variants.[27]

Ultimately, it is hoped that such characterizations of
semisynthetic analogues will lead to an increased understanding both of detailed structure-function relationships
for the specific proteins studied as well as of rules for
the function of amino acids in proteins in general.

References

1. Marglin, A., and R. B. Merrifield. Ann. Rev. Biochem.
   39, 841 (1970).
2. Geiger, R. Angew. Chem. Internat. Edit. 10, 152 (1971).
3. Taniuchi, H., and C. B. Anfinsen. J. Biol. Chem. 243,
   4778 (1968).
4. Richards, F. M., and P. J. Vithayathil. J. Biol. Chem.
   234, 1459 (1959).
5. Finn, F. M., and K. Hofmann. J. Amer. Chem. Soc. 87,
   645 (1965).
6. Finn, F. M., J. P. Visser, and K. Hofmann.   In
   Peptides 1968, Bricas, E., ed. (Amsterdam: North-
   Holland, 1968), p 330.
7. Rocchi, R., F. Marchiori, L. Moroder, A. Fontana, and
   E. Scoffone. Gazz. Chim. Ital. 96, 1537 (1966).
8. Scoffone, E., F. Marchiori, L. Moroder, R. Rocchi, and
   A. Scatturin. In *Peptides 1968*, Bricas, E., ed
   (Amsterdam: North-Holland, 1968) pp 325-329.
9. Ontjes, D. A., and C. B. Anfinsen. J. Biol. Chem.
   224, 6316 (1969).
10. Chaiken, I. M., and C. B. Anfinsen. J. Biol. Chem.
   246, 2285 (1971).
11. Chaiken, I. M. J. Biol. Chem. 247, 1999 (1972).
12. Merrifield, R. B. J. Amer. Chem. Soc. 85, 2149 (1963).
13. Merrifield, R. R. Science 150, 178 (1965).
14. Ontjes, D. A., and C. B. Anfinsen. Proc. Nat. Acad.
   Sci. U.S.A. 64, 428 (1969).
15. Chaiken, I. M., and C. B. Anfinsen. J. Biol. Chem.
   245, 2337 (1970).
16. Chaiken, I. M. J. Biol. Chem. 246, 2948 (1971).
17. Chaiken, I. M., and G. R. Sanchez. J. Biol. Chem.
   In press.
18. Cotton, F. A., C. J. Bier, V. W. Day, E. E. Hazen, Jr.,
   and S. Larsen. Cold Spring Harbor Symp. Quant. Biol.
   36, 243 (1972).
19. Taniuchi, H., D. R. Davies, and C. B. Anfinsen. J.
   Biol. Chem. 247, 3362 (1972).
20. Potts, J. T., D. M. Young, and C. B. Anfinsen. J.
   Biol. Chem. 238, 2593 (1963).
21. Hofmann, K., F. M. Finn, M. Limetti, J. Montibeller,
   and G. Zanetti. J. Amer. Chem. Soc. 88, 3633 (1966).

22. Chaiken, I. M.   Fed. Proc. <u>30</u>, 1274 (1971).
23. Chaiken, I. M., M. H. Freedman, J. R. Lyerla, Jr., and J. S. Cohen.   J. Biol. Chem.   In press.
24. Horsley, W., H. Sternlicht, and J. S. Cohen.   J. Amer. Chem. Soc. <u>92</u>, 680 (1970).
25. Tong, J. H., C. Petitclerc, A. D'Iorio, and N. L. Benoiton.   Can. J. Biochem. <u>49</u>, 877 (1971).
26. Wyckoff, H. W., D. Tsernoglou, A. W. Hanson, J. R. Knox, B. Lee, and F. M. Richards.   J. Biol. Chem. <u>245</u>, 305 (1970).

SYMPOSIUM DISCUSSIONS

Summarized by Johannes Meienhofer

THE MAJORITY OF PAPERS in this session on solid-phase
peptide synthesis dealt with new methodological developments. In the discussions, however, several comments
pertained to the standard procedure. Attention was drawn
to the observation of small amounts of ninhydrin-positive
side products when dicyclohexylcarbodiimide is added to
solutions of *t*-butyloxycarbonylamino acids. Among these
side products the free amino acid, dipeptide and polymerized
compounds have been identified. No decomposition of this
kind was observed when DCC was added to benzyloxycarbonyl-
amino acids or with active esters of Boc-amino acids, but
caution was expressed that 2-(*p*-biphenylyl)isopropyloxy-
carbonylamino acids (Bpoc-amino acids) might also be
effected by DCC.

Several remarks were made about losses of dipeptides
from peptide resins through the formation of diketopiperazines either during the neutralization step or the following
(longer lasting) coupling stage (see also pp 227 to 230
This troublesome side reaction can occur whenever the
second amino acid residue from the *N*-terminal (position 2)
is either a proline or an *N*-methylamino acid residue (see
pp 59 to 65), and it has also been observed with a Gly-
Gly-resin, as presented in detail in the forum discussion
(pp 89 to 92).

It has been generally recognized that the less crosslinked polystyrene-1% divinylbenzene resin is a superior
support compared to the 2% crosslinked material (see pp
221 to 224, 225 to 226). Unfortunately, the degree of
crosslinking of commercially available resins is never
accurately defined. Moreover, the chloromethylation reaction or prolonged storage in the presence of residual

261

catalyst can cause additional crosslinking. An approximate estimation of the degree of crosslinking could be obtained from the extent of resin swelling in solvents.

The incorporation of Boc-His(Bzl) by DCC mediated condensation is known to be accompanied by racemization.[1] This has been confirmed (pp 227 to 230); and ways to suppress this racemization were outlined in the discussion. Although addition of *N*-hydroxysuccinimide[1,2] lowers racemization, it can lead to the known β-alanyl peptide formation.[1,3] Good results were obtained by the addition of hydroxybenzotriazole.[4,5] Alternatively, racemization can be reduced to a few tenths of a percent by reducing the imidazole side chain basicity by tosyl or 2,4-dinitrophenyl protection. However, the tosyl group of Boc-His(Tos) is not very stable, and the thiol reagents employed for the *N*im-Dnp group cleavage can in residual trace amounts interfere with subsequent hydrogenolytic reactions by poisoning the catalysts.

The development of the ring-halogenated benzyl protecting groups for more acid-stabile side chain blocking (pp 191 to 195, 197 to 202) apparently followed literature guidelines, such as σ-ρ diagram values.[6] Cleavage rates were determined with 50% trifluoroacetic acid in methylene chloride, and it was pointed out that cleavage by other acids should also be examined. A troublesome side reaction occurs with *O*-protected tyrosine during the final removal of the completed peptide from the solid support by treatment with HBr in TFA or with HF-anisole (1:1). *O*-Benzyltyrosine gives 13% of 3-benzyltyrosine by intramolecular rearrangement. The extent of this undesired rearrangement is lower (4%) with *O*-(2,6-dichlorobenzyl)tyrosine, but it could not be suppressed by modifications of the cleavage reagent or by addition of various scavengers (inorganic iodides, organic sulfides.)

The successful application of Mukaiyama's oxidation-reduction condensation[7] to solid-phase synthesis (pp 209 to 211) raised the question whether it would be suitable for peptide cyclization. This seemed to have not yet been investigated. The opinion was presented that the reaction might proceed *via* intermediate mercaptopyridine esters. These anchimerically catalyzed compounds do not suffer from strong overactivation; and the Boc group cleavage observed with DCC activation (*vide supra*) would be avoided in oxidation-reduction condensations.

The report on resin-bound *N*-ethyloxycarbonyl-2-ethyloxy-1,2-dihydroquinoline (EEDQ) as a regenerable solid-phase coupling reagent (pp 217 to 219) evoked the comment that an ethyloxycarbonylamino derivative has been isolated as a by-product from an EEDQ-mediated peptide condensation[8] in solution. The formation of this type of side product is consistent with the intermediate mixed anhydride mechanism proposed by Belleau and Malek,[8] and a warning was sounded that this side reaction might also occur with the use of solid-phase EEDQ.

The presentations of three papers on the application of the solid-phase method to the preparation of biologically active peptides (lysozyme, LH-RH/FSH-RH, and valine gramicidins B and C; pp 231 to 250) were followed by arguments about alleged disparities between experimental results and the titles of some papers. One discussant placed the responsibility squarely upon the editor. My own opinion,[9] briefly, is:

Titles do imply claims of priority and achievement (especially since entire literature information systems are based on titles). The art and virtue of understatement should be practiced more often (compare titles in references 10, 11).

Synthesis, *i.e.* assembly of a molecule *identical* to a known structure, when done without characterization of each intermediate, as in solid-phase synthesis, requires meticulous examination of the final product to prove identity with the target structure.

Characterization of small peptides perhaps up to eicosapeptides, can in most cases be attained by a combination of amino acid and elemental analysis, thin layer chromatography in several solvent systems, tests for optical purity by ORD and enzymatic digestion, and demonstration of full biological activity, preferably by four-point assays of several different activities (if applicable). Additional data (IR, UV, NMR, mass spectrum) are desirable.

Large peptides or protein preparations necessarily require all the above tests but these are not sufficient. The rules for demonstrating homogeneity of a protein call for the examination of a minimum of five independent criteria (see *e.g.* [12]), such as peptide mapping, end group determination by dansylation and carboxypeptidase digestion, gel electrophoresis, isoelectric focusing in pH gradients[13] or in polyacrylamide gel,[14] rotating free zone electrophoresis,[15] sedimentation, distribution by countercurrent and chromatographic techniques, affinity chromatography, immunochemical tests, or functional purification by limited enzymatic digestion of enzyme-substrate complexes.[16]

In conclusion, it should be good to remember that researchers, not trained in peptide synthesis, are probably not aware of the problems. We have a responsibility of meticulously accurate reporting, and we must insist that the quality of work always conforms to the highest possible standards that can be attained with the newest available methods.

References

1. Windridge, G. C., and E. C. Jorgensen. In *Intra-Science Chem. Reports*, Kharash, N., ed. (Santa Monica: Intra-Science Research Foundation, 1972) pp 375-380.
2. Wünsch, E., and F. Drees. Chem. Ber. 99, 110 (1966); Weygand, F., D. Hoffmann, and E. Wünsch. Z. Naturforsch. 21b, 426 (1966).
3. Gross, H., and L. Bilk. In *Peptides 1968*, Bricas, E., ed. (Amsterdam: North-Holland, 1968) pp 156-158; Weygand, F., W. Steglich, and N. Chytil. Z. Naturforsch. 23b, 1391 (1968).
4. Windridge, G. C., and E. C. Jorgensen. J. Amer. Chem. Soc. 93, 6318 (1971).
5. König, W., and R. Geiger. Chem. Ber. 103, 788 (1970).
6. Blaha, K., and J. Rudinger. Coll. Czech. Chem. Commun. 30, 585, 599 (1965).
7. Mukaiyama, T., R. Matsueda, and H. Maruyama. Bull. Chem. Soc. Jap. 43, 1271 (1970).
8. Belleau, B., and G. Malek. J. Amer. Chem. Soc. 90, 1651 (1970).
9. Meienhofer, J. In *Hormonal Proteins and Peptides*, Vol. II, Li, C. H., ed. (New York: Academic Press, 1973) in press.
10. du Vigneaud, V., C. Ressler, J. M. Swan, C. W. Roberts, P. G. Katsoyannis, and S. Gordon. "The synthesis of an octapeptide amide with the hormonal activity of oxytocin." J. Amer. Chem. Soc. 75, 5879 (1953).
11. Schwarz, H., F. M. Bumpus, and I. H. Page. "Synthesis of a biologically active octapeptide similar to natural isoleucine angiotonine octapeptide." J. Amer. Chem. Soc. 79, 5697 (1957)
12. Meienhofer, J., and H. Zahn. In *Hoppe-Seyler/Thierfelder, Handbuch Physiol. Pathol.-Chem. Anal.*, Lang, K., and E. Lehnartz, eds. (Berlin: Springer, 1960) pp 1-115.
13. Haglund, H. In *Methods of Biochemical Analysis*, Vol. 19, Glick, D., ed. (New York: Interscience, 1971) pp 1-104.

14. Finlayson, G. R., and A. Chrambach. Anal. Biochem. <u>40</u>, 292 (1971).
15. Hjerten, S. In *Methods of Biochemical Analysis*, Vol. 18, Glick, D., ed. (New York: Interscience, 1970) pp 55-79.
16. Chaiken, I. M., and C. B. Anfinsen. J. Biol. Chem. <u>245</u>, 4718 (1970); Gutte, B., and R. B. Merrifield. J. Biol. Chem. <u>246</u>, 1922 (1971).

SECTION IV

NEW SYNTHETIC ADVANCES

Session Chairmen

Évangélos Bricas and Joseph S. Fruton

SOLID-PHASE SYNTHESIS OF RIBONUCLEASE T_1

*N. Izumiya, M. Waki, T. Kato, M. Ohno, H. Aoyagi,
N. Mitsuyasu.* Laboratory of Biochemistry, Faculty
of Science, Kyushu University, Fukuoka, Japan

SUMMARY--Several preliminary experiments were carried out
prior to the solid-phase synthesis of ribonuclease T_1.
For example, the usefulness of hydrogen chloride in formic
acid for the cleavage of N-t-butyloxycarbonyl groups has
been demonstrated in a solid-phase synthesis of a
tryptophan-containing model peptide. The Merrifield
solid-phase method was used to prepare protein through
104 steps of amino acid incorporation following the target
sequence of ribonuclease T_1. Purification of the protein
preparation isolated from the protected peptide resin
raised the specific nuclease activity up to 53% for yeast
RNA and 41% for 2',3'-cyclic guanylic acid.

INTRODUCTION--The conventional methods of peptide synthesis
have made many important and principal contributions to the
solid-phase method of peptide synthesis (SPS). For example,
many protecting groups and coupling reagents used in SPS
are originating from conventional methods, such as t-
butyloxycarbonyl (Boc), benzyloxycarbonyl (Z), benzyl ester
(OBzl; including the peptide to resin linkage), S-benzyl,
N^{im}-benzyl, N^{im}-(2,4-dinitrophenyl) (Dnp), dicyclohexylcarbodiimide (DCC), DCC plus N-hydroxysuccinimide (HOSu),
etc. Contributions from SPS to conventional methods seem
to be smaller, but the 2-(p-biphenylyl)isopropyloxycarbonyl
group (Bpoc) and the liquid hydrogen fluoride reagent are
notable contributions.

It is generally accepted that the solid-phase technique developed by Merrifield cannot at present be used for the confirmation of a proposed primary structure of a protein by synthesis.[1] However, it is well recognized that SPS possesses very attractive and useful features such as speed, simplicity, automation, *etc.* The purpose of our present investigations is outlined as follows.

(1) We intend to collect some information which might help to achieve a "clean" synthesis of a protein, perhaps by a conventional method or by a solid-phase fragment synthesis.[2] We plan to gain such information from the experimental results obtained during a stepwise solid-phase synthesis of a polypeptide with the target sequence of a known protein.

(2) New features which are developed during stepwise SPS will, it is hoped, contribute to improvements of conventional methods.

We have carried out stepwise solid-phase syntheses of basic pancreatic trypsin inhibitor (BPTI),[3] cobrotoxin (CT),[4] and ribonuclease T_1 (RNase T_1)[5] with the intentions mentioned above. In this presentation we describe the features of the solid-phase synthesis of RNase T_1 and also some experiments for improved protection of amino acid side chain functions.

Protection of Several Functional Side Chains

Although, at present, several combinations of protecting groups for functional side chains are available for SPS, we attempted to introduce new more effective protecting groups for some side chain functions.

Indole moiety in tryptophan*

It is well known that tryptophan in a peptide chain being synthesized on a polymer support undergoes oxidation during treatment with HCl in AcOH for the cleavage of Boc groups. In such cases, β-mercaptoethanol (βME) or dithiothreitol[6] have been used as a scavenger to protect tryptophan. Previero *et al.*[7] observed that bubbling of HCl into formic acid (HCOOH) solutions of tryptophan results in formylation of the indole nitrogen (N^I).   We

---

*RNase T_1 and CT (possessing 62 amino acid residues) have only one tryptophan residue each in positions 59 and 29, respectively, and BPTI (58 residues total) has no Trp.

found that tryptophan and N^I-CHO-Trp-OH are stable in 0.1-1.0 N HCl-HCOOH; the Boc group is cleaved rapidly with a 2-10 fold molar excess of HCl in HCOOH, and the N^I-formyl group is removed by 0.1 N aqueous piperidine or hydrazine in DMF although this group is not influenced by triethylamine in DMF or by HF. These observations led us to examine HCl-HCOOH as a reagent for cleaving Boc groups in SPS of tryptophan-containing peptides.

As an example, a 6-fold molar excess of HCl (0.1 N) in HCOOH was used in the synthesis of a model peptide, H-Lys-Ala-Gly-Leu-Gly-Trp-Leu-OH, which was built up by SPS in the usual way.[8] In parallel, two peptides with the same sequence were prepared using 1 N HCl-AcOH in the presence and absence of βME (2%).

One half of each protected peptide-resin was cleaved by hydrazinolysis, and the three products corresponding to Boc-Lys(Z)-Ala-Gly-Leu-Gly-Trp-Leu-NHNH₂ were isolated; they were designated as Ia (HCl-HCOOH), Ib (HCl-AcOH-βME), and Ic (HCl-AcOH), respectively. The UV spectrum of Ia was virtually identical with that of Z-Trp-OH. Extinction coefficients of Ib and Ic at 282 nm were 34% and 45% more intense than that of Ia, respectively. Such enhancement at 282 nm should be ascribed to oxidation of the indole nucleus although the side products formed have not yet been identified.

The other halves of the protected peptide-resins were treated with HF, and the peptide (IIa) produced *via* the HCl-HCOOH procedure was further treated with 0.1 N aqueous piperidine to remove the N^I-formyl group. Each product was chromatographed on a column of Sephadex G-25; IIb (synthesized *via* the HCl-AcOH-βME procedure) gave a complex elution profile due to its heterogeneity, whereas that from IIa gave a simple profile with one major peak. To ascertain whether the major peak from IIa is based only on pure peptide, IIa was treated with 2-nitro-4-carboxyphenylsulfenyl chloride (Ncps-Cl) in 80% HCOOH. The modified peptide was again chromatographed on the same column, and gave one peak at a different, delayed position. This showed that the modified peptide containing 2-thio-(2-nitro-4-carboxyphenyl)-tryptophan possessed, as expected, an increased absorptivity to the gel compared to the unmodified peptide. It is apparent from these results that the original peptide in the major peak from IIa contains only pure (unoxidized) tryptophan. Comparison of the two elution diagrams before and after reaction of IIb with Ncps-Cl indicates that IIb contains some material in which the tryptophan residue had suffered changes from the repeated HCl-AcOH-βME treatments.

We think that more experiments are required before the HCl-HCOOH system can be applied to SPS of macropeptides such as RNase T_1. We are performing the SPS of the naturally occurring decapeptide LH-RH which contains one tryptophan, as a model experiment to gain further experience with the HCl-HCOOH procedure. In the SPS of RNase T_1 reported herein we still used the usual 1 N HCl-AcOH-βME (1%) reagent and Boc-Trp-OH under complete replacement of air by N_2 gas. With respect to our intentions in the present studies, the HCl-HCOOH system developed during the SPS experiments is now being routinely applied in conventional syntheses in our laboratory. For example, crystalline H-Orn(Z)-Leu-OEt·HCl has been obtained from Boc-Orn(Z)-Leu-OEt by treatment with 0.1 N HCl (1.2 equiv) in HCOOH for 10–20 min; the Z group is more stable in 0.1 N HCl-HCOOH than in HCl-dioxane or CF_3COOH.

Phenol moiety in tyrosine

In SPS, Boc-Tyr(Bzl)-OH has been used usually, however, the phenolic benzyl ether linkage is relatively resistant to HF.[9] Stewart and Young suggested in their monograph[10] the use of p-methoxybenzyl ether which might be removed easily by HF. We prepared crystalline Boc-Tyr(MeOBzl)-OH *via* the following sequence of reactions. Tyrosine copper complex was treated with p-methoxybenzyl chloride which was prepared from anisalcohol and PCl_3. The Tyr(MeOBzl)-Cu complex was decomposed with EDTA and the resulting H-Tyr(MeOBzl)-OH was treated with Boc-N_3. We found, however, that the p-methoxybenzyl ether linkage of Boc-Tyr(MeOBzl)-OH is more resistant to cleavage by hydrogen fluoride than the benzyl ether (see Table I). Therefore, Boc-Tyr(Bzl)-OH was used in the SPS of RNase T_1. Conditions were developed to achieve practically complete cleavage of the benzyl ether linkage, as described later.

Imidazole moiety in histidine

In the SPS of histidine-containing peptides the N^α-Boc derivatives of N^{im} unprotected His, His(Z), His(Dnp), His(Boc),[6] His(Tos)[11] *etc.* have been applied. We used Boc-His(Boc)-OH in the SPS of RNase T_1 because of its greater solubility in organic solvents and its ease of preparation, although Boc-His(Tos)-OH seemed to be a promising starting material as well.

Table I

Cleavage of Protecting Groups of Tyrosine
by Liquid Hydrogen Fluoride

Compound	0°C		20°C
	0.5 hr	*1 hr*	*1 hr*
Boc-Tyr(Bzl)-OH	---	70-75%	95-98%
Boc-Tyr(MeOBzl)-OH	30%	33-40%	90%

Besides stepwise SPS of RNase T₁, we are also attempting solid-phase fragment condensation, but the present work is of very preliminary status. We prepared three Boc-peptide acids (III, IV, V)[12] and try to assemble them on the polymer support by DCC plus HOSu, or by EEDQ. For the preparation of V, Boc-Val-Ile-Thr(Bzl)-His-Thr(Bzl)-Gly-OH (VI) was tosylated to give crystalline V in a good yield, whereas benzyloxycarbonylation or dinitrophenylation of VI resulted in a poor yield or in the formation of colored material.

Boc-Val-Glu(OBzl)-Cys(MeOBzl)-OH   (III)
   101   103

Boc-Ala-Ser(Bzl)-Gly-Asn-Asn-Phe-OH   (IV)
   95   100

Boc-Val-Ile-Thr(Bzl)-His(Tos)-Thr(Bzl)-Gly-OH   (V)
   89   94

*Preliminary Experiments Prior to the Solid-Phase
Synthesis of Ribonuclease T₁*

RNase T₁ was isolated from *Aspergillus oryzae* by Egami in 1957,[13] and its amino acid sequence of 104 residues was determined by Takahashi in 1965,[14] Figure 1. The enzyme is an acidic protein (pI 2.9) and has two intramolecular disulfide bonds. Hofmann has been working on the total synthesis of this protein by conventional methods.[15] We are carrying out a synthesis of protein with the target sequence of RNase T₁ by stepwise SPS,[5] besides syntheses

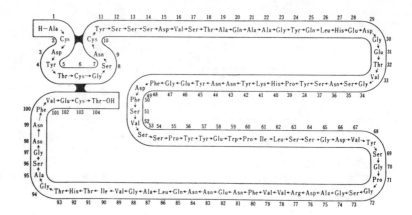

Figure 1:   The primary structure of ribonuclease T_1.

of several short fragments through conventional methods[16] and making arrangements for fragment SPS.[12]   Prior to the stepwise SPS, we performed the following experiments.

Treatment of natural RNase T_1 with liquid hydrogen fluoride

Since it is planned to treat the final, fully protected peptide-resin in the SPS with HF, it is important to examine the stability of natural RNase T_1 to HF treatment.   The natural enzyme* (220 µg, 0.02 µmol) was treated with HF (0.5 ml) in the presence and absence of tryptophan.   After certain time intervals, the solution was evaporated and the residue was tested for RNase T_1 activity using yeast RNA as a substrate.   The results are summarized in Table II and show recovery of 80-90% activity after 1 hr at 0° in the presence of added tryptophan and about 75% after 3 hr.

Treatment of some Boc-amino acids with liquid hydrogen fluoride

For the incorporation of tyrosine residues, Boc-Tyr(Bzl)-OH was selected for the SPS of RNase T_1, as mentioned before. Since the N^ε-Z group of Boc-Lys(Z)-OH is cleaved gradually

---

*Obtained from Sankyo Company.

Table II

Residual Activity (%) of RNase T$_1$ After Treatment
with Liquid Hydrogen Fluoride

| *Trp added,* | *0°C* | | | *20°C* | |
in equiv	*1 hr*	*3 hr*	*4 hr*	*1 hr*	*2 hr*
---	77	57		38,33	43
125	92		47		
250	82	76		35	

by successive treatments with HCl-AcOH or CF$_3$COOH, we chose
Boc-Lys(Dipmoc)-OH. Sakakibara introduced the Dipmoc
(diisopropylmethyloxycarbonyl) group into SPS. This group
is cleaved by HF but very stable to 1 N HCl-AcOH.[17] Our
results with HF are summarized in Table III.

Table III

Cleavage (%) of Side Chain Protecting Groups
by Liquid Hydrogen Fluoride

| *Compound* | *0°C* | | *20°C* |
	1 hr	*3 hr*	*1 hr*
Boc-Tyr(Bzl)-OH	70-75	95-97	95-98
Boc-Lys(Dipmoc)-OH	78-80	98-100	~100

From the results indicated in Tables II and III, we
selected as optimal conditions of HF treatment a temperature
of 0°C and a duration of 3 hr for the deprotection of a
fully protected peptide-resin. However, it should be noted
that some protecting groups in the protected peptide-resin
may be incompletely cleaved by HF even after 3 hr, possibly,
because some may be buried inside and cannot contact freely
with HF.

The Stepwise Solid-Phase Synthesis
of Ribonuclease T₁

We carried out the stepwise SPS of RNase T_1 as usual.
Since the procedure is similar to that employed in solid-
phase syntheses of BPTI[3] and CT,[4] we describe the course
and the present status of our experiments only briefly.

SPS procedure

Commercial 2% cross-linked polystyrene was chloromethylated in the usual manner, and a resin containing
1.46 mmol/g of Cl was obtained. This was converted to
HCl·H-Thr(Bzl)-resin with 0.21 mmol/g of threonine. Two
parallel runs of the SPS of manual way were made for syntheses of Boc-peptide-resin of 104 amino acid residues,
one with the sequence by natural RNase T_1 and the other
with that of an analog, [Tyr59]-RNase T_1. The schedule
of a cycle for the incorporation of each Boc-amino acid
was almost the same as described before[3,4] with the exception of the use of Lys(Dipmoc) and His(Boc). 1.3 N HCl-
AcOH was used as a cleaving reagent for Boc groups. After
the incorporation of tryptophan in position 59, 1% βME was
added to the 1.3 N HCl-AcOH reagent and also to the AcOH
washing solvent, and air in the reaction vessel was
completely replaced by N_2 gas. Yields of the fully protected peptide-resins are summarized in Table IV along
with the results of the BPTI and CT preparations.

Table IV

Yield of Protected Peptide-resins

Synthetic object	Starting HCl·amino acid-resin	Completed protected peptide-resin	Yield (%)		
			Weight increase	Cl titra-tion	Amino acid analysis
RNase T_1	3.8 g	10.3 g	53	56	55
[Tyr59]-RNase T_1	1.9 g	4.8 g	49	49	51
Trypsin Inhibitor (BPTI)	1.0 g	1.3 g	20		
Cobrotoxin	2.4 g	4.4 g	39		

Isolation of peptide from
protected peptide-resin

Cleavage of the polypeptides from the solid support together with the removal of all protecting groups was achieved by treatment with HF in the presence of anisole. The procedure of Exp. B (see Table V) is described as follows. After evaporation of HF and drying *in vacuo*, the

Table V

Treatment of Peptide-resin with Liquid Hydrogen Fluoride*
at 0°C and Yields of Crude Peptides

	RNase T₁			*Basic Pancreatic Trypsin Inhibitor*	*Cobrotoxin*
		Exp.			
	A	*B*	*C*		
Protected Peptide-resin	0.163 g	1.0 g	0.5 g	0.5 g	0.5 g
Time of HF treatment*	1 hr	3 hr with Trp		1 hr	1 hr
Extraction solution	Tris-HCl, pH 7.0	NH₄HCO₃, pH 8.0		10% AcOH	5% AcOH
Amount of crude peptide	6 mg	180 mg	78 mg	31 mg	55 mg
(Yield)	(8%)	(39%)	(34%)	(31%)	(36%)

*Anisole was added.

residue was washed with acetone containing 2% 1N HCl to remove anisole. To the residue, a mixture (pH 8.0) of 0.05 M NH₄HCO₃ and 0.1 M NaCl was added. The whole mixture was subjected to air oxidation for generating disulfide bonds. The filtrate from the resin was lyophilized and dialyzed. The solution was passed through a column of Sephadex G-50, and fractions containing proteins were lyophilized. The yield of crude peptide was 180 mg (39% from peptide-resin). Yields of crude peptides in three experiments are summarized in Table V.

Purification of crude peptide

The product of Exp. A was purified as follows. The crude peptide (6 mg; see Table V) was chromatographed on a column (0.9 x 38 cm) of Sephadex G-75 with 0.02 M NH_4HCO_3 (pH 7.5) as an eluant. A major peak gave 2.4 mg of a protein with a specific activity (SA) of 0.4%. This protein was treated again in similar manner; yield 0.8 mg, SA 1.1%. Further chromatography gave 0.45 mg protein with SA 2.2%.

The protein (0.45 mg), treated three times with G-75, was further chromatographed on a column (0.9 x 8 cm) of DEAE cellulose in a manner of linear-gradient elution with 0.005 M Na_2HPO_4-0.25 M $Na_2H_2PO_4$, 0.25 M NaCl. A main peak afforded 0.16 mg of a protein with SA 23%.

The protein (0.16 mg) obtained was incubated with chymotrypsin (0.016 mg) at pH 7.0 and 37°C for 15 hr. After salts in the incubation mixture were removed by Sephadex G-50 chromatography, the eluate was chromatographed on a column of DEAE cellulose; the yield was 0.09 mg and SA was 53%[5] for yeast RNA and 41% for guanosine 2',3'-cyclic phosphate.

Conclusion

Many parts of our studies have apparently not yet been completed. Furthermore, it is questionable whether a protein with 53% or 41% activity of RNase T_1 contains molecules identical to natural RNase T_1. Nevertheless, our experiments may provide some useful information for a future "clean" synthesis of RNase T_1 and its analogs.

Acknowledgment

The authors express their thanks to Mr. S. Terada and Mr. S. Matsuura for technical assistance.

References

1. Wünsch, E. Angew. Chem. Internat. Edit. <u>10</u>, 786 (1971).
2. Izumiya, N., K. Kato, M. Waki, N. Mitsuyasu, K. Noda, S. Terada, and O. Abe. In *Progress in Peptide Research*, Vol. 2, Lande, S., ed. (New York: Gordon and Breach, 1972) pp 69-78.
3. Noda, K., S. Terada, N. Mitsuyasu, M. Waki, T. Kato, and N. Izumiya. Naturwissenschaften <u>58</u>, 147 (1971).
4. Aoyagi, H., H. Yonezawa, N. Takahashi, T. Kato, and N. Izumiya. Biochim. Biophys. Acta <u>263</u>, 827 (1972).

5. Waki, M., N. Mitsuyasu, S. Terada, S. Matsuura, T. Kato, and N. Izumiya. J. Jap. Biochem. Soc. 43, 472 (1971) Japanese.

6. Blake, J., and C. H. Li. J. Amer. Chem. Soc. 90, 5882 (1968); Li, C. H., and D. Yamashiro, J. Amer. Chem. Soc. 92, 7608 (1970).

7. Previero, A., M. A. Coletti-Previero, and J.-C. Cavadore. Biochim. Biophys. Acta 147, 453 (1967).

8. Ohno, M., S. Tsukamoto, and N. Izumiya. Chem. Commun. 663 (1972).

9. Sakakibara, S., Y. Shimonishi, Y. Kishida, M. Okada, and H. Sugihara. Bull. Chem. Soc. Jap. 40, 2164 (1967).

10. Stewart, J. M., and J. D. Young. *Solid Phase Peptide Synthesis* (San Francisco, Calif.: W. H. Freeman and Co., 1969) p 23.

11. Fujii, T., and S. Sakakibara. *The 7th Symposium on Peptide Chemistry*. Tokyo, Nov. 1969.

12. Terada, S., N. Mitsuyasu, K. Noda, M. Waki, T. Kato, and N. Izumiya. *The 8th Symposium on Peptide Chemistry*. Osaka, Nov. 1970.

13. Sato, K., and F. Egami. J. Biochem. (Tokyo) 44, 753 (1957).

14. Takahashi, K. J. Biol. Chem. 240, PC4117 (1965).

15. Camble, R., G. Dupuis, K. Kawasaki, H. Romovacek. N. Yanaihara, and K. Hofmann. J. Amer. Chem. Soc. 94, 2091 (1972).

16. Mitsuyasu, N., M. Waki, T. Kato, S. Makisumi, and N. Izumiya. Bull. Chem. Soc. Jap. 43, 1556 (1970).

17. Sakakibara, S., T. Fukuda, Y. Kishida, and I. Honda. Bull. Chem. Soc. Jap. 43, 3322 (1970).

ON THE SYNTHESIS OF A PORCINE GASTRIC INHIBITORY POLYPEPTIDE

R. Camble. Imperial Chemical Industries Limited,
Pharmaceutical Division, Alderley Park, Macclesfield,
Cheshire, England.

"GASTRIC INHIBITORY POLYPEPTIDE" (GIP) was isolated from
commercially available, partially purified cholecystokinin.[1]
It is reported to be a potent inhibitor of both histamine
and pentagastrin stimulated gastric acid and pepsin secretion in the dog;[2] as such, it could be the "enterogastrone"
released from the duodenum by fat. The peptide is a single
chain of 43 amino acid residues containing no disulphide
bridges.[3] The N-terminal region shows considerable homology
with the sequences of glucagon and secretin. Dr. Brown
kindly provided us with details of the structure prior to
publication and as part of a continuing investigation into
natural inhibitors of gastric secretion[4] we undertook a
synthesis of GIP. This communication is a report of our
progress towards that goal. Our approach involved the
preparation by stepwise elongation of seven protected
fragments (Figure 1). These are to be linked by azide or
mixed anhydride reactions to give the complete molecule.

Preparation of the fragments

The fragments were chosen for obvious chemical reasons
and from a desire to pursue any indication of a biologically
active core. Fragments C, D, E, F and G were built up with
t-alkyl side chain protection and the benzyloxycarbonyl
group was used for α-amino group protection. In fragment
B the N-terminal methionine residue was introduced as its
N-*o*-nitrophenylthio derivative. For fragment A, side
chains were blocked as their **benzyl** derivatives and the
t-butoxycarbonyl group was employed for α-amino protection.

281

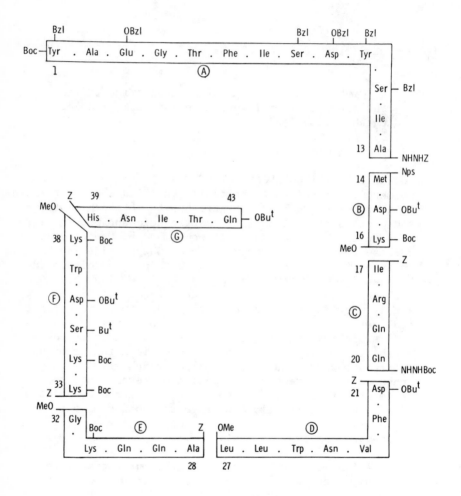

Figure 1: Sequence and protected peptide fragments for the
synthesis of gastric inhibitory polypeptide.

The presence of up to six side chain benzyl groups during the synthesis of fragment A made the coupling products very insoluble and tlc was of little use as a check for purity after the octapeptide stage. The BOC-tridecapeptide hydrazide, obtained after prolonged hydrogenolysis of fragment A over spongy palladium or 5% palladium on charcoal, exhibited four major components on tlc. These appear to arise from incomplete removal of the benzyl protecting groups. (Resistance of the serine O–benzyl derivative to hydrogenolysis has been observed in work on LRF synthesis in our laboratory[5]). We hope to employ this material for further coupling and to remove any remaining benzyl groups by acid treatment at the end of the synthesis.

The o–nitrophenylthio group was removed from fragment B with 80% thioglycollic acid[6] and the product was purified by silica gel and Bio-Rex 70 (H^+ cycle) column chromatography.

To minimize the risk of rearrangement of the N-terminal aspartic acid residue of fragment D[7] the β–t–butyl ester was removed before treatment with hydrazine. Hydrazinolysis of fragments D and F in DMF solution was slow even with large excesses of hydrazine.[8] Much faster reaction rates were obtained by using DMF–n–butanol (1:1) as solvent.[9]

When the well-characterized fragment E was treated with alkali in aqueous DMF a complex mixture was always obtained. The desired product was isolated in about 35% yield after chromatography on AG1–X2 (acetate cycle) anion exchange resin;[10] from their elution pattern the impurities appeared to be more acidic than the product and emerged at two distinct concentrations of acetic acid. NMR, hydrogenolysis and amino acid analysis results on the impurity fractions suggested that, in addition to any possible rearrangement and deamidation of the glutamine residues, the N-terminal benzyloxycarbonylalanine residue must have rearranged extensively to the corresponding hydantoin.

The five t–butyl derivatives were not removed cleanly from fragment F by aqueous trifluoroacetic acid. An acceptable product was obtained after reaction with HBr-trifluoroacetic acid. This observation was disappointing as it probably nullifies the idea behind the approach used for the synthesis of fragment A.

Fragment couplings and assessment of purity

The task of joining up the fragments has progressed to the point shown in Figure 2. The properties of the products were such that, until the heptacosapeptide stage, purification was limited to selective precipitation and chromatography

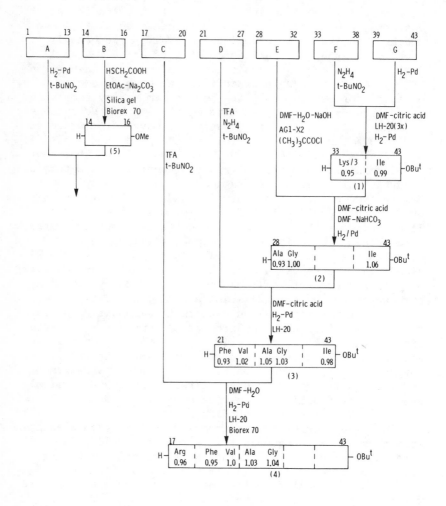

Figure 2: Fragment condensation steps toward the synthesis of gastric inhibitory polypeptide.

on LH-20 Sephadex in DMF solution. Purity was assessed by tlc and the ratios between 'diagnostic' amino-acid residues.[11] Tlc was of limited use with the heptacosapeptide and arginine was the only diagnostic residue in the N-terminal portion. At this point we began to exploit the non-quantitative but extremely sensitive dansylation technique[12] as an additional test of purity. About 4 nanomoles of the almost ninhydrin negative heptacosapeptide derivative (4) gave a very intense dansyl-isoleucine spot but no dansyl-aspartate or ε-dansyl-lysine. Dansylation of (4) following treatment with 90% trifluoroacetic acid generated both dansyl-isoleucine and ε-dansyl-lysine after hydrolysis. Backtracking, the hexadecapeptide derivative (2) gave only dansyl-alanine. In spite of an excellent ratio of diagnostic amino-acids, the tricosapeptide derivative (3) produced a faint but definite trace of dansyl-alanine as well as the expected dansyl-aspartate. The technique was also employed to confirm that the purified tripeptide (5) contained no free lysine ε-amino group after the thioglycollic acid treatment used to remove the *o*-nitrophenylthio group.

The value of diagnostic amino-acid ratios is critically dependent upon evidence, usually provided by tlc, that one of the two components in the coupling has been completely eliminated. Dansylation offers an additional and extremely sensitive tool with which to attack this problem of evaluating the purity of large synthetic peptide intermediates.

The synthetic team involved in this work comprised R. Camble, R. Cotton, A. Dutta, J. J. Gormley, C. F. Hayward, J. S. Morley and M. J. Smithers. Dansylation experiments were carried out by Mrs. B. M. Preston. Amino acid analyses were performed by M. W. Earlam.

References

1. Brown, J. C., V. Mutt, and R. A. Pederson. J. Physiol. (London) 209 , 57 (1970).
2. Pederson, R. A., and J. C. Brown. Gastroenterology 62, 393 (1972).
3. Brown, J. C., and J. R. Dryburgh. Can. J. Biochem. 49, 867 (1971).
4. Gregory, H. Amer. J. Digest. Diseases 15, 141 (1970).
5. Gormley, J. J. unpublished work.
6. Morley, J. S. J. Chem. Soc., C, 2410 (1967).
7. Bajusz, S., T. Lazar, and Z. Paulay. Acta Chim. Acad. Sci. Hung. 41, 329 (1964).

8. Strachan, R. G. *et al.*   J. Amer. Chem. Soc. <u>91</u>, 503 (1969).
9. See Ferren, R. A., J. G. Miller, and A. R. Day. J. Amer. Chem. Soc. <u>79</u>, 70 (1957).
10. Hofmann, K., J. P. Visser, and F. M. Finn. J. Amer. Chem. Soc. <u>91</u>, 4883 (1969).
11. Hofmann, K. *Peptides 1969*, Scoffone, E., ed. (Amsterdam: North Holland Publishing Co., 1971) p 130.
12. Hartley, B. S. Biochem. J. <u>119</u>, 805 (1970).

CHEMISTRY AND BIOLOGY OF PEPTIDES

ON THE PROINSULIN SYNTHESIS

Noboru Yanaihara, Chizuko Yanaihara, Tadashi Hashimoto, Masanori Sakagami, Naoki Sakura. Laboratory of Bioorganic Chemistry, Shizuoka College of Pharmacy, Shizuoka, Japan

PORCINE PROINSULIN[1,2] is a single chain polypeptide which consists of eighty-four amino acid residues, and three disulfide bridges are present in the molecule. It has been demonstrated[3,4] that the reduced proinsulin reconstitutes the proper disulfides efficiently to restore the characteristic properties of native proinsulin. The connecting peptide segment of this prohormone connects the carboxy terminus of the B chain to the amino terminus of the A chain of the insulin molecule.

We have achieved previously the syntheses of [33-glutamine,62-formyllysine]porcine proinsulin 31-63[5] and [62-formyllysine]-porcine proinsulin 31-63[6] with the amino acid sequence of porcine connecting peptide segment. These synthetic tritriacontapeptides were found to possess immunological properties identical to those of natural connecting peptide fragment.[6] Synthesis of these polypeptides was performed essentially by the fragment condensation according to the modification of the azide method developed by Honzl and Rudinger.[7]

In a previous communication,[6] we have described briefly the synthesis of a partially protected linear octahexaconta-peptide (I), Figure 1, possessing the 17 - 84 amino acid sequence of porcine proinsulin, together with its immunological properties.

The present article describes the strategies and current investigations on our studies aiming at the synthesis of porcine proinsulin. As reported previously,[6] the azide

H-Leu-Val-Cys(Ec)-Gly-Glu-Arg-Gly-Phe-Phe-Tyr-Thr-Pro-Lys(CHO)-Ala-Arg-
17 18 19   20 21 22 23 24 25 26 27 28 29   30 31

Arg-Glu-Ala-Glu-Asn-Pro-Gln-Ala-Gly-Ala-Val-Glu-Leu-Gly-Gly-Gly-Leu-Gly-
32 33 34 35 36 37 38 39 40 41 42 43 44 45 46 47 48 49

Gly-Leu-Gln-Ala-Leu-Ala-Leu-Glu-Gly-Pro-Pro-Gln-Lys(CHO)-Arg-Gly-Ile-Val-
50 51 52 53 54 55 56 57 58 59 60 61 62   63 64 65 66

Glu-Gln-Cys(Ec)-Cys(Ec)-Thr-Ser-Ile-Cys(Ec)-Ser-Leu-Tyr-Gln-Leu-Glu-Asn-
67 68 69   70   71 72 73 74   75 76 77 78 79 80 81

Tyr-Cys(Ec)-Asn-OH
82 83   84

Figure 1: Synthetic partially protected porcine proinsulin 17–84, octahexacontapeptide. Ec, Ethylcarbamyl.

Table I

Protected Peptide Fragments used for the Preparation of
Proinsulin Fragments II and III

(A) Z-Leu-Val-Cys(Ec)-Gly-NHNHBoc (positions 17-20)

(B) Z-Glu(OBu^t)-Arg-Gly-Phe-Phe-Tyr-Thr-Pro-Lys(CHO)-Ala-NHNHBoc (positions 21-30)

(C) Z-Arg(NO$_2$)-Arg-Glu-Ala-Glu(OBu^t)-Asn-Pro-Gln-Ala-Gly-NHNHBoc (positions 31-40)

(D) Z-Ala-Val-Glu(OBu^t)-Leu-Gly-Gly-Leu-Gly-NHNHBoc (positions 41-49)

(E) Z-Gly-Leu-Gln-Ala-Leu-Ala-Leu-Glu(OBu^t)-Gly-NHNHBoc (positions 50-58)

(F) Z-Pro-Pro-Gln-Lys(CHO)-Arg-Gly-Ile-Val-Glu(OBu^t)-Gln-NHNHBoc (positions 59-68)

(G) H-Cys(Ec)-Thr-Ser-Ile-Cys(Ec)-Ser-Leu-Tyr-Gln-Leu-Glu-Asn-Tyr-Cys(Ec)-
Asn-OH (positions 69-84)

condensation of protected tritriacontapeptide hydrazide (II)
(positions 17–49) with partially protected pentatriaconta-
peptide (III)(positions 50–84), followed by treatment with
hydrogen bromide in trifluoroacetic acid, resulted in the
formation of I. The seven peptide fragments shown in Table
I were employed for the construction of II and III. The
benzyloxycarbonyl group was used as α-amino protecting group;
ε-Amino groups of two lysine residues (positions 29 and 62)
were protected with the formyl function, and the side chains
of glutamic and aspartic acids were protected as t-butyl
ester. The side chain of cysteine was blocked by the
ethylcarbamyl(Ec) function.[8] The terminal carboxyl functions
of six protected peptide fragments A to F were protected with
t-butoxycarbonylhydrazide. On introduction of arginine into
a peptide chain, the guanido function of this amino acid was
protected with the nitro group. Syntheses of these protected
peptide fragments were performed according to a strategy
similar to that used for the synthesis of the peptide derivatives related to ribonuclease T_1.[9-12]

A variety of protected peptide derivatives terminating
with glutamine-t-butoxycarbonylhydrazide were prepared as
intermediates for preparation of the desirable peptide fragments. Examples are Z-Gly-Leu-Gln-NHNHBoc (positions 50-52),
Z-Pro-Pro-Gln-NHNHBoc (positions 59-61), Z-Ile-Val-Glu(OBu^t)-
Gln-NHNHBoc (positions 65-68) and Z-Cys(Ec)-Ser-Leu-Tyr-Gln-
NHNHBoc (positions 74-78). Using the hydrazides derived
from these protected intermediates, the glutamine moiety
was introduced *via* azide coupling into the peptide chain
smoothly without risk of lactam formation. The porcine
proinsulin molecule contains eleven glycine residues, of
which ten are located in the 17-84 sequence. Four glycine
moieties were used as carboxy terminal amino acids of
benzyloxycarbonyl peptide t-butoxycarbonylhydrazides.
Glycine-terminating protected di- or tripeptides such as
Z-Glu(OBu^t)-Arg(NO$_2$)-Gly-OH (positions 21-23), Z-Leu-Gly-
OH (positions 44-45), Z-Gly-Gly-OH (positions 46-47) and
Z-Lys(CHO)-Arg(NO$_2$)-Gly-OH (positions 62-64) were also
prepared and their mixed anhydrides served as acylating
agents.

For removal of benzyloxycarbonyl groups catalytic
hydrogenolysis was conducted. This process also permitted
simultaneous removal of nitro group on arginine residues.
Benzyloxycarbonyl groups of cysteine-containing peptides
were removed by treatment with hydrogen bromide in
trifluoroacetic acid and anisole. By this treatment,
t-butyl groups were also removed, while ethylcarbamyl and
formyl groups were unaffected.

The final coupling in the preparation of I was accomplished by the formation of a linkage between positions 49 and 50, at both of which positions glycine moieties are located. In the synthesis of porcine connecting peptide,[5] we carried out the azide coupling of the hydrazide derived from fragment D (positions 41-49) with a tetradecapeptide (positions 50-63) and obtained a protected tricosapeptide (positions 41-63) in 63% yield. This indicates that the azide coupling of a peptide terminating with glycine at position 49 with a glycyl peptide such as pentatriaconta-peptide III (positions 50-84) may take place efficiently under the reaction condition employed. Furthermore, we have planned to characterize our synthetic polypeptides by utilizing immunological properties specific to porcine connecting peptide and proinsulin, in addition to the usual analytical methods. Our studies[6] in cooperation with Drs. R. E. Chance and M. A. Root of the Lilly Research Laboratories have indicated clearly that the antigenic determinant in the porcine connecting peptide segment is located in the 41-54 sequence, -Ala-Val-Glu-Leu-Gly-Gly-Gly-Leu-Gly-Gly-Leu-Gln-Ala-Leu-. All synthetic and natural connecting peptide fragments containing this 41-54 sequence cross-reacted with the purified anti-porcine proinsulin antiserum on essentially equimolar basis, while the smaller synthetic nonapeptide (positions 41-49), pentapeptide (positions 50-54), and an equimolar mixture of these two peptides did not cross-react with the antiserum at all. In addition, it was found with the antiserum elicited by our synthetic connecting peptide[6] that synthetic connecting peptide is displaced by natural porcine proinsulin on an equimolar basis, while synthetic nonapeptide (positions 41-49) and tetradecapeptide (positions 50-63) do not react. Thus, none of the polypeptide fragments used for the preparation of I embodies the amino acid sequence that corresponds to the antigenic determinant, the 41-54 sequence. If the coupling of II with III results in the formation of the desired linkage between positions 49 and 50, the ensuing polypeptide embodies the 41-54 sequence and must be immunologically reactive.

Indeed, our synthetic partially protected polypeptide I cross-reacted with anti-synthetic porcine connecting peptide antiserum. The cross-reactivity of I was essentially as high as that of natural porcine proinsulin or synthetic porcine connecting peptide, indicating the successful formation of the desired linkage between positions 49 and 50. The amino acid compositions of I were those predicted by theory. Purification of I, II and III were conducted exclusively by gel filtration, because of the low solubility of these polypeptides in buffers for column chromatography.

Synthesis of protected hexadecapeptide hydrazide (positions 1-16) has been described.[6] Continuous investigation for the total synthesis of porcine proinsulin is under way.

Acknowledgment

The authors express their gratitude to Drs. T. Kaneko and H. Oka, the First Department of Internal Medicine, University of Tokyo, Faculty of Medicine, Tokyo, for their performing the radioimmunoassays.

References

1. Chance, R. E., R. M. Ellis, and W. W. Bromer. Science 161, 165 (1968).
2. Chance, R. E. *Proceedings of the VII Congress of the International Diabetes Federation, Buenos Aires, 1970.* Excerpta Medica International Congress Series No. 231, p 292.
3. Steiner, D. F., and J. L. Clark. Proc. Nat. Acad. Sci. U.S. 60, 622 (1968).
4. Bromer, W. W. BioScience 20, 701 (1970).
5. Yanaihara, N., T. Hashimoto, C. Yanaihara, and N. Sakura. Chem. Pharm. Bull. (Tokyo) 18, 417 (1970).
6. Yanaihara, N., T. Hashimoto, C. Yanaihara, M. Sakagami, and N. Sakura. Diabetes 21 (suppl. 2) 476 (1972).
7. Honzl, J., and J. Rudinger. Coll. Czech. Chem. Commun. 26, 2333 (1961).
8. Guttmann, St. Helv. Chim. Acta 49, 83 (1966).
9. Yanaihara, N., C. Yanaihara, G. Dupuis, J. Beacham, R. Camble, and K. Hofmann. J. Amer. Chem. Soc. 91, 2184 (1969).
10. Hofmann, K. In *Peptides 1969*, Scoffone, E. ed. (Amsterdam: North-Holland, 1971) pp 130-137.
11. Beacham, J., G. Dupuis, F. M. Finn, H. T. Storey, C. Yanaihara, N. Yanaihara, and K. Hofmann. J. Amer. Chem. Soc. 93, 5526 (1971).
12. Camble, R., G. Dupuis, K. Kawasaki, H. Romovacek, N. Yanaihara, and K. Hofmann. J. Amer. Chem. Soc. 94, 2091 (1972).

CRYSTALLIZATION OF SYNTHETIC POLYPEPTIDES WITH TRIPLE-HELICAL STRUCTURE

Shumpei Sakakibara, Yasuo Kishida, Saburo Aimoto.
Peptide Institute, Protein Research Foundation,
Minoh, Osaka, Japan

TWO OF US (S.S. AND Y.K.) AND OTHERS[1] have previously reported the synthesis of a new kind of sequential polypeptide of the structure (Pro-Pro-Gly)$_n$ (n = 10 or 20) using the fragment condensation technique on the Merrifield resin. These polypeptides are homogeneous in molecular weight and they have physical properties similar to collagen. The polymers are soluble in aqueous acetic acid and aqueous ethanol, but less soluble in distilled water. In dilute aqueous acetic acid these polypeptides show an optical rotation change, as shown in Figure 1, which accompanies a three fold change in molecular weight.[2] Thus, this rotation change was attributed to a transition between triple-helix and coil. Below 10°C and in high concentrations those triple-helical molecules form microcrystalline segments similar to the fibrous long–spacing segments of collagen; this phenomenon was confirmed by electron microscopy.[3]

Recently, we have found that the polypeptide (Pro–Pro–Gly)$_{10}$ gave single crystals under specific conditions.[4] We report herein the procedure for the crystallization and an attempt to incorporate a bromine atom into each molecule in order to facilitate X-ray analysis of the structure.

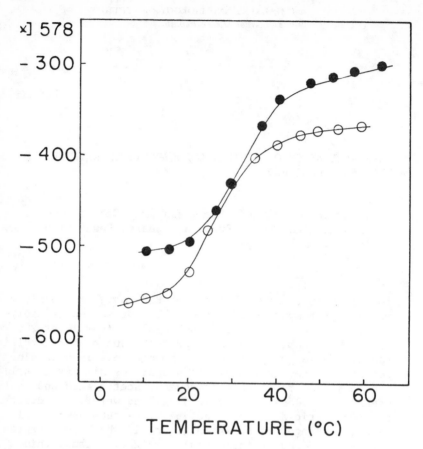

Figure 1: Temperature-dependence of specific rotation in
10% acetic acid at 578 mμ:
($\bigcirc$) (Pro-Pro-Gly)$_{10}$;
($\bullet$) Pro(4Br)-Pro-Gly-(Pro-Pro-Gly)$_9$.

Crystallization of (Pro-Pro-Gly)$_{10}$

A 5% solution of the amorphous polymer in 10% aqueous
acetic acid was sealed in a cellophane tube and dialyzed
against distilled water at a temperature below 10°C. The
dialysis was carried out in a tight cylinder with a stop-
cock at the bottom to control the speed of the release of
acetic acid. When the pH of the outside solution reached
4.7, the flow rate of distilled water was adjusted to about
one drop per second. Fine crystals appeared on the inner
surface of the cellophane bag when the pH of the effluent

reached 5.2. Dialysis was continued overnight under the
same conditions and crystals with a maximum edge of about
0.25 mm were collected by filtration with 60% recovery.
Crystallization could also be induced when amorphous polymer
was dissolved in distilled water at 40°C followed by lowering
the temperature to 10°C and seeding, but almost all crystals
so obtained were imperfect for X-ray analysis. Since these
molecules are distributed as individual triple-helices in
aqueous acetic acid at 10°C, crystallization by the dialysis
method proceeds smoothly with decreasing acetic acid concentration. In the cooling method, however, both triple-
helix formation and aggregation of helices have to take
place during the same process; therefore, packing of helical
molecules in crystals may not be as regular.

One crystal obtained by the dialysis method was examined
by X-ray diffraction, and it was confirmed that the crystal
is orthorhombic with unit cell dimensions of a=26.9, b=26.4
and c (fiber axis)=100.4 Å. The amount of absorbed water
in crystals was determined to be about 50% by measuring
the decrease of the weight during drying at 80°C over P_2O_5
in vacuo for 5 hours. On the basis of the observed water-
content and crystal density (1.31 $g \cdot cm^{-3}$), X-ray crystallog-
raphers of our group suggested that the unit cell consists
of twelve chains of $(Pro-Pro-Gly)_{10}$ which may be arranged
in four triple-helical structures. They arrived at a
tentative electron density map projected along the c-axis
as shown in Figure 2; the main peak may be a triple-helix
of $(Pro-Pro-Gly)_{10}$. Details of the X-ray study will be
published elsewhere.

Incorporation of bromine atom in polypeptide chain

In order to facilitate the X-ray analysis of the
crystal structure, incorporation of a bromine atom in
each N-terminal prolyl residue was attempted. First,
4-bromo-L-proline was synthesized as follows: N-benzyloxy-
carbonyl-L-hydroxyproline benzyl ester was treated with
freshly prepared PBr_5 at 0°C in methylene chloride. Then,
the brominated product was treated with anhydrous HF under
standard condition to remove the protecting groups.[5]
Finally, 4-bromo-L-proline was isolated from the reaction
mixture by counter-current distribution using the solvent
system *n*-butanol-acetic acid-water (4:1:5,v/v/v). Total
yield of purified 4-bromo-L-proline was about 30%; m.p.
201.5°C. (decomp.), $[\alpha]_D^{24}$ - 19.3° (c 1.1, water). This
amino acid was converted to the *t*-amyloxycarbonyl

Figure 2: A tentative electron density map projected along
   the *c*-axis.

derivative, which was then coupled with Pro–Gly–(Pro–Pro–
Gly)$_9$–Resin prepared from an intermediate for the synthesis
of (Pro–Pro–Gly)$_{10}$.[1] The product was isolated from the
resin using the HF method as in the case of the synthesis
of (Pro–Pro–Gly)$_{10}$. The physical properties of the
bromine-containing polymer in aqueous solution were found
to be similar to those of the original polymer; the optical
rotation change in aqueous acetic acid is shown in Figure 1.
     Crystallization of this polymer was attempted by the
dialysis method, and it was found that crystallization of
the brominated polymer was even better than that of the
original polymer. Crystals obtained were as big as 0.8 x
0.8 x 0.3 mm^3, (Figure 3). X-Ray studies with these
crystals are under way.

Figure 3: Single crystals of Pro(4Br)-Pro-Gly-(Pro-Pro-Gly)9 viewed under the polarized microscope. A maximum edge is about 0.8 mm long.

Acknowledgment

The authors express their gratitude to Professor Masao Kakudo of Osaka University and his colleagues for their skillful cooperation in X-ray analysis of these crystals.

References

1. Sakakibara, S., Y. Kishida, Y. Kikuchi, R. Sakai, and K. Kakiuchi. Bull. Chem. Soc. Japan 41, 1273 (1968).
2. Kobayashi, Y., R. Sakai, K. Kakiuchi, and T. Isemura. Biopolymers 9, 415 (1970).
3. Olsen, B. R., R. A. Berg. S. Sakakibara, Y. Kishida, and D. J. Prockop. J. Mol. Biol. 57, 589 (1971).
4. Sakakibara, S., Y. Kishida, K. Okuyama, N. Tanaka, T. Ashida, and M. Kakudo. J. Mol. Biol. 65, 371 (1972).
5. Sakakibara, S. *Chemistry and Biochemistry of Amino Acids, Peptides, and Proteins,* Weinstein, B., ed., Vol. I (New York: Marcel Dekker, 1971) pp 51-85.

A NOVEL METHOD FOR THE SYNTHESIS OF LONGER PEPTIDES INCLUDING THE CORRECTED STRUCTURE OF HUMAN ACTH AND ITS FRAGMENTS

L. Kisfaludy, M. Löw, T. Szirtes, I. Schön, M. Sárközi.
Chemical Works of Gedeon Richter Ltd., Budapest, Hungary

S. Bajusz, A. Turan, A. Juhász, R. Beke, L. Gráf.
Pharmaceutical Research Institute, Budapest, Hungary

K. Medzihradszky. Institute of Organic Chemistry,
Eötvös University Budapest, Budapest, Hungary

THE SYNTHESIS OF THE CORRECTED human ACTH and its fragments offered an excellent possibility to extend the applicability of pentafluorophenylesters (OPfp).[1] As is known, Swiss[2] and Hungarian[3] researchers have found that both in porcine and human ACTH Asn is present instead of Asp in the position 25 and in human ACTH the correct sequence is Gly-Ala in the position 26-27 instead of Ala-Gly. This fact has again raised the well-known problems in the synthesis of Asn-peptides, such as the formation of beta-cyanoalanin and succinimido-derivatives as by-products.[4] In fact, preparing Z-Asn-OPfp by the usual way, a reaction mixture was formed from which the desired product could not be isolated. However, working at 0°C and with short reaction time, Z-Asn-OPfp was obtained with 93% yield; moreover it was possible to recrystallize it from hot ethyl acetate obtaining fine needles with a melting point of 150°C. In a DMF solution this compound completely converts into the corresponding succinimido-derivative within 12-15 hr, Figure 1, confirming the easy formation of the cyclic derivative of the Asn moiety.

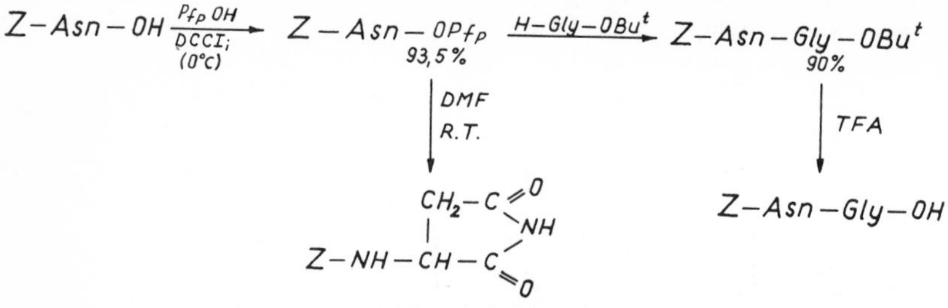

$$Z-Asn-OH \xrightarrow[\substack{DCCI; \\ (0°c)}]{P_{fp}\ OH} Z-Asn-OP_{fp} \xrightarrow{H-Gly-OBu^t} Z-Asn-Gly-OBu^t$$
$$\underset{93,5\%}{\phantom{Z-Asn-OP_{fp}}} \qquad \underset{90\%}{}$$

Figure 1: Preparation, use for synthesis and decomposition (Z-aminosuccinimide formation) of benzyloxycarbonyl-asparagine pentafluorophenyl ester.

Z-Asn-Gly-OH was obtained by reaction of this compound and H-Gly-OBu^t followed by TFA treatment. It is worth mentioning, that no example can be found in the literature for acylation with this dipeptide. We used it in the acylation reactions shown in Figure 2.

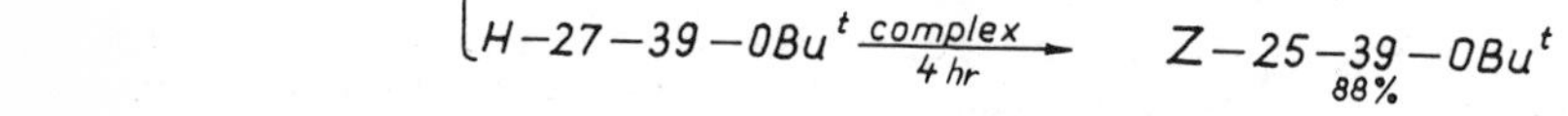

$$Z-Asn-Gly-OH + \begin{cases} H-27-31-OBu^t \xrightarrow[DCCI;\ 3hr]{P_{fp}\ OH} & \underset{80,5\%}{Z-25-31-OBu^t} \\[2ex] H-27-32-OBu^t \xrightarrow[DCCI;\ 3\ hr]{P_{fp}\ OH} & \underset{89\%}{Z-25-32-OBu^t} \\[2ex] H-27-39-OBu^t \xrightarrow[4\ hr]{complex} & \underset{88\%}{Z-25-39-OBu^t} \end{cases}$$

Figure 2: Use of Z-Asn-Gly-OH for fragment condensations.

Equimolar amounts of starting materials were used and the yields ranged from 80% to 90%. The active ester formation took place in a few minutes, as indicated by tlc, and the coupling reactions required a couple of hours. We prefer the use of the DCCI-pentafluorophenol complex because of its easy recrystallisation. During these coupling reactions we never used any excess of base which was sometimes

necessary in the case of pentachlorophenyl esters. This
fact led us to the determination of the p_K values of the
two phenols which were reported in the literature to be
the same.[3,5] We have found that in DMF solution the p_K
value of pentachlorophenol (PcpOH) is 5,05 and that of
pentafluorophenol (PfpOH) is 6,35. In other words, the
latter is present in a less dissociated state. Therefore,
protonation of the amino component and concomitant slow
down of reaction are less likely to occur.

A further advantage is the easy removal of free PfpOH
from the reaction mixture which does not always succeed
with PcpOH and particularly with its salts. If a small
amount of PcpOH is present at a catalytic hydrogenation
after a coupling reaction the HCl formed might cleave acid
sensitive protecting groups.

In the synthesis of longer peptides usually elevated
temperatures and/or an excess of acylating components are
applied to obtain higher yields and purer products.[5] In
Figure 3 condensations of larger ACTH fragments are shown.
In these steps of the synthesis equimolar amounts of
starting components were treated with only a slight excess
of DCCI-PfpOH complex at room temperature for 24 hr.

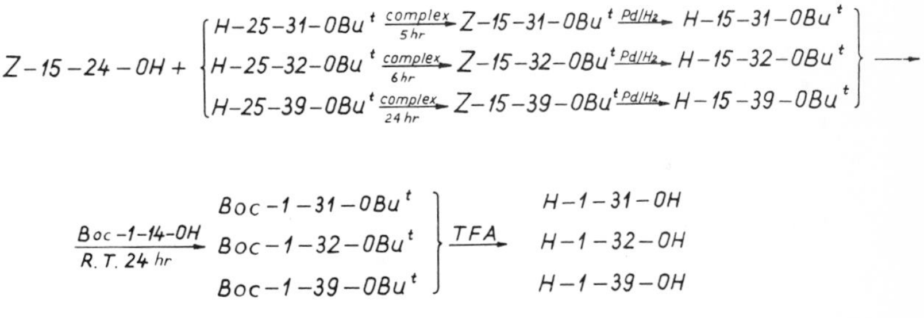

Figure 3: Fragment condensation scheme in syntheses of
corrected human ACTH and of lower homologs by the
pentafluorophenol technique.

The protected peptides thus obtained, were treated in
the usual way with TFA followed by purification on CM-
cellulose columns. The free, chromatographically homogenous
peptides showed full biological activity.

An alternative route for the synthesis of the 15-24 sequence is demonstrated in Figure 4. This route has the advantage that it avoids to expose the whole C-terminal

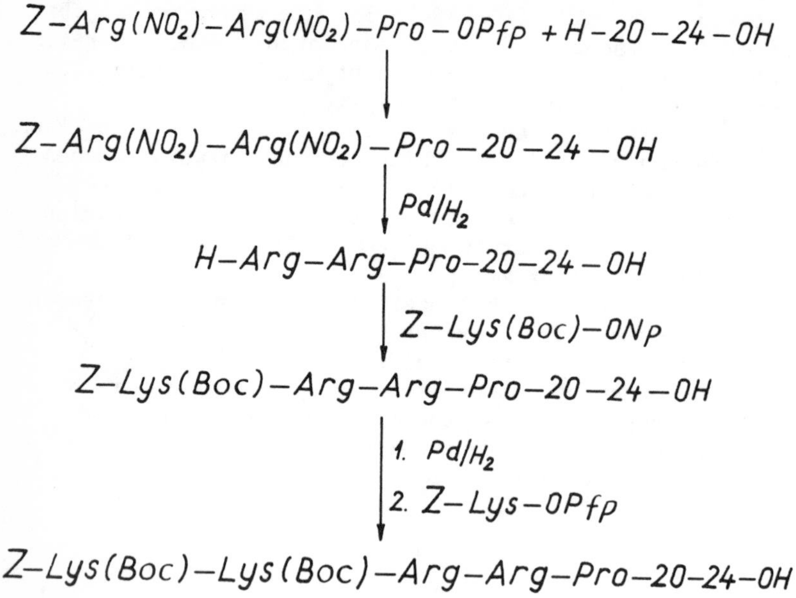

$$Z-Arg(NO_2)-Arg(NO_2)-Pro-OPfp + H-20-24-OH$$

$$\downarrow$$

$$Z-Arg(NO_2)-Arg(NO_2)-Pro-20-24-OH$$

$$\downarrow Pd/H_2$$

$$H-Arg-Arg-Pro-20-24-OH$$

$$\downarrow Z-Lys(Boc)-ONp$$

$$Z-Lys(Boc)-Arg-Arg-Pro-20-24-OH$$

$$\downarrow \begin{array}{l} 1.\ Pd/H_2 \\ 2.\ Z-Lys-OPfp \end{array}$$

$$Z-Lys(Boc)-Lys(Boc)-Arg-Arg-Pro-20-24-OH$$

Figure 4: Alternative pathway for synthesis of ACTH-(15-24)-decapeptide.

part, containing many acid sensitive protecting groups, to the time consuming catalytic hydrogenation in acetic acid solution which is needed for the removal of arginine nitro groups. This synthetic pathway presents an example for the acylation of a C-unprotected free peptide by a penta-fluorophenyl ester without using a base. In these steps we used first Z-Lys(Boc)-ONp instead of the pentafluoro-phenyl ester to avoid the formation of an acetyl-derivative which is a real danger in the case of highly activated esters.[6]

In summary, syntheses of corrected human ACTH and its species specific fragments (1-31) and (1-32) were performed without protecting the carboxamide function of asparagine. The pentafluorophenol technique was used in every important coupling step.

References

1. Kisfaludy, L., M. Q. Ceprini, B. Rákóczy, and J. Kovács. In *Peptides*, Beyerman, H. C., W. Maassen van den Brink, and A. van de Linde, eds. (Amsterdam: North-Holland Publ., 1967) p 25; Kovacs, J., L. Kisfaludy, and M. Q. Ceprini. J. Amer. Chem. Soc. <u>89</u>, 183 (1967); Kisfaludy, L., J. E. Roberts, R. H. Johnson, G. L. Mayers, and J. Kovacs. J. Org. Chem. <u>35</u>, 3563 (1970).
2. Riniker, B., P. Sieber, W. Rittel, and H. Zuber. Nature New Biology <u>235</u>, 114 (1972).
3. Gráf, L., S. Bajusz, A. Patthy, E. Barat, and G. Cseh. Acta Biochim. Biophys. Acad. Sci. Hung. <u>6</u>, 415 (1971).
4. Schröder, E., and K. Lübke. *The Peptides*, Vol. I (New York: Academic Press, 1965) p 203; Ondetti, M. A., A. Deer, T. Sheehan, J. Pluscec, and O. Kocy. Biochemistry <u>7</u>, 4069 (1968).
5. Schwyzer, R., and P. W. Schiller. Helv. Chim. Acta <u>54</u>, 897 (1971); Wünsch, E., G. Wendlberger, and P. Thamm. Chem. Ber. <u>104</u>, 2445 (1971).
6. Löw, M. Unpublished data.

SYNTHESIS OF THE HYPOTHALAMIC LH- AND FSH-RELEASING DECAPEPTIDE

G. Flouret, * *W. Arnold, W. Cole, R. Morgan, W. White, M. Hedlund, R. Rippel.* Division of Antibiotics and Natural Products, Abbott Laboratories, North Chicago, Illinois 60064

THE STRUCTURE OF THE hypothalamic luteinizing hormone-releasing hormone/follicle-stimulating hormone-releasing hormone (LH-RH/FSH-RH) has been described as that of the decapeptide[1] <Glu-His-Trp-Ser-Tyr-Gly-Leu-Arg-Pro-Gly-NH$_2$ (I), for the porcine species.

In order to provide independent confirmation of the validity of structure I for the hormone,[2] we synthesized this decapeptide and attained as well a second goal of providing a convenient route to substantial amounts of decapeptide for the more extensive studies which are required to ascertain its biological role.

In one approach solution methods were employed. Thus, glycinamide was coupled with Z-Pro-ONp to give the dipeptide Z-Pro-Gly-NH$_2$ (II). Deprotection of II with HBr-AcOH[3] and coupling with Z-Arg(NO$_2$) and DCC[4] yielded Z-Arg(NO$_2$)-Pro-Gly-NH$_2$ (III). Deprotection of III with HBr-AcOH and coupling with Boc-Leu-ONp afforded Boc-Leu-Arg(NO$_2$)-Pro-Gly-NH$_2$ (IV). Successive removal of the Boc groups with trifluoroacetic acid (TFA)-CH$_2$Cl$_2$ (1:1)[5] and coupling with appropriate Boc-amino acid active esters led to Boc-Trp-Ser(Bzl)-Tyr(Bzl)-Gly-Leu-Arg(NO$_2$)-Pro-Gly-(NH$_2$) (VIII). The Boc group of VIII was selectively removed with TFA-CH$_2$Cl$_2$ (1:1) containing 1% mercaptoethanol and the triprotected amino-octapeptide amide was coupled with Boc-His by means of DCC to yield Boc-His-Trp-Ser(Bzl)-Tyr(Bzl)-Gly-Leu-Arg(NO$_2$)-Pro-Gly-NH$_2$ (IX). After the

*Present Address: Northwestern University, The Medical School, Department of Physiology, 303 East Chicago Avenue, Chicago, Illinois 60611

removal of the Boc group of IX, the triprotected amino-
nonapeptide amide was coupled with pentachlorophenyl
pyroglutamate[6] to yield the desired triprotected decapeptide <Glu-His-Trp-Ser(Bzl)-Tyr(Bzl)-Gly-Leu-Arg(NO$_2$)-Pro-
Gly-NH$_2$ (X).

All intermediates were purified by column chromatography
on silica gel. Most of them were not crystalline, but pure
enough for characterization. Where possible, NMR spectra
were determined to corroborate the molecular structure.
All intermediates had consistent elemental analyses and gave
one spot on tlc. The yields at each step were usually high
(70-95%), the overall yield of X was (10%), based on the
starting glycinamide.

The tri-protected decapeptide (X) was also assembled by
the solid-phase method of Merrifield,[7] with Boc-Gly-Resin
being the starting material. An aliquot of the peptide-
resin was retained as octapeptide-resin (VIIIa), and the
remainder of the material was converted to the decapeptide-
resin (Xa). The completed Xa was ammonolyzed to give a
high yield (80-100%) of peptide material. Chromatography
of this product on silica gel with combinations of MeOH-
CHCl$_3$ as the eluent led to pure tri-protected decapeptide
X in good yield (35-40%). The melting point, NMR spectrum,
optical rotation and tlc pattern of this material were
identical to those of the material obtained by solution
methods. Thus, the solid-phase method appeared to be a
convenient method for obtaining X rapidly and in good yield.

Ammonolysis of VIIIa and chromatography led to pure
VIII comparable to the octapeptide made by solution methods
according to the criteria of mp, optical rotation, NMR
spectra and tlc analysis.

A fragment-condensation method was also employed for
making VIII. This was accomplished by a condensation involving the C-terminal tetrapeptide IV and Boc-Trp-Ser(Bzl)-
Tyr(Bzl)-Gly (XI). Starting with Gly-OMe, XI was assembled
by the stepwise active ester method of Bodanszky[8] employing
the appropriate Boc-amino acid-ONp. Saponification of the
tetrapeptide methyl ester gave XI. Coupling of both XI and
IV in the presence of DCC led to VIII comparable to the
material made either by the stepwise method or the solid-
phase method, according to the criteria of melting point,
NMR and tlc analysis.

The removal of the protecting groups from X was accomplished by treatment with HF-anisole.[9] The product was
treated with AG 1-X2 (Acetate) and subjected to a two-step
Sephadex gel filtration[10] purification. The lyophilized
final product was a fluffy powder obtained in 25-30% yield

from X. Elemental analysis data of synthetic LH-RH/FSH-RH were consistent with a diacetate trihydrate. Amino acid analysis[11] gave the expected values. The optical rotation, $[\alpha]_D^{24}$ = -50.5° (*c* 1, 1% AcOH), was similar to that reported by Geiger.[12] Two dimensional thin-layer electrophoresis (0.1*N* pyridine acetate, pH 6.5) and chromatography (*n*-BuOH-AcOH-H₂O, 4:1:1) showed one component.

The decapeptide I was tested biologically by *in vitro* incubation with male rat hemipituitaries. Release of both FSH[13] and LH[14] were measured and compared with the activity of a synthetic standard shown previously at Abbott and in the laboratories of A.V. Schally[15] to be equal to purified natural hormone. The results of the present *in vitro* experiments demonstrate that I has LH and FSH releasing activities which equal those of a synthetic standard.

Acknowledgment

The authors thank Mrs. J. Hood for elemental analyses, Dr. R. Egan for NMR spectra, Dr. O. Walasek for amino acid analyses, and Dr. A. V. Schally for communicating the structure of the decapeptide I.

References

1. Schally, A. V., A. Arimura, A. K. Kastin, H. Matsuo, Y. Baba, T. W. Redding, R. M. G. Nair, L. Debeljuk, and W. F. White. Science 173, 1036 (1971).
2. Schally, A. V., A. Arimura, Y. Baba, R. M. G. Nair, H. Matsuo, T. W. Redding, and W. F. White. Abstracts 53rd meeting, The Endocrine Society, San Francisco, California, June 1971, Abstract No. 55.
3. Ben-Ishai, D., and A. Berger. J. Biol. Chem. 17, 1564 (1952).
4. Sheehan, J. C., and G. P. Hess. J. Amer. Chem. Soc. 77, 1067 (1955).
5. Gutte, B., and R. B. Merrifield. J. Amer. Chem. Soc. 91, 501 (1969).
6. Flouret, G. J. Med. Chem. 13, 843 (1970).
7. Merrifield, R. B. J. Amer. Chem. Soc. 85, 2149 (1963).
8. Bodanszky, M. Nature 175, 685 (1955).
9. Sakakibara, S., and Y. Shimonishi. Bull. Chem. Soc. Japan 38, 141 (1965); Sakakibara, S., M. Shin, M. Fujino, Y. Shimonishi, S. Inouye, and N. Inukai. Bull. Chem. Soc. Japan 38, 1522 (1965); Sakakibara, S., Y. Shimonishi, Y. Kishida, M. Okada, and H. Sugihara. Bull. Chem. Soc. Japan 40, 2164 (1967).

10. Porath, J., and P. Flodin. Nature 183, 1310 (1959).
11. Spackman, D. H., W. H. Stein, and S. Moore. Anal. Chem. 30, 1190 (1958).
12. Geiger, R., W. König, H. Wissmann, K. Geisen, and F. Enzmann. Biochem. Biophys. Res. Commun. 45, 767 (1971). These authors reported a positive optical rotation for I possibly through a typographical error.
13. Steelman, S. L., and F. M. Pohley. Endocrinology 53, 604 (1953).
14. Parlow, A. F. in *Human Pituitary Gonadotropins*, Albert, A., ed (Springfield, Ill.: Charles C. Thomas, 1961) pp 300-310.
15. Schally, A. V., T. W. Redding, A. Matsuo, and A. Arimura. Endocrinol. In press.

SYMPOSIUM DISCUSSIONS

Summarized by Johannes Meienhofer

THE LUCID REPORT[1] of Dr. Klaus Hofmann on the latest progress in the synthesis of ribonuclease T_1 stimulated a very
lively discussion. Several inquiries were directed at
potential formation of side products during fragment condensation. In reply it was pointed out that detection and
identification of small amounts of side products, even of
those with known nature, becomes increasingly difficult
with progressively larger peptides. Moreover, in a project
of the magnitude of an enzyme synthesis, the time factor
renders it prohibitive to examine all side products in
mother liquors or other secondary fractions.

The inherent question of fragment condensation *versus*
incremental (stepwise) chain elongation drew several comments. Limited stability of certain side chain functions
as in S-ethylcarbamylcysteine (RNase T_1)[1] or tyrosine-O-
sulfate (cholecystokinin-pancreozymin)[2] makes incremental
(stepwise) synthesis very impractible and requires the
use of fragment condensation. Intermediates can be tested
for homogeneity with presently available analytical techniques. On the other hand, incremental chain elongation
allows more readily the use of large excesses of acylating
component to increase both yields and coupling rates; thus
diminishing the occurrence of certain unimolecular side
reactions (rearrangements). In this context it was observed
that "recoupling," *i.e.* the subsequent addition of more
excess acylating component to a reaction mixture, does
usually not produce increased yields, for reasons unknown.
This reflects the fact that we still know very little
about the kinetics of peptide condensations, and the low
solubility of larger fragments further complicates the
situation.

α→γ Rearrangements, in particular *via* succinimide formation in aspartyl peptide bonds[3] are of concern. The detection of ω-peptides by enzymatic tests would appear to be impaired since it was observed by several discussants that even the α-bonds in Asp-Asp, Asp-Gly, Asp-Thr, and Glu-Glu are not normally digested by leucine amino peptidase or aminopeptidase M. A new aminopeptidase was mentioned (E. Bricas) that seems to cleave α-peptide bonds between aminodicarboxylic acids.

It was inquired whether chromatographic purification in 50% formic acid as a solvent might cause some formylation of free amino end groups. This was not observed especially since the procedure was done at 4°C. Sephadex is considered stable against 50% HCOOH even at room temperature although traces of carbohydrate might be found in the eluate. A discussant recommended a new lipophilized gel, Sephadex LH-60, for peptide purification. An octadeca- and a nonapeptide had been successfully separated from a heptacosapeptide on LH-60.

The protection of tryptophan residues from oxidation during solid-phase synthesis continues to be a problem. The addition of mercaptoethanol[4] or dithiothreitol[5] and the use of HCl in formic acid (pp 269 to 279) for the cleavage of Boc groups suppressed Trp degradation considerably, especially when working under nitrogen. However, side products are still observed (pp 269 to 279), and it was suggested that mercaptoethanol might give rise to some sulfenylhalide formation that could lead to the known thioether formation with tryptophan.[6]

Attention was drawn to the observation that acetic acid is difficult to be completely washed out of the polystyrene resin after the use of HCl-acetic acid and HCl-dioxane was instead recommended.

The question was raised whether C-terminal fragments of the porcine gastric inhibitory polypeptide might possess biological activity. This has not yet been examined, but it was agreed that it would be useful if this large molecule (43 amino acid residues) would have a smaller active core, as *e.g.* ACTH. However, the steroidogenic activities of active fragments of ACTH approached only in isolated adrenal cell preparations the activity of the intact hormone on a molar basis. In the hypophysectomized rat the naturally occurring 1-39 sequence is still the most potent hormone (not considering synthetic preparations containing D-amino acid or other unnatural residues). Thus, the C-terminal part of ACTH seems to have a role, perhaps in the survival of the hormone.

References

1. Various commitments prevented Dr. Hofmann from preparing a manuscript. Recent publications about this project are: Storey, H. T., J. Beacham, S. F. Cernosek, F. M. Finn, C. Yanaihara, and K. Hofmann. J. Amer. Chem. Soc. 94, 6170 (1972); Storey, H. T., and K. Hofmann. In *Peptides 1971*, Nesvadba, H., ed. (Amsterdam: North-Holland Publ. Co., 1972) in press; Beachem, J., G. Dupuis, F. M. Finn, H. T. Storey, C. Yanaihara, N. Yanaihara, and K. Hofmann. J. Amer. Chem. Soc. 93, 5526 (1971); Hofmann, K. In *Peptides 1969*, Scoffone, E., ed. (Amsterdam: North-Holland Publ. Co., 1971) pp 130-137; Yanaihara, N., C. Yanaihara, G. Dupuis, J. Beacham, R. Camble, and K. Hofmann. J. Amer. Chem. Soc. 91, 2184 (1969).
2. Ondetti, M. A., J. Pluščec, E. F. Sabo, J. T. Sheehan, and N. Williams. J. Amer. Chem. Soc. 92, 195 (1970).
3. Ondetti, M. A., A. Deer, J. T. Sheehan, J. Pluščec, and O. Kocy. Biochemistry 7, 4069 (1968).
4. Marshall, G. R. Advan. Exp. Med. Biol. 2, 48 (1968); Wang, S.-S., and R. B. Merrifield. Int. J. Protein Res. 1, 235 (1969).
5. Li, C. H., and D. Yamashiro. J. Amer. Chem. Soc. 92, 7608 (1970).
6. Wieland, T., and R. Sarges. Justus Liebigs Ann. Chem. 658, 181 (1962).

SECTION V

PROGRESS IN SYNTHETIC PROCEDURES

Session Chairmen

Iphigenia Photaki and John C. Sheehan

SOME NOVEL AMINE PROTECTING GROUPS

Daniel F. Veber, Stephen F. Brady, Ralph Hirschmann.
Merck Sharp & Dohme Research Laboratories, Division
of Merck & Co., Inc., Rahway, New Jersey 07065

IN PUBLICATIONS FROM THESE laboratories dating back to
1966[1] we showed that fewer protecting groups were required
than had generally been employed in peptide synthesis. We
have concluded that protection is required only for the
α-amino groups, the ε-amino group of lysine and the thiol
of cysteine. We employed this strategy in the synthesis
of S-protein choosing to protect the α-amino groups with
Boc, the ε-amino group of lysine with Z and the thiol with
Acm.[2] Although the previous literature indicated that this
choice of protecting groups would not present problems, we
became aware of two deficiencies in our choices for amine
protection. The first of these was the partial loss of
Boc during isolation procedures when using 50% aqueous
acetic acid as solvent. The second was the partial loss
of Z during removal of the Boc protecting group. Recently,
significant advances have been made in enhancing the
specificity of the removal of the Boc protecting group in
the presence of Z.[3] Although such studies have been of
great value and have been successful in model cases, other
factors such as the poor solubility of starting materials
or product may cause problems. Furthermore, nucleophilic
scavengers are often required to trap cations generated
during the removal of protecting groups. Since the removal
of the Z group shows considerable SN_2 character while the
removal of the Boc group is largely SN_1, the presence of
such a scavenger reduces the selectivity. Such effects
are exemplified in Table I. The rate of removal of the
benzyloxycarbonyl group is significantly enhanced by the

Table I

Half-Times (Min) for Protecting Group Removal

	Formic Acid	Formic Acid-DMS* (4:1)	Formic Acid-CH$_2$Cl$_2$ (4:1)
Boc-Phe	4	26	21

	TFA[†]	TFA-DMS (4:1)	TFA-Benzene (4:1)
Z-Phe	300	60	600

*DMS, Dimethylsulfide.
[†]TFA, Trifluoracetic acid.

presence of dimethyl sulfide, a good nucleophile, but slowed
by the presence of an inert, nonpolar solvent such as benzene. On the other hand the removal of the Boc protecting
group which proceeds largely by an SN$_1$ mechanism is slowed
by the presence of dimethyl sulfide as well as by methylene
chloride. This is presumably due to the reduced solvent
polarity.

We wish to report a solution to the problem of selectivity
which involves the use of the novel isonicotinyloxycarbonyl
group (iNOC) for the protection of the ε-amino group of
lysine. Due to the positive charge on the pyridine ring
under acidic conditions, this protecting group is completely
stable in liquid HF or TFA and is only slowly removed by
HBr/acetic acid. The iNOC protecting group is, however,
removed smoothly either from a protein or from a smaller
peptide by zinc dust in 50% aqueous acetic acid. Catalytic
hydrogenation has also been found useful for the removal
of the protecting group from peptides. To show that the
chemical reduction can be applied to proteins, 6.5 out of
a possible 9-tritium labelled iNOC groups were first introduced into acetamidomethylated, reduced ribonuclease
S-protein.[4] Complete removal of the tritium label from
the protein was then accomplished by treatment with zinc
dust in 50% aqueous acetic acid at room temperature.

Peptides containing ε-iNOC-lysine have been synthesized
both by classical and solid-phase techniques. Subsequent
removal of the protecting group proceeded smoothly. An
additional advantage of the protecting group is in the
observation that it confers increased solubility both in
aqueous and organic solvents, on peptides. This protecting
group should also give advantages in purification by ion

exchange chromatography as observed by Young for the 4-picolyl ester protecting group.[5] ε-Isonicotinyloxycarbonyl-lysine (II) was prepared from lysine-copper complex by reaction with isonicotinyl succinimidoyl carbonate (Ia) or isonicotinyl p-nitrophenyl carbonate (Ib).

Lysine-copper complex + N⟨═⟩—CH₂OC(=O)—R ⟶ (II)

$$\text{Lysine-copper complex} + \text{N}\langle\!\!\!\rangle\text{-CH}_2\text{O}\overset{\text{O}}{\overset{\|}{\text{C}}}\text{-R} \longrightarrow \begin{array}{c} \text{HN-}\overset{\text{O}}{\overset{\|}{\text{C}}}\text{OCH}_2\text{-}\langle\!\!\!\rangle\text{N} \\ | \\ (\text{CH}_2)_4 \\ | \\ \overset{+}{\text{H}_3}\text{N-CH-COO}^- \end{array}$$

(I)   (II)

a) R, OSu : mp 106–107.5°,   mp 234°
   ethyl acetate–hexane   water–ethanol

b) R, ONp : mp 105–106°,
   ethyl acetate–hexane

The α–Boc derivative (mp 82–83.5° from isopropanol–hexane) was prepared by treatment of II with t-butyl succinimidoyl carbonate and by treatment of α–Boc-lysine with Ia or Ib. Ia and Ib were prepared by treatment of succinimidoyl chloroformate and bis-p-nitrophenyl carbonate, respectively, with 4-pyridine carbinol in the presence of N-methylmorpholine.

To find an α-amino blocking group which is stable in 50% acetic acid, we have examined several alkoxycarbonyl derivatives which we hoped might be more stable than Boc in this solvent system and yet be sufficiently labile under acidic conditions to be useful in peptide synthesis. We also wanted the new protecting group to be at least comparable to Boc in its effect on peptide solubility.

In order to predict the rate of removal of various urethane-type protecting groups under acidic conditions we have used the rates of solvolysis of various derivatives (tosylate, chloride and nitrobenzoate) of the corresponding alcohols as a guide. The feasibility of such an approach had been suggested by Bláha and Rudinger.[6] This approach cannot be used to precisely predict the stability of the protecting groups for several reasons. First, rates of solvolysis for a variety of derivatives are reported under varying solvolysis conditions, making direct comparisons impossible. Secondly, the degree of SN_1 character is not constant but is dependent on the nature of the alcohol. Finally, the rate of solvolysis of the tosylate of t-butanol,

needed to serve as a reference point for Boc, is not known.
A somewhat empirical approach was therefore required.

The following derivatives of phenylalanine were prepared: cyclopropylcarbinyloxycarbonyl, 1-cyclopropyletho-
xycarbonyl, 1-methylcyclohexyloxycarbonyl and 1-methylcyclo-
butyloxycarbonyl. Each of these was characterized by
physical means and shown to be a single component by tlc.
These alkoxycarbonyl-amino acids were prepared from the
alcohols via the chloroformate derivatives which were not
purified except to remove excess phosgene *in vacuo*. The
rate of removal of these groups has been studied in
trifluoroacetic acid and formic acid (Table II). It was

Table II

Half-Times (Min) for Amine Protecting Group Removal

Protecting Group (X)	t 1/2 (X-Phe) 25° (min)		t 1/2 (X-Phe-Ala-OMe) (min)	
	TFA	Formic acid	TFA	Formic acid
cyclopropylcarbinyloxycarbonyl	40	–	75	–
1-methylcyclobutyloxycarbonyl	2	–	3	180
t-Butyloxycarbonyl	*	4	†	10
1-methylcyclohexyloxycarbonyl	*	2	–	–
1-cyclopropylethoxycarbonyl	–	1.5	–	–

*Complete in 1 min.
†Complete in 1-2 min.

found that the 1-methylcyclobutyloxycarbonyl (McBoc) group
had essentially the desired stability properties. It could
be removed completely with trifluoroacetic acid in less
than 30 min at 20°, yet it was stable in 50% acetic acid
for 48 hours. A sample of Boc-Phe stored in 50% acetic
acid showed about 10-15% loss of the protecting group in
48 hours. The 1-methylcyclobutyloxycarbonyl group should
find use primarily for the protection of the amino terminus
of a relatively large peptide which is to be purified without loss of protection in an acidic solvent system such as
50% acetic acid. Some of the other protecting groups may

also prove of value when special lability or stability relative to the Boc group is required. It should be pointed out that for a given protecting group the rate of acid catalyzed cleavage from an amino acid is about twice that of a peptide due to the reduced basicity of the urethane function in a protected peptide.

The use of these two novel protecting groups, iNOC and McBoc, in conjunction with Boc and Acm fill the requirements for the protection of the three functionalities which require protection. In addition to the favorable properties of each of these protecting groups individually, the removal of iNOC, Acm and Boc/McBoc is based on "chemical" selectivity rather than kinetic differences in the rate of removal. We conclude that tactics based on such chemical selectivity should reduce side reactions in the synthesis and purification of large peptides and thus introduce a greater degree of certainty.

References

1. Denkewalter, R. G., H. Schwam, R. Strachan, T. E. Beesley, D. F. Veber, E. F. Schoenewaldt, H. Barkemeyer, W. J. Paleveda, Jr., T. A. Jacob and R. Hirschmann. J. Amer. Chem. Soc. 88, 3163 (1966).
2. Denkewalter, R. G., D. F. Veber, F. W. Holly, and R. Hirschmann. J. Amer. Chem. Soc. 91, 502 (1969).
3. Schnabel, E., H. Klostermeyer, and H. Berndt. Justus Liebigs Ann. Chem. 749, 90 (1971).
4. Veber, D. F., S. L. Varga, J. D. Milkowski, H. Joshua, J. B. Conn, R. Hirschmann, and R. G. Denkewalter. J. Amer. Chem. Soc. 91, 506 (1969).
5. Camble, R., R. Garner, and G. T. Young. J. Chem. Soc. (C) 1911 (1969).
6. Bláha, K., and J. Rudinger. Coll. Czech. Chem. Comm. 30, 599 (1965).
7. Veber, D. F., J. Milkowski, S. L. Varga, R. G. Denkewalter, and R. Hirschmann. J. Amer. Chem. Soc. 94, 5456 (1972).

THE 4,5-DIPHENYL-4-OXAZOLIN-2-ONE GROUP AS A PROTECTING GROUP IN PEPTIDE SYNTHESIS

Frank S. Guziec, Jr., John C. Sheehan. Department of Chemistry, Massachusetts Institute of Technology, Cambridge, Massachusetts

OCCASIONALLY IN PEPTIDE SYNTHESIS it would be advantageous to block a nitrogen function by replacing both hydrogens of a primary amine. Previous protecting groups of this type include the phthalimido,[1] 2-hydroxyarylidine,[2] and others[3-6] which have not attained widespread use due to solvolytic instability. We have succeeded in incorporating the amine nitrogen of primary amino acids into the extremely stable and unreactive 4,5-diphenyl-4-oxazolin-2-one ring system[7] while devising convenient methods for the removal of this protecting group.

Treatment of a mixture of benzoin (I) and phosgene in benzene with N,N-dimethylaniline, followed by cyclization of the unstable chloroformate (II) in refluxing benzene gives the unsaturated carbonate (III) in 65-70% yield (Equation 1). The carbonate is a crystalline compound which may be stored at room temperature for long periods without decomposition.

Treatment of a suspension of an amino acid tetramethylammonium salt in dimethylformamide (DMF) with the carbonate (Equation 2) gives an intermediate urethane (IV) which cyclizes under reaction conditions to the diastereomeric mixture of hydroxyoxazolidinones (V). The mixture may be clearly and quantitatively dehydrated to the desired 4,5-diphenyl-4-oxazolin-2-one derivative (VI), for which we propose the abbreviation "Ox," in overall yields of 75-85%. The "Ox" compounds are generally highly crystalline solids which fluoresce under ultraviolet light ($\lambda_{max} \sim 400$ mμ).

321

Equation 1

$$
\begin{array}{c}
\text{Ph} \quad \text{Ph} \\
\text{C}\!-\!\text{CH} + \text{COCl}_2 \\
\parallel \quad | \\
\text{O} \quad \text{OH}
\end{array}
\xrightarrow[\text{PhNMe}_2]{\text{Benzene}}
\begin{array}{c}
\text{Ph} \quad \text{Ph} \\
\text{C}\!-\!\text{CH} \\
\parallel \quad | \\
\text{O} \quad \text{OCOCl}
\end{array}
\xrightarrow{\substack{\text{Benzene} \\ \text{reflux}}}
\begin{array}{c}
\text{Ph} \quad \text{Ph} \\
\text{C}\!=\!\text{C} \\
\text{O} \quad \text{O} \\
\text{O}
\end{array}
$$

(I)   (II)   (III)

Equation 2

(III) $+ \text{H}_2\text{NCHRCOO}^-$ $\xrightarrow{\text{DMF}}$ $\left[\begin{array}{c}\text{Ph} \quad \text{Ph} \\ \text{C}\!-\!\text{CH} \\ \parallel \quad | \\ \text{O} \quad \text{OCO-NHCHRCOO}^-\end{array}\right]$ $\longrightarrow$ $\begin{array}{c}\text{Ph} \qquad \text{Ph} \\ \text{H} \qquad \text{OH} \\ \text{O} \qquad \text{NCHRCOOH} \\ \text{O}\end{array}$

(IV)   (V)

$\begin{array}{c}\text{Ph} \qquad \text{Ph} \\ \text{O} \qquad \text{NCHRCOOH} \\ \text{O}\end{array}$ $\xleftarrow{\text{TFA}}$

(VI)

Equation 3

$\begin{array}{c}\text{Ph} \quad \text{Ph} \\ \text{O} \quad \text{NCHRCOR}' \\ \text{O}\end{array}$ $\xrightarrow{\text{Reduction}}$ $\begin{array}{c}\text{Ph} \quad \text{Ph} \\ \text{H} \quad \text{H} \\ \text{O} \quad \text{NCHRCOR}' \\ \text{O}\end{array}$ $\xrightarrow{\text{Reduction}}$ $\begin{array}{c}\text{H}_2\text{NCHRCOR}' \\ + \\ \text{PhCH}_2\text{CH}_2\text{Ph} + \text{CO}_2\end{array}$

(R', peptide)

Equation 4

$\begin{array}{c}\text{Ph} \quad \text{Ph} \\ \text{O} \quad \text{NCHRCOR}' \\ \text{O}\end{array}$ $\xrightarrow{\text{Oxidation}}$ $\left[\begin{array}{c}\text{Ph} \quad \text{Ph} \\ \text{HO} \quad \text{OH} \\ \text{O} \quad \text{NCHRCOR}' \\ \text{O}\end{array}\right]$ $\xrightarrow{\text{Hydrolysis}}$ $\begin{array}{c}\text{H}_2\text{NCHRCOR}' \\ + \\ \text{PhCOCOPh} + \text{CO}_2\end{array}$

(R', peptide)   (VII)

Compounds of the general structure (VI) are stable to a variety of rigorous conditions: aqueous alkali, refluxing ethanolic hydrazine, ethanolic hydrogen chloride, hydrogen bromide in acetic acid, refluxing trifluoroacetic acid (TFA), and anhydrous hydrogen fluoride.

The 4,5-diphenyl-4-oxazolin-2-ones may be considered "protected" N-carbobenzoxy-N-benzyl derivatives. Thus, the "Ox" protecting group may be removed by a series of reductions (Equation 3). Low pressure (Parr) hydrogenation over palladium on charcoal in solvents containing an equivalent of aqueous acid most conveniently frees the amine function in nearly quantitative isolated yield.

Similarly the group may be removed by reduction with sodium in liquid ammonia.

Alternatively the vinyl oxygen, vinyl nitrogen moieties of the "Ox" group may be considered potential carbonyl functions. Oxidation to a dihydroxy compound (VII) followed by mild hydrolysis would free the amine (Equation 4). Oxidation of "Ox" derivatives in trifluoroacetic acid with excess *m*-chloroperbenzoic acid, followed by a hydrolytic workup frees the amine in 70% yield.

Simple "Ox" dipeptide derivatives have been prepared without difficulty using the water soluble 1-ethyl-3-(3-dimethylamino)propyl carbodiimide hydrochloride, and pure dipeptides isolated in high yield upon deprotection and hydrolysis.* No racemization has been observed in the preparation of "Ox" derivatives, stability studies, coupling or deprotection reactions. Investigations continue concerning the applicability of "Ox" protection in more complex peptides, especially in protection of the ε-amino group of lysine.

Acknowledgment

Frank S. Guziec, Jr. gratefully acknowledges support under National Institutes of Health Predoctoral Fellowship 5F01 GM 43911-03 from the General Medical Sciences Unit.

References

1. Sheehan, J. C., and V. S. Frank. J. Amer. Chem. Soc. 71, 1856 (1949).
2. Sheehan, J. C., and V. J. Grenda. J. Amer. Chem. Soc. 84, 2417 (1962).
3. Sheehan, J. C., and E. J. Corey. J. Amer. Chem. Soc. 74, 4555 (1952).
4. Bezas, B., and L. Zervas. J. Amer. Chem. Soc. 83, 719 (1961).
5. Gazis, E., B. Bezas, G. C. Stelekatos, and L. Zervas. In *Peptide Symposium*, Young, G. T., ed. (New York: Macmillan, 1963) p 17.
6. Carpino, L. A. J. Org. Chem. 29, 2820 (1964).
7. Filler, R. In *Advances in Heterocyclic Chemistry*, Katritzky, A. R., ed. Vol. IV (New York: Academic Press, 1965) pp 103-106 and references.

---

*Hydantoin formation has been noted in the alkaline hydrolysis of Ox-dipeptide esters.

CHEMISTRY AND BIOLOGY OF PEPTIDES

AMIDE PROTECTION IN CLASSICAL PEPTIDE SYNTHESIS

S. Bajusz, A. Turán, I. Fauszt. Research Institute
for Pharmaceutical Chemistry, Budapest, Hungary

APART FROM WELL KNOWN side known side reactions as trans-
peptidation or deamidation, the classical synthesis of free
amide group-containing peptides may involve difficulties
owing to hydrophilic intermediates which are hard to purify.
Peptide derivatives wherein more than one glycine, proline
or another amino acid with a hydrophilic side chain like
arginine, histidine, serine, or threonine reside in the
immediate vicinity of the amide groups of asparagine,
glutamine or terminal amide are particularly hydrophilic.

It was found that extreme hydrophilicity of such
intermediates could be eliminated by linking a hydrophobic
substituent to the free amide group, *i.e.* by amide protection. As a consequence, the peptide derivatives
obtained could be easily purified by simple procedures.

This second function of amide blocking may be demonstrated by the synthesis of intermediate products of salmon
calcitonin listed in Table I as in this case no side reactions were observed even in preparations without amide
blocking. These peptide derivatives and their intermediates
were purified merely by extractions with aqueous solutions
or by crystallization. All syntheses with amide protection
using Geiger's 4.4'-dimethoxy-benzhydryl group (Mbh)[1] proved
to be more efficient than those without it.[2] The only exception was the preparation of tetrapeptide sequence Cys-
Ser-Asn-Leu wherein, however, one serine residue is found.
The expression of efficiency, E_f, was based on the utilization
of amino acid derivatives applied in the coupling reactions
instead of amino acids proposed by Rydon:[3]

$$E_f = \frac{n \cdot m_{pp} \cdot 100}{m_{Ad}}$$

Table I

Efficiencies E_f of Syntheses of Protected Peptides
Containing Blocked or Free Amide Groups

Peptide Derivatives	E_f	
	X = H	X = Mbh
1. Boc–Cys(Acm)–*Ser*–Asn(X)–Leu–NHNH$_2$	47	53
2. Boc–Cys(Acm)–*Ser*–Asn(X)–Leu–*Ser*–*Thr*–Cys(Acm)–Val–Leu–Gly–OMe	43*	73*
3. Boc–Cys(Acm)–*Ser*–Asn(X)–Leu–*Ser*–*Thr*–Cys(Acm)–Val–Leu–Gly–OH	–†	83#
4. Z–Arg(NO$_2$)–*Thr*–Asn(X)–*Thr*–*Gly*–OH	21	43
5. Z–*Arg*–*Thr*–Asn(X)–*Thr*–*Gly*–OH	–†	50
6. H–*Ser*–*Gly*–*Thr*–Pro–NH–X	–†	55

*Yield in the azide coupling of components 1–4 and 5–10.

†Highly hydrophilic in character, pure product could be
 obtained by chromatography only.

#Yield of saponification.

where n is the number of amino acid residues in the peptide
chain, m_{pp} is the number of moles of pure peptide obtained
and m_{Ad} is the total number of moles of amino acid derivatives used in the coupling reactions.

     The intermediates (listed in Table II) containing only
one neighbouring serine or threonine or protected hydrophilic
side chains, *e.g.* Glu(OBu^t), Lys(Boc), were prepared without
any isolation or purification problems.

     The synthesis[4] of TRH as shown in Figure 1 may be another
example. Amide protection in the synthesis of this tripeptide, <Glu–His–Pro–NH$_2$, afforded directly a pure, homogeneous
end-product, without using any chromatographic procedure for
the purification of intermediates, *i.e.* H–His–Pro–NH–Mbh,
Z–<Glu–His–Pro–NH–Mbh, or TRH itself. The efficiency of
the synthesis was 52% being equal with that of the best performed classical synthesis of TRH published by Flouret,[5]
wherein column chromatography was used for purifying an
intermediate and the end-product.

Table II

The Intermediates for the Synthesis of Salmon Calcitonin
Prepared Without Isolation Problems

Intermediates	E_f
1. Boc-Cys(Acm)-Ser-Asn-Leu-NHNH$_2$	47
2. Z-Lys(Boc)-Leu-Ser-Gln-Glu(OBu^t)- Leu-NHNH$_2$	38
3. Z-His(Z)-Lys(Boc)-Leu-Gln-Thr- Tyr-Pro-OH	26

In the synthesis of peptide amides containing acid-sensitive residue(s), however, amide blocking is not an advantageous tool because acidolysis is the only process for removing any of the known amide masking groups. In such cases difficulties, mentioned above, can also be overcome by blocking the hydrophilic third function in the neighbourhood of amide bearing residues. Occasionally salt formation of peptide amides with some hydrophobic acid, like pentachlorophenol, can also be effective. A third possibility to facilitate purification or isolation of intermediate products in the synthesis of peptides amides

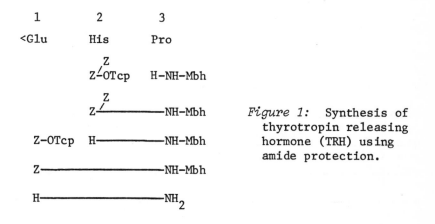

Figure 1: Synthesis of thyrotropin releasing hormone (TRH) using amide protection.

may be condensation of free amide group-containing amino
acids with longer hydrophobic fragments. The considerations
discussed above can, of course, not be generalized. Nevertheless, they have proved to be good for the synthesis of
luteinizing hormone-releasing hormone (LH-RH).

As shown in Figure 2, the sequence of the decapeptide
amide, <Glu-His-Trp-Ser-Tyr-Gly-Leu-Arg-Pro-Gly-NH$_2$, was
built up by the condensation of glycine-peptide 1-6 and
tetrapeptide component 7-10 to avoid the risk of racemization.

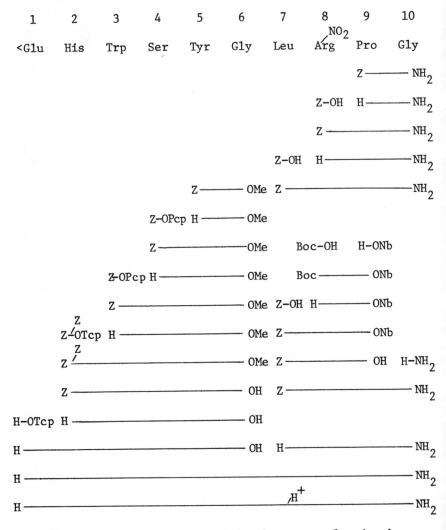

Figure 2: Synthesis of luteinizing hormone-releasing hormone
(LH-RH) with and without use of amide protection

The synthesis of peptide 1-6 was accomplished in a stepwise manner starting from glycine methyl ester and using chlorophenyl esters for acylation. The C-terminal methyl ester group was saponified at the pentapeptide stage and the free pentapeptide was coupled with either carbobenzoxy-glutamine, carbobenzoxy-pyroglutamic acid or pyroglutamic acid via activated esters. The latter gave the best result and the hexapeptide could be obtained in a yield over 90%.

In the synthesis of the tetrapeptide component, both the free guanidino group containing intermediates and carbobenzoxy-nitro-arginyl-prolyl-glycine amide are hydrophilic. It is of interest, however, that the pentachlorophenol salts of prolyl-glycine amide and of leucyl-nitro-arginyl-prolyl-glycine amide could be crystallized from water. Therefore the second version of the synthesis of protected tetrapeptide was chosen coupling glycine amide to the partially protected tripeptide 7-9. Coupling of the two components was accomplished by the dicyclohexylcarbodiimide-pentachlorophenol method.

Purification by chromatography could be omitted again. The hexapeptide 1-6, the nitro-decapeptide amide and LH-RH were purified by precipitation from aqueous solution with acetone. The synthesis presented here was rather efficient as the pure LH-RH was obtained in a yield of 30% based on the amino acid derivatives used in the coupling reactions.

References

1. König, W., and R. Geiger. Chem. Ber. <u>103</u>, 2041 (1970).
2. Turán, A., and S. Bajusz. Acta Chim. Acad. Sci. Hung. In press.
3. Bodánszky, M., in *Peptides: Chemistry and Biochemistry* Weinstein, B., and S. Lande, eds. (New York: Marcel Dekker, 1970) pp 1-15.
4. Bajusz, S., and I. Fauszt. Acta Chim. Acad. Sci. Hung. In press.
5. Flouret, G. J. Med. Chem. <u>13</u>, 843 (1970).
6. Bajusz, S., and A. Turán. Acta Chim. Acad. Sci. Hung. In press.

THE PREPARATION AND USE OF CARBOXAMIDE PROTECTED ASPARAGINE
AND GLUTAMINE DERIVATIVES

Victor J. Hruby, F. A. Muscio, W. Brown, Peter M. Gitu.
Department of Chemistry, University of Arizona, Tucson,
Arizona 85721

THE PRESENCE OF ASPARAGINE or glutamine in peptides can
provide several undesirable side reactions in the course of
synthesis including dehydration to nitrile, pyroglutaminyl
formation, hydrolysis, and imide formation. The rather
labile bis(2,4-dimethoxybenzyl),[1,2] and 2,4-dimethoxybenzyl
groups[3] have been suggested as possible carboxamide protecting groups to eliminate some of these side reactions.
  We have been examining the potential of the diphenylmethyl, 2,4-dimethylbenzyl, 3,4-dimethoxybenzyl, 4-methoxy-
benzyl, 2,3-dimethoxybenzyl, 2-methoxybenzyl, 4-methylbenzyl,
and benzyl groups as carboxamide protecting groups. The
mono acetamide derivatives of each of these groups (*e.g.* N-
N-diphenylmethylacetamide, N-2,4-dimethylbenzylacetamide,
etc.) were prepared. All of these compounds are stable in
trifluoroacetic acid (TFA) at room temperature. The rates
of solvolysis of these amides were studied at 51° in TFA,
and the relative labilities of these protecting groups were
studied in 2.5 N HBr-HOAc and liquid HF at room temperature.
The ease of solvolysis of these groups follow the order:
diphenylmethyl > 2,4-dimethylbenzyl ≳ 4-methoxybenzyl ≳
3,4-dimethoxybenzyl ≳ 2,3-dimethoxybenzyl > 2-methoxybenzyl
>> 4-methylbenzyl > benzyl. The product of the reaction
was acetamide--no acetic acid was detected.
  Asparagine and glutamine derivatives possessing the
diphenylmethyl, 2,4-dimethylbenzyl, the 3,4-dimethoxybenzyl,
and the 4-methoxybenzyl carboxamide protecting groups were
synthesized. The results of a quantitative study of the

331

solvolysis of these derivatives in liquid HF for 75 min
at 20° using amino acid analysis is given in Table I. When
the protecting groups were removed only asparagine and
glutamine were obtained--no aspartic acid or glutamic acid
was detected.

Table I

HF Cleavage of Asparagine and Glutamine Carboxamide
Protecting Groups

Compound	% Protecting Group Removed
Boc-Asn(diphenylmethyl)-OBzl	100
Boc-Gln(diphenylmethyl)-OBzl	100
Boc-Asn(2,4-dimethylbenzyl)-OBzl	100
Boc-Gln(2,4-dimethylbenzyl)-OBzl	25
Boc-Asn(4-methoxybenzyl)-OBzl	>90
Boc-Gln(4-methoxybenzyl)-OBzl	30
Boc-Asn(3,4-dimethoxybenzyl)-OBzl	100
Boc-Gln(3,4-dimethoxybenzyl)-OBzl	10

The diphenylmethyl group was used for asparagine carboxamide protection in the synthesis of [2-phenylalanine,
4-leucine]-oxytocin by the solid phase method using a
benzhydrylamine resin. The Boc-asparagine(diphenylmethyl)
was coupled quantitatively to the growing peptide chain by
dicyclohexylcarbodiimide. The oxytocin derivative was
obtained in high yield from the protected resin peptide
by treatment with liquid HF followed by oxidation and
purification. Other peptides are being synthesized using
various amide protecting groups.

Acknowledgment

Supported in part by U.S. Public Health Service Grant
AM 13411. Peter M. Gitu has been recipient of an AFGRAD
Fellowship since 1970.

References

1. Weygand, F., W. Steglich, J. Bjarnason, R. Akhtar, and N. Chytil. Chem. Ber. 101, 3623 (1968).
2. Weygand, F., W. Steglich, and J. Bjarnason. Chem. Ber. 101, 3642 (1968).
3. Pietta, P. G., P. Cavallo, and G. R. Marshall. J. Org. Chem. 36, 3966 (1971).

THE SYNTHESIS OF THE POLYMERIC A-CHAIN DISULFIDE OF SHEEP
INSULIN

Waleed Ohan Danho. Medical Research Center, University
of Baghdad, Baghdad, Iraq

Ernie Engels. Deutsches Wollforschungsinstitut an
der Technischen Hochschule Aachen, Aachen, Germany

THE PROBLEMS OF PROTECTING sulfhydryl(-SH) function in
cystein containing peptides and proteins are well-known.
An ideal sulfhydryl protecting group, which resists all
intermediate steps of peptide synthesis and which can be
easily removed at the final step is still to be found.[1]
Zahn and co-workers[2] described the synthesis of the
polymeric B-chain of insulin by using a new tactic in
order to overcome the problems of sulfhydryl protection.
They renounced the usual sulfhydryl protection by synthesizing symmetrical cystine peptides and converted them
into cysteine peptides in the final stage of the synthesis.
This principle of using cystine peptides instead of S-
protected cysteine peptides was hardly popular and previously was used only for preparing shorter peptide
fragments.[3]
In the case of larger cystine peptides, decomposition
of disulfide bonds was believed to occur. In addition,
cystine peptides having twice the molecular weight of
cysteine peptides should become quite insoluble.
The large cystine peptides, with the amino acid
sequences in the B-chain of bovine insulin (B 1-16)$_2$ and
(B 17-30)$_2$ were readily prepared by using conventional
methods of peptide synthesis.[2] No decomposition of cystine
sulfur occurred during the main steps of the synthesis.
Coupling of these large cystine peptides yielded the

335

polymeric B-chain which was easily converted to 7,19-*S*-
sulfonate via oxidative sulfitolysis. The insulin forming
potency of this preparation was higher than that of a previously published *S*-benzyl-cyst<u>e</u>ine protected B-chain.

    This success encouraged us to use this principle for a
new synthesis of the A-chain of sheep insulin. However,
it was rather difficult to use this tactic for the whole
of the A-chain of insulin. Since the A-chain contains
four cyst<u>e</u>ine residues, one has to synthesize four cyst<u>i</u>ne
peptides (Figure 1 C). However, if any two of them would

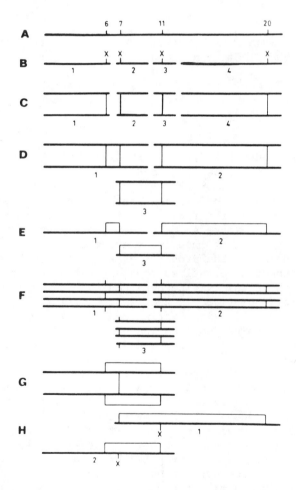

Figure 1: Schematic representation of possible disulfide-
    containing intermediates for insulin A-chain synthesis.

be coupled, polymers (Figure 1 F) would already be obtained at this intermediate stage, instead of the desired cystine peptides shown in Figure 1 D. Coupling of intermediate polymers (Figure 1 F) to form polymeric A-chain would appear to be extremely difficult. A possible approach is the coupling of cystine peptides in monomeric cyclic form (Figure 1 E).

Finally, it should be possible to work with cystine peptides in which some other cysteine residues have been S-protected, as in the cyclic cystine peptides shown in Figure 1 H.

Results

The present work describes the synthesis of the polymeric A-chain disulfide of sheep insulin in which two cysteine residues namely A_7 and A_{20} were protected by using cystine and the remaining cysteine residues in A_6 and A_{11} were protected by using the trityl-protecting group,[4] which can be easily removed by HBr-CF$_3$COOH.[5] Two large cystine peptides with the amino acid sequences of (A 1-9)$_2$ and ($\overline{A}$ 10-21)$_2$ were prepared by condensing the fragments A 1-4 with (A 5-9)$_2$ and A 10-12, A 13-16 with (A 17-21)$_2$ respectively using conventional methods of peptide synthesis.

The plans for the synthesis of these fragments are outlined in Figure 2 and Figure 3. The protected peptides were characterized by elemental and amino acid analysis, and by thin layer chromatography in several solvent systems. Physical data are shown in Table I. Coupling of the large cystine peptides (A 1-9)$_2$ and (A 10-21)$_2$ by the azide method[6] (Figure 4) should give a mixture of A-chain dimer and A-Chain polymer. The A-chain polymer was the main product (optical rotation - 31.6°, c=1, HPT). The following numbers of the amino acid residues per molecule were found in acid hydrolysate (6 N HCl, 48 hours, 110°C): 2.29 Asp, 0.80 Ser, 4.00 Glu, 2.00 Gly, 0.95 Ala, 1.20 Val, 0.20 Ile, 1.97 Leu, 1.36 Tyr.

Table I

Physical Characteristics of Protected Cystine Peptides of Sheep Insulin A-Chain

Compound		State Melting Point (Solvent, °C)	tlc* R_f A	tlc* R_f B	$[\alpha]_{22}^{D}$
[Z-Cys-Ala-Gly-OMe]$_2$	I	Crystalline (Isopropanol, 208–210)	0.78	0.25	-88.0° (c 1, DMF)[+]
[Boc-Cys(Trt)-Cys-Ala-Gly-OMe]$_2$ #	II	Crystalline (Methanol–Water, 178–180)	0.80	0.20	-32.8° (c 1, DMF)
[Boc-Gln-Cys(Trt)-Cys-Ala-Gly-OMe]$_2$	III	amorphous (212–215)	0.70	0.10	-46.4° (c 1, DMF)
[Boc-Gly-Ile-Val-Glu(OBu^t)-Gln-Cys(Trt)-Cys-Ala-Gly-OMe]$_2$	IV	amorphous (240, dec.)	0.75	0	-33.6° (c 1, HPT)√
[Boc-Gly-Ile-Val-Glu(OBu^t)-Gln-Cys(Trt)-Cys-Ala-Gly-N$_2$H$_3$]$_2$	V	amorphous (240, dec.)	----	----	-45.5° (c 1, HPT)
[Boc-Cys-Asn-OBzl]$_2$	VII	Crystalline (Ethanol, 199–201)	0.78	0.25	-106.0° (c 1, DMF)
[Boc-Tyr(Bzl)-Cys-Asn-OBzl]$_2$	VIII	Crystalline (DMF–Water, 208–209)	0.83	0.10	-33.0° (c 1, DMF)

Compound	No.	Form (Solvent, m.p.)			$[\alpha]$
[Boc-Asn-Tyr(Bzl)-Cys-Asn-OBzl]$_2$	IX	amorphous (Methanol, 223-225)	0.78	0.15	-80.0° (c 1, DMF)
[Boc-Glu(OBzl)-Asn-Tyr(Bzl)-Cys-Asn-OBzl]$_2$	X	amorphous (Methanol, 239-241)	0.75	0.12	-75.5° (c 1, DMF)
[Trt-Leu-Tyr(Bu^t)-Gln-Leu-Glu(OBzl)-Asn-Tyr(Bzl)-Cys-Asn-OBzl]$_2$	XI	amorphous (Methanol, 235-240)	0.70	0	-65.0° (c 1, DMF)
[Trt-Val-Cys(Trt)-Ser-Leu-Tyr-Gln-Leu-Glu(OBzl)-Asn-Tyr(Bzl)-Cys-Asn-OBzl]$_2$	XII	amorphous (Methanol, 240, dec.)	0.80	0	-50.0° (c 1, DMF)

*Solvent System: A, *sec.* Butanol-formic acid-water (75:13.5:11.5); B, Chloroform-methanol-acetic acid (95:5:3).

†DMF, Dimethylformamide.

#Trt, Trityl.

√HPT, Hexamethyl phosphoric acid triamide.

Z-Ala-Gly-OMe
| (a) 4 N HBr-CH$_3$COOH
| (b) DMF-Et$_3$N
H-Ala-Gly-OMe
66% | [Z-Cys-ONp]$_2$
[Z-Cys-Ala-Gly-OMe]$_2$ (I)
| (a) 4 N HBr-CH$_3$COOH
| (b) DMF-Et$_3$N
| (c) Boc-Cys(Trt)-OH, DCCI
| (d) CCD, toluene, K = 0.02
[Boc-Cys(Trt)-Cys-Ala-Gly-OMe]$_2$ (II)
| (a) 1 N HCl-Methanol
53% | (b) Ethyl acetate-aqueous Na$_2$CO$_3$
| (c) HCl-Ether
[HCl·H-Cys(Trt)-Cys-Ala-Gly-OMe]$_2$
| (a) DMF-Et$_3$N
61% | (b) Boc-Gln-ONp
[Boc-Gln-Cys(Trt)-Cys-Ala-Gly-OMe]$_2$ (III)
| (a) 1 N HCl-Methanol
70% | (b) Chloroform, Methanol-aqueous Na$_2$CO$_3$
| (c) HCl-Ether
[HCl·H-Gln-Cys(Trt)-Cys-Gly-OMe]$_2$
| (a) DMF, HPT-Et$_3$N
43% | (b) Boc-Gly-Ile-Val-Glu(OBu^t)-N$_3$
[Boc-Gly-Ile-Val-Glu(OBu^t)-Gln-Cys(Trt)-Cys-Ala-Gly-OMe]$_2$ (IV)
50% | N$_2$H$_4$
[Boc-Gly-Ile-Val-Glu(OBu^t)-Gln-Cys(Trt)-Cys-Ala-Gly-N$_2$H$_3$]$_2$ (V)
1 2 3 4 5 6 7 8 9

Figure 2: Outline of the synthesis of (1-9)-nonapeptide derivative

Boc-Asn-OBzl
| (a) CF$_3$COOH
47% | (b) DMF-Et$_3$N
| (c) [Boc-Cys-ONp]$_2$ (VI)
[Boc-Cys-Asn-OBzl]$_2$ (VII)
| (a) CF$_3$COOH
70% | (b) DMF-Et$_3$N
| (c) Boc-Tyr(Bzl)-ONp
[Boc-Tyr(Bzl)-Cys-Asn-OBzl]$_2$ (VIII)
| (a) CF$_3$COOH
61% | (b) DMF-Et$_3$N
| (c) Boc-Asn-ONp
[Boc-Asn-Tyr(Bzl)-Cys-Asn-OBzl]$_2$ (IX)
| (a) CF$_3$COOH
61% | (b) DMF-Et$_3$N
| (c) Boc-Glu(OBzl)-ONp
[Boc-Glu(OBzl)-Asn-Tyr(Bzl)-Cys-Asn-OBzl]$_2$ (X)
| (a) CF$_3$COOH
73% | (b) DMF-Et$_3$N
| (c) Trt-Leu-Tyr(Bu^t)-Gln-Leu-N$_3$
[Trt-Leu-Tyr(Bu^t)-Gln-Leu-Glu(OBzl)-Asn-Tyr(Bzl)-Cys-Asn-OBzl]$_2$ (XI)
| (a) CF$_3$COOH
62% | (b) DMF-Et$_3$N
| (c) Trt-Val-Cys(Trt)-Ser-N$_3$
[Trt-Val-Cys(Trt)-Ser-Leu-Tyr-Gln-Leu-Glu(OBzl)-Asn-Tyr(Bzl)-Cys-Asn-OBzl]$_2$ (XII)
10 11 12 13 14 15 16 17 18 19 20 21

Figure 3: Outline of the synthesis of (10-21)-dodecapeptide derivative

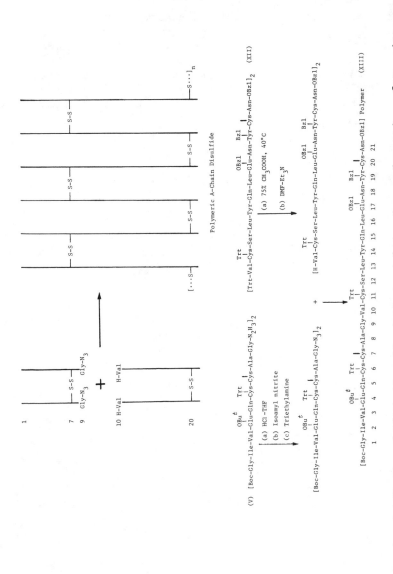

Figure 4: **Synthesis of sheep insulin A-chain polymer by condensation of two intermediate cystine peptides**

References

1. Lübke, K., and H. Klostermeyer. Advan. Enzymol. <u>33</u>, 445 (1970).
2. Zahn, H., and G. Schmidt. Tetrahedron Letters (London) 5095 (1967).
3. Ressler, C., and V. du Vigneaud. J. Amer. Chem. Soc. <u>76</u>, 3107 (1954).
4. Amiard, G., R. Heymes, and L. Velluz. Bull. Soc. Chem. Francais, 698 (1956).
5. Zahn, H., W. Danho, H. Klostermeyer, H. G. Gattner, and J. Repin. Z. Naturforschg. <u>24b</u>, 1127 (1969).
6. Honzl, J., and J. Rudinger. Coll. Czech. Chem. Commun. <u>26</u>, 2333 (1961).

CHEMISTRY AND BIOLOGY OF PEPTIDES

NEW CATALYSTS IN PEPTIDE SYNTHESIS

Wolfgang König, Rolf Geiger. Farbwerke Hoechst AG, D-623 Frankfurt-Höchst, Germany

WE REPORTED PREVIOUSLY[1,2] on the 1-hydroxybenzotriazole ester (OBt) and on the possibility of decreasing racemization by the addition of 1-hydroxybenzotriazole (HOBt) to dicyclohexylcarbodiimide-mediated peptide condensations. To compare the reactivity of these new esters with that of other active esters, we synthesized Z-Phe-Val-*o*-nitranilide and determined reaction rates spectrophotometrically. Figure 1 shows a comparison of reactivities of the 2,4,5-trichlorophenyl ester (OTcp), the *N*-hydroxysuccinimide ester (ONSu), and the OBt ester in dimethylformamide (DMF) and tetrahydrofuran (THF). It is known that *p*-nitrophenyl esters react faster in DMF than in nonpolar solvents.[3] The OTcp esters behave analogously, Figure 1. Kemp suggested[4] that the ONSu esters react more slowly in DMF than in THF, a fact which we could not confirm in this test. However, the reactivity of the OBt ester is much faster than that of the other active esters. Probably a complex (I) is formed between the amine and the OBt ester which, due to the strong polarization of the carbonyl group, reacts rapidly to give the amide.

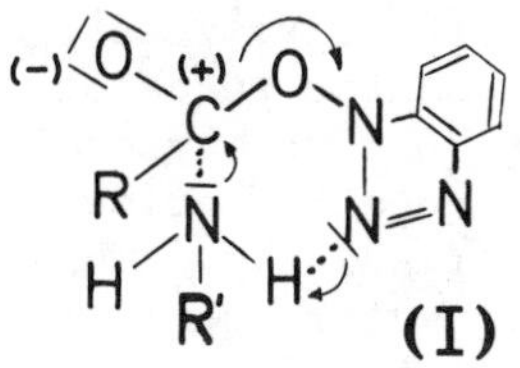

(I)

Recently we discovered a new property of HOBt. It catalyzes the aminolysis of *p*-nitrophenyl or trichlorophenyl esters, particularly in DMF or dimethylacetamide solution. The moderate catalysis of thiophenyl and *p*-nitrophenyl esters by imidazole is well known.[5,6] More active are the "bifunctional catalysts,"

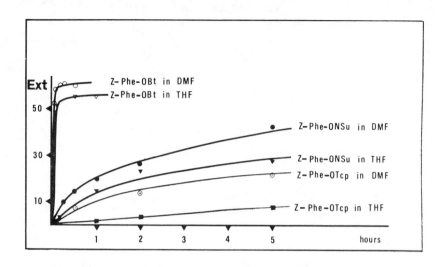

Figure 1: Active ester synthesis of Z–Phe–Val–*o*–nitranilide, monitored spectrophotometrically at 350 nm in ethyl acetate [*c* valine–*o*–nitranilide, $5.10^{-3}M$; *c* Z–Phe–active ester, $1.10^{-2}M$]. Samples were extracted with NaHCO$_3$ and 2 *N* HCl prior to reading. (OBt ester samples were treated with NaOH prior to extraction.)

e.g. certain pyrazole derivatives, 2–hydroxypyridine and 1,2,4–triazole.[7,8] However, these catalysts are effective in nonpolar solvents only and not in DMF.[3] The strong catalytic effect of HOBt on the aminolysis of Z–Phe–OTcp in DMF can be seen in Figure 2. In THF, however, HOBt inhibits the aminolysis of the trichlorophenyl ester. Figure 3 shows that the aminolysis of *N*–hydroxysuccinimide esters can also be catalyzed by HOBt in DMF, although not as effectively as that of the 2,4,5–trichlorophenyl esters.

As the reaction curves of the OBt esters and those of the HOBt–catalyzed OTcp esters are virtually identical, the question arises as to whether OBt esters are formed as intermediates. This appears to be unlikely since Z–Phe–OBt and *p*–nitrophenol actually form Z–Phe–ONp in DMF solution.

A systematic investigation of the catalytic properties of other *N*–hydroxy compounds revealed that those with an acidity similar to acetic acid are suitable as catalysts. For rapid appraisal of catalytic potency we measured the effects of the additives on the half-time of the following reaction: Z–Val–ONp + cyclohexylamine → Z–Val–cyclohexylamide + *p*–nitrophenol. The absorbance of the liberated *p*–nitrophenol

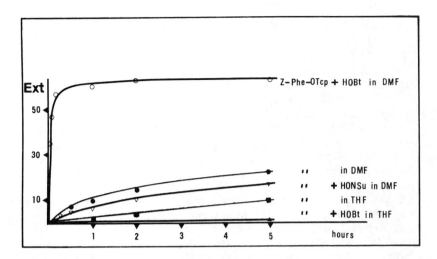

Figure 2: Reaction of Z-Phe-OTcp with H-Val-*o*-nitranilide (yield determined by UV ext at 350 nm).

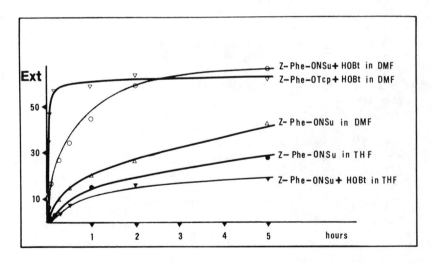

Figure 3: Reaction of activated esters of Z-Phe with H-Val-*o*-nitranilide (yield determined by UV ext at 350 nm).

was determined at 330-300 nm at regular time intervals.
Only those *N*-hydroxy compounds which possess approximately
the pK of acetic acid have a catalytic effect (Table I).
More basic compounds as 1-hydroxypiperidine do not effect
catalysis. If the additives are more acidic than HOAc *e.g.*
1-hydroxybenzotriazoles with a substituent having a minus-I-
effect, the catalytic activity becomes progressively weaker.

Catalysis of various weakly activated esters by HOBt
was examined by means of the Z-Phe-Val-OMe model synthesis.
The disappearance of H-Val-OMe was observed by tlc. 1-
Hydroxypiperidine esters, cyanomethyl esters, phenyl esters
and *p*-chlorophenyl esters could not be catalyzed. 2,4-
Dichlorophenyl, 2-chloro-4-nitrophenyl and pentachlorophenyl
esters were catalyzed, though not to the same degree as the
2,4,5-trichlorophenyl and *p*-nitrophenyl esters.

The effectiveness of these new catalysts in cases of
extreme steric hindrance was tested in the synthesis of
Nps-Ile-Cys(Trt)-Ser-Leu-OH, a tetrapeptide of the A-chain
of human insulin. Excess active ester of Nps-Ile is reacted in DMF with H-Cys(Trt)-Ser-Leu-OH. The disappearance
of the tripeptide is examined by tlc. As can be seen from
Table II, the reaction is not completed within a foreseeable
period of time both with the OTcp and the ONSu ester.
While the reaction rate of the ONSu ester is not accelerated
with 3-hydroxy-4-oxo-3,4-dihydro-quinazoline, that of the
OTcp ester is more or less catalyzed depending on the added
compound. The reactivity of the 2-chloro-4-nitrophenyl
ester is catalytically enhanced while aminolysis of the 2,4-
dinitrophenyl ester is inhibited by the same compound.
According to this test the following are particularly
suitable catalysts: 1-hydroxybenzotriazole, 3-hydroxy-4-
oxo-3,4-dihydro-quinazoline and 1-hydroxy-2-pyridone. However, 1-hydroxybenzotriazole should not be used with Nps-
amino acids or peptides, as the Nps group is partly split
off, as was first observed by J. Rudinger (personal
communication).

The basicity of the amine plays also an important part.
We failed completely in our attempt to prepare nitranilides
in this way. The catalysis described is apparently dependent on four prerequisites: only a correct combination of
solvent, activated ester, *N*-hydroxy compound and amine
results in a successful, rapid peptide synthesis. The
powerful catalytic effect on 2,4,5-trichlorophenyl esters
and *p*-nitrophenyl esters in DMF on the one hand, and the
inhibition of these esters in THF on the other, and furthermore, the inhibition of more highly activated esters in
DMF, suggest that the *N*-hydroxy compounds might form two

Table I

Effectiveness of *N*-Hydroxy Compounds to Catalyze
p-Nitrophenyl Ester Synthesis in Dimethylformamide*

Compound	pK	$t_{1/2}$*(min)
-----	-	112.0
N-hydroxypiperidine	5.9	125.0
1-hydroxy-4-methyl-6-isopropyl-2-pyridone	4.14	19.0
3-hydroxy-2-methyl-4-oxo-3,4-dihydro-quinazoline	4.11	1.9
3-hydroxy-4-oxo-3,4-dihydro-quinazoline	4.10	0.25
1-hydroxy-3,4,6-trimethyl-2-pyridone	4.09	14.8
1-hydroxy-4,6-dimethyl-2-pyridone	4.09	5.9
1-hydroxy-4-methyl-2-pyridone	4.08	3.2
1-hydroxy-2-pyridone	4.08	1.3
3-hydroxy-4-methyl-2,3-dihydro-thiazole-2-thione	4.08	1.5
1-hydroxy-3,5-dichloro-4,6-dimethyl-2-pyridone	4.05	4.0
1-hydroxy-2-oxo-2,3-dihydro-6-chloro-indole	4.05	1.5
acetic acid	4.05	60.0
N-hydroxysuccinimide	4.04	4.8
1-hydroxy-5,6-dimethyl-benzotriazole	4.02	2.1
1-hydroxy-5-methyl-benzotriazole	4.02	2.8
1-hydroxy-6-methoxy-benzotriazole	4.00	2.0
1-hydroxy-5-methoxy-benzotriazole	4.00	3.6
1-hydroxy-4-methyl-benzotriazole	4.00	3.0
1-hydroxy-benzotriazole	4.00	3.5
3-hydroxy-4-oxo-3,4-dihydro-1,2,3-benzotriazine	4.00	18.0
1-hydroxy-6-bromo-benzotriazole	3.91	8.9
1-hydroxy-6-chloro-benzotriazole	3.90	10.6
1-hydroxy-5-chloro-benzotriazole	3.89	13.9
1-hydroxy-6-trifluoromethyl-benzo-triazole	3.72	39.0
1-hydroxy-5,6-dichlorobenzotriazole	3.70	35.0
1-hydroxy-6-nitro-benzotriazole	3.51	very slow

*Half-time of the reaction Z-Val-ONp + cyclohexylamine +
catalyst → Z-Val-cyclohexylamide + *p*-nitrophenol in DMF
at 21°C (*c*, 1 μmol/ml) and pK values of the catalysts,
measured in diethyleneglycol dimethylether - water (6:4)
at 30°C (*c*, 0.05 mmol/ml).

Table II

Effectiveness of *N*-Hydroxy Compounds to Catalyze Active
Ester Condensations of Sterically Hindered Components
in Dimethylformamide*

Compound	Active Ester	Reaction Time
–	ONSu	>5 weeks
3-hydroxy-4-oxo-3,4-dihydro-quinazoline	ONSu	>5 weeks
–	OTcp	>5 weeks
3-hydroxy-4-oxo-3,4-dihydro-quinazoline	OTcp	15 hours
1-hydroxy-benzotriazole	OTcp	15 hours
1-hydroxy-5,6-dimethyl- benzotriazole	OTcp	20 hours
1-hydroxy-2-pyridone	OTcp	20 hours
1-hydroxy-4-methyl-2-pyridone	OTcp	20 hours
1-hydroxy-2,4-dimethyl-2-pyridone	OTcp	4 days
3-hydroxy-2-methyl-4-oxo-3,4-dihydro-quinazoline	OTcp	7 days
1-hydroxy-2-oxo-2,3-dihydro-6-chloro-indole	OTcp	>4 weeks
–	$OPh(2\text{-}Cl,4\text{-}NO_2)$[†]	8 days
–	$OPh(2,4\text{-}NO_2)$[#]	8 days
3-hydroxy-4-oxo-3,4-dihydro-quinazoline	$OPh(2\text{-}Cl,4\text{-}NO_2)$	2 days
3-hydroxy-4-oxo-3,4-dihydro-quinazoline	$OPh(2,4\text{-}NO_2)$	>8 days

*Synthesis of Nps-Ile-Cys(Trt)-Ser-Leu-OH using active
esters of Nps-Ile and *N*-hydroxy compounds as catalysts
(*c*, 0.1 mmol/ml, DMF, 21°C).
[†]2-Chloro-4-nitrophenyl ester
[#]2,4-Dinitrophenyl ester.

possible complexes with the active ester and the amino component (Figure 4). In the catalyzed reactions complex (I) might be formed in which the amino group is located close to the carbonyl group of the active ester. In the inhibited reactions complex (II) might be formed in which the amino group, although bound in a complex, is so unfavorably located as to inhibit aminolysis.

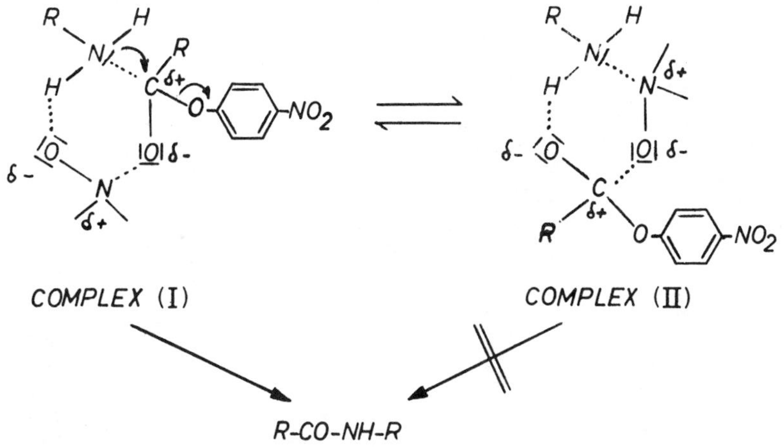

Figure 4: Proposed complexes among *p*-nitrophenyl ester, *N*-hydroxy compound and amine.

Acknowledgment

We thank our co-workers, P. Pogoda, P. Pokorny, and D. Lagner for valuable assistance.

References

1. König, W., and R. Geiger. In *Peptides 1969*, Scoffone, E., ed. (Amsterdam: North-Holland Publ., 1971) pp 17–22.
2. König, W., and R. Geiger. Chem. Ber. 103, 788 (1970).
3. Wieland, Th., and W. Kahle. Justus Liebigs Ann. Chem. 691, 212 (1966).
4. Kemp, D. S. Proc. 11th European Peptide Symposium, 1971, in press.
5. Wieland, Th., H. Determann, and W. Kahle. Angew Chem. 75, 209 (1963).

6. Mazur, R. H.   J. Org. Chem. <u>28</u>, 2498 (1963).
7. Beyerman, H. C., and W. Maassen van den Brink.
   Proc. Chem. Soc. 266, (1963).
8. Beyerman, H. C., W. Maassen van den Brink, F. Weygand,
   A. Prox, W. König, L. Schmidhammer, and E. Nintz.
   Rec. Trav. Chim. Pas-Bas <u>84</u>, 213 (1965).
9. Weygand, F., A. Prox, and W. König. Chem. Ber. <u>99</u>,
   1451 (1966).
10. König, W., and R. Geiger. Chem. Ber. <u>103</u>, 2024
   (1970).

ON THE REPETITIVE EXCESS MIXED ANHYDRIDE METHOD FOR THE
SEQUENTIAL SYNTHESIS OF PEPTIDES. SYNTHESIS OF THE
SEQUENCE 1-10 OF HUMAN GROWTH HORMONE

H. C. Beyerman. Laboratory of Organic Chemistry,
Technische Hogeschool, Julianalaan 136, Delft, The
Netherlands

GIVING HISTORICAL INTRODUCTIONS at this meeting was dis-
couraged, but I feel that I ought to give a short
introduction because the work that we have done isn't
completely new. It's just another step forward, we think,
of what has been done by other people.

The method which we call REMA-method, repetitive excess
mixed anhydride method, as you know, is based on work done
in 1951 when three groups, headed by Boissonnas,[1]
Wieland,[2] and Vaughan,[3] started to use mixed anhydrides of
the carbonates (Table I). However, these were found to
give considerable racemization, and it was only after
Anderson[4] in the United States and Wieland[5] in Germany
found conditions for racemization-free coupling that these
anhydrides could be used with more advantage. The late
Friedrich Weygand in Germany then used an excess of symmetrical anhydrides[6,7] in order to force reactions to
completion. This principle of excess has been used with
success by other people, notably by Bodanszky, *et al.*[11]
with active esters. However, symmetrical anhydrides are
difficult to prepare, and it was Weygand's pupil Tilak[8,9]
who in the United States, when using excess mixed anhydrides,
found that the relative small excess could be destroyed by
potassium hydrogen carbonate. The yields with small peptides were so high that the method could be applied without
purification of the intermediate products, that is, in a
repetitive way.

Table I

The *Repetitive Excess Mixed Anhydride* (REMA) Method
for the Stepwise Synthesis of Peptides

History

1951	Boissonnas[1]	
	Wieland and Bernhard[2]	Mixed anhydrides of *N*-protected amino acids and monoalkyl *carbonates*
	Vaughan, Jr.[3]	
1967	Anderson, *et al*.[4]	Conditions for *racemization-free* coupling
1968	Wieland, *et al*.[5]	
1967	Weygand, *et al*.[6]	*Excess symmetrical* anhydrides
1969	Weygand, *et al*.[7]	
1970	Tilak[8]	*Excess mixed* anh. (*repetitive*)
1972	Tilak, *et al*.[9]	
1972	Floor, deLeer, Beyerman[10]	Human Growth Hormone 1-10 (also *fragment* coupling)

I would like to report here, as an example of a larger peptide, the REMA-synthesis of a sequence of human growth hormone, HGH 1-10. We used with success fragment coupling in order to circumvent a very disagreeable side reaction that will be discussed later.

Equation (1) is just to show for those who don't know how one prepares mixed anhydrides (I) of *N*-protected amino acids and monoalkyl carbonates, followed by coupling with an amino acid or peptide ester (II). I would like to stress that an 0.5 molar excess of mixed anhydride is used in the coupling with the amine component (II). Activation time for making the anhydride is less than one minute in order to minimize racemization. We used ethyl acetate, tetrahydrofuran, and occasionally, dimethylformamide as solvents for preparing the anhydride. With the amino component, which will grow in our case to a octapeptide, all kinds of solvents can be used to obtain solutions, *e.g.* dimethylformamide.

The unwanted side reactions are shown in Equations 2 and 3. You can get, at about 0°, disproportionation

$$
\text{X-NH-}\underset{\underset{R_1}{|}}{\overset{\overset{H}{|}}{C}}\text{-}\overset{\nearrow O}{\underset{\searrow OH}{C}} \quad \xrightarrow{\text{(Alk)}_3N} \quad \text{X-NH-}\underset{\underset{R_1}{|}}{\overset{\overset{H}{|}}{C}}\text{-C}\overset{O}{\underset{O}{\diagup}} \quad \xrightarrow{\text{H}_2\text{N-}\underset{R_3}{\text{CH}}\text{-COOR}_4 \ \ \text{(II)}}
$$

$$
\underset{\text{R}_2\text{O-}\overset{|}{\underset{}{C}}\text{=O}}{\overset{+}{\underset{\text{Cl}}{}}} \qquad\qquad \text{(I) R}_2\text{O-C}\overset{\diagup}{\underset{\diagdown O}{}}
$$

$$
+ \quad \text{(Alk)}_3\text{N} \cdot \text{HCl}
$$

$$
\text{X-NH-}\underset{\overset{|}{R_1}}{\text{CH}}\text{-CO-NH-}\underset{\overset{|}{R_3}}{\text{CH}}\text{-COOR}_4
$$

$$
+
$$

$$
\text{R}_2\text{OH} + \text{CO}_2
$$

Equation 1

$$
\text{X-NH-}\underset{\overset{|}{R_1}}{\text{CH}}\text{-C}\overset{\diagup O}{\underset{\diagdown O}{}}\overset{O}{\diagdown}\text{C-OR}_2 \quad \xrightarrow{\text{Temp} \sim 0^\circ} \quad \begin{array}{l}\text{X-NH-}\underset{}{\text{CH-C}}\overset{R_1}{}\overset{\diagup O}{\diagdown O} \\[4pt] \text{X-NH-}\underset{\overset{|}{R_1}}{\text{CH}}\text{-C}\overset{\diagup}{\diagdown O}\end{array} \ + \ \begin{array}{l}\overset{O}{\diagup}\text{C-OR}_2 \\[4pt] \overset{}{O}\text{C-OR}_2\end{array}
$$

(I)   (III)   (IV)

Equation 2

$$
\begin{array}{l}\text{X-NH-}\underset{\overset{|}{R_1}}{\text{CH}}\text{-C}\overset{\diagup O}{} \\[6pt] \quad\quad\quad\quad\quad O \\[4pt] \text{R}_2\text{O-C}\overset{}{\diagdown O}\end{array} + \ \underset{\underset{\text{H}\ R_3}{}}{\overset{\text{H}}{\text{N-CH-COOR}_4}} \quad \longrightarrow \quad \text{X-NH-}\underset{\overset{|}{R_1}}{\text{CH}}\text{-COOH} \ + \ \text{R}_2\text{O-CO-NH-}\underset{\overset{|}{R_3}}{\overset{\overset{H}{|}}{C}}\text{-COOR}_4
$$

(I)   (II)   (V) permanently blocked

Equation 3

(Equation 2) of the mixed anhydride (I) into symmetrical
anhydride (III) and dialkylpyrocarbonate (IV). The
dialkylpyrocarbonate will react with and block permanently
the amine component. This disproportionation is negligible
if one works at a temperature at or below minus 15°, and
that is what we do in principle routinely. The formation
of a symmetrical anhydride (III) is not all that bad
because it gives the correct reaction, only you lose 50%.

Second comes a complication that is more troublesome
(Equation 3): coupling on the wrong side of the mixed
anhydride which will result in a permanently blocked urethane derivative (V). This reaction, in contrast to the
disproportionation of the mixed anhydride, is not much
temperature dependent; it depends mainly on steric factors.
We circumvented this, as shown in Figure 1, by using
fragment coupling.

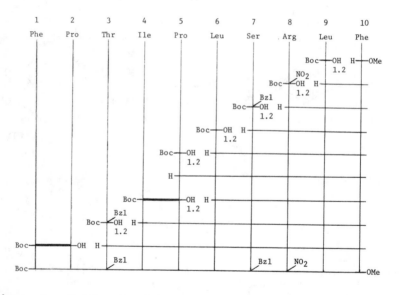

Figure 1: Synthesis of human growth hormone sequence 1-10
(HGH 1-10).

In our synthesis of the sequence 1-10 of HGH, all sidechains are protected: nitro on the arginine, benzyl on
the serine and threonine. First we made the pentapeptide

6-10 by adding stepwise the amino acid, decomposing the excess anhydride with a $KHCO_3$ - solution, and precipitating or extracting the peptide, followed by drying. One of the advantages of the method is that one can check on the purity of every intermediate product. The yield with isolation of all intermediates of the pentapeptide was 65%, and it went up to 75% with the repetitive procedure, that is, just precipitating, drying, and going on. This is not a very high yield, 75%, and this is mainly because coupling of Boc-nitroarginine gives minor by-products. After this we went on with the synthesis and found that with the isoleucine onto proline, and phenylalanine onto proline, it didn't work, we got coupling on the wrong side of the anhydride. We then coupled with the peptides drawn in heavy lines in Figure 1. These fragment couplings have the advantage inherent in proline-peptides of producing no racemization *via* the oxazolone mechanism. Without the isolation of intermediates it worked fine, and the overall yield of the pure decapeptide was 59%. I want to stress the fact that all the peptides shown in Figure 1 have been isolated too; all have correct elemental analysis for carbon, hydrogen, nitrogen; all optical rotations have been determined.

Finally it remained to prove that we really had made the decapeptide in optically pure form. We synthesized it, therefore, in another way, by a "classical" Merrifield synthesis on the polymer and split the decapeptide from the polymer by methanolysis.[11] Because for other reasons there was a Ztf protecting group on the threonine, these two decapeptides could not be compared directly. We therefore had to remove the protecting groups. Hydrogen bromide in trifluoroacetic acid gave a mixture of products. Liquid hydrogen fluoride gave much better results, and we could compare the decapeptide made by the two methods. The Merrifield peptide gave a correct amino acid analysis; its optical rotation was $[\alpha]_D^{24} -61°$ (*c* 0.8, 95% AcOH). The REMA-made HGH 1-10 showed $[\alpha]_D^{24} -50°$ *c* 0.8, 95% AcOH).

I would like to conclude that the repetitive mixed anhydride procedure is a very convenient method. It compares in the order of magnitude in speed with the Merrifield procedure; one can add to the growing peptide chain about one amino acid a day. The speed-limiting factor is mainly the drying time of the intermediate peptide, but complete dryness may not be absolutely necessary. The advantage,

like the repetitive syntheses by, for instance, Bodanszky
et al.[12] and Morley[13] is that you can isolate and check on
the properties of every intermediate. The disadvantage:
it is still a lot of work. We haven't yet a machine to do
the operations. And then, of course, you would ask, what
is the limit of the method? We stopped because we started
with the residue 10 for HGH, but being at residue 1, we
felt we could have gone on more easily. In fact, in the
beginning of the synthesis this gave some difficulties,
because then you have small and rather soluble peptides
that are not easily precipitated. In a later stage, after
about three amino acids, then it goes much better. Frankly,
I don't know, and at the moment we are trying to see how
big a peptide we can make.

Acknowledgments

The investigation was carried out under the auspices
of the Netherlands Foundation for Chemical Research (S.O.N.)
and with support from the Netherlands Organization for the
Advancement of Pure Research (Z.W.O). The experiments were
performed by J. Floor and E. W. B. de Leer.

References

1. Boissonnas, R. A.   Helv. Chim. Acta $\underline{34}$, 874 (1951).
2. Wieland, T., and H. Bernhard. Justus Liebigs Ann.
    Chem. $\underline{572}$, 190 (1951).
3. Vaughan, Jr., J. R.   J. Amer. Chem. Soc. $\underline{73}$, 3547
    (1951).
4. Anderson, G. W., J. E. Zimmerman, and F. M. Callahan.
    J. Amer. Chem. Soc. $\underline{89}$, 5012 (1967).
5. Determann, H., and T. Wieland. Justus Liebigs Ann.
    Chem. $\underline{670}$, 136 (1963).
6. Weygand, F., P. Huber, and K. Weiss. Z. Naturforsch.
    $\underline{22B}$, 1084 (1967).
7. Weygand, F., and C. Di Bello. Z. Naturforsch. $\underline{24B}$,
    314 (1969).
8. Tilak, M. A.   Tetrahedron Lett. 849 (1970).
9. Tilak, M. A., M. L. Hendricks, and D. S. Wedel. In
    Progress in Peptide Research, Vol. 2, Lande, S., ed.
    (New York: Gordon and Breach, 1972), pp 351-359.
10. Beyerman, H. C., J. Floor, and E. W. B. de Leer.
    Recl. Trav. Chim. Pays-Bas $\underline{92}$ (1973), in press.
11. De Leer, E. W. B., and H. C. Beyerman. Recl. Trav.
    Chim. Pays-Bas $\underline{92}$ (1973), in press.

12. Bodanszky, M., and V. du Vigneaud. J. Amer. Chem. Soc. 81, 5688 (1959); Bodanszky, M., J. Meienhofer, and V. du Vigneaud. J. Amer. Chem. Soc. 82, 3195 (1960); Bodanszky, M., and N. J. Williams. J. Amer. Chem. Soc. 89, 685 (1967); Bodanszky, M., M. A. Ondetti, S. D. Levine, and N. J. Williams. J. Amer. Chem. Soc. 89, 6753 (1967).
13. Morley, J. S. J. Chem. Soc. (C) 2410 (1967).

STUDIES ON THE RATE OF RACEMIZATION AND COUPLING OF N-BENZYLOXYCARBONYLAMINO ACID AND PEPTIDE ACTIVE ESTERS

J. Kovacs, E. J. Davis, R. H. Johnson, H. Cortegiano, J. Roberts. Department of Chemistry, St. John's University, Jamaica, New York 11432

DURING THE SECOND AMERICAN PEPTIDE SYMPOSIUM, rates of racemization through α-hydrogen abstraction of several active ester derivatives of cysteine, glutamic acid and aspartic acid together with the rates of coupling of these active esters with valine methyl ester were reported.[1]

In order to establish the influence of the side chain of amino acids active esters on the rate of racemization and coupling, additional amino acids, namely phenylalanine, alanine, tryptophane and serine active ester derivatives were studied. The results of the rate studies are reported in Table I.

From the previous[1,2] and present kinetic data it was concluded that the side chain of a N-benzyloxycarbonylamino acid active ester has a significant effect on the rate of racemization under the conditions studied. This is indicated in Table II, which shows the ratios of racemization rate constants of N-benzyloxycarbonyl-S-benzyl-L-cysteine active esters, which is the fastest racemizing amino acid investigated so far, and N-benzyloxycarbonylamino acid active esters. The following order of decreasing rates was determined: cysteine > serine > aspartic acid > phenylalanine > glutamic acid > tryptophane > alanine. The racemization rates of the above N-benzyloxycarbonylamino acid active esters is in the following decreasing order: N-hydroxysuccinimidyl > pentafluorophenyl > 2,4,5-trichlorophenyl > p-nitrophenyl > pentachlorophenyl. This order is independent of the structure of the amino acid side chain.

Table I

The Racemization and Coupling Rate Constants* for the
Reaction of *N*-Benzyloxycarbonylamino Acid Active Esters
with Triethylamine[†] and L-Valine Methyl Ester,[#]
Respectively

Compound	$k_{rac} \times 10^6$ $l.mol^{-1}\ sec^{-1}$		$k_c \times 10^2$ $l.mol^{-1}\ sec^{-1}$		k_c/k_{rac}
Z-Phe-R where R is:					
OSu	139.4	± 0.05**	4.04	± 0.41**	290
OPfp	82.0	± 1.1**	11.91	± 1.7[††]	1450
OTcp (2,4,5)	12.0	± 1.7**	0.204	± 0.004**	170
ONp	9.2	± 1.2[††]	0.032	± 0.001**	35
OPcp	3.3	± 0.1**	0.289	± 0.003**	875
Z-Ala-R where R is:					
OSu	17.7	± 0.5**	7.67	± 0.5[##]	4330
OPfp	8.37	± 0.04**	19.3	± 1.1**	23060
OTcp (2,4,5)	1.82	± 0.04**	0.299	± 0.004**	1600
ONp	1.38	± 0.02**	0.152	± 0.009**	1100
OPcp	0.825	± 0.025**	0.506	± 0.032[††]	6135
Z-Ser-R where R is:					
OPfp	297	± 25.0**	19.2	± 1.3**	645
OPcp	26.3	± 0.8**	0.737	± 0.03**	280
Z-Trp-R where R is:					
OPfp	10.7	± 0.7**	12.38	± 1.3**	11575
OPcp	0.716	± 0.014**	0.320	± 0.021**	4470

*All reactions were run in tetrahydrofuran at 23 ± 1°.
[†]Three different concentrations of Et$_3$N were used, 0.05 *M*,
0.357 *M*, and 1.74 *M* (1, 7 and 35 equivalents respectively),
with 0.05 *M* of the active ester: k_{rac} is a true second
order rate constant, since the three base concentrations
gave within experimental error the same values.
[#]The concentration of both the methyl ester and the active
esters was 0.13 *M*.
**Average of two experiments.   [††]Average of three experiments.
[##]Average of five experiments.

Table II

Ratio of Racemization Rate Constants of
N-Benzyloxycarbonyl-*S*-benzyl-L-cysteine Active
Esters and *N*-Benzyloxycarbonylamino Acid
Active Esters

Ester	$\dfrac{Cys}{Ser}$	$\dfrac{Cys}{Asp}$	$\dfrac{Cys}{Phe}$	$\dfrac{Cys}{Glu}$	$\dfrac{Cys}{Ala}$	$\dfrac{Cys}{Trp}$
OSu			35	109	280	
OPfp	16	13	40	97	405	330
OTcp(2,4,5)		19	41	205	270	
ONp		14	100	123	285	
OPcp	17	24	40	207	500	570

In contrast to racemization, the side chain of the above
amino acid active ester derivatives has no significant effect
on coupling with valine methyl ester. From Table I it can
be seen that the greatest difference in coupling rates is
between pentafluorophenyl and *p*-nitrophenyl esters; the
former couples 80 to 400 times faster than the latter. The
relative rates of coupling for the above amino acids decrease
in this order: pentafluorophenyl > *N*-hydroxysuccinimidyl >
pentachlorophenyl > 2,4,5-trichlorophenyl > *p*-nitrophenyl.

The decreasing order of coupling rate constants for the
above active esters is not the same as the decreasing order
for the racemization rate constants; this indicates that
the "activity" of the ester is not strictly parallel with
the racemization. This is best shown by the ratio of coupling
to racemization rate constants which is also presented in
Table I.

These ratios also indicate the relative extent of racemization which can be expected during coupling by the α-
hydrogen abstraction mechanism. The larger this number
the smaller the amount of racemization to be expected during
coupling. The decreasing order of the ratios is the following: Ala > Trp > Glu > Phe > Asp > Ser > Cys; this indicates
that there is a relationship between these ratios and the
structure of the side chain of amino acids. The decreasing
order of the ratios is the reverse of that for the rates of
racemization. For alanine or tryptophane the ratios for
all esters are very large and therefore the choice of active

ester for coupling or for the preparation of sequential
polypeptides is not as critical as in the case of phenylalanine or especially in the case of cysteine.

For comparison the rate of racemization of N-benzyloxy-
carbonylglycyl-L-phenylalanine active esters were also
studied under the same conditions, and the results are
given in Table III.

Table III

The Racemization and Coupling Rate Constants for the Reaction
of N-Benzyloxycarbonylglycyl-L-phenylalanine Active Esters
with Triethylamine and L-Valine Methyl Ester Respectively

Compound	k_{rac} x 10^6 $l.mol^{-1} sec^{-1}$	k_c x 10^2 $l.mol^{-1} sec^{-1}$	k_c/k_{rac}
Z-Gly-Phe-R where R is:			
OPfp*	25270 ± 1500	(35.8)[†]	(14.25)
OPcp*	431 ± 32	2.12 ± 0.19*	50
ONp*	187 ± 24	0.31 ± 0.09*	16.3

*Average of two experiments.
[†]Estimated from 90% reaction time (about 1 min); the reaction
is too fast to measure more than one point by the infrared
technique employed.

It can be seen from Table IV that the rates of racemization of two N-benzyloxycarbonyl-S-benzyl-L-cysteine active
esters are comparable[2] to those of the dipeptide active
esters. This indicates that the racemization of N-benzyl-
oxycarbonylcysteine active ester derivatives which is known
to proceed through α-hydrogen abstraction may be as serious
as the racemization of peptide active esters which is known
to occur primarily through an oxazolone intermediate.[3,4]

Previously we presented experimental data which indicated that racemization of N-benzyloxycarbonyl-S-benzyl-L-
cysteine pentachlorophenyl ester and N-benzyloxycarbonyl-L-
phenylalanine pentachlorophenyl ester proceeds via isorace-
mization[5] in a non-polar solvent in the presence of triethylamine. In order to obtain more information concerning

Table IV

Racemization Rate Constants (k_{rac} x 10^6 $1.mol^{-1}$ sec^{-1}) for
N-Benzyloxycarbonylglycyl-L-phenylalanine Active Esters
and Several L-Amino Acid Active Esters

	Z-Gly-Phe-R	*Z-Cys-R*
		     Bzl
			Z-Ser-R	*Z-Asp-R*
				     OMe

R is:				
OPfp	25270	3300	297	244
OPcp	431	414	26.3	17.6
ONp	187	394	–	27

the reaction of triethylamine with amino acid active esters,
labeled phenol exchange studies were initiated as indicated
here:

$$Z-L-Phe-ONp \xrightarrow[\text{HO-} \bigcirc \text{-NO}_2 \ [2,6 \ ^{14}C]]{\text{7 NEt}_3 \ \text{in tetrahydrofuran}} Z-Phe-ONp[2,6 \ ^{14}C]$$

The kinetic data for the incorporation of labeled p-nitrophenol
are given in Table V. A tenfold increase in labeled phenol
concentration did not increase the fraction incorporated;
however, without added base there was negligible incorporation.
Based on these preliminary studies, the rate of incorporation
seems to be about 40 times faster than the rate of racemization under the same conditions. It may be concluded that
the racemization of this active ester through α-hydrogen
abstraction does not involve a ketene intermediate (E1cB
mechanism); this was proposed earlier as a possible mechanism
for the base catalyzed racemization of thiophenyl esters[6]
of *N*-protected amino acids.

It is our conclusion that the presently available synthetic procedures are not refined enough to prepare large
polypeptides in good yields and sufficiently pure form.
Therefore, we believe that more investigations are needed
to understand the kinetics of coupling reactions as well
as the kinetics of racemization during coupling. These
investigations should include a larger variety of activating
and protecting groups as well as solvents and temperature.

Table V

Kinetic Data for the Triethylamine Catalyzed Incorporation
of ^{14}C Labeled p-Nitrophenol into
N-Benzyloxycarbonyl-L-phenylalanine p-Nitrophenyl Ester*

Time, min	% ^{14}C labeled phenol incorporated
10	27
20	41
40	52
60	61
150	78
240	90

*Concentration of Z-Phe-ONp was 0.05 M; concentration of
labeled phenol was 2.85 x 10^{-4} M.

Presently the most widely used tests for racemization
involve coupling with glycine ethyl ester; however, glycine
couples much faster than the more hindered amino acids;
consequently, more extensive racemization can be expected
during coupling with any other amino acid.

References

1. Kovacs, J., G. L. Mayers, R. H. Johnson, R. Giannotti,
    H. Cortegiano, and J. Roberts. In *Progress in Peptide
    Research*, Vol. 2, Lande, S., ed. (New York: Gordon and
    Breach, 1972) pp 185-193.
2. Kovacs, J., G. L. Mayers, R. H. Johnson, R. E. Cover,
    and U. R. Ghatak. J. Org. Chem. 35, 1810 (1970).
3. Goodman, M., and L. Levine. J. Amer. Chem. Soc. 86,
    2918 (1964); McGahren, W. J., and M. Goodman,
    Tetrahedron, 23, 2017 (1967).
4. Kemp, D. S., and J. Rebek, Jr. J. Amer. Chem. Soc.
    92, 5792 (1970).
5. Kovacs, J., H. Cortegiano, R. E. Cover, G. L. Mayers.
    J. Amer. Chem. Soc. 93, 1541 (1971).
6. Liberek, B., and Z. Grzonka. Bull. Acad. Polon. Sci.
    Ser. Sci. Chim. 12, 367 (1964).

RACEMIZATION SUPPRESSION BY THE USE OF ETHYL 2-HYDROXIMINO-2-CYANOACETATE

Masumi Itoh. Research Laboratories, Fujisawa Pharmaceutical Co., Higashiyodogawa-ku, Osaka 532, Japan

A COMBINATION OF DICYCLOHEXYLCARBODIIMIDE (DCC) with N-hydroxysuccinimide (I) has been recommended as a racemization free coupling reagent for peptides.[1,2] However, it is also known that I is not stable and that the combination of DCC and I occasionally leads to some side reactions.[3,4] Later 1-hydroxybenzotriazole (II) and its derivatives have been proposed as superior additive than I.[5] The author intended to search for more promising additives and alcoholic components of active esters, and now wishes to report the use of a strongly acidic oxime, ethyl 2-hydroximino-2-cyanoacetate (III), as a suitable additive for the same purpose.

(I)   (II)   (III)

III is a stable, acidic (pKa' 4.6) compound and is used in organic syntheses. Racemization with or without III was examined first by Anderson's test.[6] The coupling reaction of N-benzyloxycarbonylglycyl-L-phenylalanine with ethyl glycinate gave no racemate by the use of DCC and III in

tetrahydrofuran at room temperature, while the control
experiment, without III, gave 8% of racemate. A comparison
of additives, I, II, and III, was made by Bodanszky's test.[7]
Coupling of *N*-acetyl-L-isoleucine and ethyl glycinate with
DCC was carried out in dimethylformamide (DMF), in which
significant racemization occurs. A typical procedure is
as follows: ethyl glycinate hydrochloride (140 mg, 1.0
mmol) was dissolved in dry DMF (3.0 ml) and was neutralized
with triethylamine (0.14 ml, 1.0 mmol) under ice-cooling.
Then III (170 mg, 1.2 mmol), acetyl-L-isoleucine (173 mg,
1.0 mmol) and a solution of DCC (206 mg, 1.0 mmol) in DMF
(2.0 ml) were added, in that order, into the above solution
at 5°C. The mixture was allowed to react for 3 hr at 5°C
and overnight at room temperature. After the complete
evaporation of DMF the residue was dissolved in EtOAc and
water. The EtOAc layer was washed with 5% NaHCO$_3$ solution,
water, *N* HCl and water, and was dried over MgSO$_4$. Evaporation gave crude product, which was filtered with small
amount of ether-petroleum ether mixture; 178 mg (69%).

For the calculation of racemization the crude product,
ethyl *N*-acetylisoleucyl-glycinate, was hydrolyzed by 6*N*
HCl at 110°C and subjected to amino acid analysis to detect
D-alloisoleucine. About 0.7% of racemization occurred
during hydrolysis of acetyl-L-isoleucine under the same
conditions.

The results obtained are summarized in Table I, and
show that I and III suppress racemization potently.

Table I

	Ac-L-Ile-OH (mmol)	H-Gly-OEt.HCl (mmol)	Et$_3$N (mmol)	Additive (mmol)	Racemization (%)*
1	1.0	1.0	1.0	none	35
2	1.0	1.0	1.0	(I) 1.2	2.7
3	1.0	1.0	1.0	(II) 1.2	8.8
4	1.0	1.0	1.0	(III) 1.2	1.8

*Racemization (%) = $\dfrac{\text{Alloisoleucine} \times 100}{\text{Isoleucine} + \text{Alloisoleucine}}$

Unexpectedly, II is less effective than I. Although this
is incompatible with previously reported results by König
and Geiger, a comparison is difficult because the racemate
detection systems are completely different.

Apart from such a problem the combination of DCC with III seems to be a promising approach in coupling reactions.

Acknowledgment

The author expresses his deep gratitude to Dr. Miklos Bodanszky for his kind and helpful advice.

References

1. Weygand, F., D. Hoffmann, and E. Wünsch. Z. Naturforsch. <u>21b</u>, 426 (1966).
2. Wünsch, E., and F. Dress. Chem. Ber. <u>99</u>, 110 (1966).
3. Gross, H., and L. Bilk. Tetrahedron <u>24</u>, 6935 (1968).
4. Weygand, F., W. Steglich, and N. Chytil. Z. Naturforsch. <u>23b</u>, 1391 (1968).
5. König, W., and R. Geiger. Chem. Ber. <u>103</u>, 788, 2024, 2034 (1970).
6. Anderson, G. W., and F. M. Callahan. J. Amer. Chem. Soc. <u>80</u>, 2902 (1958).
7. Bodanszky, M., and L. E. Conklin. Chem. Commun. 773 (1967).

RACEMIZATION OF *N*-METHYLAMINO ACID RESIDUES DURING PEPTIDE
SYNTHESIS

John R. McDermott, N. Leo Benoiton. Department of
Biochemistry, University of Ottawa, Ottawa KlN 6N5
Canada

FURTHER TO OUR WORK ON the synthesis of derivatives of
N-methylamino acids,[1,2] we have initiated a study on the
incorporation of *N*-methylamino acids into peptides. No
systematic data are available in this area, and in particular,
the supposed resistance of *N*-methylamino acid derivatives
to racemization[3] has never been investigated quantitatively.
We present results here which show that *N*-methylamino acid
derivatives racemize as readily, and in some cases more
readily, than the corresponding amino acid derivatives under
conditions of peptide coupling and deprotection.

In preliminary experiments, it was found that Bz-MeLeu
lost 90% of its optical activity after hydrolysis of the
mixed anhydride formed during a 5-minute reaction with
Bu^{i}OCOCl. Under the same conditions Z-MeIle and Boc-Ala-Pro
were not racemized, whereas Z-Ala-MeLeu gave 27% of the
L-D isomer. The extent of racemization in these three and
subsequent experiments was determined by analysis of the
diastereomeric products,[4,5] after suitable deprotection,
with an amino acid analyzer. The several systems standardized and used in this study are described in Table I.

In the synthesis of standards, it was found that two
widely used and relatively racemization-free reactions
caused considerable racemization of *N*-methylamino acid
derivatives. Saponification of Z-Ala-MeLeu-OMe gave 11%,
and acidolysis of Z-Ala-MeLeu using 5.6 *N* HBr in acetic
acid gave 17% of the L-D peptide. In addition, it was
found that saponification of Z-MeIle-OMe gave 12% of the

Table I

Chromatographic Data for Analysis of Diastereomers*

Compound	Elution time (min)	Constant
Ala–MeLeu	43.5	1.1
Ala–D–MeLeu	34.5	1.0
Ala–MeLeu–Gly	35	5.7
Ala–D–MeLeu–Gly	41	5.2
Ala–Leu–Gly	33	24.3
Ala–D–Leu–Gly	38	21.9
Ala–Pro[†]	158	7
Ala–D–Pro[†]	142	9
MeIle[#]	63	5.2
D–*a*MeIle[#]	57	4.5

*Beckman amino acid analyzer. Aminex A–5 (15 cm) resin, eluted with 0.1 N sodium citrate, pH 4.25, at 68 ml/hr.

[†]AA–15 (50 cm) resin, eluted with pH 3.28 buffer (85 min), followed by pH 4.25 buffer.

[#]Elution at 34 ml/hr.

allo isomer, and that a prolonged treatment of Z–MeIle with HBr in acetic acid gave 34% *a*MeIle. Under the same conditions, the corresponding derivatives of Ala–Leu and Ile gave less than 1% of the diastereomers. Pure L–Ala–L–MeLeu was finally obtained from both Boc–Ala–MeLeu–OBzl and Z–Ala–MeLeu–OBu^t using trifluoroacetic acid and hydrogenation for deprotection.

The extents of racemization obtained with various coupling methods for the condensation of Z– or Boc–Ala–MeLeu with Gly–OBzl are recorded in Table II, along with similar data for couplings using Z–Ala–Leu. In each case, the neutral product was isolated and analyzed after suitable deprotection. Essentially optically pure products were obtained only with the N–hydroxysuccinimide ester and with dicyclohexylcarbodiimide–N–hydroxysuccinimide (DCCI–HONSu)

Table II

Extent of Racemization During Couplings with
Glycine Benzyl Ester*

Coupling method	Z–Ala–Leu	Z–Ala–MeLeu		Boc–Ala–MeLeu[†]
	TosOH·Et$_3$N	TosOH·Et$_3$N	HCl·Et$_3$N	–
DCCI–HONSu	0.4	2.8	11	<0.1
HONSu ester			<0.1	
EEDQ	0.5	15	7.7	0.5
Bu^{i}OCOCl–Et$_3$N[#]	2.0	7.0	8.2	6.4
DCCI	16	15	27	15
Woodward's K[√]		39		
DCCI–s–triazole		12		

*Percent of L–D peptide formed. Couplings carried out in tetrahydrofuran in the presence of the designated salt.

[†]Crystalline compound.

[#]–10°C, 90–sec activation time.

[√]N–Ethyl–5–phenylisoxazolium–3'–sulfonate in CH$_3$CN.

and *N*-ethoxycarbonyl-2-ethoxy-1,2-dihydroquinoline (EEDQ) *in the absence of salt.* The pronounced salt effect was not observed for the normal peptide coupling.

Oxazolonium salts have been proposed as intermediates in the racemization of *N*-methylamino acid derivatives,[6] and oxazolium–5–oxides (the conjugate base) have been prepared by heating *N*-substituted *N*-acylamino acids in acetic anhydride.[7] We have shown that such compounds can be formed under peptide coupling conditions (Z–Ala–MeLeu and DCCI in tetrahydrofuran for 10 min) by trapping the oxazolonium derivative in a 1,3-dipolar cycloaddition reaction[7,8] with methyl propiolate. The expected pyrrole, *N*-methyl,2-(1'-benzyloxycarbonylaminoethyl),3-methoxycarbonyl,5-isobutylpyrrole was obtained in 85% yield.

Promotion of oxazolone formation by chloride ion has been attributed to the basicity of the anion[9] and to the increased ionic strength of the solution.[10] Base catalysis is unnecessary for cyclization to an oxazolonium cation, however, an increase in ionic strength of the medium could enhance formation of this derivative.

Racemization during couplings might also proceed by direct α-proton abstraction. *N*-Substituted *N*-methylamino acids are more prone to racemize by this mechanism than amino acid derivatives, as was shown by the racemization of Z-Ala-MeLeu-OMe and Z-MeIle-OMe during saponification. In *N*-monosubstituted amino acid derivatives, the –*N*-H group is generally more acidic than the α-*C*-H and so will ionize first, thus protecting the α-*C*-H from ionization. No such effect obtains for *N*-substituted *N*-methylamino acids.* Activation of the *N*-protected dipeptide with a strongly electron-withdrawing group will further increase the lability of the α-proton, and the basic chloride or tosylate anion may then be able to cause ionization.

References

1. Coggins, J. R., and N. L. Benoiton. Can. J. Chem. 49, 1968 (1971).
2. Benoiton, N. L., and J. R. Coggins. In *Progress in Peptide Research*, Vol. II, Lande, S., ed. (New York: Gordon and Breach, 1972) pp 145-150.
3. Bodanszky, M., and M. A. Ondetti. *Peptide Synthesis* (New York: Interscience, 1966) p 139.
4. Manning, J. M., and S. Moore. J. Biol. Chem. 243, 5591 (1968).
5. Muraoka, M., and N. Izumiya. J. Amer. Chem. Soc. 91, 2391 (1969).
6. Cornforth, J. W., and D. F. Elliott. Science, 112, 534 (1950).
7. Huisgen, R., H. Gotthardt, H. O. Bayer, and F. C. Schaefer. Angew. Chem. (Int. ed.), 3, 136 (1964).
8. Huisgen, R. *Aromaticity*, Special Publication No. 21 (London: The Chemical Society, 1967) pp 51-73.
9. Williams, M. W., and G. T. Young. J. Chem. Soc. 3701 (1964).
10. Goodman, M., and W. J. McGahren. Tetrahedron, 23, 2031 (1967).

---

*It is significant that MeIle-OMe racemized very little on saponification.

SYMPOSIUM DISCUSSIONS

Summarized by Johannes Meienhofer

SEVERAL COMMENTS AND QUESTIONS concerned the newly developed
amine protecting groups (pp 315 to 323). The reductive
cleavage of the isonicotinyloxycarbonyl group by Zn in
acetic acid is easily achieved at room temperature, a system
suitable for work with proteins. Tryptophan residues in
S-protein seem not to be effected under these conditions,
but, strangely, some oxidation was observed when the reaction
mixture became aerated from too vigorous stirring. Mild
oxidation by *m*-chloroperbenzoic acid in trifluoroacetic
acid can be used to cleave the *N*-protecting 4,5-diphenyl-
4-oxazolin-2-one (Ox) group (pp 321 to 323) thus providing
a new selectivity. However, this reaction will not be
applicable to sulfur-containing or other oxidizable peptides and, of course, proline or other imino acids cannot
be protected by the "Ox" group.

A synthesis of luteinizing hormone-releasing hormone
in which fragments were made from considerations of hydrophilicity *vs*. hydrophobicity (pp 325 to 329) was complimented
by a discussant for the high yields achieved. Apparently,
thin layer electrophoresis proved to be much superior to
thin layer chromatography in assessing the homogeneity of
intermediates in this synthesis. The biological assays
were done by inducing ovulation in rats, rabbits, and
hamsters.

The proposal of a complex between amine component and
hydroxybenzotriazole ester (pp 343 to 350) in hydroxybenzo-
triazole-catalyzed active ester condensations appeared to
be suggestive for *o*-nitrophenyl esters[1] as well. However,
anchimeric assistance might also be an explanation for the
superior properties of these esters (as remarked by Miklos
Bodanszky). The *para*-nitrophenyl esters show (i) sometimes

slow reaction rates, (ii) incomplete reaction when
sterically hindered, and (iii) a high solvent dependence
(factor of 10), while the *ortho*-nitrophenyl esters show
several times higher reaction rates, complete reaction
even in some hindered conditions, and a much lower solvent
dependence (factor of 2). These advantages should become
especially apparent when ONo esters will be applied to
solid-phase synthesis.

It was commented that pentafluorophenyl esters (used
for a synthesis of human ACTH possessing the corrected
structure,[2] pp 299 to 303) react very fast (20 min including work-up) and show the least racemization in
chloroform and methylene chloride, followed by ethyl
acetate and dioxane. In other solvents the racemization
tendency is rather higher. | One should, of course, never
use excess thiethylamine or any other *tert*-amine in active
ester couplings, and the conditions used to study racemization (pp 359 to 364) are not recommended for a synthesis.

References

1. See: Bodanszky, M., and R. Bath. Chem. Commun. <u>1969</u>,
    1259.
2. Gráf, L., S. Bajusz, A. Patthy, E. Barat, and G. Cseh.
    Acta Biochim. Biophys. Acad. Sci. Hung. <u>6</u>, 415 (1971);
    Riniker, B., P. Sieber, W. Rittel, and H. Zuber.
    Nature New Biology <u>235</u>, 114 (1972).

SECTION VI

BIOLOGICALLY ACTIVE PEPTIDES

Session Chairmen

Helmut Zahn and Choh Hao Li

PROPERTIES OF ANTAMANIDE AND SOME OF ITS ANALOGUES

Theodor Wieland. Max-Planck-Institut für medizinische
Forschung, Abteilung Chemie, Heidelberg, Germany

SUMMARY--Antamanide (AA), an antitoxic component was discovered in the lipophilic fractions of an extract of the
green toadstool Amanita phalloides. It counteracts the
toxicity of phallotoxins by preventing their accumulation
in liver cells. The structure elucidation by mass spectrometry of peptide esters after partial methanolysis
revealed a cyclodecapeptide *cyclo*(-Val-Pro-Pro-Ala-Phe-Phe-
Pro-Pro-Phe-Phe) (I). Numerous analogues were synthesized
by the solid-phase method and checked for antitoxic activity.
A lipophilic side chain in position 1 is prerequisite of
antitoxic activity. Replacing phenylalanine by tyrosine
residues likewise leads to biologically active analogues
The perhydrogenated I (HAA) has no protecting effect.
    Antamanide and its biologically active analogues show
in apolar solvents and in absence of Na^+-ions well marked
negative dichroic absorption in the 220 nm region, which
is shifted to positive values on adding polar solvents,
particularly water, or Na^+-ions. The molecular change
giving rise to this phenomenon manifests itself also in a
blue shift of the UV absorption.
    From spectroscopic titration (and ultrasonic absorption)
experiments it is suggested that the molecule exists in two
conformers. These are in equilibrium with each other, the
ratio depending on the nature of the solvent, and the velocity
of interconversion seems to be extremely high. The "polar"
conformation (VIIb), which differs from the "apolar" VIIa
by the number of intramolecular hydrogen bridges (4 to 6)
forms of complexes with metals of radii of about 1 Å, thus
preferring Na^+ over K^+. Na^+-Complex stability constants

are in correlation with antitoxic activity of I and its analogues. However, since HAA also forms a strong Na^+-complex, biological activity must be based on additional features of the molecule.

THE DEADLY POISONEOUS GREEN MUSHROOM Amanita phalloides contains toxic bicyclic peptides--the families of phallo-toxins and amatoxins[1]--and, in addition, numerous lipophilic cyclopeptides. These can be extracted with ethyl acetate from aqueous solutions of strongly enriched material. Chromatographic fractionation of an analogous mixture ob-tained earlier in a different way[2] yielded a substance, which prevented the absolute lethal action of 5 mg of phalloidine per kg (LD_{50}=2.0 mg/kg) at the white mouse when given in a sufficiently high dose, simultaneously or a little earlier than the poison. Finally, this antitoxic substance could be obtained in crystalline state and showed its protecting power against 5 mg of phalloidine already at a dose of 0.5 - 1.0 mg/kg. A dose of 5 mg/kg of the anti-dote is sufficient for protection against 25 mg/kg of phalloidine. The substance has been called antamanide (AA).[3]

Biological Action

Using radioactively labelled phallotoxines and antamanide we got some insight into the mechanism of action of the antidote.[4] We found that death occurs in mice if the con-centration of toxin in the liver exceeds *ca.* 30 µg/g. When antamanide is given to the animals 2 min prior to the toxin the amounts of toxin appearing in the liver are greatly reduced, Table I. The uptake of the toxin by the organ can

Table I

Increase with Time of Concentration of ^{3}H-Desmethylphalloin (µg/g) in Livers of Mice Poisoned with 2 resp. 5 mg/kg of toxin in the Absence or Presence of 1 resp. 2 mg/kg of Antamanide Given 2 min Prior to the Toxin[4]

Min after application	2 mg/kg		5 mg/kg		Phallotoxin Antamanide mg/kg
	with-out	with 1 mg	with-out	with 2 mg	
5	18	4	50	6	
10	21	5	56	14	
30	25	10	52	20	

also be shown in an isolated rat liver preparation. Already
10 minutes after addition of the labelled toxin more than
90% have been absorbed by liver cells. The toxin is not
metabolized; after homogenization nearly the whole radioactivity can be extracted with methanol, and identified
with the unaltered compound by thin layer chromatography.[5]
As a consequence of phallotoxin poisoning K^+-ions begin to
flow out of the cell together with some enzymes, *e.g.*
β-glucuronidase. The rate of absorption of the toxin as
well as the leakage of membrane is strongly diminished by
antamanide, as shown in Figure 1. The phenomenon points

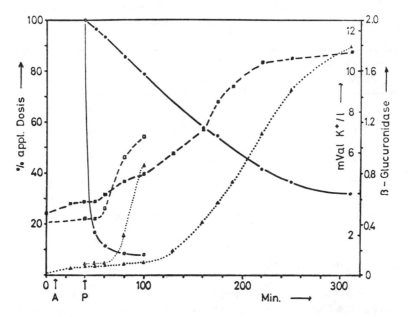

Figure 1: Inhibition of absorption of ^{3}H-desmethylphalloin
   (P, -o-), K^+ release (-□-), and efflux of β-glucuronidase
   (-Δ-) in a perfused rat liver preparation by antamanide
   (AA). 5 mg AA were added to 100 ml perfusion medium
   30 min before 250 μg (55 μCi) ^{3}H-desmethylphalloin. Open
   symbols without, filled symbols with antamanide.[4]

to a competition for one and the same receptor site of
antamanide and phallotoxin. Experiments with a [14]C-
containing derivative of antamanide, however, disproved
such an assumption. Mice were given a high dose of

phalloidin and shortly afterwards a certain amount of
labelled antamanide. Their livers did not contain less of
the label than livers of control animals, which had not
received the poison (Table II). This proves that antamanide
is not bound to the receptor sites competent for the toxin.
Therefore, its action must be sought in a rather specific
membrane tightening effect.

Table II

Amount of ^{14}C-Antamanide (µg/g) in Livers of White Mice
After *i.v.* Application of 1 mg/kg of the Antitoxin
Without, and After Pretreatment With
5 mg/kg of Phalloidine[4]

Min after application	*without*	*5 min after Phalloidine*
	µg/g	µg/g
5	3,6	3,6
10	3,4	4,1
30	3,1	2,8
60	2,5	2,4

Structure Elucidation

AA is a cyclic decapeptide containing the amino acids
L-alanine, L-valine, L-phenylalanine and L-proline in a
ratio of 1:1:4:4.[3] The structure elucidation was carried
out mainly by partial methanolysis, gas chromatographic
separation of the trifluoroacetylated peptide methyl
esters and their mass spectrometric analysis[6] and resulted
in the structural formula I.

$$
\begin{array}{ccccc}
8 & 9 & 10 & 1 & 2 \\
\multicolumn{5}{c}{\rightarrow \text{Pro–Phe–Phe–Val–Pro} \rceil} \\
\multicolumn{5}{c}{\llcorner \text{Pro–Phe–Phe–Ala–Pro} \leftarrow} \\
7 & 6 & 5 & 4 & 3 \\
\end{array} \qquad \text{I}
$$

Of some interest for peptide chemists is the observation
of "wrong" amino acid sequences among the fragments of

partial methanolysis as, *e.g.* Tfa-Phe-Phe-Phe-OMe although
in I only two phenylalanine residues are in sequence. The
presence of triphenylalanine is probably the result of an
intermediate cyclol formation in a transannular reaction.
Interaction of the NH hydrogen of Phe[9] with the carbonyl
of Phe[6] forms an orthoamide structure followed by trans-
peptidation to generate the sequence of 3 phenylalanines,
Scheme I.

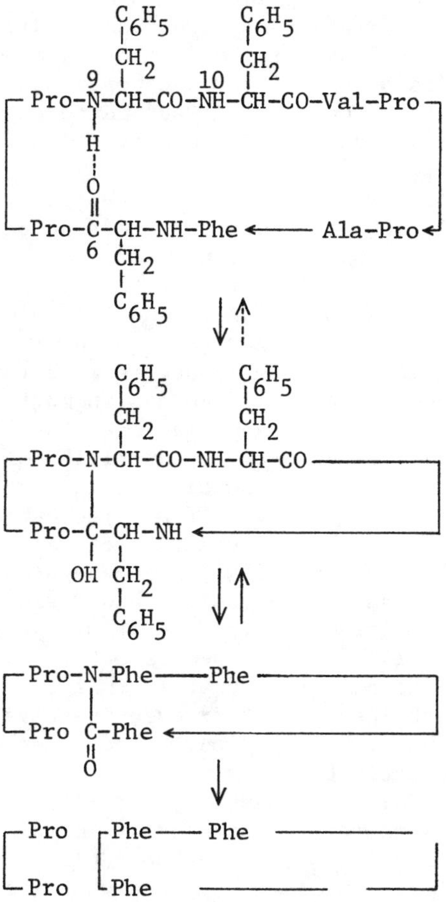

Scheme I

Syntheses of Antamanide and
Some of its Analogues

The synthesis of I can be carried out by cyclization
of ten different linear decapeptides. We have chosen
mostly the sequence 6,7--→4,5 and prepared the decapeptide
Phe-Pro-Pro-Phe-Phe-Val-Pro-Pro-Ala-Phe at first by classical methods,[7] and then[8] by the solid-phase technique of
Merrifield. The decapeptide 5 → 4 has been synthesized
by others.[9] Cyclization with the use of dicyclohexylcarbodiimide and *N*-hydroxysuccinimide gave 30–40% of I.
Gram quantities of the linear decapeptides can now be
prepared within 30 hours in an automated peptide synthesizer
(Schwarz BioResearch) using the conventional or a novel
reactor system.[10]

The relatively easy access to the cyclic decapeptide
made it possible to synthesize a great number of analogues
of antamanide to investigate the structural details necessary for the biological action of the molecule. At first
the single amino acids 1-valine and 4-alanine have been
substituted by several other amino acids[11] including the
unusual constituent L-α-aminobutyric acid (Abu).[12] The
results of these investigations are summarized in Table
III and in Figure 2. It appeared that the lipophilic
nature of the side chain of residue 1 is most essential.
Valine can be replaced by leucine or isoleucine without
loss of antitoxic activity. If a three-carbon chain is
present in position 1, the structure of residue 4 is not
very critical since the [Gly⁴] analogue still possesses
appreciable activity. The amino acid residue in position
4 can even be omitted and the resulting cyclic nonapeptide
(XXVI in Table IV) still possesses activity. However,
reduction of the number of C atoms in the side chain of
Val¹ gives much less potent or even inactive analogues.
Benzyl side chains in positions 1 and 4 apparently abolish
the antitoxic activity.

The [Ala¹,Val⁴] analogue, L-*retro*-AA (No. XII in Table
III) could not be tested, because it immediately precipitated from its solution in DMSO on adding the required
amount of water. The same happened with its mirror
antipode, D-*retro*-AA, which was synthesized in my laboratory
by B. Penke.[13] The difficulty was overcome by introduction
of a solubilizing group as will be mentioned later. D-Ant-
amanide, the enantiomer consisting of the D-amino acids
in the correct sequence has been synthesized in Yu. A.
Ovchinnikov's laboratory and, independently, in ours. It
has about 10% of the antitoxic activity of AA.

Table III

Protecting Doses Against 5 mg/kg of Phalloidine for
the White Mouse of Some 1,4-Variants of Antamanide

No.	Amino acid in position 1	4	Protecting dose (mg/kg)	
I	Val	Ala	0,5	(Antamanide)
II	Val	Abu	1	
III	Val	Val	1	
IV	Val	Gly	2,5	
V	Leu	Ala	0,5	
VI	Ile	Ala	0,5	
VII	Abu	Ala	2,5	
VIII	Abu	Abu	15	
IX	Ala	Ala	10	
X	Ala	Gly	15	
XI	Gly	Ala	10	
XII	Ala	Val	–	insoluble L-*retro*-AA
XIII	Gly	Gly	>20	
XIV	Val	Phe	>20	
XV	Phe	Phe	>20	
XVI	Val	Phe(Val^6)	1,5	C_2 Symmetry

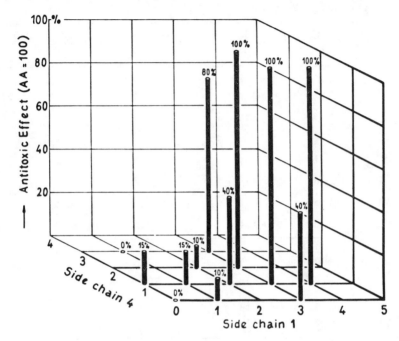

Antitoxic effects of derivatives of antamanide
varied at 1- and 4-side chains

Figure 2: Influence of side chains length of amino acids
in positions 1 and 4 of antamanide on antitoxic effectivity

The almost symmetrical molecule of AA offers several
possibilities for conversion into entirely symmetrical
analogues. By replacing Ala[4] by Val, or Val[1] by Ala, two
molecules have been synthesized (III and IX, respectively,
in Table III) which happen to be identical with their
respective *retro*-forms. For obtaining C_2-symmetry in each
of them only an exchange of amino acids 4 for 6 was necessary. Since the [Val[4]] analogue (III) is about 10 times
more active than IX, we synthesized analogue XVI and found
that it had relatively high protecting potency, at a dose
of only 1.5 – 2 mg/kg.[14]

Pro-Phe-Phe-Val-Pro
Pro-Phe-Phe-Val-Pro

Analogue III

Pro-Phe-Phe-Val-Pro
Pro-Val-Phe-Phe-Pro

C_2 symmetric XVI

Perhydro-AA (HAA) was obtained by hydrogenation of AA over Pt-catalyst in glacial acetic acid in Ovchinnikov's group and, simultaneously, in our laboratory.[15] The compound which contains four residues of cyclohexylalanine instead of phenylalanine has an extremely low solubility in solvents containing water and, perhaps as a consequence, exhibits no antitoxic activity.

Very active analogues of AA are obtained by replacing one of the four phenylalanines by tyrosine. The [Tyr^6] analogue XVIII, Table IV, has been synthesized by Ch. Rietzel[16] *via* classical methods of peptide chemistry using Boc-Tyr(OBzl). The other tyrosine-containing analogues were prepared by Ch. Birr using the solid-phase technique.[17]

Table IV

Tyrosine-Containing Analogues of Antamanide (I), of Some Ile and Gly Containing Analogues of the des-Ala^4-Homologue

No.		*Amino Acids in Position* 1	2	3	4	5	6	7	8	9	10		*Protecting Dose*
I	*cyclo*(	-Val	-Pro	-Pro	-Ala	-Phe	-Phe	-Pro	-Pro	-Phe	-Phe-	)	0,5
XVII	*cyclo*(					Tyr						)	0,5
XVIII	*cyclo*(						Tyr					)	0,5
XIX	*cyclo*(									Tyr		)	2
XX	*cyclo*(										Tyr	)	5
XXI	*cyclo*(					Tyr-	Tyr			Tyr-	Tyr	)	10
XXII	*cyclo*(	-Ile					Tyr					)	1
XXIII	*cyclo*(	-Ile				Tyr					Tyr	)	5
XXIV	*cyclo*(	-Gly			Gly-	Tyr						)	>20
XXV	*cyclo*(	-Gly			Gly		Tyr					)	>20
XXVI	*cyclo*(				des					Tyr		)	5-10

The [Tyr⁵] and [Tyr⁶] compounds, XVII and XVIII, as
well as the [Ile¹,Tyr⁵] analogue XXII have equal protecting
effectiveness as AA (I).   [Tyr⁹]-AA and [Tyr¹⁰]-AA (XIX and
XX) are markedly less active; and activity is also reduced
by introduction of more than one Tyr (XXIII).   Analogue
XXI which contains four tyrosine residues is not protective
even with doses up to 10 mg/kg.[18]   The tyrosine-containing
[Gly¹,Gly⁴] analogues, XXIV and XXV, exhibit (as XIII) no
antitoxic effect.

The phenolic OH groups are in different micro environ-
ments in each of the four mono-Tyr analogues which show
different spectral behaviour in UV in alkaline methanolic
solution, Figure 3.   The phenolate curves of [Tyr⁵]- and

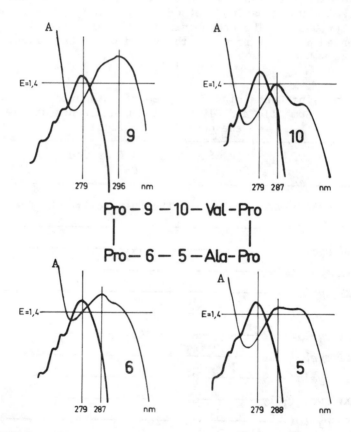

Figure 3: UV absorption spectra in neutral and alkaline
  (A) methanolic solution of the four different Tyr analogues
  of antamanide

[Tyr[6]]-AA have still some similarity, but differ markedly from [Tyr[9]]- and [Tyr[10]]-AA which again are different from each other in shape, maxima, and extinction.

Using the phenolic hydroxyls as a handle it was possible to introduce groups into the molecule, which lend good water-solubility to the otherwise very slightly soluble compounds.[19] The monoesters of sulfuric or phosphoric acid of XVII and XVIII or their O-carboxymethyl- or O-3-sulfo-propyl derivatives form water soluble neutral alcali metal salts, which have antitoxic effects comparable to the mother compounds. Strikingly, only in the sulfopropyl ether of XVII the activity is reduced by 90%.

It was this solubilizing effect, which enabled us to test also a D-*retro* species of AA.[13] [D-Tyr[6]]-all-D-*retro* AA which is nearly insoluble in the biological testing system was solubilized as Na[+] salt of its sulfuric acid ester, which was antitoxic at a protecting dose of 2 mg/kg. Assuming that it is also the poor solubility of perhydro-AA (HAA), which causes its ineffectivity as an antitoxin, the carboxymethylated [Tyr[6]]-AA was hydrogenated over Pt to obtain a water soluble derivative. The desired product, [cyclohexylalanyl[5,9,10],4-carboxymethoxycyclohexylalanyl[6]]-AA could indeed be isolated.[20] Its NH$_4$[+]-salt was readily soluble in water, but had also no antitoxic effectivity up to 10 mg/kg.

[O-Carboxymethyl-Tyr[6]]-AA   Product of perhydrogenation (Cha = cyclohexylalanyl)

Metal Complexes of Antamanide and Analogues

AA interacts with Na and K ions. This was shown in Shemyakins and in our laboratories by mass spectrometry (occurrence of an ion AA·Na[+]), IR spectroscopy (increase of carbonyl absorption at 1630 cm[-1] in the presence of Na[+]), potential measurements with ion specific glass electrodes, vapor pressure osmometry, ORD-spectrometry and decrease of electrical conductivity of ethanolic NaCl solutions in the presence of AA.[21] We have now extended these investigations by comparing the reaction of several metal ions with AA and some of its structural analogues.[22] We also used the solvent extraction method of Pedersen,[23] in which the amount of metal picrates extracted from aqueous solutions by AA dissolved in CH_2Cl_2 corresponds to complex formation. For alcali ions and alcaline earth ions we found the values given in Table V. Evidently, ions of *ca.* 1 Å radius are

Table V

Alcali and Alcaline Earth Picrates Extracted from
Aqueous Solution by Antamanide in
Methylene Chloride

Metal	$r[Å]$	% picrate extracted based on AA	Metal	$r[Å]$	% picrate extracted based on picric acid
Li	0.6	0.8	Mg	0.65	0.0
Na	0.97	14.0	Ca	0.99	9.7
K	1.33	0.7	Sr	1.13	1.2
Rb	1.48	0.3	Ba	1.35	0.7
Cs	1.67	0.2			
NH_4		0.5			

much preferred by AA. Analogues devoid of antitoxic activity extracted little or none of the Na picrate. More quantitative spectroscopic assays on complexation capacity with several cations were also performed. The ORD spectrum of AA in 96% ethanol has two rather large negative Cotton effects between 200 nm and 220 nm, whose magnitude and shape depend on the nature of the solvent and the presence of cations.[21] This pertains also to the CD spectra where AA exhibits a very strong negative dichroic absorption at 224 nm in dioxane and in other non-polar solvents, whereas a positive band is observed at about the same wave length region in the presence of water or of Na ions.[24] Methanol also causes a positive shift, but not to the same extent as water (Figure 4). Thus the positive shift of ellipticity at 224 nm produced by several metal ions could be utilized as a measure of complexing activity. In Figure 5, the

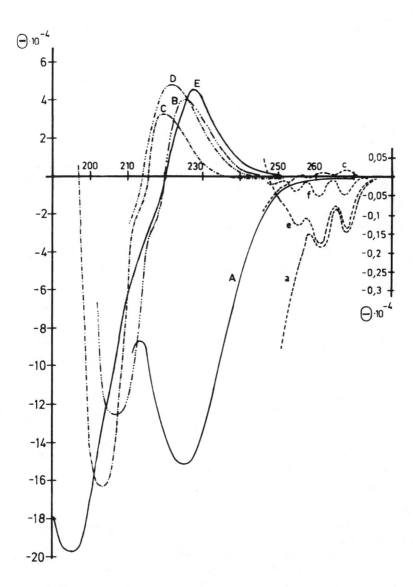

Figure 4: CD spectra of antamanide in various solvents, and in the presence of Na+ ions. a, A, in dioxane; e, E, in methanol-water (1:1). B, Antamanide and Na+ (1:15) in dioxane; c, C, in methanol, D in acetonitril. f, AA·NA+ (1:1) in acetonitrile.

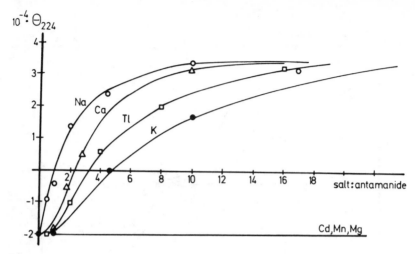

Figure 5: Positive shift of molar ellipticity of antamanide
(Θ) at 224 nm ($4,35 \cdot 10^{-3}M$ in methanol) caused by increasing
amounts of different metal ions.

positive shift of molar ellipticity of AA is depicted in
response to increasing concentrations of several cations.
Here again Na and Ca formed the most stable complexes.
Half complexation in methanol was found in $3.4 \cdot 10^{-3}M$ AA
solution at a molar ration of Na/AA = 1.5:1, for Ca at 3:1,
for K at 7:1, and surprisingly also for Tl at 4.5:1. No
spectral changes were caused by Cd (r, 0.97 Å), Mn(II)
(r, 0.80 Å) and Mg (r, 0.66 Å).

Complexation of the AA molecule manifests itself also
in UV spectra, because the interaction of the complexing
cation with the n electron of the peptide carbonyl oxygen
shifts the $n - \pi^*$ transitions to shorter wave lengths. This
effect can be seen also in the 250 nm region of the spectrum
where the absorption decreases with increasing ion concentration. The complex stability constants of AA with Na^+,
K^+ and Ca^+ were determined in solvents of different polarity
by this spectrophotometric titration method and compared
with analogous values obtained by vapor pressure osmometry
and ion-selective glass electrodes. Some of the values
obtained are summarized in Table VI. High values of K
(strong complexing), are found in the more lipophilic solvents like acetonitril or ethanol. In methanol- or water-
containing acetonitrile or ethanol the values are lower by
1-2 orders of magnitude. Solvents with high affinity to
the cations and to the complexing carbonyl oxygen atoms,

Table VI

Stability Constants (K) of AA Complexes in
Different Solvent Systems

	CH_3CN	$CH_3CN \cdot H_2O$ (96:4)	$CH_3CN \cdot H_2O$ (92:8)	$C_2H_5OH \cdot H_2O$ (96:4)	$C_2H_5OH \cdot H_2O$ (30:70)	CH_3OH
Na	$3 \cdot 10^4$	$2,6 \cdot 10^3$	$1,2 \cdot 10^3$	$2,0 \cdot 10^3$	0	$5,0 \cdot 10^2$
K	$2,9 \cdot 10^2$	$2 \cdot 10^1$	$2,8 \cdot 10^2$	$1,8 \cdot 10^2$	-	10^1
Ca	$1,0 \cdot 10^5$	-	-	-	-	$3 \cdot 10^1$

especially water and low alcohols, diminish complex stabilities; thus in 70% water-containing ethanol no complexation at all was observed. Indeed, it is in lipophilic environment, that the most stable Na-complex is formed. The crystalline perchlorate of $AA \cdot Na^+$ could be obtained by adding $NaClO_4$ to a 0,1% solution of AA (I) in methanol.[22]

The selectivity of AA for Na^+ over K^+, which is essentially the consequence of a steric adaption of the cyclopeptide to a certain ion size is maintained throughout the different solvents, but it changes quantitatively due to the different solvation energies. One finds values of *ca.* 100 for the ratio K_{Na}/K_K in acetonitrile, but decreased selectivity in more polar solvents (*e.g.* 10 in 4% H_2O-containing ethanol).

As mentioned above, addition of water (or methanol) to AA in non-polar solvents caused changes in CD spectra that are quite similar to those caused by Na^+-complexation. This similarity also applies to UV spectra, in which not only complex forming ions (see above), but also strong polar agents, particularly water, bring about a blue shift of the carbonyl adsorption, presumably by formation of H-bonds. This manifests itself in a decrease of absorbancy in the region of the phenyl absorption, which could be used for a spectrometric titration of 248 nm of AA with water. In Figure 6 the extinction of 248 nm of AA in 1,4-dioxane or acetonitril is plotted against increasing water concentration and a parallel (positive) change of molar ellipticity at 224 nm. The sigmoidal shape of the curves points to the transformation of one conformer more stable in lipophilic solvents to a second one which is stable in presence of water. The situation reminds us of the reaction

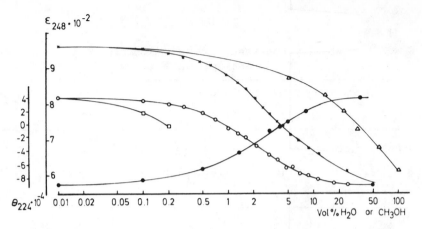

Figure 6: Decrease of absorption at 248 nm and increase
in molar ellipticity (Θ) at 224 nm of antamanide ($10^{-3}M$)
in different solvents on addition of water or methanol;
Δ methanol in 1,4-dioxane, x water in 1,4-dioxane, •
water in acetonitrile (• = Θ), □ water in CH_2Cl_2.

of AA with Na$^+$-ions. A conformation equilibrium of only
two species dependent on polarity has already been suggested
from the existence of an isosbestic point in the ORD spectra
by Ivanov *et al.*,[24] and was recently confirmed by ultrasonic
absorption measurements.[25] It was suggested that the
interconversion of the two conformers is a very fast process
which happens within microseconds while the complexation
reaction takes milliseconds.

Structural Analogues

Some of the structural analogues mentioned earlier have
been investigated with respect to the magnitude of the negative
Cotton effect[26] and their complexing behavior (Table VII).[22]
In all analogues so far studied, negative rotation, complexation capacity and biological activity are attributes
of the molecules which parallel each other. This argues
in favor of an ionophoric mechanism of AA action. From
perhydrogenated AA (HAA), however, we learned that Na
complexation can not be the only prerequisite of antitoxic
activity. HAA forms a Na complex as stable as AA itself,
but does not protect against phallotoxins.[16] Since its CD
spectrum also resembles that of AA in all solvent mixtures
its molecular structure is assumed to be not very different

Table VII

Molar Rotation Φ_{240} and Na^+-Complex Stability Constants (K_{Na}) in Ethanol-Water (96:4) as Related to Protective Doses of Analogues of Antamanide (see also Table III)

Analogue	$\Phi_{240} \cdot 10^{-4}$	K_{Na} *(1/mole)*	*Protective dose against 5 mg phalloidine (mg/kg)*
AA	-8,5°	2000	0,5
[Leu^1]-	-8,6°	1000	0,5
[Ile^1]-		2300	0,5
[Ala^1]-	-5,3°	150	10
[Gly^1]-	-5,8°	180	10
[Ala^1,Gly^4]-	-4,5°	120	15
[Gly^1,Gly^4]-	-2,0°	100	> 20
[Abu^1]-		1000-2000	2,5
[Tyr^6]-		2000	0,5
HAA	-9,0°	~ 2000	> 20

from AA. The argument of slight solubility in biological systems as an explanation for the lack of antitoxic activity seems invalid since a soluble O-carboxymethyl compound likewise was not protective. Therefore we compared the selectivity of AA and HAA for Na^+ over K^+ and found that HAA is a relatively good complexing agent also for K^+-ions: K_{Na}/K_K in acetonitrile-H_2O (96:4) is *ca.* 100 for AA and only 10 for HAA. Differences in affinities to water as compared with Na^+ may thus play a role in the mechanism of biological action.

Conformation of Antamanide

The conformation of the Na-complex of AA has been described by the Moscow Academy group,[24] who used ORD, NMR and IR data for the structure analysis in solution. Recently, minimum energy calculations by Tonelli *et al.*[27] based on NMR and CD led to a proposal for the conformation

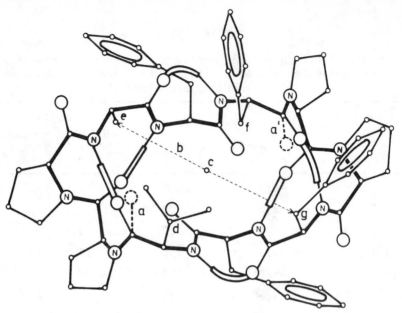

Figure 7a: Conformation of antamanide in nonpolar solvents.[22]

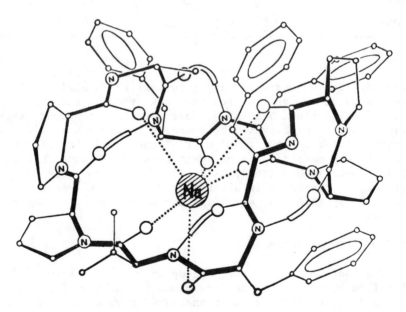

Figure 7b: Conformation of the Na$^+$-complex of antamanide
(independent of nature of solvent). [From Ivanov *et al.*[24]]

of uncomplexed AA in nonpolar solvents. From NMR studies
of exchange with deuterated alcohols in CDCl$_3$ solution it
was concluded that the peptide NH resonances were not
hydrogen bonded and accordingly a conformation was suggested. The spectral changes seen in AA solutions when
polar solvents, particularly water, are added, were
interpreted as solvent effects. It has been pointed out,
however, on page 392 that from spectroscopic titration
experiments with water and ultrasonic absorption measurements a conformational transition rather than solvent
effects have to be considered.[28] This transition process,
occurring already in the presence of tiny amounts of water
or methanol, is very fast. Thus, the high exchange rate
of some protons which led to the disproval of H bridges
and the sharp average signals actually found in the spectrum[27]
may find an explanation.

The unusually high negative dichroism at about 230 nm
of AA in nonpolar solvents (Figure 4) is attributed[28] to n→π*
transition of two transoid tertiary amide groups distorted
out of plane. These are the bonds between Pro^3-Ala^4 and
Pro^8-Phe^9 (Figure 7a). Two additional intramolecular H-
bridges are formed between NH of Phe^9 and CO of Phe^6 and
between NH of Ala^4 and CO of Val^1 in favor of out of plane
deformations of 20-30° of the two carbonyls concerned. By
H$^+$ containing solvents or by complexing metal ions these
bridges will readily be abolished thus converting the molecule to the structure of Figure 7b, in which, according to
Ivanov *et al*.[24] the Na$^+$ ion is held in an octahedral O-complex.
For *cis*-prolyl-peptide bonds in AA, as ascertained by ^{13}C-nmr
Fourier transformation confer the paper of F. A. Bovey (p 3).

References

1. Wieland, Th., and O. Wieland. In *Microbial Toxins*,
   Vol 8 (New York: Academic Press, 1972) pp 249–280.
2. Wieland, Th., and J. X. de Vries. Justus Liebigs Ann.
   Chem. **700**, 174 (1966).
3. Wieland, Th., G. Lüben, H. Ottenheym, J. Faesel, J. X.
   de Vries, W. Konz, A. Prox, and J. Schmid. Angew.
   Chem. Internat. Edit. **7**, 204 (1968).
4. Wieland, Th., H. Faulstich, W. Jahn, M. V. Govindan,
   H. Puchinger, Z. Kopitar, H. Schmaus, and A. Schmitz.
   Hoppe-Seyler's Z. Physiol. Chem., in press.
5. Puchinger, H., and Th. Wieland. Europ. J. Biochem.
   11, 1 (1969).
6. Prox, A., J. Schmid, and M. Ottenheym. Justus Liebigs
   Ann. Chem. **722**, 179 (1969).
7. Wieland, Th., J. Faesel, and W. Konz. Justus Liebigs
   Ann. Chem. **722**, 197 (1969).

8. Wieland, Th., Ch. Birr, and F. Flor. Justus Liebigs Ann. Chem. 727, 130 (1969).
9. König, W., and R. Geiger. Justus Liebigs Ann. Chem. 727, 125 (1969).
10. Birr, Ch., and W. Lochinger. Synthesis 1971, 319.
11. Wieland, Th., L. Lapatsanis, J. Faesel, and W. Konz. Justus Liebigs Ann. Chem. 747, 194 (1971).
12. Wieland, Th., Ch. Birr, and A. V. Dungen. Justus Liebigs Ann. Chem. 747, 207 (1971).
13. Wieland, Th., B. Penke, and Ch. Birr. Justus Liebigs Ann. Chem. 759, 71 (1972).
14. Wieland, Th., A. V. Dungen, and Ch. Birr. FEBS Letters 14, 299 (1971).
15. Faulstich, H., H. Trischmann, and Th. Wieland. Unpublished results; Wieland, Th. in *Peptides 1971, Proceeding of the XIth European Peptide Symposium,* Nesvadba, H., ed. (Vienna: North Holland Publ. Comp.) in press; Ovchinnikov, Yu. A., V. T. Ivanov, A. I. Miroshnikov, K. KH. Kalilulina, and N. N. Uvarova. Khim. Prir. Soedin 4, 469 (1971).
16. Wieland, Th., and Ch. Rietzel. Justus Liebigs Ann. Chem. 754, 107 (1971).
17. Wieland, Th., Ch. Birr under assistance of R. Frodl, W. Lochinger, and G. Stahnke. Justus Liebigs Ann. Chem. 757, 136 (1972).
18. Birr, Ch. Private communication.
19. Wieland, Th., Ch. Rietzel, and A. Seeliger. Justus Liebigs Ann. Chem. 759, 52 (1972).
20. Trischmann, H. Unpublished results.
21. Wieland, Th., H. Faulstich, W. Burgermeister, W. Otting, W. Möhle, and M. M. Shemyakin, Yu. A. Ovchinnikov, V. T. Ivanov, and G. G. Malenkov. FEBS Letters 9, 89 (1970).
22. Wieland, Th., H. Faulstich, and W. Burgermeister. Biophys. Biochem. Res. Commun. 47, 984 (1972).
23. Pedersen, C. J. Fed. Proc. 27, 1305 (1968).
24. Ivanov, V. T., A. I. Miroshnikov, N. D. Abdullaev, L. B. Senyavina, S. F. Arkhipova, N. N. Uvarova, K. Kh. Khalilulina, V. F. Bystrov, and Yu. A. Ovchinnikov. Biochem. Biophys. Res. Commun. 42, 654 (1971).
25. Burgermeister, W. Dissertation Universität Heidelberg, 1972.
26. Lapatsanis, L. Dissertation Frankfurt a.M. 1970; Wieland, Th., Jahrbuch der Max-Planck-Gesellschaft, 146 (1970).
27. Tonelli, A. E., D. J. Patel, M. Goodman, F. Naider, H. Faulstich, and Th. Wieland. Biochemistry 10, 3211 (1971).
28. Faulstich, H., W. Burgermeister, and Th. Wieland. Biochem. Biophys. Res. Commun. 47, 975 (1972).

ON THE SYNTHESIS OF SCOTOPHOBIN: A SPECIFIC BEHAVIOR-INDUCING
PEPTIDE

Wolfgang Parr, Günther Holzer.   Chemistry Department,
University of Houston, Houston, Texas 77004.

SUMMARY--Seryl-asparagyl-asparaginyl-asparaginyl-glutaminyl-
glutaminyl-glycyl-lysyl-seryl-alanyl-glutaminyl-glutaminyl-
glycyl-glycyl-tyrosylamide was synthesized via the solid
phase method. In biological and analytical tests, the peptide is identical with scotophobin, a specific behavior
inducing pentadecapeptide, that was isolated from the brain
of rats which were trained to have fear of dark.

IN 1968 UNGAR *ET AL*.[1] SHOWED that dark avoidance could be
induced in mice or rats by injection of a brain extract
which was taken from rats that were trained to fear the
dark. Using this behavior as basis for a bioassay an active
material, named scotophobin, was isolated from the brains
of 4000 trained rats.[2,3] Scotophobin has been formulated
as a pentadecapeptide with the sequence I given in Figure 1.
The structure determination was done by means of mass
spectrometry[4] however the functionalities of the amino
acids in position 2, 5 and 11 remained uncertain. Since
no more natural material was available only the synthesis
of the possible analogs could clarify the structure of
scotophobin. Sequence IV was synthesized by Ali *et al*.[5]
This peptide had 2.5% (75 U/mg) biological activity and
its R_f values on thin layer plates did not agree with those
of natural scotophobin.
    We first synthesized structure III,[6] peramidoscotophobin,
which possessed 10-12% (300 U/mg) of the activity of the
natural material. However, its analytical data differed

I   Ser-Asx-Asn-Asn-Glx-Gln-Gly-Lys-Ser-Ala-Glx-Gln-Gly-Gly-TyrNH$_2$

II   Ser-Asp-Asn-Asn-Gln-Gln-Gly-Lys-Ser-Ala-Gln-Gln-Gly-Gly-TyrNH$_2$

III   Ser-Asn-Asn-Asn-Gln-Gln-Gly-Lys-Ser-Ala-Gln-Gln-Gly-Gly-TyrNH$_2$

IV   AcSer-Asp-Asn-Asn-Glu-Gln-Gly-Lys-Ser-Ala-Glu-Gln-Gly-Gly-TyrNH$_2$

       1   2   3   4   5   6   7   8   9   10   11   12   13   14   15

Figure 1: Synthetic analogs of scotophobin. I represents
   the amino acid sequence deduced by mass spectrometry.
   Amino acids in position 2, 5, and 11 are uncertain in
   terms of their functionality. Sequences II, III, and IV
   were synthesized. Biological and chemical properties
   of sequence II are identical with those of the natural
   peptide.

from those of natural scotophobin. Both structures IV and
III could now be rejected.

In our second synthetic approach we chose structure II[7]
in which Asn in position 2 is replaced by Asp. The protected
pentadecapeptide resin ester Boc-Ser(Bzl)-Asp(OBu^t)-Asn-Asn-
Gln-Gln-Gly-Lys(Z)-Ser(Bzl)-Ala-Gln-Gln-Gly-Gly-Tyr(Bzl)-
resin ester was synthesized by the method of Merrifield[8]
using an automatic peptide synthesizer.

The t-butyloxycarbonyl (Boc) group was used to protect
the α-amino group of all amino acids except for Asp which
was protected by the 2-(p-biphenyl) isopropyloxycarbonyl
(Bpoc) group.[9] The following derivatives of the protected
amino acids were used: Tyr(Bzl), Asp(OBu^t), Lys(Z), Ser
(Bzl). Coupling of Gly, Ala, Ser, Lys and Asp was achieved
at room temperature with dicyclohexylcarbodiimide (DCC).
Methylene chloride was used as solvent. Gln and Asn residues were coupled as p-nitrophenyl esters in dimethylformamide (DMF). Boc-amino acids and Boc-amino acid nitrophenyl
esters were used in six fold excess. For the incorporation
of the amino acids in position 12(Gln), 11(Gln), 9(Ser),
8(Lys), 6(Gln), 5(Gln), 4(Asn), 3(Asn), 2(Asp) and 1(Ser)
the coupling reaction was repeated. Reaction times were
3-5 hr for DCC condensations and 8-12 hr for p-nitrophenyl
ester couplings. Three different deblocking reagents were
used for effective removal of the Boc group. Residues 15,
14, 13, 10, 9 were deblocked with 3N HCl in dioxane-CH$_2$CL$_2$
(1:1), residues 8, 7, 3 with HCl saturated CH$_2$Cl$_2$-TFA (1:1)
and residues 12, 11, 6, 5, 4 with CH$_2$Cl$_2$-TFA (1:1). The

Bpoc group of Asp was deblocked with acetic acid–formic acid–H_2O (7:1:2). Reaction times for the removal of Boc groups varied between 30 min and 60 min. The removal of the Bpoc group was achieved in 3 hr. The amount of free amino groups were determined by titration after deblocking[10] and after coupling.[11] Unreacted amino groups were acetylated after each coupling step in order to prevent the formation of failure sequences.[12] A graphical representation of yields in the coupling steps is given in Figure 2.

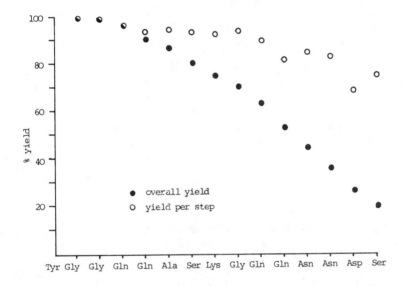

Figure 2: Yields per step and overall yields in the synthesis of sequence II. After each coupling step the unreacted amino groups were determined with pyridine–HCl in methylene chloride. The free chloride was titrated according to the method of Volhard. The amount of Tyr bonded to the resin was taken as 100%.

Starting with 4.5 mmol Boc–Tyr(Bzl) resin ester (the chloromethylated beads were purchased from Schwarz/Mann, New York, capacity: 3.7 mmol/g), the following cycles were used to build the pentadecapeptide. Removal of the α-amino protection group, 12 washing steps, neutralization with Et_3N–$CHCl_3$ (1:9) for 10 min, 9 washing steps, coupling in

CH_2Cl_2 or DMF, 10 washing steps, determination of non-reacted amino groups with $0.1N$ pyridine HCl in CH_2Cl_2 for 20 min, 12 washing steps, neutralization with Et_3N-$CHCl_3$ (1:9) for 10 min, 8 washing steps, acetylation (if the coupling step was not repeated) with 15 ml DMF, 1.2 ml acetic acid anhydride, 0.5 ml Et_3N, 8 washing steps. Washing solvents were used in 30 ml portions and reagents in 15 ml portions. A three way solenoid valve activated by the relay control directed the waste to a collecting vessel in which the titrations were done. The cleavage of the protected pentadecapeptide from the resin was achieved by direct ammonolysis.[13] The t-butyl ester provides an ammonia resistant protection for Asp under the conditions applied [liq. NH_3-DMF (1:1), room temperature, 7 days]. The crude protected pentadecapeptide amide was treated for 2 hr with HF-TFA at 20°C. After evaporation of the solvent 3.8 g of crude peptide was obtained. A portion of this material (900 mg) was purified by gel filtration on Sephadex LH-20 using MeOH as solvent. Further purification was achieved by preparative thin layer chromatography on silica gel plates. Three ninhydrin positive spots were obtained. The spot at R_f 0.57 was eluted with pyridine-acetic acid buffer. By this procedure 180 mg of pure peptide was obtained. Amino acid analysis gave the following molar ratios: Asp 2.85, Ser 1.80, Glu 3.85, Gly 3.00, Ala 1.00, Tyr 1.10, Lys 0.95, NH_3 7.20.

In Table I chromatographic and biological data of synthetic peptide II are compared with those of natural scotophobin. Both peptides showed similar R_f values on thin layer chromatography (tlc). After microdansylation[14] the synthetic peptide II and the natural material appeared as single spots on tlc and had corresponding R_f values. For further identification a tryptic digestion was performed. The resulting two fragments had similar R_f values on tlc. The biological activity of the synthetic peptides was tested on mice. Only animals which spent 90 sec out of 180 sec in the dark box were selected for injections. In these tests peptide II is identical with the natural material. Furthermore, the dose response shows no significant difference in activity between synthetic peptide II and natural scotophobin. The potency ratio was found to be 1.11 and the slope ratio 1.09. These results have since been confirmed by other laboratories.[15-18] On the basis of these analytical and biological data it is highly probable that scotophobin is a peptide of structure II.

Table I

Comparison of Chromatographic Properties and
Biological Activities of Natural Scotophobin
and Synthetic Peptide II

	Natural Scotophobin (I)	Synthetic Peptide (II)	Solvent System
Rf Values	0.58	0.57	n-Butanol-Ethanol-Acetic Acid-Water 80:20:10:30
Dansyl-derivatives			
1. Dimension	0.16	0.16	Formic Acid-Water 1.5:100
2. Dimension	0.16	0.18	Benzene-Acetic Acid 9:1
Tryptic Fragments			
T_1	0.26	0.30	n-Butanol-Ethanol-
T_2	0.40	0.37	Acetic Acid-Water 80:2-:10:30
Biological Activities	3000.U/mg*	3000.U/mg	

*One unit of activity is the amount of material that reduces the mean time spent in the dark from 130 to 60 sec.

Acknowledgment

The authors are indebted to Dr. G. Ungar, Baylor College of Medicine, Houston, Texas for providing the bioassay. This work was supported by a grant from the Robert A. Welch Foundation.

References

1. Ungar, G., L. Galvan, and R. H. Clark. Nature 217, 1259 (1968).
2. Ungar, G., I. K. Ho, L. Galvan, and D. M. Desiderio. Proc. Western Pharmacol. Soc. 13, 149 (1970).

3. Ungar, G., D. Desiderio, and W. Parr. Nature 238, 198 (1972).

4. Desiderio, D. M., G. Ungar, and P. A. White. Chem. Commun. 9, 432 (1971).

5. Ali, A., J. H. R. Fraesel, D. Sarantakis, D. Stevenson, and B. Weinstein. Experientia 27, 1138 (1971).

6. Parr, W., and G. Holzer. Nature, in press.

7. Parr, W., and G. Holzer. Hoppe-Seyler's Z. Physiol. Chem. 352, 1043 (1971).

8. Merrifield, R. B. Biochemistry 3, 1385 (1964).

9. Sieber, P., and B. Iselin. Helv. Chim. Acta 51, 614 (1968); 51, 622 (1968).

10. Bayer, E., and H. Hagenmaier. Tetrahedron Lett. 2037 (1968).

11. Dorman, L. C., and E. C. Britton. Tetrahedron Lett. 2319 (1969).

12. Bayer, E., H. Eckstein, K. Hägele, W. A. König, W. Brüning, H. Hagenmaier, and W. Parr. J. Amer. Chem. Soc. 92, 1735 (1970).

13. Parr, W., C. C. Yang, and G. Holzer. Tetrahedron Lett. 1, 101 (1972).

14. Neuhoff, V., and F. Kiehl. Arzneimittelforschung 19, 1898 (1969).

15. Domagk, G. F. Symposium on Memory and Transfer of Information, Göttingen, Germany, May 1972.

16. Bryne, W., and R. C. Bryant. Symposium on Memory and Transfer of Information, Göttingen, Germany, May 1972.

17. Guttman, H. N., G. Matwyshyn, and G. H. Warriner III. Nature 235, 6 (1972).

18. McConnell, J. V. Symposium on Memory and Transfer of Information, Göttingen, Germany, May 1972.

TUFTSIN, THREONYL-LYSYL-PROLYL-ARGININE, THE PHAGOCYTOSIS
STIMULATING MESSENGER OF THE CARRIER CYTOPHILIC γ-GLOBULIN
LEUCOKININ

*Paul S. Satoh, Andreas Constantopoulos, Kenji Nishioka,
Victor A. Najjar.* Division of Protein Chemistry,
Department of Molecular Biology and Microbiology,
Tufts University School of Medicine, 136 Harrison
Avenue, Boston, Massachusetts 02111

IN PREVIOUS COMMUNICATIONS from this laboratory a naturally
occurring[1,2] phagocytosis stimulating tetrapeptide, tuftsin,
was shown to be covalently bonded to a specific γ-globulin
fraction, leucokinin.[3] It is released by a specific enzyme
present on the outer surface of the polymorphonuclear (PMN)
neutrophil.[1] This peptide is responsible for all the
phagocytosis stimulating activity of leucokinin. No other
peptide shows that effect.[1,2] Trypsin also releases tuftsin
from leucokinin rich phosphocellulose fraction (PC) IV.[2,4,5]
Tuftsin activity is detectable only in the H chain of leucokinin. The tetrapeptide has recently been identified in the
Fc fragment of the H chain.[6]

The Isolation of Tuftsin

This has been accomplished by leucokininase derived
from human, dog and rabbit neutrophil membrane preparations
as well as by crystalline trypsin.[1,2] The amino acid analysis yielded unit values for threonine, lysine, proline and
arginine and its sequence proved to be Thr-Lys-Pro-Arg.
This was accomplished through the use of leucine aminopeptidase (LAP) and the dansyl method[7] for the amino-terminal
end and carboxypeptidase B (CP-B) and tritium exchange[8] for

the carboxy-terminal. Tuftsin was synthesized by the solid phase method[9] and proved to be identical with the natural product in all physical, chemical and biological parameters.[2]

The Covalent Bond Between Tuftsin and the Carrier γ-Globulin

Tuftsin is at the carboxy-terminal end of either the nicked segment of it, or a branch off that chain. The evidence was derived from the use of LAP and CP-B, both of which destroy the biological activity of free tuftsin.[2] Six samples of fraction PC IV prepared from Cohn II and six samples from fresh human γ-globulin were reacted with LAP and with CP-B. Each sample was then treated with leucokininase to release tuftsin. Tuftsin activity was then tested by its ability to stimulate phagocytosis. The phagocytosis values calculated for (a-f) samples as percent of reagent controls, were as follows: CP-B treated Cohn PC IV values were (a) 85, (b) 105, (c) 110, (d) 90, (e) 95, (f) 103 – averaging 98; while the LAP treated samples were 200, 200, 185, 190, 205, 205 respectively – averaging 198. Similarly, CP-B treated fresh γ-globulin PC IV were (a) 100, (b) 98, (c) 87, (d) 100, (e) 115, (f) 95 – averaging 99; while the same samples treated with LAP were 220, 205, 190, 200, 195, 210, respectively – averaging 203. These results indicate that the carboxy-terminal arginine is free and the tetrapeptide is bound at the amino-terminal. This conclusion was further strengthened by the liberation of the arginine residue by CP-B treatment.

Until pure leucokinin is obtained, the exact nature and location of tuftsin on the Fc portion of the H chain can not be defined with certainty. It appears that tuftsin is linked, at least in part, through the hydroxyl of the amino-terminal threonine. Several lines of evidence favor the ester bond linkage. (a) γ-globulin was thoroughly dialyzed in five changes of 40 volumes of 8 M urea followed by 0.15 M NaCl and water. It was then treated with 2 M hydroxylamine for one hour at room temperature.[10] After deproteinization, the hydroxylamine salt was separated on Sephadex G-10. It was then chromatographed on G-25 in 1 M acetic acid. The effluent between Kav 0.34-0.40 was subsequently chromatographed on Aminex.[2] The peptide yielded approximately unit ratios of Thr 0.98, Lys 1.0, Pro 0.86, Arg 1.1. It was subsequently chromatographed on silica gel. The peptide gave the exact R_f values as the enzymatically isolated tuftsin in two chromatographic systems. Sequence analysis also yielded the same tetrapeptide. CP-B digestion liberated arginine. The remaining tripeptide incorporated ^{3}H from ^{3}H$_2$O only in proline, 532 counts per minute per n mole. For further confirmation, the tetrapeptide was dansylated and yielded identical R_f values in three systems as authentic tuftsin. On hydrolysis it yielded N^α-dansyl-threonine and N^ϵ-dansyl-lysine. CP-B treatment of the dansylated peptide followed by dansylation of the product, yielded dansyl-arginine and a dansylated polypeptide with the same R_f as synthetic N^α-dansyl-threonyl-N^ϵ-dansyl-lysyl-proline. Tritium incorporation from ^{3}H$_2$O into this polypeptide was found only in proline, 547 counts per minute per n mole. Thus the sequence is Thr-Lys-Pro-Arg. The yield on hydroxaminolysis amounted to 10% of the tetrapeptide present. (b) Thoroughly dialyzed γ-globulin in 6 M guanidine HCl, was dansylated[7] and then subjected to hydroxaminolysis as above. Dansyl tuftsin was readily identified on thin layer chromatography with a yield of 18%. Acid hydrolysis yielded N^α-dansyl-threonine and N^ϵ-dansyl lysine. CP-B treatment yielded arginine as carboxy-terminal. The low yield in both instances (a and b) is to be expected because of the O → N shift under the conditions of high pH. Similar low yields were obtained by hydroxylamine treatment of serine and threonine peptidyl esters of rearranged protamine. These were shown to result from O → N shift.[11] (c) Trypsin readily released tuftsin in the presence of 50% 2-chloro-ethanol. Under these conditions, peptidase activity of trypsin is destroyed and only esterase activity is preserved.[12]

The existence of tuftsin as a molecular entity is suggested (a) by the specificity of the carrier molecule and the presence of a specific enzyme for its release. (b) Post-splenectomy leucokinin is devoid of activity.[4] (c) Human mutants have been identified in several families who are defective in tuftsin activity.[1,5]

It is of particular interest that the complete sequence of γG$_1$ (EU) myeloma protein shows the same tetrapeptide in the Fc fragment of the heavy chain, residues No. 289-292.[13] Assuming its presence in normal γG$_1$, it is unlikely that this is the direct source of the tetrapeptide. (a) Tuftsin is a carboxy-terminal peptide of, or within, the Fc. The reported activity of Fab[10] and its absence in Fc was due to leucokinin contamination of Fab and the lack of binding of Fc to the cell membrane. (b) Not all γG$_1$ carry tuftsin activity.[4,5] PC I-PC III contain about 94% of the γG$_1$ of the serum and these are inactive.[4,5] The only active fraction, PC IV, contains a small quantity of γG$_1$. (c) Splenectomized humans

possess γG_1 in the normal range, 10 mg per ml, yet all
fractions, including PC IV are devoid of activity.[4,5]
(d) γG_1 levels in cases of the tuftsin deficiency syndrome
are within the normal range averaging 10.07 mg per ml of
serum.[4,5] (e) Myeloma protein of the γG_1 type was devoid
of even traces of activity. (f) Once activity is removed
by trypsin, the carrier molecule does not release further
activity upon treatment with leucokininase.[2] (g) Tuftsin
is released by NH_2OH. (h) Incubation of fresh serum at
37° for 1-2 hours destroys tuftsin[4] presumably because of
the presence of CP-B like enzyme.[14] Trypsin in 50% 2-
chloroethanol loses its peptidase, but retains its esterase
activity, yet is still capable of quick release of tuftsin.
These data put together make it unlikely that tuftsin de-
rives from the tetrapeptide of the Fc fragment of γG_1 as
an intact part of the polypeptide chain unless it is
ruptured[15] at the carboxy-terminal arginine by a specific
enzyme. All the data are consonant with the presence of a
carboxy-terminal tetrapeptide fragment and probably esterified at the amino-terminal.

   If indeed the tetrapeptide is present both within some,
and terminal in other H chains, it would not be unique in
serum proteins. Two types of low molecular weight human
kallidinogens exist, one has the kinin, methionyl-lysyl-
bradykinin, at its carboxy-terminal, and the other has the
same kinin fragment in its internal structure.[16] More
relevant is the demonstration that in high molecular weight
bovine kininogen-I methionyl-lysyl-bradykinin moieties are
located both at the carboxy-terminus and inside of the
polypeptide chain.[17]

   Should it turn out that the same tetrapeptide is present
within the Fc fragment of normal γG_1 species, it may indeed
have biological significance in furthering enhanced phagocytosis and pinocytosis. Tuftsin, as a carboxy-terminal
peptide of leucokinin, could be released by leucokininase
as the trigger mechanism that initiates the act of phagocytosis. γG_1 of the plasma bearing the tetrapeptide within
the Fc piece, would be taken up in the phagocytic vacuole
and provide another source of tuftsin within the cell. This
may explain the observation that "once a leucocyte has in-
gested one or more particles, it becomes phagocytically more
active than a resting cell." In keeping with this fact is
the well-known tendency of a relatively small proportion of
the granulocytes in an acute bacterial lesion to accomplish
most of the phagocytosis.[18]

Acknowledgment

Victor A. Najjar, American Cancer Society Professor of Molecular Biology (MA Div.). We are grateful to Dr. Peter H. Schur, Department of Medicine, Harvard Medical School, Robert B. Brigham Hospital, Boston, Massachusetts, for the assay of γG_1 subclass. Supported by Grant AI-09116, National Institutes of Health, U.S. Public Health Service, and Grant GB-31535X, National Science Foundation.

References

1. Najjar, V. A., and K. Nishioka. Nature 228, 672 (1970).
2. Nishioka, K., A. Constantopoulos, P. S. Satoh, and V. A. Najjar. Biochem. Biophys. Res. Commun. 47, 172 (1972).
3. Fidalgo, B. V., and V. A. Najjar. Biochemistry 6, 3386 (1967).
4. Najjar, V. A., and A. Constantopoulos. J. Reticuloen-dothel. Soc. 12, 197 (1972).
5. Constantopoulos, A., and V. A. Najjar. J. Pediat. 80, 564 (1972).
6. Nishioka, K., J. F. X. Judge, and V. A. Najjar. Unpublished.
7. Seiler, N. *Methods of Biochemical Analysis*, Glick, D., ed. (New York: Interscience Publishers, 1970) Vol. 18, pp 259-337.
8. Matsuo, H., and Y. Fujimoto. Biochem. Biophys. Res. Commun. 22, 69 (1966).
9. Merrifield, R. B. Biochemistry 3, 1385 (1964).
10. Lahiri, A. K. and V. A. Najjar. Arch. Biochem. Biophys. 141, 602 (1970).
11. Iwai, K., and T. Ando. Methods in Enzymology XI (Academic Press, N.Y., 1967) pp 263-270.
12. Coletti-Previero, M. A., A. Previero, and E. Zuckerkandl. J. Mol. Biol. 39, 493 (1969).
13. Endelman, G. M., B. A. Cunningham, E. W. Gall, P. D. Gottlieb, U. Rutishauser, and M. J. Waxdal. Proc. Nat. Acad. U. S. 63, 78 (1969).
14. Erdös, E. G., I. M. Wohler, M. I. Levine, and M. P. Westerman. Clin. Chim. Acta. 11, 39 (1965).
15. Connell, G. E., and R. R. Porter. Biochem. J. 24, 53 (1971).
16. Pierce, J. V., and M. E. Webster. *Hypotensive Peptides*, Erdös, E. G., N. Back, and Sicuteri, eds. (W. Germany: Springer-Verlag, 1966) pp 130-138.
17. Yano, M., S. Nagasawa, and T. Suzuki. Biochem. Biophys. Res. Commun. 40, 914 (1970).

18. Davis, B. D., R. Dulbecco, H. N. Eisen, S. Ginsberg, and W. B. Wood. *Microbiology* (New York: Harper and Row, Inc., Publishers, 1970) p 629.

BIOLOGICALLY ACTIVE POLYPEPTIDES FROM THE LUNG

Sami I. Said, Satoshi Kitamura. University of Texas
(Southwestern) Medical School and Veterans Administration
Hospital, Dallas, Texas

Viktor Mutt. Karolinska Institute, Stockholm

THE WORK TO BE PRESENTED HERE was prompted by the finding
that extracts of fresh lung contained soluble vasoactive
materials, some of which appeared to be protein or polypeptide in nature.[1]
   Our efforts to isolate a vasoactive polypeptide from
hog lungs have led to the extraction and partial purification
of two biologically active peptide fractions, which are
dilators of peripheral systemic and pulmonary vessels, and
have activity on non-vascular smooth muscle.
   The extraction work was carried out at the Karolinska
Institute, and the biological characterization was done at
the V.A. Hospital, Dallas, and The University of Texas
(Southwestern) Medical School.

Extraction and Purification

   Fresh lung was collected in the abbatoir from hogs which
had just been killed. The lungs were kept on ice during
quick transfer to the laboratory, where they were sliced
into thin slices and submerged in boiling H_2O for ten minutes,
to destroy proteolytic enzymes. The boiled lungs were then
minced and extracted in dilute acetic acid and the extract
was filtered until clear. Concentration of the peptides
from this extract was accomplished by a sequence of procedures
adapted from those used in the isolation of gastrointestinal
polypeptides.[2,3] These procedures were: adsorption on

alginic acid and subsequent elution with 0.2 M HCl, precipitating out the peptides by saturating the eluate with
NaCl, extraction of this salt precipitate with methanol at
neutral pH, followed by precipitation with ether at a low
pH, and a second procedure of salting out.

   To isolate the peptides in question from the methanol
extract, two further purification steps were carried out:
1) chromatography on a column of Sephadex G-25 fine,
developed in 0.2 M acetic acid buffer (Figure 1); and

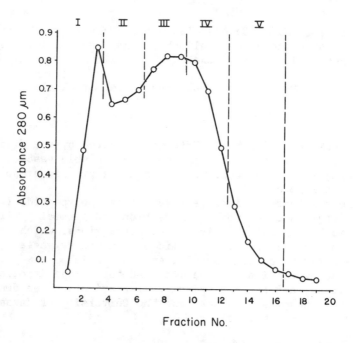

Figure 1: Chromatography of methanol extract of hog lung
   on Sephadex G-25, fine, in 0.2M acetic acid. Column
   dimensions: 95 x 3 cm. Fractions (11 ml) collected
   q 5 min. after void volume of 167 ml.

2) ion-exchange chromatography, on a column of carboxymethyl cellulose, developed in 0.02 M and 0.2 M NH$_4$HCO$_3$.
   The last chromatography resulted in the elution of at
least two peptide peaks (Figure 2), (A) which was eluted
with the weaker buffer, and (B) which was eluted with the

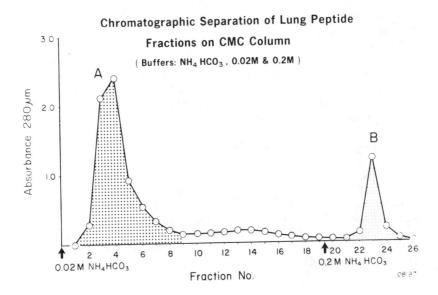

Figure 2: Chromatography of fraction IV from Sephadex column on CMC, in 0.02 *M* and 0.2 *M* NH₄HCO₃. Column dimensions: 25 x 1. Fractions (4 ml) collected q 10 min. after void volume of 5.5 ml.

stronger buffer. A third fraction, emerging in between, was not investigated further. Approximately 10 kg of fresh lung was required for the extraction of 1 mg of fraction B.

Bioassay

Throughout the purification steps, the bioassay we employed was based on peripheral vasodilator activity, measured as the ability of each preparation to increase the femoral arterial blood flow in anesthetized dogs. The flow was measured by a non-cannulating probe (Carolina Medical Electronics), placed around one femoral artery; saline solutions of each fraction were infused into a superficial branch of the same artery. Systemic (aortic) blood pressure was recorded simultaneously, to permit estimation of changes in femoral vascular resistance.

Biological Activity

We examined three aspects of biological activity of the lung peptides: systemic vasoactivity, evaluated by measuring femoral arterial blood flow and aortic blood pressure; pulmonary vasoactivity, determined by infusing the peptide fractions into isolated lobes of dog lung, perfused with Krebs-dextran at constant flow, and measuring perfusion pressure; and activity on non-vascular smooth muscle, tested by adding the peptide fractions to several isolated smooth-muscle organs, superfused with Krebs solution.

1. Systemic Vasodilator Effect

Infusions of approximately 6 µg/kg of the peptide frac-tions for one minute caused peripheral vasodilation, as evidenced by considerable increases in femoral arterial blood flow, with little or no fall in systemic arterial blood pressure. In some instances, the blood pressure actually *increased* slightly (5-10 mm Hg) while femoral flow increased (Figure 3). The femoral vasodilation was

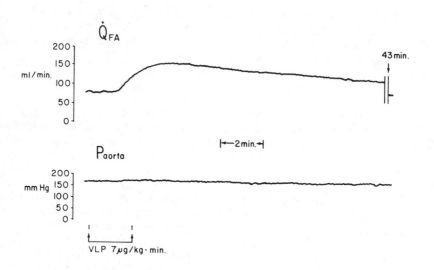

Figure 3: Simultaneous records of mean femoral arterial blood flow (Q_{FA}) and mean aortic blood pressure (P_{aorta}) in anesthetized dog, showing the effect of infusion of lung peptide (Sephadex fraction II). Infusion was given for two minutes.

prolonged (at least 40 minutes) with fractions II and III from the Sephadex column, but relatively brief (3-6 min) with fractions IV and V.

The absence of systemic hypotension with the increase in peripheral blood flow distinguishes this peptide from the recently isolated Vasoactive Intestinal Polypeptide (and from most other vasodilators), and suggests that other vascular beds may be constricted while extremity vessels are dilated.

2. Pulmonary Vasodilator Effect

In six experiments on isolated, perfused lung, lung peptide fractions A and B, given in doses of 100 µg/min, produced about the same fall (2 mm Hg) in perfusion pressure as 10 µg of isoproterenol or of acetylcholine. However, the vasodilation caused by either peptide fraction was several times as prolonged, lasting for up to 25 minutes, as opposed to 3 or 4 minutes for the other compounds.

3. Activity on Non-Vascular Smooth Muscle Organs

When tested on isolated guinea-pig trachea, continually superfused with Krebs solution, both fraction A and fraction B relaxed the trachea.[4] When we added another smooth muscle tissue, rat stomach strip, it became evident that fractions A and B contained *two distinct* principles, since A contracted rat stomach while B relaxed it (Figure 4).

Possible Role in Normal Function or Disease

The extraction of biologically active polypeptides from the lung naturally raises the questions: What role, if any, do they have in the normal regulation of the airways and pulmonary vessels? Are they normal local "tissue" hormones or could they be released into the circulation? How effective may they prove to be as therapeutic dilators of bronchi and of pulmonary vessels? These and other questions must await full purification of the active peptides and the development of sensitive assays for them.

Acknowledgments

This investigation was supported by the Swedish Medical Research Council (Grant B73-13X-1010-08A), the Albert och Gerda Svenssons stiftelse, the Research Funds of the Karolinska Institute, the National Tuberculosis and

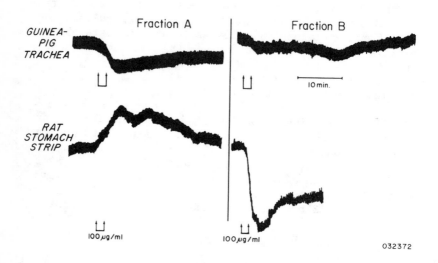

Figure 4: Responses of isolated, superfused guinea pig
trachea, above, and rat stomach strip, below, to CMC
fraction A (left) and to fraction B (right) of lung
peptide.

Respiratory Disease Association, and a Center Award from
the National Heart and Lung Institute.

We are grateful for technical assistance to Mr. Juha
Sääksi, Mr. M. Rifaat, Mrs. Eva Papinska, Mrs. Louise Melin
and Mrs. Karin Thermaenius.

References

1. Said, S., H. L. Estep, M. E. Webster, and H. A. Kontos.
   J. Clin. Invest. 47, 85a (1968).
2. Mutt, V. Arkiv Kemi 15, 69 (1959); Jorpes, J. E., and
   V. Mutt. In *Ciba Found. Symp. on the Exocrine Pancreas.*
   de Reuck, A. V. S. and M. P. Cameron, eds (London: J.and
   A. Churchill, Ltd., 1962) pp 150–154.
3. Said, S. I., and V. Mutt. Europ. J. Biochem., in press.
4. Said, S. I., V. Mutt, and S. Kitamura. Fed. Proc. 31,
   308 (1972) (abstract).

L-(N^5-PHOSPHONO)METHIONINE-S-SULFOXIMINYL-L-ALANYL-L-ALANINE, AN ANTIMETABOLITE OF L-GLUTAMINE PRODUCED BY A STREPTOMYCETE

James P. Scannell, David L. Pruess, Thomas C. Demny, Helen A. Ax, Florence Weiss, Thomas Williams, Arthur Stempel. Chemical Research Department, Hoffmann-La Roche Inc., Nutley, New Jersey 07110

IN RECENT YEARS, largely as a result of a deliberate search for antimetabolites,[1-3] several biologically active di- and tripeptides have been isolated from fermentation broths.[4-10] We wish to report here the isolation and structural elucidation of a new member of this series, L-(N^5-phosphono)methionine-S-sulfoximinyl-L-alanyl-L-alanine, I.

The substance is produced by an unclassified streptomycete species and inhibits the growth of *Serratia sp.* on a chemically defined minimal medium.[11] The growth inhibition is relieved by addition of L-glutamine to the medium. The fermentation conditions, microbiological assay and isolation procedure will be described in detail elsewhere.[12]

Characterization of I

After a 5 step 10,000 fold purification analytically pure I was deposited slowly as amorphous spherules from methanol-water solution; mp 192° dec. Elemental analysis gave correct data for $C_{11}H_{23}N_4O_8PS$. The NMR spectrum is similar to that of the dephospho compound II described below. The IR spectrum (KBr, 1 mg) contained peaks at 3000, 1650 and 1020 cm^{-1} which may be attributed to $-NH_3^+$, $-\overset{O}{\underset{}{C}}-O^-$, and $S=O$[13] groups respectively, but a peak at 1200 cm^{-1} assigned[13] to the unsubstituted S=N group in II and III was not present in I or V.

415

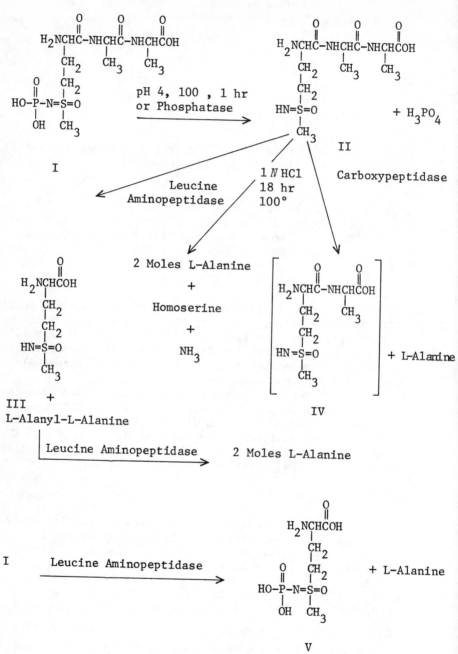

Figure 1: Degradation of L-(N^5-phosphono)methionine-S-sulfoximinyl-L-alanyl-L-alanine (I).

Chemical and Enzymatic Conversion of I to II

Dephosphorylation to II was effected by maintaining an aqueous solution of I at pH 4 and 100° for 1 hr (1/2 life-10 minutes) or by the enzymatic action of intestinal or *E. coli* alkaline phosphatase. The quantitative release of phosphate was followed by the method of Fiske and Subba Row.[14] N^5-Phosphonomethionine sulfoximine, V, prepared either synthetically[15] or by enzymatic degradation of I (see below) was hydrolyzed chemically and enzymatically at essentially the same rates. Purification of II was accomplished by absorption onto AG50WX4 (50-100 mesh in the H^+ form), elution with 10% aqueous pyridine, evaporation at reduced pressure and crystallization from methanol: mp 204-206°; NMR (20 mg, D_2O, ext TMS, both 60 and 100 Hz spectra with decoupling experiments) δ 1.79, 4.57 (A_3X, 4, J=7Hz, CH_3CH), 1.85, 4.82 (A_3X, 4, J=7Hz, CH_3CH), 3.62 (s,3, CH_3S), 3.92, 2.85, 4.75 (AA'MM'X, 5, CH_2CH_2CH). Elemental analysis gave correct data for $C_{11}H_{22}N_4O_5S$. The properties of II prepared by action of *E. coli* alkaline phosphatase were identical to those reported above for II prepared by chemical dephosphorylation.

Strong Acid Hydrolysis of I or II

Treatment of either I or II with 5.0 N HCl at 100° for 3 hr released 2 mol of L-alanine (quantitation by amino acid analyzer; structure by isolation and crystallization as described below; absolute configuration by optical rotation) and roughly equivalent amounts of NH_3 (Nesslers reagent) and methionine sulfone or sulfoxide (vpc of TMS derivative). Small amounts of alanyl-alanine were detected (vpc of TMS derivative) during the course of the reaction. In 1 N HCl at 100° up to 18 hr, alanine and NH_3 were released with alanylalanine as a transient intermediate, but the major degradation product of the methionine sulfoximine portion of the molecule was homoserine (vpc of TMS derivative of lactone and free acid).

Leucine Amino Peptidase Degradation of II

A solution consisting of 132 mg II and 1.3 mg pre-activated leucine amino-peptidase (Miles Laboratory) in 19 ml 0.01 M $MgCl_2$ was incubated at pH 8.4 and 25°C for 3 1/2 hr. The solution was then added to the top of a column (2.5 cm x 19 cm) which contained 75 ml AG50WX4

resin (200-400 mesh in the Na$^+$ form). The column was
eluted first with 700 ml 0.2 M sodium phosphate-citrate
buffer[16] pH 2.9 then with 300 ml of the same buffer adjusted
to pH 4.8. Alanine and methionine sulfoximine were detected
at elution volumes 400 to 650 and 800 to 900 ml respectively.
Each amino acid was desalted by readsorption on AG50WX4 (50-
100 mesh in the H$^+$ form) followed by elution with 10% aqueous
pyridine and evaporation at reduced pressure. Alanine was
crystallized from 80% aqueous ethanol. Yield 37 mg (52%);
mp 296°; $[\alpha]_D^{25}$ +12.4° (c 1, 5 N HCl) lit.[17] +14.6°, but we
obtained +12.53° for an authentic sample. Elemental analysis
gave correct data for $C_3H_7NO_2$.

Methionine sulfoximine, III, was crystallized from 80%
aqueous ethanol in two crops, 17 mg (23%); mp 240°; $[\alpha]_D^{25}$
+33.3° (c 1, 1 N HCl). Lit.[18] for 2(s), S(s) isomer; mp
239°; $[\alpha]_D^{25}$ +34°; for 2(s), S(R); mp 235°; $[\alpha]_D^{25}$ +39.0°.
IR (KBr) essentially identical to that of synthetic III
(racemic at S). Elemental analysis gave correct data for
$C_5H_{12}N_2O_3S$.

When the enzymatic hydrolysis was carried out with
relatively less enzyme, substantial amounts of alanylalanine
were detectable (vpc of TMS derivative and Technicon amino
acid analyzer) during the course of the reaction.

Leucine Amino Peptidase Degradation of I

A solution of 18 mg pre-activated leucine amino peptidase and 100 mg I in 37 ml 0.05 M MgCl$_2$ was incubated at pH
8.4 for 5 hr at 25°C. The solution was then applied to a
column which contained 45 ml AG50WX4 (100-200 mesh in the
H$^+$ form). N^5-Phosphonomethionine sulfoximine, V, was washed
off the column with water and after evaporation at reduced
pressure, an amorphous preparation of V was obtained which
had the same tlc mobility and essentially the same IR (in
DMSO) as synthetic V (racemic at sulfur). This compound
was synthesized by the method of Rowe *et al.*[15] modified by
the introduction of a base hydrolysis step (15 min, 60°,
0.1 M NaOH) to ensure removal of protecting groups. The
product has the properties described by Rowe *et al.*[15] except
that we obtained mp 183-185° (dec) compared to 149-152°
cited.

End Group Analysis of II

Alanine was shown (tlc and vpc of TMS derivative) to
be liberated when 800 μg II in 1 ml 0.025 M tris HCl buffer
pH 7.5 containing 0.5 M NaCl was treated with 5 μg

carboxypeptidase A (Worthington Biochemical) for 4 hr at
25°C. Prior to chromatography it was necessary to desalt
the hydrolysate by the AG50W-pyridine elution technique
described above. Another product of the reaction which
had a long retention time during vpc and which, like II,
gave a yellow reaction product with ninhydrin but which
had a tlc mobility different from II, was presumed to be
IV.

After II was subjected to dansylation and hydrolysis
by the method of Hartley and Grey,[19] the dansyl derivative
of methionine sulfoximine but not that of alanine was shown
to be present by electrophoresis and tlc.

Discussion

The composition of II follows unambiguously from (1)
the elemental analysis; (2) the release of 2 mol of L-alanine
from acid hydrolysates; (3) the recovery of L-methionine-
S-sulfoximine in the leucine amino peptidase hydrolysate;
(4) the previously known[20] strong acid hydrolysis of
methionine sulfoximine to homoserine, methionine sulfone
and methionine sulfoxide. That the amino acids are in a
linear peptide array is shown by the susceptibility
to enzymatic attack and the presence of $\overset{+}{N}H_3$ and $\overset{O}{\overset{\|}{C}}-O^-$
groups (IR, electrophoresis and method of purification).
Dimeric or polymeric structures are ruled out by the absence of required intermediates during enzymatic hydrolysis.
The peptide sequence is demonstrated by the end group
analyses and the release of alanylalanine as a transient
intermediate during acid and leucine amino peptidase
hydrolysis. The location of the phosphate group in I is
demonstrated by lack of a reasonable alternative ($\overset{O}{\overset{\|}{C}}-O^-$ and
$\overset{+}{N}H_3$ unsubstituted as shown by IR and electrophoresis) and
the direct analogies to the IR and rates of chemical and
enzymatic dephosphorylation of V.

The reversal by L-glutamine of the antibacterial
activity of I is not surprising since methionine sulfoximine
is a known inhibitor of glutamine synthetase[21] and it has
been shown[22] that in the process of inhibition methionine
sulfoximine is phosphorylated by the enzyme to the N^5-
phosphono compound which is then irreversibly bound to the
active site of the enzyme.

The remarkably similar bacterial product,[10] phosphino-
thricyl-L-alanyl-L-alanine, VI, is also an inhibitor of
glutamine synthetase.

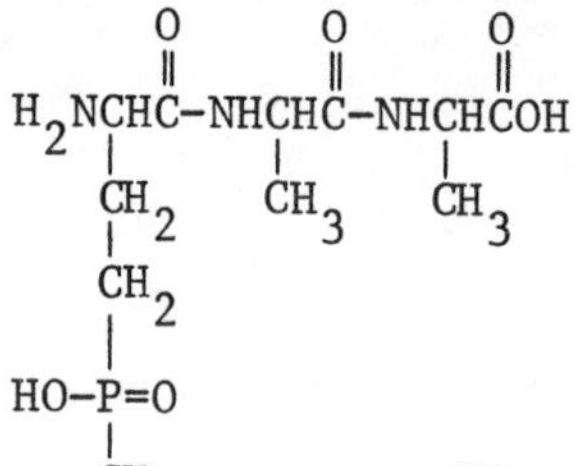

VI

References

1. Hanka, L. J. Proc. 5th Int. Congr. Chemother. B 9/2, 351 (1967).
2. Nass, G., K. Poralla, and H. Zähner. Naturwissenschaften 58, 603 (1971).
3. Scannell, J. P., D. L. Pruess, T. C. Demny, F. Weiss, T. Williams and A. Stempel. J. Antibiotics 24, 239 (1971).
4. DeVol, S. E., N. E. Rigler, A. J. Shay, J. H. Martin, F. C. Boyd, E. J. Backus, J. H. Mowat, and N. Bohonos. *Antibiotics Annual* (New York: Medical Encyclopedia, Inc.) p 730.
5. Walker, J. E., and E. P. Abraham. Biochem. J. 118, 557, 563 (1970).
6. Baggaley, K. H., B. Blessington, C. P. Falshaw, W. D. Ollis, L. Chaiet, and F. J. Wolf. Chem. Comm. 1, 101 (1969).
7. Stewart, W. W. Nature 229, 174 (1971).
8. Kondo, S., H. Yamamoto, K. Maeda, and H. Umezawa. J. Antibiotics 24, 732 (1971).
9. Molloy, B. B., D. H. Lively, R. M. Gale, M. Gorman, L. D. Boeck, C. E. Higgins, R. E. Kastner, L. L. Huckstep, and N. Neuss. J. Antibiotics 25, 137 (1972).
10. Bayer, E., K. H. Gugel, H. Hägele, H. Hagenmaier, S. Jessipow, W. A. König, and H. Zähner. Helv. Chim. Acta 55, 224 (1972).
11. Davis, B. D., and E. S. Mingioli. J. Bact. 60, 17 (1950).
12. Pruess, D. L., J. P. Scannell, T. C. Demny, H. A. Ax, F. Weiss, M. Kellett, and A. Stempel. J. Antibiotics. In press.
13. Short, L. N., and H. W. Thompson. Nature 166, 514 (1950).

14. Fiske, C. H., and Y. Subba Row. J. Biol. Chem. <u>66</u>, 375 (1925).
15. Rowe, W. B., R. A. Ronzio, and A. Meister. Biochemistry <u>8</u>, 2674 (1969).
16. McIlvaine, T. C. J. Biol. Chem. <u>49</u>, 183 (1921).
17. Greenstein, J. P. Adv. Prot. Chem. <u>9</u>, 121 (1954).
18. Christensen, B. W., and A. Kjaer. Chem. Comm. <u>4</u>, 169 (1969).
19. Gray, W. R., and B. S. Hartley. Proc. Biochem. Soc. <u>89</u>, 59P (1963).
20. Campbell, P. N., T. S. Work, and E. Mellanby. Biochem. J. <u>48</u>, 106 (1951).
21. Pace, J., and E. E. McDermott. Nature <u>169</u>, 415 (1952).
22. Ronzio, R. A., W. B. Rowe, and A. Meister. Biochemistry <u>8</u>, 1066 (1969).

A COMPARISON OF THE STRUCTURAL AND FUNCTIONAL PROPERTIES OF NERVE GROWTH FACTOR AND INSULIN

*Ralph A. Bradshaw, William A. Frazier, Ruth Hogue Angeletti.** Washington University School of Medicine, Department of Biological Chemistry, St. Louis, Missouri 63110

SUMMARY--The data presented on the similarity of the effects of insulin and nerve growth factor (NGF) on their respective target tissues and particularly the metabolic stimulation of NGF sensitive neurons by insulin lends strong support to the concept that insulin and NGF represent an example of distantly related proteins which have nevertheless retained a sufficient number of common structural features to allow them to perform related biological functions. Furthermore, the initial observations that insolubilized NGF, like insolubilized insulin, is biologically active indicate that the related function of these two proteins may well be expressed through a primary interaction with a receptor located on the surface membrane of the responsive cell.

POLYPEPTIDE MESSENGERS ARE AN integral part of the homeostatic control mechanisms of adult organisms. Unique among this class of molecules is nerve growth factor (NGF) which not only stabilizes the adult sympathetic nervous system, but also appears to direct its differentiation in the developing organism as well.[1] The wide species distribution of NGF, the specific cytotoxic effects of antiserum prepared against it[2] as well as the absolute requirement for NGF by

---

*Present Address: Laboratorio di Biologia Cellulare, CNR, Roma, Italia

sensory and sympathetic neurons *in vitro*,[1] all serve to establish the biological importance of this protein. NGF purified from male mouse submandibular glands[3] and several snake venoms[4],[5] has been shown to cause the specific hypertrophy of sympathetic ganglia *in vivo* and to stimulate the anabolic metabolism of sympathetic neurons *in vitro*.[6],[7]

Although the metabolic and morphological effects of NGF have been well documented, little is known about the mechanism by which these responses are elicited in its target cells. To provide the basis for experiments directed at this problem, detailed physical and chemical analyses of 2.5 S mouse submandibular gland NGF were undertaken. Ultracentrifugal analysis and quantitative end group determination revealed the 2.5 S molecule to be a dimer of very similar subunits.[8] Sequence analysis indicated that the primary subunit was composed of a 118 residue polypeptide chain of molecular weight 13,259 containing three intrachain disulfide bridges.[9],[10] Side chain amide assignments[10] were consistent with the previously determined isoelectric point of 9.3.[11]

In addition to more clearly defining the molecular properties of NGF, the elucidation of the sequence provided the opportunity for comparing the primary structure of NGF with that of other proteins for the purpose of identifying proteins related to NGF by plausible evolutionary events. Initial comparisons with enzymes and proteins of similar size did not yield any significant homologies.[10] However, when NGF was aligned with several proteins of the hormone class, which displayed some similarities in biological function, apparently significant relatedness was observed with portions of NGF and insulin. Thus, the known sequences of various insulins and proinsulins were compared with that of NGF and the results of these more detailed comparisons of the primary and secondary structure as well as an examination of the biological function of insulin and NGF have led us to postulate that these two proteins are not only derived from a common evolutionary precursor but also remain functionally and mechanistically related in their present versions.[12]

Structural Comparisons

The comparison of insulin (proinsulin) and NGF at the primary structure level is summarized in Figure 1. The sequence of human proinsulin (PI)[13] and guinea pig insulin[14] are positioned with NGF[9],[10] by alignment of the amino termini. Guinea pig insulin has been included in the alignment since it is the only other insulin sequence that supplies a significant number of additional similar residues. This arrangement directly yields the maximum number of identities, enclosed with solid boxes, and favored replacements, enclosed with dashed boxes. Favored amino acid replacements are defined in this comparison from the relatedness–odds matrix[15] as interchanged amino acid pairs with a value R_{ij} greater than 10, where R_{ij} is ten times the ratio of the probability that the particular amino acid substitution occurred during the evolutionary development of two related proteins divided by the probability that the substitution occurred by chance. In this alignment, residues 1-81 of NGF correspond to residues 1-86 of human PI. In order to achieve maximum similarity, five deletions were inserted in the PI structure.

A numerical analysis of this comparison[12] indicated that significant structural similarity existed in the regions corresponding to the A and B chains but not the C peptide segment. The lack of similarity in this region is not so surprising since the C peptide is quite variable in sequence as well as length even among the five proinsulin sequences which have been elucidated.[13,16-18] Of more importance is the fact that the C peptides of human (or rat) PI fit into the alignment (Figure 1) without any deletions allowing the regions of NGF which are most similar to the two insulin chains to align perfectly with their corresponding identities in the insulin sequence.

An additional feature of this alignment of the insulin molecule with NGF that must be considered is the remaining 37 residues of NGF not accounted for by the proinsulin molecule. As depicted in Figure 1, the alignment of a second B chain (designated as B') satisfies this disparity, and as described below, allows for a plausible genetic explanation. It is noteworthy that the extent of similarity, as manifested by identical residues, is very limited in this region and the statistical data[12] suggest an only slightly higher than random relationship. This situation is improved somewhat if the four additional identities, obtained by examination of all known insulin sequences[14,15] are included in the comparison. These replacements make a total of 7 identical residues out of 32 (22%) in the B' segment as compared to 11 of 21 (52%) and 9 of 30 (30%) for the A and B chain segments, respectively. It should be pointed out that this nonuniform distribution of insulin identities within the NGF sequence may be of significance in delineating regions of functional importance in the NGF molecule.

The three peptide segments contained within the PI
sequence have been shown to be arranged in the order B
chain-C peptide-A chain.[16]   Insulin is produced by proteolytic cleavage at each end of the C peptide leaving the
A and B chains joined by two of the three insulin disulfide
bridges. Thus the relation of the conserved regions to the
disulfide structure in each molecule is an essential part
of the comparison. These relationships are shown schematically in Figure 2. The consecutive line of circles
represents the linear amino acid sequence of NGF and its
relation to the corresponding residues of insulin is indicated by various degrees of shading. In order to emphasize
the functionally important A and B chains of insulin, the
NGF residues corresponding to these segments are indicated
by solid circles and the C bridge and B' segments by broken
circles. As can be seen by the distribution of the filled
circles, which represent the identical residues in the
region corresponding to the insulin A and B chains, and
the cross-hatched circles, which represent the identical
residues in the C bridge and B' segments, the highest
concentration of identities occurs in the region corresponding to residues 6-15 and 68-80 of NGF.

Three of the identities of this group are half-cystinyl
residues, two of which are bonded together in the same
manner as the proinsulin disulfide bond which connects the
B chain to the carboxyl region of the A chain (Cys B-19 to

Figure 1: The alignment of the amino acid sequence of
mouse submandibular gland NGF[9,10] with those of human
proinsulin[13] and guinea pig insulin.[14] Numbers above
the sets of lines are those of the NGF residue positions
and numbers below the sets of lines indicate the positions of the proinsulin and insulin residues. Solid
boxes enclose sets of identical residues and dashed boxes
enclose sets of residues considered to be favored amino
acid replacements, defined as those pairs of residues
with an R_{ij} value greater than the random value of 10.[15]
Asterisks denote the positions at which residues from
other insulins and proinsulins increase the number of
identities and favored replacements. Taken from reference
12. (Copyright 1972 by the American Association for the
Advancement of Science.

```
                    1                5              10              15
Mouse NGF    Ser - - Ser-Thr-His-Pro- - -Val-Phe-His-Met-Gly-Glu - - - -Phe-Ser-Val-Cys-Asp-Ser
Human PI     Phe-Val-Asn-Gln-His-Leu-Cys-Gly-Ser-His-Leu-Val-Glu-Ala-Leu-Tyr-Leu-Val-Cys-Gly-Glu
Gn.Pig Ins.  Phe-Val-Ser-Arg-His-Leu-Cys-Gly-Ser-Asn-Leu-Val-Glu-Thr-Leu-Tyr-Ser-Val-Cys-Gln-Asp
             B-1        B-5             B-10            B-15          B-20


                    20              25              30              35
Mouse NGF    Val-Ser-Val-Trp-Val-Gly-Asp-Lys-Thr-Thr-Ala-Thr-Asn-Ile-Lys-Gly-Lys-Glu-Val-Thr-Val
Human PI     Arg-Gly-Phe-Phe-Tyr-Thr-Pro-Lys-Thr-Arg-Arg-Glu-Ala-Glu-Asp-Leu-Gln-Val-Gly-Gln-Val
Gn.Pig Ins.  Asp-Gly-Phe-Phe-Tyr-Ile-Pro-Lys-Asp-[
                   B-25       B-30  C-1        C-5             C-10
                                                *


                    40              45              50              55
Mouse NGF    Leu-Ala-Glu-Val-Asn-Ile-Asn-Asn-Ser-Val-Phe-Arg-Gln-Tyr-Phe-Phe-Glu-Thr-Lys-Cys-Arg
Human PI     Glu-Leu-Gly-Gly-Gly-Pro-Gly-Ala-Gly-Ser-Leu-Gln-Pro-Leu-Ala-Leu-Glu-Gly-Ser-Leu-Gln
Gn.Pig Ins.
                    C-15            C-20            C-25            C-30
                              *


                    60              65              70              75
Mouse NGF    Ala-Ser-Asn-Pro-Val-Glu-Ser-Gly-Cys-Arg-Gly-Ile-Asp-Ser-Lys-His - -Trp-Asn-Ser-Tyr-
Human PI     Lys-Arg-Gly-Ile-Val-Glu-Gln-Cys-Cys-Thr-Ser-Ile-Cys-Ser-Leu-Tyr-Gln-Leu-Glu-Asn-Tyr-
Gn.Pig Ins.          Gly-Ile-Val-Asp-Gln-Cys-Cys-Ala-Gly-Thr-Cys-Thr-Arg-His-Glu-Leu-Gln-Thr-Leu-Tyr-
             C-35 A-1        A-5             A-10            A-15
                                      *               *


                    80              85              90              95              100
Mouse NGF    Cys-Thr-Thr-Thr-His-Thr-Phe-Val-Lys-Ala-Leu-Thr-Thr-Asp-Glu-Lys-Gln-Ala-Ala-Trp-Arg-
Human PI     Cys-Asn-COOH      NH2-Phe-Val-Asn-Gln-His-Leu-Cys-Gly-Ser-His-Leu-Val-Glu-Ala-Leu-
Gn.Pig Ins.  Cys-Asn-COOH      NH2-Phe-Val-Ser-Arg-His-Leu-Cys-Gly-Ser-Asn-Leu-Val-Glu-Thr-Leu-
             A-20       B'-1        B'-5            B'-10
                              *


                    105             110             115             118
Mouse NGF    Phe-Ile-Arg-Ile-Asp-Thr-Ala-Cys-Val-Cys-Val-Leu-Ser-Arg-Lys-Ala-Thr-Arg-COOH
Human PI     Tyr-Leu-Val-Cys-Gly-Glu-Arg-Gly-Phe-Phe-Tyr-Thr-Pro- - -Lys-Thr-Arg-Arg-C-peptide
Gn.Pig Ins.  Tyr-Ser-Val-Cys-Gln-Asp-Gly-Phe-Phe-Tyr-Ile-Pro - -Lys-Asp-COOH
             B'-20           B'-25           B'-30            B'-15
                                      *
```

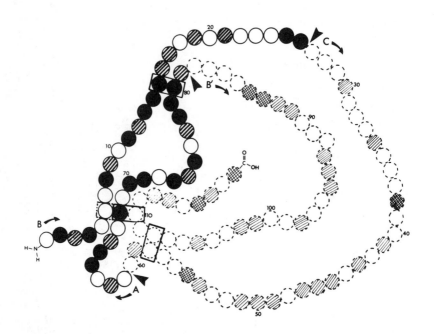

Figure 2: Schematic representation of the comparison of
the covalent structure of human proinsulin and mouse
2.5 S nerve growth factor.[9,10] Solid circles indicate
the portions of the NGF sequence which correspond to the
functional B and A chains of insulin as indicated. (NGF
1-26 and NGF 61-81). Broken circles represent the segments of NGF which correspond to the insulin C-peptide
and the repeated B chain, B'. Filled or cross-hatched
circles indicate residues identical in both sequences.
Diagonally shaded circles indicate favored amino acid
replacements.[12] The deletions introduced into the NGF
sequence (Figure 1) are not considered. The solid boxes
enclose pairs of half-cystinyl residues which form disulfide bonds in NGF. The dotted extension of the box
enclosing residues 68 and 110 in NGF indicates the
corresponding disulfide bridge in human proinsulin.
Arrowheads mark the peptide bonds for NGF which correspond to the cleavage points for the activation of
proinsulin.

Cys A-20).[19] The third conserved half-cystinyl residue is not bonded in the same manner in the two molecules. This residue, Cys 80, is linked to Cys 110 in NGF, as indicated by the solid box, while in insulin it is bonded to the residue corresponding to residue 6 of NGF (see Figure 1) as indicated by the dashed box. Thus, the partially conserved disulfide pairing strengthens the proposed relationship of the two proteins and indicates that some of the regions of NGF and insulin will have similar three-dimensional structures while other portions, particularly in the areas where the disulfide bonding is different, are probably quite dissimilar. The similarities and differences in biological action may be reflected in the same way as these structural features.

Evolution of Nerve Growth Factor

The similarities in primary and secondary structure of NGF described above suggest a structural and evolutionary relationship to proinsulin. Although the observed similarity is greatest with the functionally significant A and B chains, the portion of NGF corresponding to the repeated B chain is clearly suggestive of a gene duplication event. Two possible evolutionary routes, involving such steps and leading to the production of mouse NGF from an ancestral proinsulin-like molecule, are depicted in Figure 3. Both pathways invoke a duplication of the ancestral gene resulting in two independent genes in the manner now classical for proteins related by primary structure.[20] One gene would then have undergone a further duplication (contiguous reduplication, Figure 3) to produce a gene twice the length of the original, coding for a protein of some 170 amino acid residues, while the other evolved into proinsulin. Analogous events have been described for the human haptoglobins[21] and for the immunoglobulins.[22] From this point, two possible routes diverge. The first possibility would involve the deletion of the genetic material coding for the carboxyl-terminal C peptide and A chain regions of the double proinsulin-like peptide, so that a protein of some 120 residues would be produced. The second alternative would involve no further genetic events materially affecting the length of the gene, so that the primary gene product would thus be some 170 residues long. After translation, a proteolytic event (or events) would cleave this larger NGF molecule to yield 2.5 S NGF (118 residues), containing the regions corresponding to the first complete PI molecule plus the B' chain, and the carboxyl-terminal segment (*ca.*

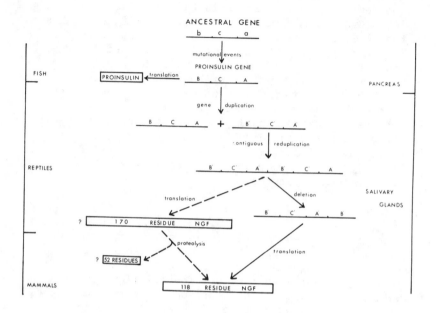

Figure 3:   A hypothetical scheme depicting the evolution
of NGF in its present form (118 residues) from an ancestral proinsulin-like molecule by a series of plausible
genetic events. Bars indicate genes, the modes dividing
them into regions corresponding to the peptide regions
of proinsulin. Boxes represent gene products (proteins)
and the dotted arrows indicate a hypothetical pathway
(see text). The scale on the left indicates an evolutionary era during which the most highly developed animal
extent was that named between the dividing marks and
during which the events depicted in the central part of
the figure are presumed to have occurred. The right hand
scale is comparable in time to that on the left, but the
organ has been listed in which the translational events
depicted in the central figure occur. Taken in part from
reference 12. (Copyright 1972 by the American Association
for the Advancement of Science).

52 residues) containing regions corresponding to the C'
peptide and A' chain.

While definitive evidence favoring one of these two
pathways is lacking, several considerations tend to support
limited proteolysis. For example, the carboxyl-terminal
residue of NGF is arginine. NGF is usually isolated from
the submandibular gland as part of a multiprotein complex,
as described by Varon *et al.*,[23] which contains as one of
its subunits a potent arginine esterase. Such an enzyme
could perform the required cleavage of the larger NGF
molecule at the carboxyl-terminus to produce the shortened
form. Furthermore, this carboxyl-terminal arginine corresponds to one of the sites at which proinsulin is cleaved.
A similar possibility based on the same structural feature
has been suggested for epidermal growth factor by Taylor
et al.[24] which is also isolated from mouse submandibular
glands.

It is interesting to note that the residues of NGF
which align with the two cleavage points of the first PI
molecule (NGF 27, 28 and NGF 60, 61) are not the argininyl
bonds which are cleaved in PI to produce insulin. This
observation raises the question as to whether they have
subsequently appeared in proinsulin or disappeared in NGF.
Although this question is not readily resolvable, it seems
probable that insulin has evolved directly as the zymogen.
This molecule probably originally possessed full activity,
which has been passed on, albeit in somewhat altered form,
to NGF which still exerts its physiological effects without
requirement for cleavage within the part of the NGF molecule
that corresponds to proinsulin. It is a direct consequence
of this hypothesis that the proinsulin precursor gene must
have acquired inactivity by mutation which has been subsequently counteracted at the phenotypic level by proteolytic
activation.

The evolutionary development of the organs of origin of
these two proteins is also indicated in Figure 3. Interestingly, the salivary glands of vertebrates parallel the
appearance of NGF while insulin and the pancreas appeared
significantly earlier, being found even in primitive fish.
Salivary glands of the vertebrate type are first found in
amphibians.[25]

The more extensive sequence data available for insulins
of different species[14,15] allows some reconstruction of the
genetic events marking the pathways illustrated in Figure 3.
These data indicate that insulins can be subdivided into
two classes--fish and mammalian. The higher degree of
similarity of mouse NGF with the mammalian insulins suggests

that the first gene duplication event (Figure 3) occurred
after the divergence of fish and higher vertebrates. There
is, however, insufficient information, particularly with
regard to the structure of other NGF molecules to allow
any more detailed conclusions.

Physical Chemical Comparisons

It has been well established that proteins of the same
class, *e.g.* the serine proteases, which display sequential
similarity, including conservation of the intrachain disulfide crosslinks, possess high degrees of similarity in
the course of folding of the main polypeptide chains.[26]
In the absence of crystallographic analyses, three-dimensional comparisons must be conducted at less specific
levels by examination of various physical and chemical
properties. One such property which is known to depend
on the three-dimensional arrangement of amino acid residues
is the solution aggregation behavior of proteins. Insulin,
proinsulin and NGF all exist as specific dimers or higher
even aggregates over a wide range of pH.[8,27-29] A unique
property of insulin and proinsulin is the polymerization
to hexamers induced by zinc which depends on the histidine
at position 10 of the insulin B chain.[30] Since this
histidine is conserved as His-8 in the NGF sequence, it
may be possible for NGF to polymerize in the presence of
zinc. Experiments in progress to test this hypothesis
have been complicated by the presence of chains missing
the first eight residues (and hence, His-8) in some preparations of NGF.[9,10]

Two parameters which also reflect the polypeptide chain
conformation are the optical rotatory dispersion (ORD) and
circular dichroism (CD) spectra of proteins. Frank and
Veros[28] have reported ORD and CD spectra for insulin and
proinsulin at neutral pH, and concluded that the spectra
are representative of approximately 25% α-helical structure
in insulin and 14% α-helix in proinsulin. They concluded
that the lower percentage in proinsulin is due to the
additional residues of the C peptide in an "unordered" or
"random" conformation (Table I). The percentage α-helix
in the insulin structure is borne out by the model deduced
from crystallographic studies.[30,31] The ORD and CD spectra
of NGF at pH 5 and pH 8.5 indicate that the molecule consists of approximately equal proportions of unordered and
β structure with a low (*ca.* 7%) α-helix content.[32] Thus,
the number of residues in the α-helical conformation is
approximately the same in insulin, proinsulin and NGF, the

Table I

The Relative Helicity of Insulin, Proinsulin and Nerve Growth Factor
as Derived from Circular Dichroic Spectra

Protein	*Position of Negative Extremum (nm)*	*Magnitude of Negative Extremum Mean Residue Ellipticity*	*Approximate % α-Helix*	*Number of Residues in Polypeptide Chain*	*Approximate Number of Residues in α-Helix*
Insulin[a]	208	14,000	25	51	13
Proinsulin[a]	207	10,000	14	84	12
Nerve Growth Factor[b]	207	6,000	7	118	8

[a]From reference 28.
[b]Per cent α-helix calculated for NGF using the depth of the trough at 208 nm.[32]

remainder of each being composed of unordered and β structure (Table I). Particularly in the case of NGF, this
conclusion is supported by the large number (45%) of helix
destabilizing residues[33] in the sequence. The conformation
of NGF as revealed by ORD and CD measurements is thus consistent with the possibility that some regions of the
three-dimensional structure of NGF may parallel portions
of proinsulin or insulin. Conclusive proof of this hypothesis, however, must await the collection of high
resolution structural data for both proinsulin and NGF.

Functional Similarities and Mechanistic Implications

The conservation of specialized function through large
spans of evolution is a well documented concept. Thus,
while a structural comparison of NGF and insulin has shown
them to be distantly related, a consideration of the functional properties of the two proteins indicates close
similarities in their effects on their respective target
tissues. For example, the spectrum of stimulatory anabolic
effects of insulin on 3T3 cells *in vitro,* referred to as
the "positive pleiotypic response,"[34] exactly parallels
the metabolic response of sensitive neurons to NGF. A
process that is in some ways distinct from the metabolic
stimulation is nerve fiber outgrowth. This morphological
"end point" of the *in vitro* NGF response, occurs in the
presence of actinomycin D [7] (but not *p*-fluorophenylalanine[6]
or cycloheximide[7]) and may thus represent post-transcriptional control of protein synthesis.[7] Insulin has been
shown to act at a post-transcriptional stage in the case
of tyrosine amino transferase induction in rat hepatoma
cells.[35] Just as protein hormones amplify tissue-specific
processes, NGF has recently been shown to increase the
activity of enzymes specific for noradrenergic function,
including tyrosine hydroxylase and dopamine-β-hydroxylase.[36]
Particularly interesting with regard to the functional
similarity of insulin and NGF is the fact that insulin can
stimulate the anabolic metabolism of NGF-sensitive sympathetic neurons.[37,38] While many agents are capable of
producing a growth response in various cell types, this
observed physiological "cross-reaction" of insulin with
NGF sensitive nerve tissue strongly suggests that the
structural similarities noted for these two proteins reflect the conservation of related function at the level
of cellular response.
As in the case of NGF, little is known about the initial
biochemical events which trigger the insulin response in

its target tissues. However, Cuatrecasas has used insoluble derivatives of insulin coupled covalently to Sepharose to demonstrate that insulin acts at the surface membrane of responsive cells.[39] Furthermore, all of the insulin binding activity of fat and liver cells resides in the plasma membrane fraction of cell homogenates[40] from which a protein retaining insulin binding activity has been purified.[41,42]

To explore the possibility that NGF may also act at the cell membrane of responsive neurons, insoluble derivatives of NGF were prepared by reacting NGF at pH 6.4 with cyanogen bromide-activated Sepharose. This procedure (essentially that of Cuatrecasas[43]) produces a highly substituted derivative containing 2 mg (0.15 µmoles) of NGF per ml of packed Sepharose beads. After extensive washing, 5 ml of NGF Sepharose was washed with one liter of water which was then lyophilized and redissolved in 1 ml; no NGF activity could be detected in this wash with the biological assay which is sensitive to less than 20 nanograms of NGF per ml.

Since native 2.5 S NGF is a tightly associated dimer,[8] the possibility existed that only one subunit of the dimer might be covalently linked to the Sepharose allowing the other subunit to dissociate during the prolonged incubations. To eliminate this possibility, NGF-Sepharose was incubated in 6 M guanidine hydrochloride, pH 5, for 24 hr at 37° followed by further washing with 6 M guanidine hydrochloride and finally with buffer. This treatment resulted in the release of very little NGF from the resin indicating that most of the molecules were bound covalently through each subunit of the dimer. The guanidine HCl treatment ensures that no noncovalently bound NGF remains in the NGF-Sepharose preparation.

The insoluble NGF-Sepharose derivatives prepared in this manner were tested for biological activity. One of the most reliable indications of NGF activity is the stimulation of nerve fiber outgrowth from primary explants of eight day chick embryo dorsal root ganglia. These ganglia explants were cultured in liquid medium and in plasma clot preparations with NGF-Sepharose and control Sepharose. Since the ganglia are only about ten times the size of a Sepharose bead (~600µ *vs* ~60µ), the number of beads that can pack around a ganglion is quite limited, and only a few cells at the surface of each ganglion can be stimulated. Thus in these experiments, which depend on the morphological response, one cannot expect to observe neurite outgrowth of the type stimulated by soluble NGF in which all responsive cells contribute to the fiber outgrowth. Nevertheless, when ganglia explants are cultured with NGF-Sepharose for

24-48 hours, tufts of nerve fibers can be seen originating
at points on the surface of the ganglia which are in close
contact with Sepharose beads. Longer term cultures maintained up to ten days in the presence of NGF-Sepharose
exhibit large numbers of surviving neurons which appear
in clumps covered with Sepharose beads. Nerve fibers,
many several millimeters long, form a network of thick
bundles which span the entire distance between the groups
of neurons which remain at the position at which the
ganglia were originally placed. Figure 4 shows a part of
such a culture maintained 9 days with NGF-Sepharose.

Figure 4:   The effect of NGF-Sepharose on 8-day old chick
   embryo dorsal root ganglia after long term culture (9 days).
   Bundles of nerve fibers can be seen to connect two sensory
   ganglia. The dark mass in the lower left is a dorsal root
   ganglion covered with Sepharose beads. The Sepharose
   beads clinging to the surface of a second ganglion are
   seen in the upper right. The distance between the two
   ganglia is approximately 1 mm. Additional nerve fibers
   which lie in other focal planes can also be seen. Control
   ganglia incubated with untreated Sepharose showed no
   fiber outgrowth over the same period (Nomarski Optics,
   ca. 150X).

Control explants cultured in the presence of untreated
Sepharose show no fiber outgrowth during any stage of culture. In these experiments, the Sepharose beads do not
adhere to the ganglia and after a few days only fibroblastic
and other nonneuronal cell types survive, distributed
randomly throughout the culture.

These observations are consistent with the hypothesis
that NGF exerts its action through interaction with the
cell membrane of responsive neurons. Furthermore, the NGF
response noted in these cultures was not due to dissociation
from the Sepharose of non-covalently bound NGF since the
results were the same whether the NGF-Sepharose had been
treated with guanidine HCl or only washed with large volumes
of buffers. There remain, however, other means by which
soluble NGF activity could be generated from the Sepharose
derivative. These include the release of enzymes by the
cells in culture which could degrade the Sepharose to yield
NGF attached to a few sugar residues or enzymes that could
proteolyze the insoluble NGF to produce active fragments.
These possibilities are unlikely since no gross degradation
of the Sepharose was noted even in long term cultures, and
experiments to generate active fragments of NGF with known
proteolytic enzymes have thus far yielded no active peptides.
Experiments to more rigorously eliminate the generation of
soluble NGF activity from the insolubilized derivatives
are presently in progress.

While the similarities in the metabolic function of NGF
and insulin described above tend to support the structural
relationship observed for the two proteins, these common
effects can be considered little more than coincidental in
the absence of information about the means by which these
actions are initiated. The experiments with insolubilized
NGF supply this type of information and indicate that, like
insulin, the primary action of NGF is its interaction with
the cell membrane. Thus the observed functional similarities
between insulin and NGF may well be manifested through a
common primary mechanism and as such may indeed reflect
the persistence through divergent evolution of common
structural features.

Acknowledgments

The authors express their appreciation to Mr. Delio
Mercanti for excellent technical assitance. They also
gratefully acknowledge the helpful discussions of Drs.
Rita Levi-Montalcini, Milton Goldstein, Peter A. Neumann,
and Garland R. Marshall. Supported by a research grant

from U.S. Public Health Service, NS-10229. Ralph A.
Bradshaw, Research Career Development Awardee of the
National Institutes of Health, AM-23968; William A.
Frazier, National Science Foundation Predoctoral Trainee.

References

1. Angeletti, P. U., and R. Levi-Montalcini. Developmental
   Biology 7, 653 (1963).
2. Levi-Montalcini, R., and P. U. Angeletti. Pharmacol.
   Rev. 18, 619 (1966).
3. Bocchini, V., and P. U. Angeletti. Proc. Natl. Acad.
   Sci. U.S. 64, 787 (1969).
4. Cohen, S. J. Biol. Chem. 234, 1129 (1959).
5. Angeletti, R. H. Proc. Natl. Acad. Sci. U.S. 65, 668
   (1970).
6. Levi-Montalcini, R. and P. U. Angeletti. Physiological
   Rev. 68, 534 (1968).
7. Partlow, L. M., and M. G. Larrabee. J. Neurochem. 18,
   2101 (1971).
8. Angeletti, R. H., R. A. Bradshaw, and R. D. Wade.
   Biochem. 10, 463 (1971).
9. Angeletti, R. H., W. A. Frazier, and R. A. Bradshaw.
   Proc. of the Third International Cong. of Neurochemistry.
   In press.
10. Angeletti, R. H., and R. A. Bradshaw. Proc. Natl.
    Acad. Sci. U.S. 68, 2417 (1971).
11. Bocchini, V. Eur. J. Biochem. 15, 127 (1970).
12. Frazier, W. A., R. H. Angeletti, and R. A. Bradshaw.
    Science 176, 482 (1972).
13. Oyer, P. E., S. Cho, J. D. Peterson, and D. F. Steiner.
    J. Biol. Chem. 246, 1375 (1971).
14. Smith, L. F. Amer. J. Med. 40, 662 (1966).
15. Dayhoff, M. O., ed. *Atlas of Protein Sequence and
    Structure 1969*, Vol. 4 (Silver Spring, Md.: Natl.
    Biomedical Res. Foundation), Chapter 9.
16. Nolan, C., E. Margoliash, J. D. Peterson, and D. F.
    Steiner. J. Biol. Chem. 246, 2780 (1971).
17. Chance, R. E., R. M. Ellis, and W. W. Bromer. Science
    161, 165 (1968).
18. Sundby, F., and J. Markussen. Eur. J. Biochem. 25,
    147 (1972).
19. Brown, H., F. Sanger, and R. Kitai. Biochem. J. 60,
    556 (1955).
20. Ingram, V.M. Nature 189, 704 (1961).
21. Dixon, G. H. "Mechanisms of Protein Evolution," Essays
    in Biochemistry 2, 147 (1966).

22. Hill, R. L., R. Delaney, R. E. Fellows, and H. E. Lebovitz. Proc. Natl. Acad. Sci. U.S. 56, 1762 (1966).
23. Varon, S., J. Nomura, and E. Shooter. Biochem. 6, 2202 (1967).
24. Taylor, J. M., S. Cohen, and W. M. Mitchell. Proc. Natl. Acad. Sci. U.S. 67, 164 (1970).
25. Junqueira, L. C. U. In *Proc. of Int. Cong. on Mechanisms of Salivary Secretion and Regulation*, Schneyer, L. and C. Schneyer, eds. (New York: Academic Press, 1967) p 286.
26. Desnuelle, P., H. Neurath, and M. Ottensen, eds. *Structure-Function Relationships of Proteolytic Enzymes* (New York: Academic Press, 1970).
27. Jeffrey, P. D. and J. H. Coates. Biochem 5, 489 (1966).
28. Frank, B. H., and A. J. Veros. Biochem. Biophys. Res. Comm. 32, 155 (1968).
29. Zuhle, H., and J. Behlke. FEBS Letters 2, 130 (1968).
30. Adams, M. J., T. L. Blundell, E. J. Dodson, G. G. Dodson, M. Vijayan, E. N. Baker, M. M. Harding, D. C. Hodgkin, B. Rimmer, and S. Sheats. Nature 224, 491 (1969).
31. Blundell, T. L., J. F. Cutfield, S. M. Cutfield, E. J. Dodson, G. G. Dodson, D. C. Hodgkin, D. A. Mercola, and M. Vijayan. Nature 231, 506 (1971).
32. Greenfield, N., and G. D. Fasman. Biochem. 8, 4108 (1969).
33. Fasman, G. D. In *Poly-α-amino Acids*, Vol. 1, Fasman, G. D., ed. (New York: Marcel Dekker, Inc., 1967) Chapter 11.
34. Hershko, A., P. Mamont, R. Shields, and G. Tomkins. Nature New Biol. 232, 206 (1971).
35. Gelehrter, T., and G. Tomkins. Proc. Natl. Acad. Sci. U.S. 66, 390 (1970).
36. Thoenen, H., P. U. Angeletti, R. Levi-Montalcini, and R. Kettler. Proc. Natl. Acad. Sci. U.S. 68, 1598 (1971).
37. Levi-Montalcini, R. *The Harvey Lectures, Series 60* (New York: Academic Press, 1966) p 217
38. Partlow, L. M. Ph.D. Thesis, The Johns Hopkins University, Baltimore, Md., 1969.
39. Cuatrecasas, P. Proc. Natl. Acad. Sci. U.S. 63, 450 (1969).
40. Cuatrecasas, P. Proc. Natl. Acad. Sci. U.S. 68, 1264 (1971).
41. Cuatrecasas, P. Proc. Natl. Acad. Sci. U.S. 69, 318 (1972).
42. Cuatrecasas, P. Proc. Natl. Acad. Sci. U.S. 69, 1277 (1972).
43. Cuatrecasas, P. J. Biol. Chem. 245, 3059 (1970).

SIGNIFICANCE AND INTERACTION OF THE AMINO ACID RESIDUES IN POSITIONS 1, 2, 3 AND 8 OF VASOPRESSINS ON CONTRACTILE ACTIVITY IN VASCULAR SMOOTH MUSCLE

Burton M. Altura. Departments of Anesthesiology and Physiology, Albert Einstein College of Medicine of Yeshiva University, Bronx, New York

UNTIL RECENTLY, VASOPRESSIN WAS given little serious thought as a useful vasoactive drug because of its well documented coronary constrictor action. However, evidence has now accumulated to suggest that at least two synthetic analogues of vasopressin namely [2-phenylalanine, 8-lysine]-vasopressin (PLV-2) and [8-ornithine]-vasopressin, may have significant antiarrhythmic properties.[1,2] Furthermore, PLV-2 has been reported to increase the relative myocardial blood flow in rats[3] and to relax a variety of isolated bovine coronary arteries.[4] Possibly more importantly, PLV-2 has been reported to be very beneficial in the treatment of various forms of experimental and clinical shock syndromes.[5-9] The anti-shock action of PLV-2 is thought to be due to its unusual microcirculatory actions.[6,7,9-12] Interestingly, several synthetic analogues of vasopressin, including PLV-2, have been shown, by direct *in vivo* microscopy, to have a predominant constrictor (or contractile) action on muscular venules in the microcirculation; *i.e.*, changes in the molecular structure of [8-arginine]--or [8-lysine]--vasopressin in positions 2, 3 and 8 can change the affinities of the altered peptide for a particular type of peripheral microscopic blood vessel (*e.g.*, arteriole, venule, metarteriole, precapillary sphincter).[9-12] In view of such surprising findings we initiated systematic pharmacologic and structure-activity studies of the neurohypophyseal hormones (NHPH), and their synthetic analogues, not only

441

at the microcirculatory level[9-12] but on various types of
isolated mammalian blood vessels[4],[12-15] since the latter
would be divorced from any *in vivo* effects of metabolism
and blood flow; factors which could compromise rat pressor
assays.

It is generally believed that the degree of basicity
of the amino acid residue in position 8 of vasopressin is
probably the single critical structural factor for optimizing the pressor or vasoconstrictor properties of the
molecules.[16],[17] This hypothesis is, however, primarily
based on crude intravenous blood pressure assays in
adrenergically blocked rats. But such *in vivo* assays may
not give accurate estimates of either the potencies or
affinities* of these peptide molecules on vascular smooth
muscle.[10-13],[15],[18] For example, we recently presented
data on a variety of isolated canine blood vessels[12-14]
which suggest that: (a) the *length* of the amino acid side
chain interacting with the basicity in position 8 of the
vasopressins may be extremely important for both affinity
and intrinsic (contractile) activity* of these peptides on
vascular smooth muscle; (b) the phenolic and aromatic
groups in positions 2 and 3, respectively, of the vasopressins may also be of importance not only for affinity but
intrinsic contractile activity as well; (c) there might
be chemical or steric differences from one target site to
another within the receptor molecules in different blood
vessels, even within a single mammalian species; and
(d) different isolated canine blood vessels from different
regional vascular beds appear to exhibit different depen-
dencies on magnesium ions for [8-arginine]-vasopressin-
induced contractions.[14]

Since our previous studies were done on isolated canine
blood vessels and most all of the structure-activity data
derived so far for the NHPH has been projected on the basis
of rat pressor assays, several possibilities could be enter-
tained to explain the discrepancies between our data and
that of others: (1) species differences, (2) *in vitro
versus in vivo*, (3) arterial *vs.* arteriolar responsiveness
(*i.e.*, heterogeneity of drug receptors which subserve contraction may exist with respect to the structure-activity
relationships of the NHPH and analogues on blood vessels
[11-15],[20]), (4) not enough analogues were examined in our
in vitro studies, and (5) complex autonomic reflex actions
may complicate rat pressor assays.[18] The present study using

---

*For a thorough discussion of affinity and intrinsic activity
see reference 19.

12 NHPH peptide analogues and direct *in vivo* microscopy on
rat mesenteric arterioles and isolated rat aortas was therefore designed to elucidate these contingencies. Delineation
of structure-activity correlations are strategic requirements
for the design of new NHPH with more precise vasomotor
selectivity, especially in view of the therapeutic potential
of such molecules in the treatment of low-flow states.[5-9]
    Isolated, helically cut thoracic rat aortic strips were
set up for isometric recording as described previously.[21]
The rat mesocecal mesenteric preparation[10] was used for the
direct *in vivo* study of arterioles. *In vivo* quantitative
microscopic observations were carried out on rat mesenteric
arterioles by means of an image-splitting television microscope recording system.[22] The latter TV system, at magnifi-
cations up to 4,000 times, allowed us to quantitatively make
rapid measurements of lumen diameters in response to topical
application of 0.1 ml volumes of the neurohypophyseal peptides. Such a system has very recently been effectively
used to make rapid *in vivo* micrometric measurements from
which complete log dose-response curves have been constructed
for drug effects on various kinds of muscular microvessels
including arterioles.[23-25]
    The present *in vitro* and *in vivo* findings on rat arterial
and arteriolar smooth muscle (Figures 1 and 2) suggest that

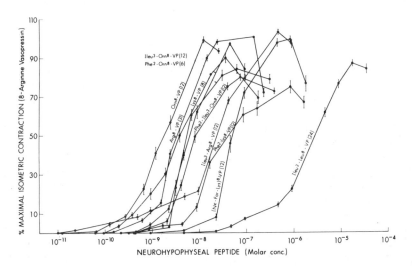

Figure 1: Comparative contractile actions of neurohypophyseal
   hormones and synthetic analogues on isolated rat aortas.
   Numbers in parentheses denote the number of different male
   rats utilized. Values are means ±S.E.M. Cumulative log
   dose-isometric response curves.

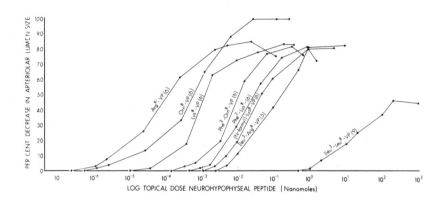

Figure 2: Graded contractile responses of rat mesenteric
   arterioles to topically applied neurohypophyseal hormones
   and synthetic analogues. Note that the abscissa is expressed in nanomoles. Each point represents the mean
   value ± S.E.M. obtained from *in vivo* measurements on
   vessels from different male rats (indicated by numbers
   in parentheses). The mean control lumen sizes for the
   arterioles range from 22 to 30μ.

an optimum interaction between the length of side chain and
basicity in position 8 is, indeed, necessary for maximal
activation (or contraction) of smooth muscle cells by the
vasopressin receptor in mammalian somatic vascular muscle.[12,13]
Maximal basicity alone in position 8 does not in itself promote optimum contractile activity (*e.g.*, [8-ornithine]-
vasopressin is more potent than either [8-arginine]--or
[8-lysine]--vasopressin). Furthermore, the present data
on rat blood vessels support our previous suggestion that
the phenolic and aromatic groups in positions 2 and 3,
respectively, are involved in both the affinity and the
intrinsic activity of the hormone on vascular muscle.[12,13]
Analogues lacking either of these functional groups (*e.g.*,
[2-phenylalanine, 8-lysine]-vasopressin, [3-isoleucine,
8-ornithine]-vasopressin, [3-isoleucine, 8-arginine]-
vasopressin) have reduced affinity and intrinsic activity
(Figures 1 and 2).

The absence of only the functional amino group also results in a marked loss of both affinity and intrinsic activity (Figure 3). This is, thus, quite different from

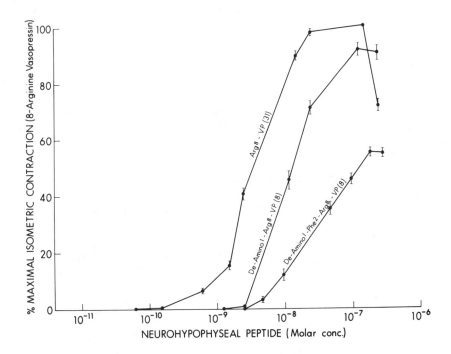

Figure 3: Comparative contractile actions of [8–arginine]–vasopressin, deamino–[8–arginine]–vasopressin and deamino–[2–phenylalanine, 8–arginine]–vasopressin on isolated rat aorta.

that which has been reported for this functional group in oxytocin on uterine smooth muscle.[26] Simultaneous absence of both the functional amino group and the phenolic hydroxyl (*e.g.*, 1–deamino–[2–phenylalanine, 8–arginine]–vasopressin) results in further losses in both affinity and intrinsic activity of the hormone on rat blood vessels (Figure 3). Thus the amino group in position 1, the phenolic hydroxyl in position 2 and the aromatic group in position 3 all are required to promote optimum hormonal contractile activity on mammalian somatic blood vessels.

The widely divergent relative affinities and intrinsic
activities seen for a variety of NHPH analogues on rat
aorta *versus* rat arterioles could not only be used to support our previous suggestion that the vasopressin receptor
may not be identical on blood vessels within a single
mammalian species,[12-14] but could aid in explaining some
or all of the discrepancies observed between our observations and crude rat pressor assays. These present data
when coupled with previous observations[4,10-15,23,25]
emphasize the importance of a comparative pharmacologic
approach in the analysis of structure-activity relationships
for NHPH in the cardiovascular system and could be used to
buttress the concept that drug molecules with *selective
regional vascular actions* can be realized.

Acknowledgments

This study was supported by research grants HL-12462
and HL-11391 from the National Institutes of Health, USPHS.
I thank Mr. C. F. Reich and Mr. R. W. Burton for their excellent technical assistance. I am deeply indebted to Dr.
B. Berde, Sandoz Ltd., and Dr. R. Walter, Mt. Sinai School
of Medicine, for generously supplying the pure, synthetic
peptides used in these studies. Burton M. Altura, recipient
of Research Career Development Award 5-K3-GM-38, 603 from
the N.I.H., USPHS.

References

1. Katz, R. L. Anesthesiology 26, 619 (1965).
2. Nielsen, O. V., and N. Valentin. Acta Obstet. Gynec.
   Scand. 49, 45 (1970).
3. Berde, B. In *Advances in Oxytocin Research*, Pinkerton,
   J. H. M., ed. (London: Pergamon Press, 1965) pp 11-35.
4. Altura, B. M. Amer. Heart J. 72, 709 (1966).
5. Altura, B. M., R. Hsu, V. D. B. Mazzia, and S. G.
   Hershey. Proc. Soc. Exp. Biol. Med. 119, 389 (1965).
6. Altura, B. M., S. G. Hershey, and V. D. B. Mazzia.
   Amer. J. Surgery 111, 186 (1966).
7. Hershey, S. G., and B. M. Altura. Schweiz. Med. Wschr.
   96, 1467 (1966), German.
8. Cohn, J. N., F. E. Tristani, and I. M. Khatri. Circulation 38, 151 (1968).
9. Altura, B. M., S. G. Hershey, and B. T. Altura. Adv.
   Exp. Med. Biol., Vol. 8: *Bradykinin and Related Kinins*
   (New York: Plenum, 1970) pp 239-247.

10. Altura, B. M., S. G. Hershey, and B. W. Zweifach. Proc. Soc. Exp. Biol. Med. 119, 258 (1965).
11. Altura, B. M., and S. G. Hershey. Angiology 18, 428 (1967).
12. Altura, B. M. In *Physiology and Pharmacology of Vascular Neuroeffector Systems*, Bevan, J. A., R. F. Furchgott, R. A. Maxwell, and A. P. Somlyo, eds. (Karger, Basel, 1971), pp 274-290.
13. Altura, B. M. Amer. J. Physiol. 219, 222 (1970).
14. Altura, B. M. Experientia 26, 1089 (1970).
15. Altura, B. M., D. Malaviya, C. F. Reich, and L. R. Orkin. Amer. J. Physiol. 222, 345 (1972).
16. Walter, R., J. Rudinger, and I. L. Schwartz. Amer. J. Med. 42, 653 (1967).
17. Bodanszky, M., and G. Lindeberg. J. Med. Chem. 14, 1197 (1971).
18. Krejčí, I., B. Kupková, I. Vávra, and J. Rudinger. European J. Pharmacol. 13, 65 (1970).
19. Ariëns, E. J. *Molecular Pharmacology*, Vol. 1 (New York: Academic Press, 1964).
20. Altura, B. M., and B. T. Altura. European J. Pharmacol. 12, 44 (1970).
21. Altura, B. M., and B. T. Altura. Microvascular Res., in press.
22. Baez, S. J. Appl. Physiol. 211, 299 (1966).
23. Altura, B. M. Microvascular Res. 3, 361 (1971).
24. Altura, B. M. Proc. Soc. Exp. Biol. Med. 138, 273 (1971).
25. Altura, B. M. Proc. Soc. Exp. Biol. Med. 140, 1270 (1972).
26. Chan, W. Y., and N. Kelley. J. Pharmacol. Exp. Ther. 156, 150 (1967).

DIRECTED BIOSYNTHESIS OF ANTIBIOTIC PEPTIDES WITH ISOLEUCINE STEREOISOMERS AND DL–PIPECOLIC ACID

Edward Katz, Joseph V. Formica, Takehiko Yajima, Mary Ann Grigg.* Department of Microbiology, Georgetown University Schools of Medicine and Dentistry, Washington, D.C.

STREPTOMYCES CHRYSOMALLUS produces a number of actinomycins which contain D-valine, D-alloisoleucine or both amino acids, whereas those synthesized by *S. antibioticus* contain D-valine (Figure 1).[1] *N*-Methyl-L-valine is also a normal constituent of the actinomycins. We have carried out studies to determine the effect of the four stereoisomers of isoleucine on actinomycin peptide formation.

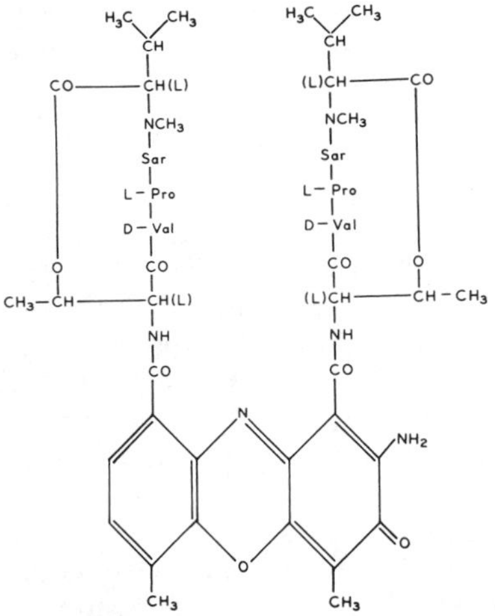

Figure 1: The structure of actinomycin D.

*Present Address: Department of Microbiology, Medical College of Virginia, Richmond, Virginia.

The influence of various concentrations of each of the four stereoisomers of isoleucine on the production of actino- mycin mixtures by *S. antibioticus* was examined. D-Alloiso- leucine and D-isoleucine were found to be quite inhibitory to antibiotic formation, whereas little effect was observed with L-isoleucine and L-alloisoleucine. The latter compound did inhibit antibiotic production, to some extent, for a 24-hour period.

Amino acid hydrolysates of actinomycin mixtures were examined by high voltage electrophoresis (4% formate, 4800 V, 3 hr) in one dimension, followed by paper chromatography (butanol-acetic acid-water, 4:1:5) in the second dimension. By this procedure, it was ascertained that isoleucine and *N*-methylalloisoleucine (but not *N*-methylisoleucine) were present in actinomycin preparations synthesized in the pre- sence of L- or D-isoleucine or L-alloisoleucine. Quantitative data of amino acids in actinomycin hydrolysates (*S. antibioticus*) were obtained with the Beckman Model 120 C amino acid analyzer. Control mixtures contained *N*-methyl-valine predominantly; but, in addition, small amounts of *N*-methylalloisoleucine were found. D-Valine also predominates, but trace amounts of isoleucine, alloisoleucine and even leucine were detected. Significant levels of *N*-methylalloisoleucine, isoleucine and even alloisoleucine were found with each of the four isomers of isoleucine supplied in the medium. Similar results were obtained in experiments with *S. chrysomallus*.

Isoleucine and *N*-methylalloisoleucine were isolated and purified from hydrolysates of the actinomycin mixture formed in the presence of D-isoleucine by *S. antibioticus*. Homogeneity and identity of the amino acids were established with standard amino acids by high voltage electrophoresis, paper and column chromatography.

Optical configuration of the amino acids was determined by optical rotatory dispersion.[2] *N*-Methyl-L-alloisoleucine exhibited a positive Cotton effect; by contrast, isoleucine showed a negative Cotton effect denoting that the isoleucine has the D-configuration. L-Leucyl dipeptides were prepared with the four stereoisomers of isoleucine and with the iso- leucine isolated from actinomycin mixtures by the method of Manning and Moore.[3] It was established that the dipeptide syn- thesized with the isoleucine isolated from actinomycin hydroly- sates had the same retention time as the standard of L-leucyl-D- isoleucine. Isoleucine purified from hydrolysates also was incubated with dialyzed preparations of L- or D-amino acid oxidase. The amino acid was destroyed after D-amino acid oxidase, but not after L-amino acid oxidase treatment, pro- viding further evidence that isoleucine has the D-configuration.

The mechanism of biosynthesis of D-isoleucine and *N*-methyl-L-alloisoleucine from the various stereoisomers of isoleucine remains to be established with cell-free systems. The interconversion of the four stereoisomers probably involves both enzymatic and non-enzymatic reactions with the keto acids (L-α-keto- and D-α-keto-β-methylvaleric acid) playing a key role in the process.[4] Presumably, the enzymatic methylation of the appropriate L-amino acid (*e.g.*, L-valine normally or L-alloisoleucine) involving S-adenosylmethionine leads to *N*-methylamino acid synthesis. Based on recent studies concerning the ATP-dependent racemization of L-phenylalanine to D-phenylalanine during gramicidin S and tyrocidine formation by *Bacillus brevis*,[5],[6] it is postulated that a single racemase with rather broad specificities for the L-form of the branched chain amino acids catalyzes an energy-dependent biosynthesis of the D-amino acids, D-valine, D-alloisoleucine and D-isoleucine. The data reveal that *N*-methyl-L-alloisoleucine and D-isoleucine substitute for *N*-methyl-L-valine and D-valine, respectively, in actinomycin peptides.

The actinomycins synthesized by *S. antibioticus* differ solely in the imino acid site of the antibiotic molecule.[1] When DL-pipecolic acid, the higher analogue of proline, was supplied during antibiotic formation, several new actinomycins were synthesized by *S. antibioticus*.[7] These antibiotics, isolated and purified from mixtures, were designated actinomycins Pip 2, Pip 1 alpha, Pip 1 beta, Pip 1 gamma, Pip 1 delta and Pip 1 epsilon. The amino acid composition of the various actinomycins was established qualitatively by a combination of high voltage electrophoresis and paper chromatography. Quantitative data reveal that the new actinomycins contain two residues each of D-valine, *N*-methyl-L-valine, sarcosine and L-threonine, but differ in the number of the imino acid residues (Pip 2: pipecolic acid – 2; Pip 1 beta: proline – 1, pipecolic acid – 1; Pip 1 alpha: pipecolic acid – 1, 4-oxopipecolic acid – 1; Pip 1 delta: proline – 1, 4-oxopipecolic acid – 1; Pip 1 gamma: pipecolic acid – 1, 4-hydroxypipecolic acid – 1; Pip 1 epsilon: proline – 1, 4-hydroxypipecolic acid – 1).

The presence of pipecolic acid and 4-oxopipecolic acid was confirmed by cochromatography and coelectrophoresis with authentic standards of these amino acids. Also, a comparison of hydrolysates of actinomycins Pip 1 alpha and Pip 1 delta with hydrolysates of Vernamycin B alpha and Ostreogrycin, peptide antibiotics known to contain 4-oxopipecolic acid, revealed the presence of the oxoimino acid in the actinomycin components. Further, oxopipecolic acid was reduced with sodium borohydride to a mixture of *cis*-4-hydroxypipecolic

acid and *trans*-4-hydroxypipecolic acid as reported by
Clark-Lewis and Mortimer.[8] The identity of the imino acid
in actinomycins Pip 1 gamma and Pip 1 epsilon as *trans*-4-
hydroxypipecolic acid was established also by cochromatography
and coelectrophoresis with authentic *cis*- and *trans*-4-
hydroxypipecolic acid and with 3- and 5-hydroxypipecolic
acid.

The effect of the six pipecolic acid-containing
actinomycins on the *E. coli* DNA-dependent RNA polymerase
reaction was also studied. The data correlate well in that
actinomycins containing at least one proline residue were
found to be more inhibitory than those containing a pipecolic
acid residue in place of proline. Moreover, the actinomycins
that contain hydroxypipecolic acid (Pip 1 epsilon, Pip 1
gamma) exhibited the lowest activity for each series of
compounds. The order of activity: actinomycin IV = Pip
1β > Pip 1 delta > Pip 1 epsilon > Pip 2 > Pip 1 alpha >
Pip 1 gamma.

Normally, *S. antibioticus* synthesizes proline from
glutamic acid during actinomycin production.[1] Proline is
incorporated directly into actinomycin peptides or is
modified via hydroxylation or oxidation reactions to give
trans-4-hydroxy-L-proline (actinomycin I) or 4-oxo-L-
proline (actinomycin V), respectively.[9,10] When supplied
with the proline analogue, pipecolic acid, the organism
apparently catalyzes similar types of biochemical reactions.
The utilization of these compounds results in the synthesis
of novel actinomycins with quantitatively differing biological
properties.

Acknowledgment:

The investigation was supported by research grant
CA-06926 from the National Cancer Institute, U.S.P.H.S.

References

1. Katz, E. Biosynthesis Vol. II: *Antibiotics* (New York:
   Springer Verlag, 1967) pp 276-341.
2. Jennings, J. P., W. Klyne, and P. M. Scopes. J. Chem.
   Soc. 294 (1965).
3. Manning, J. M., and S. Moore. J. Biol. Chem. 243,
   5591 (1968).
4. Meister, A. J. Biol. Chem. 195, 813 (1952).
5. Kurahashi, K., M. Yamada, K. Mori, K. Fujikawa, M. Kambe,
   Y. Imae, E. Sato, H. Takahashi, and Y. Sakamoto.
   Cold Spring Harbor Symposium Quantitative Biology 34,
   815 (1969).

6. Kleinkauf, H., and W. Gevers. Cold Spring Harbor Symposium Quantitative Biology 34, 805 (1969).
7. Katz, E., and W. A. Goss. Biochem. J. 73, 458 (1959).
8. Clark-Lewis, J. W., and P. I. Mortimer. J. Chem. Soc. 189 (1961).
9. Katz, E., D. Prockop, and S. Udenfriend. J. Biol. Chem. 237, 1585 (1962).
10. Salzman, L. A., H. Weissbach, and E. Katz. Proc. Nat. Acad. Sci. U. S. 54, 542 (1965).

MECHANISM OF THROMBIN ACTION ON FIBRINOGEN; ACTIVITY OF
THROMBIN TOWARD HUMAN α(A)-FIBRINOGEN PEPTIDES

L. C. Dorman, R. C. Cheng, F. N. Marshall. Chemical
Biology Research, The Dow Chemical Company, Midland,
Michigan 48640

ONE OF THE KEY EVENTS involved in the clotting of blood is
the reaction of thrombin on fibrinogen.[1] Thrombin cleaves
the Arg-Gly bonds in the α(A) and β(B) chains of fibrinogen
releasing peptides A and B. Subsequently, the α and β
chains of the remaining protein, fibrin, are crosslinked
covalently by fibrin stabilizing factor forming a permanent
clot.

As far as proteins are concerned, the action of thrombin
is bond and protein specific, *i.e.*, only the Arg-Gly bonds
in the protein fibrinogen are cleaved by thrombin. Many
studies of this reaction indicate that the high specificity
of the reaction is not associated with the conformation of
fibrinogen. Instead the primary structure or amino acid
sequence around the Arg-Gly bonds have become suspect as
being responsible for this high specificity.[2] Shortly
after the 50-unit N-terminal sequence of the α(A) chain of
human fibrinogen was reported,[3] we initiated a study to
test this hypothesis.

Nonapeptide, H-Gly-Val-Arg-Gly-Pro-Arg-Val-Val-Glu-OH (*1*),
the 14-22 sequence of human α(A) fibrinogen was synthesized
by the solid-phase method.[4] When a 0.2% solution of *1* was
digested with thrombin (48 NIH units/ml)[5] and soybean
trypsin inhibitor (0.04% solution)[6] in 0.15M NH$_4$HCO$_3$, pH
8.2, at 37° for 4 hr, a rather complex cleavage pattern
was observed involving both Arg-Gly and Arg-Val peptide
bonds (Scheme I). Cleavage products were isolated by ion-
exchange and gel chromatography and identified by amino

Scheme I

Reaction of Thrombin on Synthetic Peptide *1*
[Human α(A) Fibrinogen-(14-22)-nonapeptide]

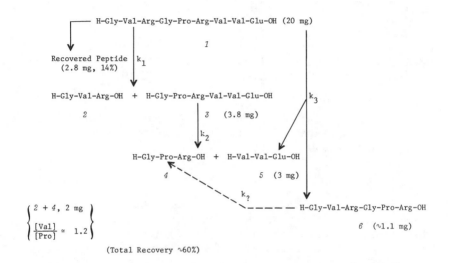

(Total Recovery ∿60%)

acid analyses and by comparison of their tlc and tle mobilities with authentic peptides.

Having demonstrated that *1* was a thrombin substrate, reactivity towards analogs of *1*, *i.e.*, *7-11*, was examined to determine the effects of structural change (cf. Table I). With regard to Arg-Gly cleavage, substitution of Pro[18] by Gly had little effect. Blombäck[3] had speculated that this proline residue aids in exposing Arg-Gly of the α(A) chain of fibrinogen to thrombin attack. Substitution of Arg[19] by Gly reduced the extent of Arg-Gly cleavage. The [Sar[17]] analog (*9*) was relatively stable to thrombin attack as was *10* in which adjacent Gly[17] and Pro[18] were reversed. Shortening of peptide *1* at the C-terminus,*i.e.*, *11* had no apparent effect on Arg-Gly cleavage. There was apparent greater specificity toward Arg-Val cleavage; Val-Val-Glu was readily formed from *1*, but very little or none was detected in the reactions of *7*, *9* and *10*. Thrombin cleavage of Arg-Val has been observed on large (43 and 50 residues) N-terminal fragments of the α(A) chain,[7,8] but the rate was reported[7] to be considerably slower than Arg-Gly cleavage.

Table I

Reaction* of Thrombin on Human α(A)
Chain Peptides and Analogs

Peptide	*Cleavage*[†]
14 15 16 17 18 19 20 21 22	
(*1*) H-Gly-Val-Arg-Gly-Pro-Arg-Val-Val-Glu-OH	+++
(*7*) H-Gly-Val-Arg-Gly-⌐Gly¬-Arg-Val-Val-Glu-OH	++
(*8*) H-Gly-Val-Arg-Gly-⌐Gly¬-Val-Val-Glu-OH	+
(*9*) H-Gly-Val-Arg-⌐Sar¬-Pro-Arg-Val-Val-Glu-OH	O
(*10*) H-Gly-Val-Arg-⌐Pro¬-⌐Gly¬-Arg-Val-Val-Glu-OH	O
(*11*) H-Gly-Val-Arg-Gly-Pro-Arg-Val-Val-⌐ ¬-OH	+++

*Digestion conditions: peptide (0.40%), bovine thrombin
 (100 units/ml) in 0.15 NH_4HCO_3 at pH 8.2 for 6 hr.
[†]As indicated by formation of Gly-Val-Arg and/or Val-Val-
 Glu.
+++=complete reaction
+=slight reaction
O=very little or no reaction

     To assess the relative potential of these peptides and
their fragments to bind thrombin, "thrombin times," *i.e.*,
inhibition of thrombin to clot fibrinogen, were determined.
Generally, thrombin times were not significantly above controls except for fragment[8] Gly-Pro-Arg (*4*), which was higher
than that of Tos-Arg-OMe (TAME), the standard thrombin
substrate. The high binding of *4* to thrombin prompted us
to look at some of its analogs and derivatives, *12-18*, presented in Table II with clotting data. These data show
that amino or carboxyl protection of Gly-Pro-Arg diminishes
thrombing binding. More significantly, a smaller protected
peptide, Boc-Pro-Arg-OBzl (*17*) displayed even higher thrombin
binding.

Table II

In Vitro Thrombin Times* for Gly-Pro-Arg and Derivatives

Peptide	$\overline{15}$ (sec)	Thrombin Times (sec) at Peptide Concentrations (mg/ml)				
		0.125 mg/ml	0.25	0.5	1.0	2.0
Control	15 (sec)	0	0	0	0	0
4 Gly-Pro-Arg·AcOH·H_2O		27 (sec)	32	46	76	191
12 Boc-Gly-Pro-Arg·TosOH		—	—	—	46	46
13 Boc-Gly-Pro-Arg-OBzl·TosOH		23	29	42	56	75
14 Gly-Pro-Arg-OBzl·TosOH·CF_3CO_2H		19	26	44	68	135
15 Pro-Arg·AcOH		+	—	+	+	+
16 Boc-Pro-Arg·TosOH·1/2 H_2O		—	—	—	47	48
17 Boc-Pro-Arg-OBzl·TosOH·MeOH		40	44	73	91	159
18 Pro-Arg-OBzl·TosOH·CF_3CO_2H		19	23	31	46	69
Tos-Arg-OMe (TAME)		24	30	39	68	87

*Clotting mixture: 0.2 ml oxalated dog plasma, 0.1 ml barbital buffer containing peptide and 0.1 ml thrombin (1 unit/ml)

†Not significantly above control value.

References

1. Laki, K.   In *Fibrinogen*, Laki, K., ed. (New York: Marcel Dekker, 1968) pp 1-22.
2. Gladner, J. A.   In *Fibrinogen*, Laki, K., ed. (New York: Marcel Dekker, 1968) p 102; Liem, R. K. H., R. H. Andreatta, and H. A. Scheraga.   Arch. Biochem. Biophys. 147, 201 (1971); Andreatta, R. H., R. K. H. Liem, and H. A. Scheraga.   Proc. Nat. Acad. Sci. U.S. 68, 253 (1971).
3. Blombäck, B., M. Blombäck, B. Hessel, and S. Iwanaga. Nature 215, 1445 (1967).
4. Merrifield, R. B.   In *Advances in Enzymology*, Vol. 32 (New York: Interscience, 1969) pp 221-297; Dorman, L. C. and J. Love.   J. Org. Chem. 34, 158 (1969); Markley, L. D. and L. C. Dorman.   Tetrahedron Lett. 1787 (1970); Dorman, L. C.   Tetrahedron Lett. 2319 (1969).
5. Roberts, P. S., R. K. Burkat, and W. E. Braxton. Thromb. Diath. Haemorrh. 21, 103 (1969).
6. Marden, V. J., and N. R. Shulman.   J. Biol. Chem. 44, 2120 (1969); Lanchantin, G. F., J. A. Friedmann, and D. W. Hast.   J. Biol. Chem. 244, 865 (1969).
7. Blombäck, D., M. Blombäck, A. Henschen, B. Hassel, S. Iwanaga, and K. R. Woods.   Nature 218, 130 (1968).
8. Iwanaga, S., P. Wallen, N. J. Gröndahl, A. Henschen, and B. Blombäck.   Eur. J. Biochem. 8, 189 (1969).

STUDIES ON BRADYKININ-POTENTIATING PEPTIDES

Duane A. Tewksbury. Marshfield Clinic Foundation for
Medical Research and Education, Marshfield, Wisconsin.

BRADYKININ-POTENTIATING PEPTIDES have been isolated from
snake venoms.[1,2,3] Peptides with this activity have also
been obtained from the plasmin hydrolysis of fibrinogen and
fibrin,[4] from the trypsin hydrolysis of plasma[5,6] and from
the trypsin hydrolysis of albumin.[6] This paper reports the
study of the peptides obtained from the enzymatic digestion
of four different proteins with five different proteolytic
enzymes.

Materials and Methods

Reagents

The proteins that were used as substrates were albumin
(Mann, human fraction V), casein (Sigma, technical grade),
fibrin obtained by clotting outdated blood bank plasma with
thrombin and Bence Jones protein of the λ type isolated from
human urine by ammonium sulfate fractionation. The enzymes
used were trypsin (Worthington, TPCK treated, 200 units/mg),
plasmin (KABI, human in 50% glycerol), pronase (Calbiochem,
B grade, 45,000 PUK units/g), pepsin (Worthington, 3,000
units/mg) and chymotrypsin (Worthington, alpha chymotrypsin
3x crystalized).

Enzymatic Hydrolysis

Hydrolyses were carried out at 37°C for 3 hr with 1%
substrate solutions. E:S was 1:1,000 except the concentration of plasmin which was 10 units/100 ml. For the pronase

hydrolysis of casein E:S was 1:100. The following buffers were used: 0.008M sodium phosphate, pH 7.5, for plasmin hydrolysis; 0.01M HCl for pepsin hydrolysis; 0.08M tris-HCl, pH 7.8, for chymotrypsin hydrolysis; and 0.05M tris-HCl, pH 8.1, for trypsin and pronase hydrolysis except for the trypsin and pronase hydrolysis of casein where 0.05M sodium phosphate, pH 8.1, was used.

Fractionation of Hydrolysates

The hydrolysates were centrifuged and then fractionated on Amicon ultrafilters at 4°C. The digests were first ultrafiltered using a PM 10 membrane which retains material with a molecular weight greater than 10,000. The material which passes through this membrane was again ultrafiltered using the UM 2 membrane which retains material with a molecular weight greater than 1,000. After four volumes of water have been passed through the ultrafilter the material retained by the UM 2 membrane was freeze-dried. This material constitutes the peptide fraction.

Bioassay

The bradykinin-potentiating activity of each peptide fraction was determined on the isolated guinea pig ileum as previously described.[7]

Results and Discussion

It is to be noted from Table I that bradykinin-potentiating activity was demonstrated in 19 out of the 20 peptide fractions. The peptide fraction obtained from the plasmin digest of fibrin did not exhibit appreciable bradykinin-potentiating activity but since such hydrolysates have previously been reported to possess bradykinin-potentiating activity[4] it is suggested that the active peptides may have been of a different size than those presently studied. Hamberg and Stelwagen suggested that the bradykinin-potentiating activity of tryptic peptides was due to the basic C-terminal amino acid residue. The fact that peptides produced by the action of pronase, pepsin and chymotrypsin also have bradykinin-potentiating activity would suggest that this activity is due to the amino acid sequence of the peptide rather than to its having a basic C-terminal residue.

Table I

Bradykinin-Potentiating Activity
of the Peptide Fractions

Enzyme	*Amount Required for 2-fold Potentiation (µg/ml)**			
	Bence Jones	*Albumin*	*Fibrin*	*Casein*
Trypsin	300	100	150	300
Pronase	350	400	300	200
Pepsin	500	100	200	80
Plasmin	300	450	N.A.	65
Chymotrypsin	500	450	150	100

N.A. – Not active at 500 µg/ml.

*Bradykinin-potentiating activity is expressed as the
concentration of peptide, µg/ml of bath solution, required
to increase the effect of a dose (x) of bradykinin to match
the effect of a double dose (2x). Thus the potentiating
activity is inversely proportional to the dose required
for 2-fold potentiation.

References

1. Ferreira, S. H., D. C. Bartelt, and L. J. Greene.
   Biochemistry 9, 2583 (1970).
2. Kato, H., and T. Suzuki. Biochemistry 10, 972 (1971).
3. Ondetti, M. A., N. J. Williams, E. F. Sabo, J. Pluscec,
   E. R. Weaver, and O. Kocy. Biochemistry 10, 4033 (1971).
4. Malofiejew, K. B., and M. Czokalo. Bull. Acad. Pol.
   Sci. 14, 193 (1966).
5. Hamberg, U., P. Elg, and P. Stelwagen. Scand. J. Clin.
   Lab. Invest. 24, Suppl. 107, 21 (1969).
6. Aarsen, P. N. Brit. J. Pharmacol. Chemother. 32, 453
   (1968).
7. Tewksbury, D. A., and M. A. Stahmann. Arch. Biochem.
   Biophys. 112, 453 (1965).

SYMPOSIUM DISCUSSIONS

Summarized by Johannes Meienhofer

A DISCOURSE AROSE about interpreting the large negative
rotation of antamanide (pp 377 to 396), *e.g.* in dioxane,
in terms of bent amide bonds. This interpretation was
based on examinations of *N*-acetylprolinamide derivatives
and on comparisons with CD spectra of enniatine, but other
interpretations might also be possible. Shifts of negative
CD absorption to a positive absorption indicate in proline
peptide bonds *trans* to *cis* isomerization, and this could
be suggestive but not conclusive evidence for the type of
transformation occurring in antamanide when changing from
the uncomplexed to the complexed form.

The complete structure elucidation of the pentadecapeptide amide scotophobin still requires the determination
of acid or amide functions in positions 2 (Asx), 5 and 11
(Glx). Two of eight possible isomers were prepared by
solid-phase synthesis, and one was found to exhibit high
biological activity (pp 397 to 402). An argument about
the validity of scotophobin assays arose which could not
be resolved, but it was agreed that even full biological
activity does not *per se* establish identity with natural
scotophobin because one or more of the other not yet prepared analogs might also be active. It was suggested
that the mass spectral analysis of the synthetic products
might provide further information about their relationship
to natural scotophobin. With respect to the biological
role of scotophobin, the question arose whether the information for fear of darkness might be contained in the
peptide or whether the information might be present in
the brain and the peptide effects its release.

Several suggestions were made for further chemical
studies on the interesting phagocytosis stimulating tetrapeptide tuftsin (pp 403 to 407), but further work will have

465

to await the isolation of more material. This will be
laborious because tuftsin occurs only in nanomolar quantities and elicits its activity at levels of 0.05 µg/ml. It
is assumed that cyclic AMP plays a role in its mechanism
of action. Tuftsin is linked to the heavy chain of a γ-
globulin carrier by an ester bond between a protein carboxyl
and the hydroxyl group of tuftsin's *N*-terminal threonine.
In response to an inquiry about the danger of an *O→N* shift,
the audience learned that this shift does indeed occur
during isolation of the tuftsin-globulin complex to an
extent that the recovery of ester is in the range of 15-18%.

The interesting comparison of nerve growth factor (NGF)
with insulin (pp 423 to 439) stimulated a lively discussion
about bioassay procedures, about the observation that NGF
has an almost equal tendency to aggregate as insulin, and,
therefore, commonly occurs as a dimer, and about the amount
of cross-reactivity of the two proteins in radioimmunoassays
which amounts to 5-10%. Concerning attachment of proteins
through their amino groups to Sepharose by cyanogen bromide,
it was pointed out that at pH values above 5 release of
protein was observed at rates of nanomoles/min to micromoles/
min. The breakage was found to occur at the attachment site
and was probably effected by neighboring hydroxyl groups.
Another discussant revealed that concern about this problem
led to the use of crosslinked agarose. If the cyanogen
bromide coupling is carried out at pH 9-10 instead of at
previously employed pH 11, one can obtain very stable
bonding and no release of protein occurs. The application
of statistical analysis to the examination of homologies
between NGF and insulin (pp 423 to 439) was welcomed by a
discussant, because serious studies should confine comparisons between proteins to those parts of the sequences
that have statistical significance.

In a discourse following the paper on the comparative
contractile actions of vasopressin analogs on vascular
smooth muscle (pp 441 to 447) auto inhibitory effects,
frequently observed in dose-response curves above an optimal
concentration, were mentioned. Apparently such effects
vary between different blood vessels in the dog and the
same analog may show no auto inhibition at all at some
vessels but an even increased effect in others. These
varying responses among several analogs might be construed
as evidence for a second type of receptor which subserves
relaxation. It was also observed that several analogs
with lower intrinsic activities possessed much higher
affinities. These phenomena are not surprising, a discussant pointed out, if one assumes a mechanism of action

in which the process of binding is not identical with the process of stimulus initiation, perhaps involving separate molecular regions. V. du Vigneaud reminisced on how some pharmacologists at a Federation Meeting in the early fifties argued passionately in favor of changing the name "vaso-pressin" to "antidiurethin."

SECTION VII

ANGIOTENSIN

Session Chairmen

Robert Schwyzer and F. Merlin Bumpus

PHYSIOLOGICAL ROLES OF ANGIOTENSIN

Michael J. Peach. Department of Pharmacology,
University of Virginia, School of Medicine,
Charlottesville, Virginia

SUMMARY--The proteolytic enzyme renin splits an α_2-globulin
to produce angiotensin I. Angiotensin II is a product
formed from angiotensin I by converting enzyme. Acute
injections of angiotensin II produce an increase in systolic/
diastolic blood pressure. The peptide can induce an elevation
of blood pressure for hours or days when administered by
infusion. Angiotensin causes marked vasoconstriction in
splanchnic, coronary, hepatic and cutaneous vascular beds
while blood flow in the uterus, brain and skeletal muscle
may increase. In pressor structure-activity-studies, amino
acid residues #4, 6, 7 and 8 represent important activity
sites. Angiotensin II causes a positive chronotropic cardiac response via stimulation of the sympathetic innervation
of the myocardium. Angiotensin-induced increases in myocardial contractile force represents a direct effect of the
peptide on cardiac muscle. Structure-activity-relationship
studies in the papillary muscle indicate that proline is
not required as the #7 amino acid residue. Angiotensin
also interacts with the nervous systems. A central neurogenic vasoconstrictor response to angiotensin has been well
documented. These central effects are mediated via increased
efferent sympathetic activity. Angiotensin also stimulates
the adrenal medulla and autonomic ganglia and facilitates
peripheral sympathetic nervous transmission. Structural
requirements established for angiotensin-induced adrenal
catecholamine release indicate that a C-terminal aromatic
amino acid is not necessary for stimulus-secretion coupling.
Autonomic ganglia also do not discriminate C-terminal

471

structural changes that markedly alter smooth muscle contractile responses. Angiotensin stimulates the biosynthesis
of aldosterone in the adrenal cortex. The peptide appears
to stimulate the first step in the biosynthesis of aldosterone; the conversion of cholesterol to pregnenolone. Preliminary structure-activity studies in the adrenal cortex
indicate that a metabolite of angiotensin II, des-asp[1]-
angiotensin II (heptapeptide); is at least as active as
the octapeptide in stimulating the synthesis and release
of aldosterone. Analogs of angiotensin II with C-terminal
aliphatic amino acids substituted for phenylalanine inhibit
smooth muscle, papillary muscle, adrenal medullary,
autonomic ganglia and adrenal cortical responses to
angiotensin II.

FORMATION AND DEGRADATION--The name of this small, remarkably
active peptide describes the activity for which it is best
known (angio=vessel, tensin=tension). The peptide, angiotensin II, is the product of 2 enzymatic reactions. Renal[1,2]
and possibly other tissues (*i.e.* placenta, uterus, salivary
gland, brain)[3-6] contain a proteolytic enzyme called renin.
Renin substrate is an α_2-globulin and the product of this
reaction is a decapeptide designated angiotensin I. Angiotensin I is the substrate for a dipeptidyl carboxypeptidase
(converting enzyme) and this reaction yields an octapeptide,
angiotensin II, and histidyl-leucine.[7]
   Endogenous or exogenous angiotensin II is rapidly degraded by plasma and possibly tissue enzymes. Three plasma
enzymes have been described. (1) Angiotensinase A$_1$, is an
aminopeptidase specific for N-terminal asparagine. This
enzyme is inhibited by EDTA, has a pH optimum of 7.4 and
is stable at 60°C for 30 min.[8,9] (2) Angiotensinase A$_2$ is
an aminopeptidase specific for N-terminal aspartic acid.
It is also inhibited by EDTA, has a pH optimum of 6.8 and
is heat labile.[9] (3) Angiotensinase B is an endopeptidase
that splits the molecule into 2 tetrapeptides. This enzyme[10]
is inhibited by DPF and has no activity at physiological pH.
A tissue carboxypeptidase that metabolizes angiotensin II
has been isolated from liver and kidney and is called
angiotensinase C.[11,12] A summary of this brief presentation
is shown in Figure 1. All work in this area has been reviewed recently by Page and McCubbin[13] and Fisher.[14]

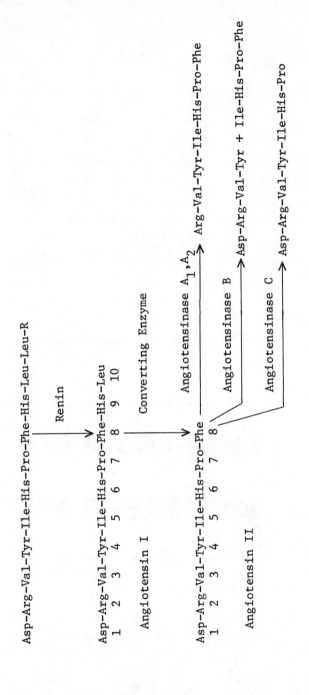

Figure 1: Pathways for formation and degradation of angiotensin II
R, α_2-globulin.

Effects on the Cardiovascular System

Arterial Blood Pressure

The response to intravenous injections of angiotensin II in all species studied is characterized by about a 20 sec lag period followed by a sharp rise in systolic-distolic pressure. This increase reaches maximum in 1-2 min and declines to the initial level in 3-5 min.[15] The response to small doses is constant and repeatable while large doses produce tachyphylaxis.[16] Intraarterial injections of angiotensin II produce less of an effect on blood pressure than intravenous doses.[17] Angiotensin pressor responses induce a reflex bradycardia which is mediated by the vagus[18],[19] and is blocked by treatment with atropine. Representative arterial blood pressure and heart rate responses to angiotensin II before and after treatment with atropine are presented in Figure 2.

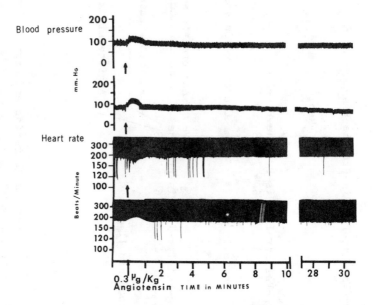

Figure 2: Representative femoral arterial blood pressure and heart rate responses to 0.3 μg/kg intravenous angiotensin II in the rabbit. The upper blood pressure recording is control and the lower recording is from the same animal after treatment with atropine, 0.5 mg/kg. Heart rate recordings also depict responses before (upper tracing) and after (lower tracing) atropine treatment.

Angiotensin can induce an elevation of blood pressure
for hours or days when administered by intravenous infusion.
Very low, acutely subpressor doses of angiotensin produce
a slow progressive rise in arterial pressure[20],[21] while
doses of peptide that cause an immediate pressor response
tend to lose their effect over a period of days unless the
dose is continuously increased.[22],[23]

Regional Blood Flow

Blood flow responses to angiotensin differ qualitatively
and quantitatively in different vascular regions. Antio-
tensin reduces mesenteric blood flow in the cat, dog and
man apparently by a constrictor effect on small arteries.[24],[26]
In studies with perfused hind-limbs of rabbit, rat, cat and
dog, angiotensin produces a vasoconstriction.[27-29] Coronary
vasoconstriction occurs in isolated perfused hearts,[30] heart-
lung preparations[31] and *in vivo* in the cat[24] and dog.[32]
There is an increase in blood flow in the uterus of dogs
and sheep following angiotensin administration. When the
peptide is administered intravenously in dogs and cats there
is an initial increase in skeletal muscle blood flow.[24-34]
In the cat this vasodilatation is inhibited by β-adrenergic
blockade and adrenal ligation.[34],[35] Cutaneous temperature
and blood flow are reduced by angiotensin due to a vasoconstrictor effect on cutaneous vessels.[26],[36] Cerebral blood
flow in the rat is increased by the administration of angiotensin.[37] In summary, angiotensin causes marked vasoconstriction in splanchnic,[3] coronary, heptatic and cutaneous
areas while blood flow in the uterus, brain and skeletal
muscle may increase. In regional vascular beds where blood
flow increases, this may simply reflect an elevation in
tissue perfusion pressure and not any direct effect of
angiotensin on the vasculature of these tissues.

Pressor Structure-Activity Relationships

Structure-activity relationship studies with pressor
and oxytocic assays have recently been reviewed.[13] The
following is a summary of the structural requirements of
angiotensin for activity in vascular and uterine smooth
muscle: (1) There must be at least 6 amino acids (residues 3-8 of angiotensin II) in the peptide chain; (2) The
presence of 8 amino acid residues (angiotensin II) yields
maximum activity; (3) A tyrosine residue is required in
position 4 of angiotensin II; (4) The imidazole ring of
histidine is required in position 6; (5) Proline must be

adjacent to position 8; (6) An aromatic amino acid with
free carboxyl must be in position 8.

Heart

Chronotropic Effects

In vivo administration of angiotensin produces a reflex
bradycardia.[18-19] If vagal tone is inhibited, angiotensin
induces a tachycardia. Tachycardia is markedly reduced by
β-adrenergic blockade, cardiac denervation or pretreatment
with reserpine.[38,39] This positive chronotropic effect
appears to be due to the release of norepinephrine from
sympathetic neurons in the heart and not dependent on release of adrenal medullary catecholamines.[40,41] Some
studies indicate that the tachycardia is due to angiotensin-
induced stimulation of the stellate[42] or caudal cervical
ganglia[43] while other reports suggest a direct effect on
adrenergic nerve endings.[44,45]
      In isolated perfused rabbit heart[46] or spontaneously
beating atria[47] angiotensin has no effect on cardiac rate.
This supports the suggestion that the tachycardia induced
by angiotensin is dependent on an intact sympathetic innervation of the myocardium. Large doses of angiotensin prevent or reverse ventricular fibrillation induced by
chloroform, epinephrine[48] and toxic doses of g-strophanthin.[49]

Inotropic Effects

Angiotensin has a positive inotropic effect in dog heart-
lung preparation,[31] perfused cat heart,[50] isolated cat
papillary muscle[51,52] and isolated atria from cat, guinea
pig and rabbit.[47,53,54] In most of these preparations
responses are obtained with 10^{-10} to 10^{-9} M angiotensin.
The positive inotropic effect of the peptide was not altered
by lowering the Ca^{2+} concentration from 2.54 to 0.63 mM,
however, the effect was decreased at 100 mM Na^+ concentration and increased at 160 mM Na^+.[55] Angiotensin-induced
increases in contractile force are not affected by reserpine
pretreatment[31,47,52,55] β-adrenergic blockade,[51] and extrin-
sic cardiac denervation[50] indicating the peptide has a direct
effect on the myocardium. Hypoxia does not depress responses
to angiotensin II in the cat papillary muscle.[56]
      In vivo angiotensin probably exerts both this direct
inotropic effect and indirect inotropic activity via stimulation of the sympathoadrenal system.[34,43,57]

Structure-Activity-Studies in Papillary Muscle

The effects of angiotensin I, angiotensin II, angiotensin II-amide, des-(Asp^1)-angiotensin II and des-(Asp^1, Arg^2)-angiotensin II on contractility of the papillary muscle are presented in Figure 3. All these peptides, with

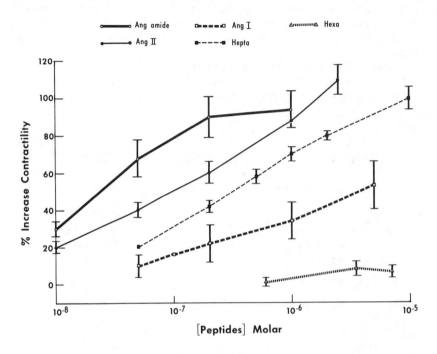

Figure 3: Comparison of the effects of angiotensin II-amide, angiotensin II, angiotensin I, hepta- and hexapeptides on myocardial contractility. Each point represents the mean ± S.E.M. of 8-12 experimental observations. Isolated papillary muscles were used in the study. Muscles were maintained at 35°C and paced by field stimulation at 60 beats/min with twice threshold voltage. Isotonic contractions were recorded from a base of 1 gram resting tension. Doses of each peptide are shown as molar concentration for ease of comparison.

the exception of hexapeptide, produced dose-dependent responses. Angiotensin II-amide was more potent than angiotensin II at concentrations from 10^{-9} to $3 \times 10^{-7} M$; however, there was no difference between these 2 octapeptides at concentrations of $10^{-6}M$ or greater. The heptapeptide [des-(Asp[1])-angiotensin II] was not as potent as angiotensin II (about 60% relative activity at any concentration studied) but did exert marked positive inotropic activity. Angiotensin I produced an increase in contractility at concentrations about $5 \times 10^{-8}M$. Angiotensin I was about 30% as active as angiotensin II. This positive inotropic effect of angiotensin I was markedly reduced (80%) by the addition of *Bothrops jararaca* pentapeptide, an inhibitor of converting enzyme.

Data obtained from the papillary assay using analogs of angiotensin with single substitutions of amino acid residues 1 through 8 are shown in Table I. Of particular interest

Table I

Comparison of Activities of Angiotensin Analogs
in Uterus and Heart

Peptide	Per Cent Activity Relative to Angiotensin II	
	Uterus	*Papillary Muscle*
[Ile[1]]-A II	20	40
[Arg[1]]-A II	55	75
(Poly-*O*-acetylseryl)-A II	40	90
[Ala[3]]-A II	30	100
[Tyr(OMe)[4]]-A II	1	8
[Ala[5]]-A II	5	30
[Ala[6]]-A II	0.1	1
[Ala[7]]-A II	0.1	50
[Hyp[7]]-A II	5	50
[Ile[8]]-A II	1	1

is the remarkable activity displayed by 3 of these analogs, (poly-*O*-acetyl-seryl)-angiotensin II, [Ala[7]]-angiotensin II and [Hyp[7]]-angiotensin II. The poly-*O*-acetyl-seryl analog has a molecular weight of approximately 27,000 and should not cross a cell membrane, however, on a molar comparison it was not different from angiotensin II. The two

7-substituted analogs appear to be specific stimulants of myocardial contractility.

The positive inotropic effects of angiotensin II are antagonized by [Ile^8]-angiotensin II. This data is presented in Table II. Other studies have reported that

Table II

Inhibition of Angiotensin II-Induced Inotropic Responses
by [Ile^8]-Angiotensin II

M Concentration	Per Cent Increase in Contractile Force		
	Control	*[Ile^8]-A II* $3x10^{-8}M$	*[Ile^8]-A II* $1x10^{-7}M$
3×10^{-9}	5 ± 1	0*	0*
6×10^{-9}	10 ± 2	0*	0*
1×10^{-8}	20 ± 3	0*	0*
6×10^{-8}	40 ± 4	7 ± 3*	0*
3×10^{-7}	60 ± 6	25 ± 5*	5 ± 1*
1×10^{-6}	80 ± 5	40 ± 6*	10 ± 4*
6×10^{-6}	110 ± 8	55 ± 6*	20 ± 4*

*$p < 0.01$.

[Ile^8]-angiotensin II blocks the myotropic effects of angiotensin II in smooth muscle.[58-59] An effective blockade in the papillary muscle is achieved with $3x10^{-8}M$ [Ile^8]-angiotensin II.

The structural studies in the papillary muscle indicate that proline is not required in position 7, which is a requirement in smooth muscle. The minimum chain length that is effective is 7 amino acid residues instead of 6 residues. It has been reported that the heptapeptide, des-(Asp^1)-angiotensin II, is much more active in hypoxic papillary muscle than it is in muscle at normal oxygen tension.[56]

Effects on the Nervous System

Central Nervous System

Bickerton and Buckley[60] first reported a central neurogenic vasoconstrictor response to angiotensin in 1961. This study involved cross-perfusion experiments in which a donor animal supplied blood to the head of a recipient. The head of the recipient received no blood from its own cardiovascular system but the nervous system of the recipient was intact. When angiotensin II was injected into the circulation of the donor or into the aterial supply to the head of the recipient a systemic pressor response resulted in the recipient. Since this systemic response was blocked by an α-adrenergic blocking agent, they concluded that angiotensin induced a central hypertensive effect mediated via the sympathetic nervous system. It was reported later that the central cardiovascular effects of angiotensin were greatly potentiated by β-adrenergic blockade with pronetholol.[61] The entire response was shown to be only partially dependent on the sympathetic nervous system since acute surgical sympathectomy reduced but did not abolish the response. These early experiments were not fully appreciated because of the large unphysiological doses of angiotensin required. Further studies revealed that when angiotensin was infused into the vertebral circulation of the unanesthetized rabbit hypertension resulted.[62] The dose of angiotensin that produced a centrally mediated pressor response had no effect when administered systemically. Several studies have now been reported from different laboratories and they all indicated that the central effects of angiotensin are mainly mediated by increased efferent sympathetic activity.[63-66] Fukyjama *et al.*[67] and Sweet *et al.*[68] have reported that the central hypertensive effect of angiotensin persisted when the peptide was administered continuously for 7 days. The studies of Ferrario *et al.*[69] indicated that these centrally induced pressor responses are due to an increase in total peripheral resistance. Microinjections of angiotensin into the area postrema on the caudal medulla resulted in a systemic pressor response.[70] Bilateral ablation of the area postrema in the dog completely abolished pressor responses to infusions of angiotensin into the vertebral arteries.[71,72] Recently it was reported that ablation of the area postrema also attenuated pressor responses to *i.v.* infusion of angiotensin.[73]

In the cat, angiotensin-induced central pressor responses occur when the peptide is administered intraventricularly.[63, 74,75] Deuben and Buckly[76] reported that the central site of action of angiotensin in the cat was the subnucleus medialis or nucleus mesencephalicus profundus.

Sympathoadrenal System

Sympathetic Nervous System

Studies in the early 1960's with cross-perfusion experiments,[60] ganglionic stimulation[77] and acute, surgical sympathectomy[29] suggested an interrelationship between angiotensin and the sympathetic nervous system. In 1963, McCubbin and Page[78] reported that angiotensin potentiated responses to drugs or reflexes that induced the release of norepinephrine. Benelli *et al.*[79] reported that angiotensin potentiated responses to nerve stimulation in the guinea pig vas deferens and cat spleen. Angiotensin-induced enhanced responses have also been reported in perfused mesenteric blood vessels,[80,81] perfused spleen following tyramine administration[82,83] and perfused rabbit ear with norepinephrine and tyramine.[84] Responses to tyramine in isolated aortic strips[85] and rabbit atria[47] are also potentiated by angiotensin administration. Angiotensin-induced facilitation can be blocked by α-adrenergic receptor blockade,[85,86] inhibition of norepinephrine release with bretylium[85] or depletion of norepinephrine stores with reserpine pretreatment.[85]

These effects of angiotensin on the peripheral adrenergic system occur with concentrations of angiotensin that have no direct effects on the assay organs. Angiotensin stimulates the biosynthesis of norepinephrine[87,88] in sympathetic neurons presumably by interfering with end-product inhibition of tyrosine hydroxylase, the rate limiting enzyme in the synthesis of adrenergic transmitter.

There are essentially two theories proposed to explain the mechanism of this interaction of angiotensin with peripheral sympathetic neurons. Evidence has been presented that angiotensin partially inhibits the neuronal uptake of norepinephrine.[41,46,81,89,90] Since uptake inhibition potentiates responses to norepinephrine this could explain facilitation induced by angiotensin. Structure-activity-relationship studies on uptake inhibition of norepinephrine in perfused heart[46] led us to postulate at least two receptors for angiotensin. This suggestion was based on activities displayed by several 8-substituted angiotensin analogs that did not correlate with activities from pressor/oxytocic assays. These studies also initiated a revitilization of studies with angiotensin analogs.

The second mechanism proposed to explain the interaction of angiotensin with sympathetic nerves is that angiotensin releases norepinephrine or sensitizes the neuron so that more norepinephrine is released per stimulus frequency.[79,91-94] Regardless of the exact mechanism, the evidence is overwhelming that angiotensin does modulate peripheral sympathetic activity resulting in an increase in the concentration of norepinephrine at the effector organ.

Adrenal Medulla

Release of adrenal catecholamines by angiotensin was first demonstrated in 1940, by Braun-Menendez *et al.*[95] using a crude peptide preparation. In 1959, Haas and Goldblatt[96] showed that infusions of a ganglionic stimulant, dimethyl-phenylpiperazinium iodide (DMPP), potentiated cardiovascular responses to angiotensin II. This DMPP-induced facilitation of responses to angiotensin II was abolished by bilateral adrenalectomy or administration of an α-adrenergic blocking agent phentolamine.[77]

Adrenalectomy in rats was reported to decrease the vasopressor action of angiotensin II, and direct perfusion of rat adrenals with angiotensin II caused an increase in catecholamine output.[97] Feldberg and Lewis (1964) demonstrated that injections of angiotensin II into the central stump of the celiac artery in the eviscerated cat induced marked adrenal catecholamine release.[98] In a subsequent study the intravenous administration of renin was reported to induce the secretion of adrenal catecholamines.[99]

In vivo studies in the dog estimating plasma catecholamines indicated that angiotensin II (*i.v.* infusion 0.05 μg/kg/min or *i.v.* injections of 0.5 μg/kg) evoked adrenal catecholamine release.[100] Other *in vivo*[101-102] *in situ*[103] and *in vitro*[104-105] studies on adrenal chromaffin stimulation by angiotensin II are in agreement with the findings of Feldberg and Lewis.[98]

Angiotensin II-induced adrenal medulla stimulation is not inhibited by bilateral splanchnicotomy, spinal block with a local anesthetic or the administration of hexamethonium, pentolinium, morphine, 5-hydroxytryptamine, histamine, bradykinin, kalliden and eledoisin.[99,102,106] In isolated adrenal medullary cells of the gerbil, angiotensin II has been shown to depolarize the chromaffin cell membrane.[107] This depolarizing effect was not altered by the administration of hexamethonium and atropine in doses that completely inhibited acetylcholine-, nicotine- and pilocarpine-induced membrane depolarization. All these studies indicate that angiotensin has a direct stimulatory effect on adrenal chromaffin tissue.

In the *in situ* perfused adrenal of the cat, Poisner and Douglas (1966) reported that Ca^{2+} was required for release of catecholamines by angiotensin II. This requirement for extracellular Ca^{2+} was confirmed in the isolated dog[104] and cat adrenal.[108] It would appear that angiotensin depolarizes the chromaffin cell membrane which then results in a net transfer of extracellular Ca^{2+} into the cell. It is not known if angiotensin has the ability to effect the translocation of intracellular Ca^{2+} and induce secretion-coupling without depolarizing the cell membrane; however, this would seem unlikely.

Analogs of angiotensin have been studied in the adrenal medulla and their activities relative to angiotensin have been determined.[95,102,105,108] Structural requirements established for stimulation of medullary catecholamine secretion are as follows: (1) there must be at least 6 amino acids in the peptide chain (residues 3-8 of angiotensin II); (2) a tyrosine residue is required in position 4 of the angiotensin II molecule; (3) the imidazole ring of histidine must be in position 6; and (4) proline must be adjacent to position 8.

Stimulation of adrenal chromaffin tissue by angiotensin can be blocked by several 8-substituted analogs, as [Leu[8]]-angiotensin II, [Val[8]]-angiotensin II, [Ile[8]]-angiotensin II, [8-cyclohexylalanine]-angiotensin II. Inhibition of angiotensin-induced secretion of adrenal catecholamines by [Ile[8]]-angiotensin II is shown in Figure 4. This 8-substituted analog is a competitive inhibitor of angiotensin in the adrenal medulla.

Parasympathetic Nervous System and Autonomic Ganglia

Part of the contractile response induced by angiotensin in isolated intestinal strips is due to the release of acetycholine from the enteric nerve plexus in the intestine.[109,110] The contraction of guinea pig ileum in response to angiotensin is a composite of two effects.[111] There is a fast component manifested as a rapid rise in tension and subsequent partial relaxation. The second effect is a slow, progressive contraction that reaches maximum in about 2 min. The initial rapid component can be blocked by atropine and morphine and potentiated by acetycholinesterase inhibitors.[112,113]

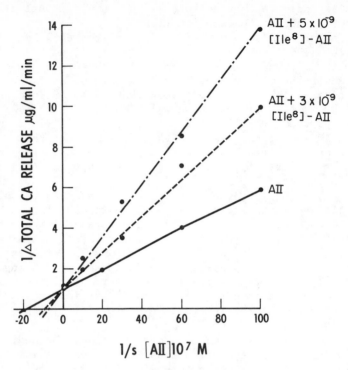

Figure 4: Inhibition of angiotensin II-induced adrenal
   catecholamine release by [Ile8]-angiotensin II. Studies
   were carried out in the isolated retrograde perfused
   adrenal of the cat. Each point represents the mean
   determined from 6-10 experimental observations. The
   8-substituted analog is a competitive inhibitor of
   angiotensin II.

In 1965, Lewis and Reit[114] first reported that close
intraarterial injections of angiotensin to the superior
cervical ganglion of cats produced a contraction of the
nictitating membrane. This angiotensin-induced response
was prevented by sectioning the postganglionic nerve but
was unaffected by chronic decentralization. Ganglionic
responses to angiotensin were unimpaired by treatment with
hexamethonium and atropine. Further studies showed that
morphine[115] and depolarization phase nicotine blockage
prevented[115,116] angiotensin-induced ganglionic stimulation.
During non-depolarization phase nicotine blockade of the

ganglion[115],[116] or following a burst of high frequency preganglionic stimuli,[115],[116] ganglionic responses to angiotensin were markedly potentiated. Angiotensin had very weak activity on the superior cervical ganglion of rabbits and no activity in dogs.[115] Angiotensin can also facilitate responses to preganglionic stimulation in the superior cervical ganglion of the cat.[117]–[119] This facilitation is felt to be due to increased release of acetylcholine from the preganglionic neuron.[120] Angiotensin has also been shown to stimulate the stellate ganglion in the cat.[42],[121] This stimulatory effect in the stellate is only on the adrenergic ganglion cells resulting in cardioacceleration.[121] Angiotensin-induced ganglionic stimulation leading to cardiac stimulation has also been reported in the dog,[39],[43] however, the ganglion involved was the caudocervical instead of the stellate.

There have been two structure-activity studies in autonomic ganglia, one, in the superior cervical ganglion[115] and, the other, in the enteric plexus of the intestine.[122] Ganglionic activities of one *N*-terminal substituted analog and two peptides with changes in position 4 correlated well with activities displayed in the adrenal medulla and on blood pressure.[115] Khairallah *et al.*[122] studied ten analogs of angiotensin in the guinea pig ileum. With substitutions of residues 1 to 7, they reported a parallelism between pressor responses and both the direct myotropic and acetylcholine-mediated responses of the ileum. Substitutions in position 8 yielded analogs with activities that did not parallel pressor responses induced by these peptides. The smooth muscle of the ileum appeared to discriminate between phenylalanine, tyrosine and *p*-methoxy-tyrosine substitutions in position 8 but the parasympathetic neurons did not. [Ala^8]-angiotensin II completely blocked both components of angiotensin-induced contraction of the ileum. This represented the first indication that aliphatic amino acids substituted for phenylalanine yield competitive inhibitors of angiotensin II.

Adrenal Cortex

Stimulation of Aldosterone

In cross circulation studies, blood from dogs with secondary hyperaldosteronism[123] or from sodium-depleted sheep[124] produced secretion of aldosterone in normal recipients. Further experiments showed that the kidney was involved in aldosterone production induced by acute

hemorrhage.[125] It was also reported that the administration
of angiotension II stimulated aldosterone secretion *in
vivo.*[126] Stimulation of aldosterone was specific for
angiotensin, and not for other pressor agents, with maximum
stimulation occurring at moderately pressor doses of the
peptide.[127,128] The effect of angiotensin in the adrenal
cortex was shown to be direct in studies that showed a
marked increase in aldosterone secretion in slices of bovine
adrenal[129] and isolated, perfused adrenals of dogs.[130]
Antibodies to renin were shown to block aldosterone secretion
induced by the administration of renin.[131] Renal and plasma
renin was found to be increased in dogs with caval constriction[132] or animals on low sodium diets.[133] Adrenal cortical
reponses to angiotensin can be reduced by an elevation of
plasma sodium content.[134] High sodium and/or low potassium
concentration appears to inhibit the stimulatory effect of
angiotensin on the adrenal cortex.[135-137] The rat appears
to be the one species in which angiotensin has a very weak
effect on the adrenal cortex.[138]

Angiotensin appears to stimulate the first steps in the
biosynthesis of aldosterone--probably the conversion of
cholesterol to pregnenolone.[129,139,140] In addition,
chronic sodium depletion[141] or renin administration[140] increases the conversion of corticosterone to adlosterone
without altering 11-hydroxylase activity. With acute
angiotensin administration or *in vitro* the peptide has no
effect on corticosterone conversion to aldosterone.[129,138,139,14]

Structure-Activity-Relationship Studies

Only two structure-activity-studies have been reported
with angiotensin and the adrenal cortex. The first study
by Hageman *et al.*[142] used a 6-substituted analog, [6-B-
(pyrazolyl-3)-1-alanine]-angiotensin II. Its activity
relative to the parent octapeptide showed good correlation
of induced aldosterone secretion in the dog with activity
in several other assays. The second study by Blair-West
et al.[143] determined the effects of heptapeptide, des-(Asp[1])-
angiotensin II, and hexapeptide, des-(Asp[1],Arg[2])-angiotensin
II, on aldosterone secretion *in vivo*. The heptapeptide was
as active as angiotensin II in stimulating aldosterone secretion while the hexapeptide was totally inactive. This potent
effect of the heptapeptide suggests a possible physiological
role for the metabolite of angiotensin II. In preliminary
studies (Chiu and Peach, unpublished observations) with
adrenal cortical cell suspensions of dog adrenal, the heptapeptide on a molar basis is more potent than angiotensin II

in stimulating incorporation of H^3-cholesterol into H^3-aldosterone. Stimulation of aldosterone biosynthesis can be blocked by [Ile^8]-angiotensin II (Peach unpublished observation).

In summary, one must realize that by no means does this presentation cover all the activities reported for angiotensin. The three general activities of angiotensin covered in some detail here, stimulation of the cardiovascular system, nervous system, and adrenal cortex, do represent effects of the peptide that are felt to be important to understanding the various roles of angiotensin. Angiotensin is a potent, biologically active peptide with a very broad spectrum of physiological activities. At the present time, one must consider that any one of these activities or all may represent the key to elucidating pathophysiological roles of angiotensin.

References

1. Page, I. H., and O. M. Helmer. J. Exp. Med. 71, 29 (1940).
2. Braun-Menendez, E., J. C. Fasciolo, L. F. Leloir, and J. M. Munoz. J. Physiol. 98, 283 (1940).
3. Ziegler, M., B. Riniker, and F. Gross. Biochem. J. 102, 28 (1967).
4. Ferris, T. F., P. Gordon, and P. J. Mulrow. Am. J. Physiol. 212, 698 (1967).
5. Oliver, W. J., and F. Gross. Arch. Exp. Path. Pharmacol. 255, 55 (1966).
6. Fischer-Ferraro, C., V. E. Nahmod, D. J. Goldstein, S. Finkielman. J. Exp. Med. 133, 353 (1971).
7. Skeggs, L. T., J. R. Kahn, and N. P. Shumway. J. Exp. Med. 103, 295 (1956).
8. Khairallah, P. A., F. M. Bumpus, I. H. Page, and R. R. Smeby. Science 140, 672 (1963).
9. Khairallah, P. A., and I. H. Page. Biochem. Med. 1, 1 (1967).
10. Pickens, P. T., F. M. Bumpus, A. M. Lloyd, R. R. Smeby, and I. H. Page. Circulation Res. 17, 438 (1965).
11. Johnson, D. C., and J. W. Ryan. Biochim. Biophys. Acta 160, 196 (1968).
12. Yang, H. Y. T., E. G. Erdos, and T. S. Chiang. Nature 218, 1224 (1968) London.
13. Page, I. H., and J. W. McCubbin, eds. *Renal Hypertension* (Illinois: Year Book Medical Publishers, Inc., 1968) pp 15-98.
14. Fisher, J. W. *Kidney Hormones* (New York: Academic Press, 1971) pp 93-136.

15. Page, I. H., J. W. McCubbin, H. Schwarz, and F. M. Bumpus. Circulation Res. 5, 552 (1957).
16. Bock, K. D., and F. Gross. Circulation Res. 9, 1044 (1961).
17. Akinkugbe, O. O., W. C. B. Brown, and W. I. Cranston. Clin. Sci. 30, 409 (1966).
18. Page, I. H., and F. Olmsted. Am. J. Physiol. 201, 92 (1961).
19. Westfall, T. C., and M. J. Peach. Biochem. Pharmacol. 14, 1916 (1965).
20. Dickinson, C. J., and J. R. Lawrence. Lancet I, 1354 (1963).
21. McCubbin, J. W., R. Soares De Moura, I. H. Page, and F. Olmsted. Science 149, 1394 (1965).
22. Olmsted, F., and I. H. Page. Circulation Res. 16, 140 (1965).
23. Day, M. D., J. W. McCubbin, and I. H. Page. Am. J. Physiol. 209, 264 (1965).
24. Barer, G. R. J. Physiol. 156, 49 (1961) London.
25. McCubbin, J. W., Y. Kaneko, and I. H. Page. Circulation Res. 11, 74 (1962).
26. DeBono, E., G. de J. Lee, F. R. Mottram, G. W. Pickering, J. J. Brown, H. Keen, W. S. Peart, and P. H. Sanderson. Clin. Sci. 25, 123 (1963).
27. Meier, R., J. Tripod, and A. Struder. Arch. Int. Pharmacodyn Ther. 117, 185 (1958).
28. Folkow, B., B. Johansson, and S. Mellander. Acta Physiol. Scand. 175, 50 (1960).
29. Zimmerman, B. G. Circulation Res. 11, 780 (1962).
30. Douglas, C. R., A. Ponce-Zumino, E. Ruiz-Petrich, F. Puig, and J. Talesnik. Acta Physiol. Latinoam. 14, 161 (1964).
31. Fowler, N. O., and J. C. Holmes. Circulation Res. 14, 191 (1964).
32. Marchetti, G. V., L. Merlo, and V. Noseda. Arzneimittel-Forsch. 15, 1244 (1965).
33. Assali, N. S., and A. Westersten. Circulation Res. 9, 189 (1961).
34. White, F. N., and G. Ross. Am. J. Physiol. 210, 1118 (1966).
35. Ross, G., and F. N. White. Am. J. Physiol. 211, 1419 (1966).
36. Von Capeller, D., L. K. Widmer, H. Staub. Helvet Physiol. Pharmacol. Acta 18, C-12 (1960).
37. Mandel, M. J., and L. A. Sapirstein. Circulation Res. 10, 807 (1962).

38. Krasney, J. A., F. T. Paudler, D. C. Smith, L. D. Davis, and W. B. Youmans. Am. J. Physiol. 209, 539 (1965).
39. Farr, W. C., and G. Grupp. J. Pharmacol. Exp. Ther. 156, 528 (1967).
40. Peach, M. J., and G. D. Ford. J. Pharmacol. Exp. Ther. 162, 92 (1968).
41. Peach, M. J., W. H. Cline, D. Davila, and P. A. Khairallah. Europ. J. Pharmacol. 11, 286 (1970).
42. Aiken, J. W., and E. Reit. J. Pharmacol. Exp. Ther. 159, 107 (1968).
43. Farr, W. C., and G. Grupp. J. Pharmacol. Exp. Ther. 177, 48 (1971).
44. Krasney, J. A., F. T. Paudler, P. M. Hogan, R. F. Lowe, and W. B. Youmans. Am. J.Physiol. 211, 1447 (1966).
45. Krasney, J. A., J. L. Thompson, and R. F. Lowe. Am. J. Physiol. 213, 134 (1967).
46. Peach, M. J., F. M. Bumpus, and P. A. Khairallah. J. Pharmacol. Exp. Ther. 167, 291 (1969).
47. Illanes, A., J. Perez-Olea, M. Quevedo, A. Ortiz, and M. Lazo. J. Pharmacol. Exp. Ther. 158, 487 (1967).
48. Beaulnes, A., J. Panisset, J. Brodeur, E. Beltrami, G. Gariepy. Circulation Res. 15 Suppl II, 210 (1964).
49. Turker, K. R. Experientia 21, 707 (1965).
50. Dempsey, P. J., Z. T. McCallum, K. M. Kent, and T. Cooper. Am. J. Physiol. 220, 477 (1971).
51. Koch-Weser, J. Circulation Res. 14, 337 (1964).
52. Koch-Weser, J. Circulation Res. 16, 230 (1965).
53. Beaulnes, A. Biochem. Pharmacol. 12, Suppl 181 (1963).
54. Heig, E., and K. Meng. Arch. Exp. Path. Pharmacol. 250, 35 (1965).
55. Lefer, A. M. Am. Heart J. 73, 674 (1967).
56. Kent, K. M., T. L. Goodfriend, Z. T. McCallum, P. J. Dempsey, and T. Cooper. Circulation Res. 30, 196 (1972).
57. Gross, F., D. Montague, R. Rosas, and D. H. Bohr. Circulation Res. 16, 150 (1965).
58. Regoli, D., W. K. Park, F. Rioux, and C. S. Chan. Rev. Can. Biol. 30, 319 (1971).
59. Yamamoto, M., R. K. Turker, P. A. Khairallah, and F. M. Bumpus. Eurp. J. Pharmacol. In Press (1972).
60. Bickerton, R. K., and J. P. Buckley. Proc. Soc. Exp. Biol. Med. 106, 834 (1961).
61. Vogin, E. E., and J. P. Buckley. J. Pharmaceut. Sci. 53, 1482 (1964).
62. Yu, R., and C. J. Dickinson. Lancet 2, 1276 (1965).
63. Severs, W. B., A. E. Daniels, H. H. Smookler, W. J. Kinnard, and J. P. Buckley. J. Pharmacol. Exp. Ther. 153, 530 (1966).

64. Ueda, H., Y. Uchida, K. Ueda, T. Gondaria, and S. Katayama. Jap. Heart J. 10, 243 (1969).
65. Rosendorff, C., R. D. Lowe, H. Lavery, and W. I. Cranston. Cardiovasc. Res. 4, 36 (1970).
66. Yu, R., and C. J. Dickinson. Arch. Int. Pharmacodyn. Ther. 191, 24 (1971).
67. Fukiyama, K., J. W. McCubbin, and I. H. Page. Clin. Sci. 40, 283 (1971).
68. Sweet, C. S., P. Kadowitz, and M. J. Brody. Am. J. Physiol. 221, 1640 (1971).
69. Ferrario, C. M., C. J. Dickinson, and J. W. McCubbin. Clin. Sci. 39, 239 (1970).
70. Ueda, H. Proc. 5th Eur. Congr. Cardiol. 249 (1968).
71. Joy, M. D. Clin. Sci. 41, 89 (1971).
72. Joy, M. D., and R. D. Lowe. Nature 228, 1303 (1970) London.
73. Scroop, G. C., F. Katic, M. D. Joy, and R. D. Lowe. Br. Med. J. I, 324 (1971).
74. Nashold, B. S., E. Mannarino, and M. Wunderlick. Nature 193, 1297 (1962) London.
75. Smookler, H. H., W. B. Severs, W. J. Kinnard, and J. P. Buckley. J. Pharmacol. Exp. Ther. 153, 485 (1966).
76. Deuben, R. R., and J. P. Buckley. J. Pharmacol. Exp. Ther. 175, 139 (1970).
77. Kaneko, Y., J. W. McCubbin, and I. H. Page. Circulation Res. 9, 1247 (1961).
78. McCubbin, J. W., and I. H. Page. Circulation Res. 12, 553 (1963).
79. Benelli, G., D. Della Bella, and A. Gandini. Brit. J. Pharmacol. 22, 211 (1964).
80. McGregor, D. D. J. Physiol. 177, 21 (1965).
81. Panisset, J. C., and P. Bowdois. Can. J. Physiol. Pharmacol. 46, 125 (1968).
82. Thoenen, H., A. Huerlimann, and W. Haefely. Med. Pharmacol. Exp. 13, 379 (1965).
83. Herrting, G., and J. Suko. Brit. J. Pharmacol. 26, 368 (1966).
84. Sakurai, T., and Y. Hasimoto. Jap. J. Pharmacol. 15, 223 (1965).
85. Khairallah, P. A., I. H. Page, and K. R. Turker. Circulation Res. 19, 538 (1966).
86. Scroop, G. C., and R. F. Whelan. Aust. J. Exp. Biol. Med. Sci. 46, 563 (1968).
87. Boadle, M. C., J. Hughes, and R. H. Roth. Nature 222, 987 (1969).
88. Khairallah, P. A. Pharmacologist 13, 185 (1971).

89. Palaic, D., and P. A. Khairallah. Biochem. Pharmacol. 16, 2291 (1967).
90. Palaic, D., and P. A. Khairallah. J. Neurochem. 15, 1195 (1968).
91. Lieban, H., A. Distler, and H. P. Wolff. Klin. Wchnschr. 44, 322 (1966).
92. Schumann, H., and H. Schmitt. Arch. Pharmakol. Exp. Path. 256, 169 (1967).
93. Zimmerman, B. G., and L. Whitmore. Int. J. Neuropharmacol. 6, 27 (1967).
94. Schumann, H. *New Aspects of Storage and Release Mechanisms of Catecholamines*, Schumann, H., and G. Kroneberg, eds. (Berlin: Springer-Verlag, 1970).
95. Braun-Menendez, E., J. C. Fasicolo, L. F. Leloir, and J. M. Munoz. Rev. Soc. Argent. Biol. 16, 398 (1940).
96. Haas, E., and H. Goldblatt. Am. J. Physiol. 196, 763 (1959).
97. Cession, G., and A. Cession-Fossion. C. R. Soc. Biol. 157, 1830 (1963) Paris.
98. Feldberg, W., and G. P. Lewis. J. Physiol. 171, 98 (1964) London.
99. Feldberg, W., and G. P. Lewis. J. Physiol. 178, 239 (1965) London.
100. Peach, M. J., W. H. Cline, and D. T. Watts. Circulation Res. 19, 571 (1966).
101. Staszewska-Barczak, J., and K. Konopka-Rogatko. Bull. Acad. Pol. Sci. 25, 503 (1967).
102. Staszewska-Barczak, J., and J. R. Vane. Brit. J. Pharmacol. 30, 655 (1967).
103. Poisner, A. M., and W. W. Douglas. Proc. Soc. Exp. Biol. Med. 123, 62 (1966).
104. Robinson, R. L. J. Pharmacol. Exp. Ther. 156, 252 (1967).
105. Peach, M. J., F. M. Bumpus, and P. A. Khairallah. J. Pharmacol. Exp. Ther. 176, 366 (1971).
106. Lewis, G. P., and E. Reit. Brit. J. Pharmacol. 26, 444 (1966).
107. Douglas, W. W., T. Kanno, and S. R. Sampson. J. Physiol. 188, 107 (1967).
108. Peach, M. J. Circulation Res., Suppl. II to 28, 29 11-107 (1971).
109. Khairallah, P. A., and I. H. Page. Am. J. Physiol. 200, 51 (1961).
110. Robertson, P. A., and D. Rubin. Brit. J. Pharmacol. 19, 5 (1962).
111. Godfraind, T., A. Kaba, and P. Polster. Arch. Int. Pharmacodyn Ther. 163, 227 (1966).

112. Khairallah, P. A., and I. H. Page. Ann. N.Y. Acad. Sci. 104, 212 (1963).
113. Goldenberg, M. M. Can. J. Physiol. Pharmacol. 46, 159 (1968).
114. Lewis, G. P., and E. Reit. J. Physiol. 179, 538 (1965) London.
115. Lewis, G. P., and E. Reit. Brit. J. Pharmacol. 26, 444 (1966).
116. Trendelenburg, U.. J. Pharmacol. Exp. Ther. 154, 418 (1966).
117. Haefely, W. A. A. Hurlimann, and H. Thoenen. Biochem. Pharmacol. 14, 1393 (1965).
118. Panisset, J. C., P. Biron, and A. Beaulnes. Experientia 22, 394 (1966).
119. Machova, J., and D. Boska. Europ. J. Pharmacol. 1, 233 (1967).
120. Panisset, J. C. Can. J. Physiol. Pharmacol. 45, 313 (1967).
121. Aiken, J. W., and E. Reit. J. Pharmacol. Exp. Ther. 169, 211 (1969).
122. Khairallah, P. A., A. Toth, and F. M. Bumpus. J. Med. Chem. 13, 181 (1970).
123. Yankopeulos, N. A., J. O. Davis, B. Kliman, and R. E. Peterson. J. Clin. Invest. 38, 1278 (1959).
124. Denton, D. A., J. R. Goding, and R. D. Wright. Brit. Med. J. 2, 447 (1959).
125. Davis, J. O., C. C. J. Carpenter, C. R. Ayers, J. E. Holman, and R. C. Bahn. J. Clin. Invest. 40, 684 (1961).
126. Laragh, J. H., M. Angers, W. G. Kelly, and S. Lieberman. J. Am. Med. Assoc. 174, 234 (1960).
127. Biron, P., E. Koiw, W. Nowaczynski, J. Brouillet, and J. Genest. J. Clin. Invest. 40, 338 (1961).
128. Ames, R. P., A. J. Borkowski, A. M. Sicinski, and J. H. Laragh. J. Clin. Invest. 44, 1171 (1965).
129. Kaplan, N. M., and F. C. Barlter. J. Clin. Invest. 41, 715 (1962).
130. Ganong, W. F., P. J. Mulrow, A. Boryczka, and G. Cera. Proc. Soc. Exp. Biol. Med. 109, 381 (1962).
131. Ganong, W. F., H. H. Von Brunt, T. C. Lee, and P. J. Mulrow. Proc. Soc. Exp. Biol. Med. 112, 1062 (1963).
132. Higgins, J. T. Jr., J. O. Davis. J. Urguhart, and M. J. Olichney. Am. J. Physiol. 207, 814 (1964).
133. Brown, T. C., J. O. Davis, M. J. Olichney, C. I. Johnston. Circulation Res. 18, 475 (1966).
134. Blair-West, J. R., J. P. Coghlan, D. A. Denton, J. R. Goding, M. Wintour, and R. D. Wright. Circulation Res. 17, 386 (1965).

135. Denton, D. A. Anst. Ann. Med. 13, 121 (1964).
136. Kaplan, N. M. J. Clin. Invest. 44, 2029 (1965).
137. Dufau, M. L., J. D. Crawford, and B. Kliman. Endocr. 84, 462 (1969).
138. Haning, R., S. A. Tait, and J. F. Tait. Endocr. 87, 1147 (1970).
139. Muller, J. Acta Endocr. 52, 515 (1966).
140. Aquilera, G., and E. T. Marusic. Endocr. 89, 1524 (1971).
141. Marusic, E. T., and P. J. Mulrow. J. Clin. Invest. 40, 2101 (1967).
142. Hageman, W. E., K. Hofman, R. J. Ertel, and J. P. Buckley. J. Pharmacol. Exp. Ther. 168, 295 (1969).
143. Blair-West, J. R., J. P. Coghlan, D. A. Denton, J. W. Funder, A. Scoggins, and R. D. Wright. J. Clin. Endocr. Metab. 32, 575 (1971).

FROM ANGIOTENSIN TO ANTI-ANGIOTENSIN

D. Regoli, F. Rioux, W. K. Park. Department of
Pharmacology, Centre Hospitalier Universitaire,
Sherbrooke, Canada

VARIOUS HORMOMES AND naturally occurring substances, like
angiotensin, have a polypeptide structure. It is assumed
that the groups responsible for binding a peptide to receptors or for evoking biological responses, are scattered
along the side chains. As pointed out by Dickerson and
Geis:[1] "If the polypeptide chain provides the fundamental
pattern, the ground bass of the composition, it is the side
chains that build the melody."

Identification of active groups in a peptide is the
primary step for the discovery of specific antagonists. By
definition an antagonist should possibly have the same
affinity for the receptors as the agonist has, but be
devoid of intrinsic activity.[2] A rational approach to the
identification of active groups in the side chains of the
various amino acids composing a peptide is the substitution
of the single amino acids with presumably inactive compounds,
such as Gly, Ala or unnatural amino acids with a saturated
ring, like 1-amino-cyclopentanecarboxylic acid (Acpc).

In a preliminary study, a series of analogs of the
octapeptide 5-Ile-angiotensin II (A II) were prepared by
replacing with Acpc the amino acids in positions 1 to 8.
Affinities and intrinsic activities were evaluated "in
vitro" on rat isolated stomach strips suspended in a 40 ml
bath or in a cascade superfusion system, according to Vane.[3]
Details of the methods have been described previously.[4,5]

It was found that the replacement of 4-Tyr decreases
the affinity, while the substitution of 6-His, 7-Pro and
8-Phe reduce both affinity and intrinsic activity.[4] Tests

495

for antagonism indicated that [Acpc[6]]-A II and [Acpc[7]]-A II
are inactive compounds, while [Acpc[8]]-A II is an inhibitor.[4]
When Gly and Ala were used instead of Acpc, to replace
4-Tyr, 6-His and 7-Pro, similar effects were observed with
the analogs substituted in position 4, but the intrinsic
activities of [Ala[6]]-A II and [Ala[7]]-A II were found to be
higher (see Table I) than those of analogs substituted with
Gly and Acpc. Results shown in Table I indicate that the
presence of an asymmetric α-carbon in positions 6 and 7 is
necessary to maintain the steric orientation of the phenyl
ring in position 8.

Table I

Intrinsic Activities (α^E) and Affinities (pD_2) of
Various Angiotensin II (A II) Analogs,
on the Rat Isolated Stomach Strip

1	2	3	4	5	6	7	8
Asp	- Arg	- Val	- Tyr	- Ile	- His	- Pro	- Phe

Compound	α^E	pD_2*
A II	1.0	8.0
[Phe[4]]-A II	1.0	6.8
[Acpc[4]]-A II	0.8	4.9
[Ala[4]]-A II	0.9	4.8
[Gly[4]]-A II	0.7	4.8
[Acpc[6]]-A II	0.3	---
[Ala[6]]-A II	0.9	5.0
[Gly[6]]-A II	0.5	4.0
[Acpc[7]]-A II	0.5	5.8
[Ala[7]]-A II	1.0	6.4
[Gly[7]]-A II	0.7	4.4

*Log of molar concentration

To find potent and specific antagonists of A II a series
of analogs substituted in position 8 were synthesized. The
compounds were tested "in vivo" (on the blood pressure of
nephrectomized rats anesthetized with urethane) and "in

vitro" (on the rat isolated stomach strip, suspended in a
5 ml bath, with oxygenated Krebs solution at 37° C, to
record isometric contractions).[5] Affinity of antagonists
was evaluated with the method of Schild[6] by estimating pA_2.
Residual intrinsic activity was measured by applying to the
tissues increasing concentrations of the analogs.

As shown in Table II, the lengthening of the aliphatic
chain from Ala to Leu, increases progressively the affinity

Table II

Relative Potencies of Various A II Antagonists as Determined
"In Vivo" (on the Rat Blood Pressure) and
"In Vitro" on the Rat Isolated Stomach Strip

| Compound | A II: $pD_2 = 8.00$ | | $[Leu^8]$-A II: $pA_2 = 8.00$ | |
	ID_{50} (µg/kg/min)	Relative Potency in vivo	pA_2	Relative Potency in vitro
$[Gly^8]$-A II	1.10	0.3	6.77	0.06
$[Ala^8]$-A II	1.00	0.3	6.86	0.07
$[n$-But$^8]$-A II	0.41	0.8	7.48 0	0.3
$[Val^8]$-A II	0.67	0.5	7.51	0.3
$[Ile^8]$-A II	0.36	0.9	7.90	0.8
$[Leu^8]$-A II	0.32	1.0	8.00	1.0
$[Glu^8]$-A II	10.0	0.03	6.61	0.04
$[Lys^8]$-A II*	100.0	0.003	----	----
$[Sar^1,Leu^8]$-A II	0.07	4.6	8.60	4.0
$[\beta$-Asp$^1,Leu^8]$-A II	0.26	1.2	8.20	1.4
$[Sar^1,Ala^8]$-A II	0.28	1.1	8.38	2.5

ID_{50}, Dose producing 50% of inhibition.
pA_2, Indicates -log of molar concentration.
*A preliminary analysis of this compound with infra red
spectrum in a solution cell has shown a partial cyclisation.

of the antagonist for the receptors and leads to analogs
($[Leu^8]$-A II) with a pA_2 value (8.0) equal to the pD_2 value
(8.0) of A II. These findings indicate that optimal

stimulation of the receptors requires the presence of a
phenyl ring in position 8. Binding of the phenyl ring to
the receptors may be due to its hydrophobic character,
which is shared by the side chain of Leu, but stimulation
of receptors appears to be due to the resonance of the
phenyl ring, which is obviously absent in Leu.

This conclusion is supported by the results obtained
with [Glu[8]]-A II and [Lys[8]]-A II. Replacement of 8-Phe
with an acid (Glu) or a basic (Lys) hydrophilic group
brings about almost inactive compounds.

Peptides hormones have generally a short half-life in
the circulation, because they are rapidly taken up by the
tissues and inactivated by the polypeptidases. Inactiva-
tion of a peptide can be partially prevented by substituting,
at the amino end, compounds which prevent the action of
aminopeptidases. It has been shown that replacement of
asparagine with β or β-D aspartic acid in A II increases
the potency of the analog, as well as its duration of
action.[7]

To obtain long acting inhibitors of A II double substituted (in positions 1 and 8) analogs were prepared.
Results are shown at the bottom of Table II. Replacement
of 1-Asp with Sar or with β-Asp, increases the potency
and the duration of action[8] of the antagonist. This
suggests that the compounds may be slowly degraded by
aminopeptidases. However, the results obtained "in vivo"
with [β-Asp[1]]-A II and [Sar[1]]-A II (see Table III) indicate

Table III

Relative Potencies and Pharmacological Characterization of
Angiotensin II (A II) [β-Asp[1]]-A II and [Sar[1]]-A II,
"in vivo" (Rat Blood Pressure) and "in vitro"
(Rat Isolated Stomach Strip)

Compounds	"in vivo"			"in vitro"		
	R.P. %	D.A. min.		α^E	pD_2	R.P.
A II	100	2.0 ± 0.1 (9)		1.0	8.0	100
[β-Asp[1]]-A II	172	3.5 ± 0.4 (9)		1.0	8.2	150
[Sar[1]]-A II	100	2.5 ± 0.2 (9)		1.0	8.3	200

α^E, Intrinsic activity. pD_2, Affinity.
In parentheses, the number of determinations.
R.P., Relative potency.   D.A., Duration of action.

that degradation by aminopeptidases is probably not the primary factor involved in increasing the potency of the two antagonists [Sar^1,Leu^8]-angiotensin II and [β-Asp^1,Leu^8]-angiotensin II.

Acknowledgment

This work was supported by the M.R.C. of Canada.

References

1. Dickerson, R. E., and I. Geis. *The Structure and Action of Proteins* (New York: Harper and Row, 1969) p 17.
2. Ariens, E. J. *Molecular Pharmacology* (New York: Academic Press, 1964).
3. Vane, J. R. Br. J. Pharmacol. <u>23</u>, 260 (1964).
4. Regoli, D., and W. K. Park. Can. J. Physiol. & Pharmacol. <u>50</u>, 99 (1972).
5. Rioux, F., D. Regoli, and W. K. Park. Revue Can. Biol. <u>30</u>, 333 (1971).
6. Schild, H. O. Br. J. Pharmacol. <u>2</u>, 189 (1947).
7. Regoli, D., B. Riniker, and H. Brunner. Biochem. Pharmacol. <u>12</u>, 637 (1963).
8. Regoli, D., F. Rioux, and W. K. Park. Revue Can. Biol. <u>31</u>, 73-77 (1972).

COMPARATIVE PHARMACOLOGY OF ANGIOTENSIN ANTAGONISTS

Philip Needleman, Eugene M. Johnson, Jr., William Vine, Everett Flanigan, * *and Garland R. Marshall.* Departments of Pharmacology and of Physiology and Biophysics, Washington University School of Medicine, St. Louis, Missouri 63110

THE *IN VIVO* CONVERSION OF the decapeptide Asp^1, Val^5-angiotension I (A I) to the octapeptide angiotensin II (A II) has been shown to occur primarily in the pulmonary circula tion.[1-3] A I possesses minimal biological activity whereas A II is an extremely potent substance with a wide range of effects and of course, the renin angiotensin system is implicated in renal hypertension.[4] $[Phe^4-Tyr^8]$-A II was the first reported specific antagonist of A II[5] and there have since been a number of A II antagonists reported.[6,7] In addition, peptides initially isolated from snake venom have proven to be potent converting enzyme inhibitors while apparently having little or no effect at A II receptor sites.[8] This investigation describes a number of peptide analogs which have been assayed for their ability to act as angiotensin antagonists both *in vitro* and *in vivo* as well as in renal hypertensive rats. Materials: The angiotensin analogs were synthesized by the solid-phase procedure of Marshall and Merrifield[9] and the compounds were purified by Sephadex chromatography and characterized by amino acid analysis. The hippuryl histidyl-leucine and SQ 20881 (Glu-Tyr-Pro-Arg-Pro-Gln-Ile-Pro-Pro) were kindly supplied by D. W. Cushman of Squibb Research Institute. $[Phe^4-Tyr^8]$-A II[5] $[Ala^8]$-A II[7] and $[Ile^8]$-A II[10] are the

---

*Present Address: Department of Biology and Chemistry, Brookhaven National Laboratory, Upton, New York.

only angiotensin analogs considered in this report that have previously been described.

Results and Discussion

Oxytocic effects of peptide analogs

Peptides which were angiotensin agonists and antagonists in uterus are described in Table I and Table II. [p-fluoro-Phe[4]]-A II was equipotent to A II, whereas A I had only 2% of the oxytocic activity. Rat uterus is known to have minimal converting enzyme activity.[15] The octapeptide analogs ([Ile[8]]-A II, [Cys[8]]-A II, [Phe[4]-Tyr[8]]-A II and [p-fluoro-Phe[4]]-A II) were potent competitive antagonists of A II and A I (Table I). Of special interest is that [p-fluoro-Phe[4]]-A II represents the first antagonists which is not modified in the 8-position of angiotensin. [Ile[8]]-A II, [Cys[8]]-A II, A II, and [p-fluoro-Phe[8]]-A II appear to have the same affinity for the uterine receptor site.

Table I

Effect of Angiotensin (A) Analogs as Agonists and Antagonists on Isolated Rat Uterus Strips*

Agonists		*Potent Antagonists*	
Peptide Analogs	K_m ng/ml	Peptide Analogs	K_I ng/ml
A II	5	[Ile[8]]-A II	5
[p-fluoro-Phe[8]]-A II	5	[Cys[8]]-A II	6
A I	250	[Phe[4], Tyr[8]]-A II	240
		[p-fluoro-Phe[4]]-A II	500

*Uterine segments were suspended in 5 ml of de Jalon's solution[11] which was bubbled with O_2-CO_2 (95:5%) at room temperature. K_m and K_I were calculated from reciprocal plots of gm of uterine contraction versus dose of angiotensin analogs. Each peptide was tested on 4-8 uterine strips.

Table II

Effectiveness of Octapeptide and Decapeptide Analogs
as A II and A I Antagonists on Rat Uterine Strips*

Peptide Analogs	Conc. of antagonist µg/ml	Angiotensin ED_{50}
A II	0	5
A II + [Ile^8]-A I	100	50
A II + [Phe^4]-Tyr^8-A I	25	50–100
A II + [Ala^8]-A II	10	100
A I	0	250
A I + [Ile^8]-A I	100	2500
A I + [Phe^4, Tyr^8]-A I	25	2500–5000

*ED_{50} was the dose of A II or A I needed to produce a con-
traction which was 50% of the maximal response. Each
peptide was tested on 4–8 uterine strips.

The decapeptide analogs [Phe^4-Tyr^8]-A I and [Ile^8]-A I,
as well as the octapeptide [Ala^8]-A II were weak antagonists
which required very high concentrations to produce comparable
10–20 fold shifts in the both A II and A I dose response
curves (Table II). This data supports the conclusion that
A I and A II must act at the same receptor site. If significant conversion of the decapeptides to octapeptides was
occurring the compounds would have been more potent
antagonists.

Conversion of peptide analogs by rabbit lung extracts

Converting enzyme activity was measured in rabbit lung
acetone powder extracts by a modification[12] of the spectro-
photofluorometric method of Piquilloud, *et al* [13] using
either hippuryl-His-Leu or A I as the substrate.
Hippuryl-His-Leu (HHL) was used as a substrate to
characterize converting enzyme activity because it is not
hydrolyzed by most other tissue peptidases. The K_m for
HHL with the rabbit lung acetone powder was 2.5 mmol/liter.

The decapeptides A I, [Ile[8]]-A I, and [Phe[4], Tyr[8]]-A I were equipotent competitive inhibitors of HHL with a K_i of 1-4 μmol/liter. Incubation of the decapeptides with rabbit lung (at concentrations comparable to that needed for HHL inhibition) did not result in the appearance of assayable His-Leu (therefore less than 2% was converted). However, the K_m of the three decapeptides when studied as substrates for the rabbit lung enzyme was 50-200 μmol/liter. The K_m's of HHL and A I agree with those previously reported.[14]

In Vivo angiotensin antagonism in normotensive rats

[Ile[8]]-A II and [Cys[8]]-A II were potent antagonists of the A II induced vasoconstriction in normotensive rats. When infused at 100 μg/kg/min, these two analogs caused a 200-250 fold shift in the A II dose response curve (Table III). [Phe[4], Tyr[8]]-A II, and [*p*-fluoro-Phe[4]]-A II were less potent antagonists causing a 10-25 fold shift in the A II dose-response curve (when antagonists were infused at 100 μg/kg/min). [Phe[4], Tyr[8]]-A II and [*p*-fluoro-Phe[4]]-A II possess some agonistic activity. Park and Regoli[6] demonstrated that [Ala[8]]-A II and [Phe[4], Tyr[8]]-A II were equipotent as A II and A I antagonists *in vivo*.

A I was essentially equipotent with A II in elevating rat blood pressure, indicating rapid *in vivo* conversion of the decapeptide to the octapeptide. [Ile[8]]-A I proved to be much less potent (100 fold) as an A II antagonist than [Ile[8]]-A II, thus the decapeptide appears to be a less efficient substrate *in vivo* for the rat converting enzyme (*in vivo*) than does A I (Table III). The A II inhibitory activity of [Ile[8]]-A I was substantially reversed by the nonapeptide converting enzyme inhibitor SQ 20881 (Table III). SQ 20881 did not alter A II vasoconstriction but caused a 200 fold decrease in A I activity thereby indicating that there was some *in vivo* conversion of [Ile[8]]-A I to the octapeptide antagonist. Aiken and Vane[15] have noted that poor correlation existed between converting enzyme activity in rat isolated organs and homogenates of those tissues.

Effect of peptides on renal hypertension

Infusion of 100 μg/kg/min of [Cys[8]]-A II, [Ala[8]]-A II, [Ile[8]]-A II, [*p*-fluoro-Phe[4]]-A II, [Phe[4], Tyr[8]]-A II, as well as 50 μg/kg/min of SQ 20881 and 1 mg/kg/min of [Ile[8]]-A I and [Phe[4], Tyr[8]]-A I cause a 15-30% reduction in mean blood pressure which returned to the preinfusion level

Table III

Ability of Peptide Analogs to Lower Blood Pressure in Renal Hypertensive Rats and to Inhibit the A II Pressor Response in Normotensive Rats

Peptide analogs	Dosage μg/kg/min	A II in presence of peptide analogs ED$_{25}$ μg/kg*	Percent fall of blood pressure in renal hypertensives†
Controls (A II or A I or [p-fluoro-Phe^8]-A II		0.1 (8)#	
[Cys^8]-A II	10	1.2 (4)	
	100	20 (3)	29 ± 2 (5)#
[p-fluoro-Phe^4]-A II	100	2.5 (3)	17 ± 4 (4)
[Ala^8]-A II	100	–	26 ± 5 (3)
[Phe^4, Tyr^8]-A II	100	1.0 (5)	14 ± 4 (8)
[Phe^4, Tyr^8]-A I	1000	–	22 ± 2 (4)
[Ile^8]-A II	1	0.1 (2)	
	10	5 (3)	
	100	25 (4)	22 ± 3 (9)
[Ile^8]-A I	10	0.1 (3)	
	100	0.2 (5)	
	1000	4.5 (6)	20 ± 3 (7)
SQ 20881	50	0.1 (3)	30 ± 2 (4)
SQ 20881 + [Ile^8]-A I	50 + 1000	0.6 (3)	33 ± 5 (4)

Normotensive rats were pentobarbital anesthetized and pre-treated with phenoxybenzamine and propranolol. The analogs were infused for 15 min and throughout the A II dose response curves (given as pulse injections in the other jugular vein).

†Acute renal hypertension was produced by unclamping the left renal pedicle (contralateral kidney was left intact) which had been occluded for 4.5 hr.[16] The mean blood pressure in the phenoxybenzamine (30 mg/kg), propranolol (15 mg/kg) pentobarbital sodium (30 mg/kg), treated rats before clamp removal was 79 ± 3 and after clamp removal was 112 ± 4. The values represent the mean ± SE of blood pressure measured after 10 min of infusion.

#Numbers in parentheses represent numbers of animals used in each test.

10 min after completion of the infusion (Table III). Infusion of the combination of SQ 20881 and [Ile^8]-A I caused
no further reduction in renal hypertension than either agent
used alone. Such agents have potential therapeutic activity
for confirmatory diagnosis in renal hypertension.

Acknowledgment

The authors acknowledge the technical assistance of Mrs.
Anne H. Kauffman and Miss Patricia A. Burnheimer. This
research was supported by USPHS NIH Grants 5-R01-HE-11771;
HE-14397-01, HE-14509-01, AM-13025 and American Heart Grants
69-722 and 70-672. The authors acknowledge Established
Investigatorships of the American Heart Association (PN,
68-115 and GRM, 70-111).

References

1. Ng, K. K. F., and J. R. Vane. Nature (London) 21, 762
   (1967).
2. Ng, K. K. F., and J. R. Vane. Arch. Pharm. Exp. Path.
   259, 188 (1968).
3. Ng. K. K. F., and J. R. Vane. Nature (London) 218,
   144 (1968).
4. Page, I. B., and J. W. McCubbin. *Renal Hypertension*
   (Chicago: Year Book Medical Publishers, Inc., 1969).
5. Marshall, G. R., W. Vine, and P. Needleman. Proc.
   Natl. Acad. Sci. U.S. 67, 1624 (1970).
6. Park, W. K., and D. Regoli. Brit. J. Pharmacol. 43,
   418p (1971).
7. Turker, R. K., M. Yamamoto, P. A. Khairallah, and F. M.
   Bumpus. Europ. J. Pharmacol. 15, 285 (1971).
8. Ferreira, S. H., L. J. Greene, V. A. Alabaster, Y. S.
   Bakhle, and J. R. Vane. Nature (London) 225, 379
   (1970).
9. Marshall, G. R., and R. B. Merrifield. Biochemistry
   4, 2394 (1965).
10. Khairallah, P. A., personal communication.
11. Burn, J. H. *Practical Pharmacology* (Oxford: Blackwell
    Scientific Publishers, 1952) pp 7-16.
12. Cushman, D. W., and H. S. Cheung. Biochem. Pharmacol.
    20, 1637 (1971).
13. Piquilloud, Y., A. Reinharz, and M. Roth. Biochim.
    Biophys. Acta 206, 136 (1970).
14. Cushman, D. W. and H. S. Cheung. *International Symposium
    on the Renin-Angiotensin-Aldosterone-Sodium System in
    Hypertension, Mount Gabriel, Quebec Sept. 30-Oct. 1971,*
    (in press).

15. Aiken, J. W., and J. R. Vane. Nature <u>228</u>, 30 (1970).
16. Krieger, E. M., H. C. Salgado, C. J. Assan, L. J.
    Greene, and S. H. Ferreira. Lancet <u>269</u> (Feb. 6,
    1971).

HIGHLIGHTS OF RECENT STRUCTURE–ACTIVITY STUDIES WITH ANGIOTENSIN II

Robert R. Smeby, M. C. Khosla, F. Merlin Bumpus.
Research Division, Cleveland Clinic Foundation,
Cleveland, Ohio 44106

IN THE LAST FEW YEARS a large number of angiotensin II analogs have been prepared and these studies have shown that structural modifications in amino acid positions 1 through 7 usually resulted either in retention or destruction of pressor or myotropic activity. Assays of these analogs in widely differing biological systems gave results that were consistent from one assay system to another. Thus, amino acid changes in these positions, which destroy biological activity, either remove a group required to obtain binding between peptide and receptor site or destroy the conformation necessary for peptide–receptor site interaction.[1]

However, when structural modifications were made in position 8 of angiotensin II more interesting results were obtained. Thus, [8-alanine]-, [8-(3-amino-3'-phenylisobutyric acid)]-, and [8-(2-amino-4-phenylbutyric acid)]-angiotensin II exhibited low pressor activity but were found to be highly active in inhibiting uptake of catecholamines by sympathetic nerve endings and causing release of catecholamines from adrenal medulla. On the other hand, [8-tyrosine]-angiotensin II has high pressor activity but lower activity in inhibiting catecholamine uptake.[2] Later [8-alanine]-angiotensin II was shown to competitively inhibit the response of isolated muscle strips to angiotensin II[3] *in vitro* but did not inhibit the *in vivo* pressor response of the cat to angiotensin II. This analog has very low biological activity.

Increasing the size of the aliphatic side group of the amino acid in position 8 to give [8-isoleucine]-angiotensin II increased both the biological activity and the angiotensin II antagonistic activity. This analog is a potent competitive antagonist of the *in vitro* myotropic response to the parent hormone. The pressor response to angiotensin II in the dog, cat and rat is also inhibited by this analog. Infusion of this analog into rats with acute renal clip hypertension, both with the contralateral kidney intact (2 kidney) and with the contralateral kidney removed (1 kidney), caused a reduction in blood pressure. These data, shown in Table I, demonstrate that peptide analogs which inhibit angiotensin II will be very useful tools in the study of the etiology of hypertensive disease in animals and as a possible diagnostic aid in the treatment of patients with renal hypertensive disease.

Table I

Reduction of Blood Pressure in Renal Hypertensive Rats by Infusion of [8-Isoleucine]-Angiotensin II

Group	No. in group	Average B.P. before infusion (mm Hg)	Average B.P. after infusion (mm Hg)	Dose infused (ng/rat/min)	p value
Normotensive	7	90	94	200 and 1000	N.S.
2 Kidney Hypertensive	4	167	141	200	0.01
1 Kidney Hypertensive	7	138	115	1000	0.001

Angiotensin II increases capillary permeability in rabbit skin and this effect is also inhibited by [8-isoleucine]-angiotensin II. However, in isolated frog skin, where angiotensin II increases the rate of sodium transport, this analog is twice as active as an agonist as the natural peptide. It is not known whether this effect is due to a different type of receptor mechanism involved in the sodium transport system as compared to the muscle receptor or merely due to the change in species.

Changing the aromatic ring of phenylalanine in position 8 to a saturated ring, to give [8-cyclohexylalanine]-

angiotensin II, has yielded an analog which gives assay results difficult to interpret. The initial dose of this analog cuases a moderate pressor and myotropic response (about 25% of the parent hormone) but subsequent responses to itself or angiotensin II are abolished. It is not known whether this compound is a potent, long-lasting, competitive antagonist or whether it is making the assay system tachyphylactic. This peptide requires much further study.

Further changes in position 8 to give [8-valine]-, [8-leucine]-, and [8-norleucine]-angiotensins II have not increased the antagonistic potency of [8-isoleucine]-angiotensin II. It is interesting in this regard to note that the isoleucine side chain is very similar in size and shape to the aromatic ring of phenylalanine. Placing an unnatural amino acid in the 1 position, [1-sarcosine, 8-isoleucine]-angiotensin II increases the duration of antagonistic activity probably because the peptide is protected from the action of aminopeptidase.

Recent physical studies by Fermandjian and Fromageot[4] on some of these analogs modified in position 8 have indicated that these peptides have a conformation different from angiotensin II. It is not yet known whether the degree of conformational change can in any way be associated with the antagonistic activity. Whatever the effect of the 8-position is toward causing the specific, competitive inhibition of angiotensin II, it may not be limited to the 8-position alone because [4-phenylalanine, 8-tyrosine]-angiotensin II is an antagonist[5] while [8-tyrosine]-angiotensin II is not.

The specific, competitive inhibition produced by these analogs of angiotensin II may be highly useful for both laboratory and clinical studies of hypertensive disease. It is to be hoped that these studies will lead to future compounds which will be more potent and longer acting.

References

1. For earlier review of much of this work see: a) Bumpus, F. M. "Angiotensin: Structure-Activity Relationships," In *Structure-Activity Relationships of Protein and Polypeptide Hormones,* Margoulies, M., and F. C. Greenwood, eds. (Amsterdam: Excerpta Medica, 1971) pp 18-22. b) Bumpus, F. M., and R. R. Smeby. "Angiotensin," In *Renal Hypertension,* Page, I. H., and J. W. McCubbin, eds. (Chicago: Yearbook Medical Publishers, Inc., 1968) pp 62-98.

2. Peach, M. J., F. M. Bumpus, and P. A. Khairallah.
    J. Pharmacol. Exp. Ther. <u>176</u>, 366 (1970).
3. Turker, R. K., M. Yamamoto, P. A. Khairallah, and
    F. M. Bumpus. Eur. J. Pharmacol. <u>15</u>, 285 (1971).
4. Fermandjian, S., J. L. Morgat, and P. Fromageot.
    Eur. J. Biochem. <u>24</u>, 252 (1971); Fermandjian, S., D.
    Greff, and P. Fromageot. In *Chemistry and Biology of
    Peptides, Proc. Third Amer. Peptide Symposium,*
    Meienhofer, J., ed. (Ann Arbor, Mich: Ann Arbor Science
    Publishers, 1972) pp
5. Marshall, G. R., W. Vine, and P. Needleman. Proc.
    Nat. Acad. Sci. U.S. <u>67</u>, 1624 (1970).

RELATIONSHIPS BETWEEN PRESSOR ACTIVITY AND THE PROPERTIES
OF SIDE CHAINS OF ANGIOTENSIN II

Eugene C. Jorgensen, Graham C. Windridge, Kun-Hwa Hsieh.*
Department of Pharmaceutical Chemistry, School of
Pharmacy, University of California, San Francisco CA
94122

Thomas C. Lee. Department of Human Physiology, School
of Medicine, University of California, Davis, CA 95616

THE IMPORTANCE OF THE CARBOXYL terminal hexapeptide sequence
of angiotensin II for its myotropic and pressor effects has
been well established.[1] We have examined specific characteristics of the amino acid side chains within the 5-8
portion of this sequence by solid-phase synthesis of analogs
of [1-asparagine,5-isoleucine]-angiotensin II. The analogs
were studied for their pressor effects in the pentolinium-
treated nephrectomized rat (Table I).

The 5-Isoleucine Position

Previous analog studies[2] showed that activity equal to
that of angiotensin II could be produced by the replacement
of isoleucine by a variety of lipophilic aliphatic and
alicyclic β-branched L-α-amino acids, such as valine, alloisoleucine α-amino-β-ethylvaleric acid, α-cyclopentylglycine,
and α-cyclohexylglycine. Activity was significantly reduced
when β-branching was absent in related analogs. In order
to test the importance of lipophilic character for a side
chain in the 5-position, the more hydrophilic β-branched

*Present Address: School of Pharmacy, Virginia Commonwealth
University, Richmond, Virginia 23219.

Table I

Relative Pressor Activities of Angiotensin II Analogs
in the Rat

Peptide	Pressor Activity
[Asn[1],Val[5]]-angiotensin II	100
[Asn[1],Thr(Me)[5]]-angiotensin II*	118
[Asn[1],Thr[5]]-angiotensin II	10
[Asn[1],Ile[5],A2bu[6]]-angiotensin II[†]	0.02
[Asn[1],Ile[5],Pya[6]]-angiotensin II[#]	3
[Asn[1],Ile[5],D-Pya[6]]-angiotensin II[#]	0.1
[Asn[1],Ile[5],Leu[8]]-angiotensin II	1
[Asn[1],Ile[5],Nle[8]]-angiotensin II	6
[Asn[1],Ile[5],Ahp[8]]-angiotensin II[√]	10

*Thr(Me),O-Methylthreonine

[†]A2bu,2,4-diaminobutyric acid

[#]Pya,β-(Pyridyl-2)alanine

[√]Ahp, α-Aminoheptanoic acid

5-O-methylthreonine and 5-threonine analogs were prepared.
It was expected that the increasingly polar methyl ether
and hydroxyl containing side chain analogs would show a
regular decrease in activity with decreasing lipophilic
character. However, [1-asparagine, 5-O-methylthreonine]-
angiotensin II showed 118% activity relative to 100% for
[1-asparagine,5-valine]-angiotensin II in the rat pressor
assay. [1-Asparagine,5-threonine]-angiotensin II was 10%
as active.

The equally high activity contributed by the side chains
of a wide variety of highly lipophilic β-branched aliphatic
and alicyclic amino acids, and by the relatively polar side
chain of O-methylthreonine, indicates that the steric effect
is of primary importance. The low activity of the more
hydrophilic [5-threonine] analog may be due to its exceeding
limits of hydrophobic character. However, it is believed

more likely that the free hydroxyl group produces changes
in the conformation of the peptide through the formation
of new hydrogen bonds.

The NMR spectra of the [Thr[5]] and [Thr(Me)[5]] analogs
in D_2O were virtually identical for the protons of the
side chains, except for the presence of the OMe singlet
(3.25 ppm) in the [Thr(Me)[5]] analog. The peptide backbone
CH spectra (4.0-4.5 ppm) showed significant differences
between the analogs, although a more detailed study would
be required to identify these with specific conformational
differences.

The 6-Histidine Position

The imidazole ring of histidine possesses a number of
characteristics which have been associated with its functional role in enzymes, such as aromatic character, acid-
base behavior, and hydrogen donor and hydrogen acceptor
ability in hydrogen bond formation.

The need for aromatic character was indicated by the
low activities for the neutral [6-alanine] analog (0.8%),[3]
and for the [6-lysine] analog (0.1%)[4] in which a basic
nitrogen atom was placed at a spacing which approximates
that of the *tele*-nitrogen of histidine. We have prepared
the [6-(2,4-diaminobutyric acid)] analog in which the basic
nitrogen approximates the position of the *pros*-nitrogen of
the imidazole ring. [1-Asparagine,5-isoleucine,6-(2,4-
diaminobutyric acid)]-angiotensin II shows negligible
activity (0.02%) in the rat pressor assay.

Aromatic character alone is not sufficient as has been
shown by the low activities for the [6-phenylalanine] (1%)[5]
and [6-thienylalanine] (1%)[6] analogs. That acid-base behavior is not a critical function of the imidazole ring
was shown with the weakly basic (pK$_a$ ~ 2.5) but highly
active (57%, rat pressor) [6-β-(pyrazolyl-3)-alanine]
analog.[7] To extend this heterocyclic series we have prepared the more basic [6-β-(pyridyl-2)-alanine] analog in
which the pyridine ring has a pK$_a$ of about 6.0.
[1-Asparagine,5-isoleucine,6-β-(pyridyl-2)-alanine]-
angiotensin II showed 3% activity in the rat pressor assay.

Two correlative physical properties are present in the
N-heterocyclic series. The highly active imidazole and
pyrazole analogs have high dipolar character, and possess
tautomeric forms with a pyrrole-like NH in the *pros*-position
adjacent to the side chain. These features could facilitate
the intramolecular interaction between the imidazole ring
and phenylalanyl carboxylate ion demonstrated for the active

analog, des-Asp[1]-[Gly[2],Ile[5]]-angiotensin II,[8] and thus help
to stabilize an active conformation.

In the solid-phase synthesis, incorporation of
Boc-β-(pyridyl-2)-L-alanine in methylene chloride with
dicyclohexylcarbodiimide and 1-hydroxybenzotriazole, under
conditions which minimized the racemization of Boc-*im*-
benzyl-L-histidine,[9] led to virtual complete racemization
of this residue. Approximately equal amounts of the 6-β-
(pyridyl-2)-L-alanine and 6-β-(pyridyl-2)-D-alanine analogs
were formed and were separated by chromatography on Sephadex
G-25 using *sec*-butanol-3% NH_4OH (100:44). The individual
diasteriomeric peptides were further purified by ammonium
acetate gradient elution from carboxymethylcellulose (CM-25),
until homogenous to tlc and electrophoresis at pH 1.85.
Acid hydrolysis followed by L-amino acid oxidase digestion
and amino acid analyses served to identify the L- and D-
amino acid-containing peptides. This unusual degree of
racemization is apparently related to the presence of the
basic pyridine nitrogen atom adjacent to the amino acid
side chain. The [6-β-(pyridyl-2)-D-alanine] analog was
essentially inactive (0.1%) in the rat pressor assay.

The 8-Phenylalanine Position

Aromatic analogs of the 8-phenylalanine residue have
shown pressor activities which may be related to the
lipophilic character of their side chains. The 8-tyrosine
residue is apparently too hydrophilic (83%,[10] 10%[4]), while
the 8-*p*-bromophenylalanine residue may be too lipophilic
(50%).[11] The lower aliphatic analogs [5-isoleucine,8-
alanine]-angiotensin II[3],[12] and [1-asparagine,5-valine,
8-alanine]-antiotensin II[13] and the lower alicyclic analogs
[5-isoleucine,8-α-aminocyclopentanecarboxylic acid]-
angiotensin II[14] and [5-isoleucine,8-α-aminocyclohexane-
carboxylic acid]-angiotensin II[15] showed virtually no
agonist properties, but were active as antagonists to
angiotensin II. We have prepared analogs with increasing
aliphatic lipophilic character in the 8-position in order
to differentiate the aromatic *versus* lipophilic roles in
the agonist and antagonist responses. The [8-leucine],
[8-norleucine], and [8-α-aminoheptanoic acid] analogs
showed increasing pressor activities with increasing side
chain lipophilic character: 1.0%, 6%, and 10%, respectively.
When administered in a single dose concurrently with a
single dose of angiotensin II, these 8-position aliphatic
analogs showed additive pressor effects.

[1-Asparagine,5-isoleucine,8-leucine]-angiotensin II, when perfused at minimal pressor levels of 37.5 and 75 picomol/min, significantly reduced the pressor responses to single injections of 2, 4, and 8 picomol of [1-asparagine, 5-valine]-angiotensin II (Figure 1). In order to enhance

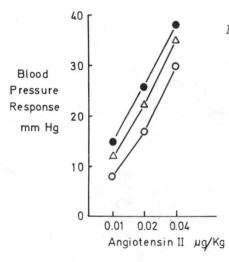

Figure 1: Antagonism to the rat pressure response to angiotensin II during infusion of [Asn[1],Ile[5], Leu[8]]-angiotensin II. ●Control; △ [Leu[8]] analog 0.2 µg/kg/min; ○ [Leu[8]] analog 0.4 µg/kg/min.

and prolong this antagonistic effect, we have prepared [1-sarcosine,5-isoleucine,8-leucine]-angiotensin II, an analog potentially resistant to aminopeptidase degradation.*

A single 400 picomol injection in a 180 gm male rat of the [Sar[1],Leu[8]] analog produced a significant inhibition of the pressor responses to single doses of 3, 6, and 12 picomol of angiotensin II, given during a 45 min period after administration of the antagonist (Figure 2).

Infusion of angiotensin II at a rate of 70 picomol/min over 90 min produced a stable rat blood pressure which was 60 mm Hg higher than normal (Figure 3). Concomitant infusion of the [Sar[1],Leu[8]] analog at a rate of 1400 picomol/ min (20/1 molar ratio) produced a 50% fall in blood pressure within 3 min and a further decrease to a total of 67% during 30 min of infusion. When infusion of the [Sar[1], Leu[8]] analog was stopped, but that of angiotensin II continued, blood pressure rose during 20 min, but was still depressed by about 40% relative to the control value. Blood pressure rapidly fell to a level close to that of

*After completion of this work we learned of a report of the long-lasting inhibitory properties of this compound.[16]

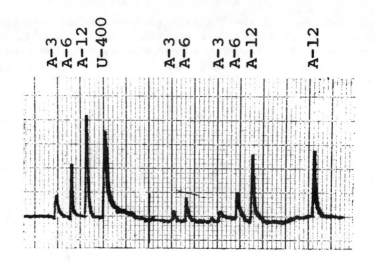

Figure 2: Record of rat pressure response to [Asn^1,Ile^5]-angiotensin II (A, x 10^{-12}mol), before and after administration of 400 picomol of [Sar^1,Ile^5,Leu^8]-angiotensin II (U, x 10^{-12}mol). Heavy lines equal 5 min.

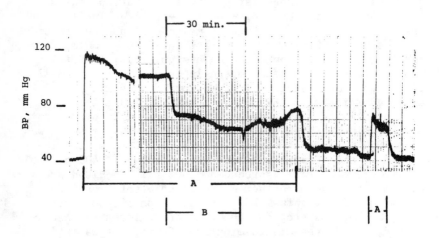

Figure 3: Record of rat pressure response to infusion of [Asn^1,Ile^5]-angiotensin II (A, 70 x 10^{-12}mol/min) and to [Sar^1,Ile^5,Leu^8]-angiotensin II (B, 1400 x 10^{-12}mol/min).

normal (40-50 mm Hg) when angiotensin II infusion was stopped during 25 min, then rose to about 40% of the normal response when infusion of angiotensin II was begun again, indicating the long lasting residual antagonistic effect of the infused [Sar[1],Leu[8]] analog.

In conclusion, specific properties of side chains have been related to the pressor effects of angiotensin II. In the 5-position, the favorable effect of β-branching is present in side chains widely varying in polarity, except that a free hydroxyl group is detrimental.

The high dipolar character of an N-heterocyclic ring system, and presence of a tautomeric form which places a pyrrole-like NH group adjacent to the amino acid side chain, appear to be characteristics for maximal pressor response in the 6-histidine position.

In the 8-position, the lipophilic character of the unsubstituted phenylalanine aromatic ring is associated with maximal pressor response, although higher aliphatic residues also produce moderately active analogs. Aliphatic or alicyclic residues of low molecular weight are ineffective as agonists, but are effective as antagonists to angiotensin II.

Acknowledgments

This investigation was supported in part by Research Grant AM 08066 from the National Institute of Arthritis and Metabolic Diseases, and by Research Funds of the University of California Academic Senate Committee on Research. We thank Dr. S. R. Rapaka for preliminary studies on the synthesis of [Asn[1],Thr[5]]-angiotensin II.

References

1. Schröder, E., and K. Lübke. *The Peptides*, (New York: Academic Press, 1966), Chapter 1, Vol. 2.
2. Jorgensen, E. C., S. R. Rapaka, G. C. Windridge, and T. C. Lee. J. Med. Chem. <u>14</u>, 899 (1971).
3. Khairallah, P. A., A. Toth, and F. M. Bumpus. J. Med. Chem. <u>13</u>, 181 (1970).
4. Schröder, E., and R. Hempel. Ann. Chem. <u>684</u>, 243 (1965).
5. Schröder, E. Ann. Chem. <u>680</u>, 142 (1964).
6. Bumpus, F. M., R. R. Smeby, and P. A. Khairallah. In *Peptides: Chemistry and Biochemistry*, Weinstein, B., and S. Lande, eds. (New York: Marcel Dekker, 1970) pp 127-150.

7. Andreatta, R., and K. Hofmann. J. Amer. Chem. Soc. 90, 7334 (1968).
8. Weinkam, R. J., and E. C. Jorgensen. J. Amer. Chem. Soc. 93, 7038 (1971).
9. Windridge, G. C., and E. C. Jorgensen. J. Amer. Chem. Soc. 93, 6318 (1971).
10. Sivanandaiah, K. M., R. R. Smeby, and F. M. Bumpus. Biochemistry 5, 1224 (1966).
11. Schwyzer, R. Helv. Chim. Acta 44, 667 (1961).
12. Türker, R. K., M. Yamamoto, P. A. Khairallah, and F. M. Bumpus. Europ. J. Pharmacol. 15, 285 (1971).
13. Pals, D. T., F. D. Masucci, F. Sipos, and G. S. Denning, Jr. Circulation Research 29, 664 (1971).
14. Regoli, D. and W. K. Park. Canad. J. Physiol. Pharmacol. 50, 99 (1972).
15. Park, W. K., and D. Regoli. Brit. J. Pharmacol. 43, 418P (1971).
16. Regoli, D., F. Rioux, and W. K. Park. Revue Canadienne de Biologie 31, 73 (1972).

PROPERTIES OF ANGIOTENSIN RECEPTORS IN SMOOTH MUSCLE: A PROPOSED MODEL

John Morrow Stewart, Richard J. Freer. Department of Biochemistry, University of Colorado School of Medicine, Denver, Colorado 80220

DATA FROM EXPERIMENTS ON the effects of changes in pH and Ca^{++} concentration on the response of smooth muscles to angiotensin II, (A II) from a study of the effect of amino acid substitutions at position 6 in A II and results of experiments with alkylating (affinity labeled) peptide derivatives have been combined to provide the basis for a hypothetical model of the topography of the A II receptor of smooth muscle, and to suggest a mechanism for the production of tachyphylaxis at low pH and Ca^{++} concentration.

The response of rat uterus to A II in low Ca^{++} (0.18 mM) de Jalon's solution is maximal at pH 8.0, and decreases at higher and lower pH. The decreased activity is most marked below pH 7, due to the production of a rapid and severe tachphylaxis. This observation suggested that the tachyphylaxis might be due to protonation of the imidazole of the histidine residue (pK 6.8). A series of 6-substituted analogs of A II was synthesized and assayed to test this hypothesis.[1] These [Asp[1], Ile[5]]-angistensins II contained thienylalanine, phenylalanine, methionine, leucine, ornithine, arginine, or glutamic acid in position 6. In addition, [Asp[1], Val[5], pyrazolylalanine[6]]-angiotensin II, a gift of Dr. Klaus Hofmann, was tested. The results showed that for maximum A II activity the side chain at position 6 must have both aromatic and nucleophilic character, and that when a strongly basic residue was present in this position, production of tachyphylaxis was severe.[2] In contrast to the earlier work of Rocha e Silva on histamine,[3] these data suggest that the response is determined by protonation of the imidazole in the hormone, rather than in the receptor.

Development of tachyphylaxis to A II in rat uterus was dependent not only on pH of the medium but also on Ca^{++} concentration. When the concentration of Ca^{++} was 1.0 m*M*, tachyphylaxis was never seen at any pH. Recovery from tachyphylaxis in low pH, low Ca^{++} medium was very slow, requiring about one hour. However, the recovery could be greatly accelerated by raising either the pH or the Ca^{++} concentration of the medium. Guinea pig ileum responds to changes in pH and $[Ca^{++}]$ in a similar but less clear-cut manner.

Two peptides containing the nitrogen mustard chlorambucil (Chl) (*p*-[*N,N*-bis(2-chloroethyl)amino]phenylbutyric acid) were used to provide additional information about the A II receptor. [Chl^1]-A II specifically and irreversibly inhibited the action of A II on isolated guinea pig ileum, without affecting the response of the tissue to bradykinin, histamine, or ions.[4] This result suggests that the nitrogen mustard moiety of the hormone derivative is alkylating an anionic site on the receptor which normally combines with the guanidinium side chain of the Arg^2-residue. Chl-Pro-Phe-Arg^5 also produces a permanent inhibition of the response of ileum to A II, but this inhibition is quite different from that produced by [Chl^1]-A II. After treatment of the tissue with the larger derivative, no amount of A II could produce a maximal response, but after treatment with the small derivative, the dose response curves obtained had the characteristics of competitive inhibition; that is, the dose response curve was parallel to and displaced to the right of the control curve. A maximal response could always be obtained by large doses of A II. However, after washing, the inhibition returned. A suggestion for this type of inhibition[6] is that the inhibitor is bound to the tissue by alkylation of an anionic site not directly on the receptor, but adjacent to it, so that the receptor is partially occluded. High concentrations of A II would displace that part of the derivative which lay directly over the receptor, and allow a full response. Since the derivative is covalently attached to the tissue, once the high concentration of A II was washed out the peptide part of the inhibitor would return to occlude the receptor.

Figure 1 depicts schematically the structure of A II, showing those functional groups which are known to be important for biological activity. In the adjacent rectangle are suggested functional groups on the receptor which could

Figure 1   Figure 2

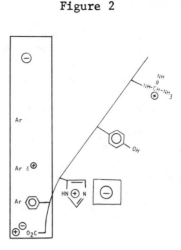

Figure 1: Suggested normal angiotensin-receptor interaction. The receptor area is indicated by the large vertical rectangle. The anionic site to the right of the receptor is normally occupied by Ca^{++}.

Figure 2: Suggested angiotensin-receptor interaction in tachyphylaxis.

most logically interact with the important groups on the peptide. The anionic site on the receptor adjacent to the locus of combination with the histidine of A II is probably normally occupied by Ca^{++}. Figure 2 depicts a suggested peptide-receptor interaction in tachyphylaxis. In low Ca^{++} medium, the anionic site would no longer be saturated with Ca^{++}, and would be available to combine strongly with the protonated form of the imidazole to anchor the peptide firmly to the receptor in an abnormal way. This would prevent normal agonist action of the adsorbed peptide, and would also inhibit entry of additional agonist to the receptor. The anionic site indicated at the top of the receptor model normally interacts with the guanidinium of the arginine residue, and may be the site alkylated by [Chl1]-A II. Further experiments are under way to seek additional confirmation of this receptor model.

References

1. Freer, R. J., and J. M. Stewart. In *Proceedings,
    Second International Symposium on Protein and Polypep-
    tide Hormones,* Part II, Margoulies, M. and F. C.
    Greenwood, eds. (Amsterdam: Excerpta Medica, in
    press).
2. Freer, R. J., and J. M. Stewart. Submitted to J. Med.
    Chem.
3. Rocha e Silva, M. J. Pharm. Pharmacol. 21, 778 (1969).
4. Paiva, T. B., A. C. M. Paiva, R. J. Freer, and J. M.
    Stewart. J. Med. Chem. 15, 6 (1972).
5. Freer, R. J., and J. M. Stewart. J. Med. Chem. 15, 1
    (1972).
6. Rocha e Silva, M. Eur. J. Pharmacol. 6, 224 (1969).

MODE OF ACTION OF PEPTIDE INHIBITORS OF ANGIOTENSIN-CONVERTING
ENZYME.-I.-

*M. A. Ondetti, J. Pluščec, E. R. Weaver, N. Williams,
E. F. Sabo, O. Kocy.* The Squibb Institute for Medical
Research, New Brunswick, New Jersey 08903

FERREIRA *ET AL.*[1] DESCRIBED IN 1970 the isolation from the
venom of *Bothrops jararaca* of several peptides capable of
potentiating the biological activities of bradykinin and
inhibiting the conversion of angiotensin I to angiotensin
II. The sequence of one of these peptides was shown to be:
<Glu-Lys-Trp-Ala-Pro (BPP$_{5a}$). Our work on the isolation of
inhibitors of angiotensin-converting enzyme from the same
venom led to the isolation and synthesis of six nona- to
tridecapeptides:[2] <Glu-Trp-Pro-Arg-Pro-Thr-Pro-Gln-Ile-Pro-
Pro (SQ 20,661); <Glu-Trp-Pro-Arg-Pro-Gln-Ile-Pro-Pro (SQ
20,881); <Glu-Asn-Trp-Pro-Arg-Pro-Gln-Ile-Pro-Pro (SQ
20,861); <Glu-Asn-Trp-Pro-His-Pro-Gln-Ile-Pro-Pro (SQ
20,858); <Glu-Ser-Trp-Pro-Gly-Pro-Asn-Ile-Pro-Pro (SQ
20,857); <Glu-Gly-Gly-Trp-Pro-Arg-Pro-Gly-Pro-Glu-Ile-Pro-
Pro (SQ 20,718).
  The studies of Cushman and Cheung[3,4] on the inhibition
of isolated angiotensin-converting enzyme with the pentapeptide BPP$_{5a}$ and longer-chain inhibitors, showed two
important differences between these two groups of inhibitors:
a) BPP$_{5a}$ is (on a molar basis) at least ten times more active
than the longer chain inhibitors; and b) the inhibitory
activity of BPP$_{5a}$ is destroyed rapidly when the pentapeptide
is incubated with the converting enzyme in the absence of
substrate and chloride ions. Under the same conditions,
the inhibitory activity of the longer chain peptides is
completely stable.

These two observations led us to undertake the synthesis of a number of analogs and derivatives of the pentapeptide BPP_{5a} with the aim of clarifying its mode of action and the cause of the instability of its inhibitory activity.

Synthesis

The strategy of stepwise condensation, starting from the C-terminal amino acid was employed for the synthesis of all peptides described in Table I. In a large number of cases, the C-terminal amino acid moiety was bound to an insoluble polystyrene divinylbenzene copolymer and the synthesis was carried out according to the solid-phase technique developed by Merrifield.[5] The t-butyloxycarbonyl group was utilized to protect the α-amino function of all amino acids, except pyroglutamic acid, and the benzyloxy-carbonyl group to protect the ε-amino function of lysine. Dicyclohexylcarbodiimide was used for all the coupling steps, except in the cases of lysine and glutamine, for which p-nitrophenyl esters were employed. The removal of the butyloxycarbonyl group was achieved with N hydrogen chloride in acetic acid, and the removal of the peptide from the resin was carried out with hydrogen bromide in acetic acid. This latter procedure led to incomplete re-moval of the benzyl ester from the γ-carboxyl of glutamic acid in peptide 8, and hydrogenolysis was needed to carry this deprotection to completion.

The synthesis of the t-butyloxycarbonyl derivative of 3-L-amino-4-phenylbutyric acid (β-homophenylalanine) re-quired for the solid-phase synthesis of peptide 20 was carried out by the Arndt-Eistert procedure, starting with t-butyloxycarbonyl-L-phenylalanine isobutylcarbonate mixed anhydride:[6]

$$\underset{\text{Boc-NH-CH-COOH}}{\overset{\text{CH}_2\text{-C}_6\text{H}_5}{|}} \xrightarrow[\underset{i\text{-BuOCOCl}}{}]{\text{NEt}_3} \underset{\text{Boc-NH-CH-COO-CO-C}_4\text{H}_9}{\overset{\text{CH}_2\text{-C}_6\text{H}_5}{|}}$$

$$\xrightarrow{\text{CH}_2\text{N}_2} \underset{\text{Boc-NH-CH-CO-CHN}_2}{\overset{\text{CH}_2\text{C}_6\text{H}_5}{|}} \xrightarrow[\underset{\text{CH}_3\text{OH}}{}]{\text{C}_6\text{H}_5\text{COOAg}} \underset{\text{Boc-NH-CH-CH}_2\text{-COO-CH}_3}{\overset{\text{CH}_2\text{C}_6\text{H}_5}{|}}$$

$$\xrightarrow{\text{OH}^-} \underset{\text{Boc-NH-CH-CH}_2\text{-COOH}}{\overset{\text{CH}_2\text{-C}_6\text{H}_5}{|}}$$

Table I

Inhibition of Angiotensin-Converting Enzyme

	Structure	Method of Synthesis*	I_{50}† (μg/ml)
BPP-5a	<Glu-Lys-Trp-Ala-Pro	A	0.05
1	<Glu-Lys-Phe-Ala-Pro	A	0.05
2	Cbc-Lys-Phe-Ala-Pro#	A	0.04
3	Cpc-Lys-Phe-Ala-Pro**	A	0.06
4	Chc-Lys-Ala-Pro††	A	0.06
5	Boc-Lys-Trp-Ala-Pro	A	0.9
6	<Glu-Nle-Phe-Ala-Pro	B	0.17
7	<Glu-Gln-Phe-Ala-Pro	B	0.36
8	<Glu-Glu-Phe-Ala-Pro	B	3.0
9	<Glu-Ala-Pro-Ala-Pro	B	31.0
10	<Glu-Lys-Pro-Ala-Pro	B	1.1
11	<Glu-Lys-Ile-Ala-Pro	B	1.6
12	<Glu-Lys-Ser-Ala-Pro	B	2.4
13	<Glu-Lys-Phe-Gly-Pro	B	0.14
14	<Glu-Lys-Phe-Pro-Pro	B	3.2
15	<Glu-Lys-Phe-Ala-N⬠	A	>100.0
16	<Glu-Lys-D-Trp-Ala-Pro	B	72.0
17	<Glu-Lys-Phe-Lac-Pro	B	0.06
18	<Glu-Lys-NH-CH-CH$_2$O-CH$_2$-CO-Pro (CH$_2$C$_6$H$_5$)	A	24.5
19	<Glu-Lys-NH-CH-CH$_2$-O-CH-CO-Pro (CH$_2$C$_6$H$_5$, CH$_3$)	A	19.0
20	<Glu-Lys-NH-CH-CH$_2$-CO-Ala-Pro (CH$_2$C$_6$H$_5$)	A	0.25

*A: Stepwise approach in solution, B: Stepwise in solid-phase.
†Concentration required to inhibit 50% of the activity of
  angiotensin-converting enzyme isolated from rabbit lung.[3]
#Cbc, cyclobutylcarbonyl.
**Cpc, cyclopentylcarbonyl.
††Chc, cyclohexylcarbonyl.

The intermediate diazoketone and methyl ester, and the final product, *t*-butyloxycarbonyl-L-β-homophenylalanine, were isolated in crystalline form, with good yields. The synthesis of this derivative by a similar procedure has been described recently.[7]

The synthesis of the depsipeptide 17 (Table I) by the solid-phase technique followed the approach described by Gisin *et al.* for the synthesis of valinomycin.[8] moiety was synthesized in solution:

$$\underset{\text{Boc-NH-CH-COOH}}{\overset{\overset{\displaystyle CH_2-C_6H_5}{|}}{}} \xrightarrow[\underset{\overset{\displaystyle |}{CH_3}}{\text{L-HO-CH-COOCH}_2C_6H_5}]{\text{Carbonyldiimidazole}} \underset{\text{Boc-NH-CH-CO-O-CH-CO}_2CH_2C_6H_5}{\overset{\overset{\displaystyle CH_2-C_6H_5 \quad CH_3}{| \qquad\quad |}}{}}$$

$$\xrightarrow{\text{H}_2/\text{Pd}} \underset{\text{Boc-NH-CH-CO-O-CH-COOH}}{\overset{\overset{\displaystyle CH_2-C_6H_5 \quad CH_3}{| \qquad\quad /}}{}}$$

Synthesis of the amino acid ether moieties required for the synthesis of peptides 18 and 19 proceeded as follows:

$$\underset{\underline{\text{I}}}{\underset{\text{Pht-N-CH-CH}_2\text{OH}}{\overset{\overset{\displaystyle CH_2C_6H_5}{|}}{}}} \xrightarrow{\text{NaH/BrCH}_2\text{COOC}_2\text{H}_5} \underset{\underline{\text{II}}}{\underset{\text{Pht-N-CH-CH}_2-O-CH_2-COOC_2H_5}{\overset{\overset{\displaystyle CH_2C_6H_5}{|}}{}}}$$

I ──→ (NaH; $\overset{\displaystyle CH_3}{\underset{|}{\underline{\underline{D}}\text{-BrCH-COOCH}_3}}$) ──→

$$\underset{\underline{\text{III}}}{\underset{\text{Pht-N-CH-CH}_2-O-CH-COO CH_3}{\overset{\overset{\displaystyle CH_2C_6H_5 \quad CH_3}{| \qquad\quad |}}{}}}$$

II ──→ (1)OH⁻ 2)H+) ──→

$$\underset{\underline{\text{IV}}}{\underset{\text{NH}_2-\text{CH-CH}_2-O-CH_2-COOH}{\overset{\overset{\displaystyle CH_2C_6H_5}{|}}{}}}$$

III ──→ (1)OH⁻ 2)H+) ──→

$$\underset{\underline{\text{V}}}{\underset{\text{NH}_2-\text{CH-CH}_2-O-CH-COOH}{\overset{\overset{\displaystyle CH_2C_5H_6 \quad CH_3}{| \qquad\quad |}}{}}}$$

The amino acid ether IV was obtained in crystalline form, but all attempts to crystallize V were unsuccessful. Both compounds showed the expected IR and NMR spectra. The corresponding *t*-butyloxycarbonyl derivatives were obtained as oils, and were purified by countercurrent distribution or through crystallization of the corresponding dicyclohexylamine salts. The configuration of the asymmetric carbon atom in the "lactyl" moiety of V is expected to be L, assuming a 100% inversion in the reaction from I to III.

In some of the peptides of Table I, the strategy of stepwise condensation was carried out in solution, starting from a benzyl ester of the C-terminal amino acid, and utilizing the active ester procedure for all coupling steps. The synthesis of peptide 15 was also carried out in the stepwise manner, starting with pyrolidine.

Structure-Activity Relationships

How the replacement of amino acids effected the inhibitory activity of BPP_{5a} on the angiotensin-converting enzyme can be summarized as follows:

a) The pyroglutamyl residue in position 1 can be replaced by an alicylic carboxylic acid of similar size without any loss of activity. Bulky alkyloxycarbonyl acyl groups are considerably less efficient in this respect (peptides 2, 3, 4, and 5, Table I).

b) The side chain amino function of the lysine residue in position 2, is not essential for the inhibitory activity (peptides 6 and 7). However, it cannot be replaced by a carboxyl function (peptide 8) nor can the aliphatic chain be shortened significantly (compare peptides 9 and 10) without considerable loss of inhibitory potency.

c) An aromatic amino acid is the best choice for position 3 (peptide 1). Replacement of tryptophan with aliphatic acyclic amino acid or cyclic amino acid residues lowers the inhibitory activity (peptides 10 and 11) and the reduction in activity is even greater if hydrophilic functional groups are present in the side chain (peptide 12).

d) The replacement of the alanine residues in position 4 by a proline residue diminishes the inhibitory activity drastically (peptide 14). However, this inhibitory activity remains unaltered when this peptide 14 is incubated with the converting enzyme in the conditions described in the introduction, as is the case with the longer-chain peptidic inhibitors that have the C-terminal sequence Ile-Pro-Pro.

e) Elimination of the carboxyl group from the C-terminal amino acid moiety of BPP_{5a} destroys the inhibitory activity completely (peptide 15).

The structure-activity relationships described above parallel rather closely the requirements for substrate specificity of the converting enzyme,[10] and support the hypothesis advanced by Cushman and Cheung[4] that BPP$_{5a}$ interacts with the enzyme at the active site, and is, under certain conditions, cleaved at the Trp-Ala bond by the converting enzyme itself.

Since substitution of Pro for Ala (peptide 14, Table I) gives inhibitors that are stable, but of decreased potency, other modifications of this bond were attempted:

1) Replacement of L-Trp by D-Trp (peptide 16);
2) Change from an amide to an ether bond (peptides 18 and 19);
3) Replacement of alanine by lactic acid (peptide 17); and
4) Substitution of β-homophenylalanine for tryptophan (peptide 20).

Only the last two modifications yielded inhibitors with potency comparable to that of BPP$_{5a}$. The inhibitory activity of the depsipeptide 17 was destroyed rapidly by the converting enzyme, but that of the homopeptide 20 remained unaltered under the same conditions.

Acknowledgment

The authors thank Dr. D. Cushman and Mr. H. S. Cheung for the assays of inhibition of the angiotensin-converting enzyme, and for the stimulating exchange of ideas.

References

1. Ferreira, S. H., D. C. Bartelt, and L. J. Greene. Biochemistry 9, 2583 (1970).
2. Ondetti, M. A., N. J. Williams, E. F. Sabo, J. Pluščec, E. R. Weaver, and O. Kocy. Biochemistry 10, 4033 (1971).
3. Cushman, D. W., and H. S. Cheung. Biochem. Pharmacol. 20, 1637 (1971).
4. Cushman, D. W., and H. S. Cheung. In preparation.
5. Merrifield, R. B. Advan. Enzymol. 32, 221 (1969).
6. Penke, B., J. Czombos, L. Baláspiri, J. Petres, and K. Kovács. Helv. Chim. Acta 53, 1057 (1970).
7. Chaturvedi, N. C., W. K. Park, R. R. Smeby, and F. M. Bumpus. J. Med. Chem. 13, 177 (1970).
8. Gisin, B. F., R. B. Merrifield, and D. C. Tosteson. J. Amer. Chem. Soc. 91, 2691 (1969).
9. Liberek, B., and C. Cupryszak. Ann. Soc. Chim. Polonorum 45, 677 (1971).

10. Elisseeva, Y. E., V. N. Orekhovich, L. V. Pavlikhina, and L. P. Alexeenko. Clin. Chim. Acta $\underline{31}$, 413 (1971).

CHEMISTRY AND BIOLOGY OF PEPTIDES

ISOTOPE DERIVATIVE ASSAY OF NANOGRAM QUANTITIES OF
ANGIOTENSIN I: USE FOR RENIN MEASUREMENT IN PLASMA

Robert I. Gregerman, Mary Ann Kowatch. Gerontology
Research Center, National Institute of Child Health
and Human Development, National Institutes of Health,
Baltimore City Hospitals, Baltimore, Maryland 21224

WE HAVE PREVIOUSLY DESCRIBED a double isotope derivative
method for the quantitation of nanogram amounts of angiotensin II and some related peptides.[1] Subsequently we
reported an assay of human renin by a procedure which uses
this method for quantitation of angiotensin I generated by
the action of the enzyme on an excess of hog renin substrate.[2] This technique has now been standardized for the
assay of renin in human plasma.

Figure 1 shows a general scheme for the double isotope
quantitation of peptides. The principles involved follow
those which have been used extensively in the field of
steroid analysis. Derivatization of the peptide is performed with labeled reagent of one isotope while recovery
is corrected by use of an indicator amount of the compound
(or its derivative) which contains a second isotope.
Varying degrees of purification of the derivatized product
are accomplished by whatever means prove necessary and
convenient for the particular application. In the following description italicized numbers refer to Figure 1. For
the angiotensins we use tritium-labeled 1-fluoro-2,4-
dinitrobenzene (^{3}H-FDNB) *1* to derivatize the peptides to
the dinitrophenyl (Dnp) derivatives. ^{14}C-Dnp-angiotensin
I, prepared separately, is used to monitor recovery *2* and
carrier Dnp-angiotensin I is added after derivatization *4*.
Thin layer chromatography *5* is used for purification of
the product after removal of excess ^{3}H-FDNB *3*. In our

533

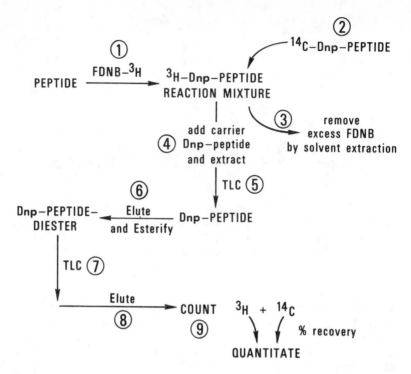

Figure 1: General scheme for the double isotope derivative
assay of peptides

procedure each tlc step is performed in two dimensions.
After the first tlc the Dnp-angiotensin I is eluted and
simultaneously further derivatized *6* to the dimethyl ester
before the repeated tlc. The extent of this purification
is determined primarily by our need for a low reagent and
plasma blank in order to measure low levels of renin in
human plasma. For some applications only a single tlc
step might be sufficient while for others, in which complex
mixtures of peptides were involved, additional purification
might be required.

After derivatization with FDNB, angiotensin I contains
4 Dnp groups: Dnp-Asp-Arg-Val-Tyr(Dnp)-Ile-His(Dnp)-Pro-
Phe-His(Dnp)-Leu. The four sites which are derivatized
are the N-terminal NH_2 of Asp, the two imidazoles of His
and the phenolic hydroxyl of Tyr. This stable derivative
is, of course, yellow. More importantly, it is strongly
UV-absorbing, which allows non-destructive localization on

tlc. The tritium-labeled reagent is also stable and relatively inexpensive. Details of the chemical manipulations have been published previously.[1,2]

The scheme for preparation of the plasma for renin assay and isolation of generated angiotensin I from the plasma incubation mixture is shown in Figure 2. Passage of a

1. PLASMA (EDTA) THRU DOWEX 50-X8

2. ADD SUBSTRATE, BUFFER, PMA

3. INCUBATE

4. PASS THRU DOWEX 50-X8

5. WASH WITH WATER

6. ELUTE WITH 2 M PYRIDINE

7. CONCENTRATE TO DRYNESS

Figure 2: Plasma handling for renin assay by double isotope derivative assay: preparation of plasma, incubation, and isolation of generated angiotensin I

2 ml plasma sample through a 1 ml Dowex bed removes any preformed angiotensin I and a number of materials which would otherwise react with the ^{3}H-FDNB reagent and so interfere with the procedure. An 0.5 ml sample of the "cleaned" plasma is then incubated with excess hog substrate in the presence of EDTA at pH 7.6. Phenylmercuric acetate (PMA) is used as a bacteriostat. These conditions are essentially those of Gould *et al.*,[3] except that in our procedure the EDTA concentration is high enough (0.03 M) to inhibit converting enzyme as well as angiotensinases. Angiotensin standards and renin (plus hog substrate) standards are run with each assay. After completion of incubation for 18 hours the plasma mixture is passed through a 1 ml Dowex bed in a disposable syringe barrel. This adsorbs the generated angiotensin I, and eliminates the bulk of the protein and salts. The angiotensin I is then eluted with pyridine directly into the derivatization tube, evaporated under an air stream, and derivatized.

As reported earlier, 1 x 10^{-4} Goldblatt unit (GU) of human renin yields approximately 100 ng of angiotensin I in 18 hours under these conditions.[2] The recovery of 1 x 10^{-4} GU of renin added per ml was 87.7 ± 8.7% (S.D.) for 24 determinations made over about 18 months. Thus, the mean recovery is close to 90%, but some samples do appear to have some residual angiotensinase activity.

Renin assays have been made by the double isotope derivative method applied to plasma from approximately 20

normal human subjects on normal and high sodium intakes.
The recumbent values have a mean of about 0.2×10^{-4} GU/ml
and approximate those reported in the literature using
similar incubation conditions and bioassay. Upright values
and post diuretic values (4 hr after 80 mg furosemide given
orally) increase about as expected to approximately 0.35
and 0.6×10^{-4} GU/ml respectively. Values for pregnant
subjects are greatly elevated (mean about 1.2×10^{-4} GU/ml)
but perhaps somewhat less so than one might have expected.[3]
Our seven anephric subjects (3 males, 4 females), whose
kidneys had been removed prior to anticipated kidney trans-
plants, all had measurable renin with values very close to
those of the normal recumbent subjects.

The literature is conflicting on the issue of whether
or not the plasma of anephric subjects contains measurable
renin. A number of workers using bioassay methods have
reported low normal or normal values[4] while recently both
a bioassay method and radio-immunoassay technique were
said to show no detectable renin.[5] If one takes the view
that our method measures material other than angiotensin I,
i.e., that the anephric plasmas represent the biological
"blank" for our method, then one must conclude that, since
recumbent normal subjects have the same values, that they
too have no detectable renin. We do not favor such an
interpretation and at present take the view that anephric
subjects have low but measurable rates of angiotensin
generation. This point is under further investigation.

We have compared the results by our isotope derivative
method on individual samples with those obtained by bioassay of the same samples performed in Dr. Gould's
laboratory. The comparison shows a reasonable but by no
means perfect correlation of the values.

This application of the double isotope approach to
peptide quantitation suggests to us that, even though
techniques such as radio-immunoassay (angiotensin I) may
be more convenient for routine use, and therefore preferable
for renin measurement, the double isotope method may well
be useful under special circumstances and for the quantitation of other peptides of biological interest.

References

1. Gregerman, R. I., and M. A. Kowatch. Biochem. Med. 3,
   13 (1969).
2. Gregerman, R. I., and M. A. Kowatch. J. Clin.
   Endocrinol. and Metab. 32, 110 (1971).

3. Gould, A. B., L. T. Skeggs, and J. R. Kahn. Lab. Invest. 15, 1802 (1966).
4. McKenzie, J. K., and J. Z. Montgomerie. Nature 223, 1156 (1969).
5. Berman, L. B., V. Vertes, S. Mitra, and A. B. Gould. New Engl. J. Med. 286, 58 (1972).

ON THE IDENTITY OF PEPSITENSIN

T. B. Paiva, A. Grandino, M. E. Miyamoto, A. C. M. Paiva. Department of Biophysics and Physiology, Escola Paulista de Medicina, 04023, São Paulo, S.P., Brazil

PEPSITENSIN, A VASOPRESSOR PEPTIDE obtained by peptic proteolysis of a plasma substrate, has biological activities that are qualitatively identical with those of angiotensin II,[1] but the two peptides were shown to be different.[2] Two different amino acid sequences have been proposed for this peptide: Franze *et al.*[3] isolated [Val5]-angiotensin I (AI) from ox plasma incubated with pepsin at pH 6 and Hochstrasser *et al.*[4] proposed the sequence Asp-Arg-Val-Tyr-Val-His-Pro-Phe-His-Leu-Leu ([Val5]-AI-Leu) for a pepsitensin obtained from peptic proteolysis of ox plasma at pH 3. These results would suggest that pepsin acts at a different peptide bond in plasma renin-substrate, according to the pH of incubation (Figure 1). In order to verify this hypothesis we have synthesized the following peptides: *(1)* [Val5]-AI-Leu; *(2)* [Ile5]-AI-Leu; *(3)* Val-Tyr-Ser; *(4)* Leu-Val-Tyr-Ser; *(5)* His-Leu-Leu-Val-Tyr-Ser. Merrifield's solid phase method was employed[5] with some modifications.[6] The activity of compounds *1* and *2* on the rat's blood pressure was, respectively, 1.28(±0.08)% and 1.42(±0.09)%, relative to angiotensin II. Compounds *1* and *2* were easily converted to fully active angiotensin II by the action of carboxypeptidase A.

Peptides *2 - 5*, [Ile5]-angiotensin I and [Ile5]-angiotensin II were compared with the products of peptic hydrolysis of synthetic [Ile5]-tetradecapeptide renin substrate (Schwarz/Mann) done at pH 3.0 and at pH 6.0. At both pHs, the main products were angiotensin I and *4*,

Asp-Arg-Val-Tyr-Val-His-Pro-Phe-His-Leu + Leu-Val-Tyr-Ser
   (Angiotensin I)

Pepsin | pH 6

Asp-Arg-Val-Tyr-Val-His-Pro-Phe-His-Leu-Leu-Val-Tyr-Ser-R
   (Renin Substrate)

Pepsin | pH 3

Asp-Arg-Val-Tyr-Val-His-Pro-Phe-His-Leu-Leu + Val-Tyr-Ser
   (Previously Proposed Pepsitensin)

Figure 1: Hypothetical production of two different
"pepsitensins" by peptic hydrolysis of renin substrate
at pH 3 and at pH 6.

shown by high voltage paper electrophoresis and tlc with
three solvent systems. *2* and *3* were not detected as
products of peptic hydrolysis at either pH. This indicates
that there is only one pepsitensin, which is identical with
angiotensin I.

Acknowledgment

This work was supported by grants from the Fundação de
Amparo a Pesquisa do Estado de São Paulo ("Projeto Bioq/
FAPESP") and the Conselho Nacional de Pesquisas, Brazil.

References

1. Alonso, O., R. Croxatto, and H. Croxatto. Proc. Soc.
    Exp. Biol. Med. 52, 61 (1943).
2. Paiva, A. C. M., T. Bandiera, and J. L. Prado. Science
    120, 611 (1954).
3. Franze, M. T., A. C. Paladini, and A. E. Delius.
    Biochem. J. 97, 540 (1965).
4. Hochstrasser, K., F. Bachhuber, and E. Werle. Hoppe-
    Seyler's Z. Physiol. Chem. 350, 1225 (1969).
5. Stewart, J. M., and J. D. Young. *Solid Phase Peptide
    Synthesis* (San Francisco, Calif.: W. H. Freeman and
    Co., 1969).
6. Paiva, T. B., A. C. M. Paiva, R. J. Freer, and J. M.
    Stewart. J. Med. Chem. 15, 6 (1972).

BIO-ISOSTERES OF A PEPTIDE RENIN INHIBITOR

M. J. Parry, A. B. Russell, M. Szelke. Research
Division, Pfizer Ltd., Sandwich, Kent, England

IT WAS REPORTED BY KOKUBU and coworkers[1] in 1968 that the
tetrapeptide ester H–Leu–Leu–Val–Phe–OMe *1* was a competitive
in vitro inhibitor of rabbit renin, and that it lacked
hypotensive activity in rats. It was also shown[1] that the
Leu[1]–Leu[2] grouping and an aromatic amino acid residue at
position 4 were essential for renin inhibitory activity
and that even conservative replacements of Leu, as in
H–Ile–Leu–Val–Phe–OMe *2* or H–Val–Leu–Val–Phe–OMe *3* produced
inactive compounds.

In the course of our investigations concerned with the
role of the renin–angiotensin system in hypertension we
found that *1* did not inhibit endogenous renin in rat plasma,
probably as a result of enzymatic breakdown and binding to
plasma proteins. Subsequently, we undertook various
chemical modifications of *1* with the aim of obtaining a
renin inhibitor also effective *in vivo* and therefore of
practical value in assessing the pathogenetic role of
renin in experimental hypertension.

In view of the specific structural requirements with
regard to the side-chains in *1*, these were retained during
modification. In order to increase stability towards
proteolytic enzymes, some or all of the α-amino acid units
in *1* were replaced by the isosteric "reduced" units
–NH–HCR–CH$_2$– or carbazic acid units –NH–NR–CO– bearing
the side-chain R of the amino acid residue that they
replace.

In an *in vitro* assay system based on the radioimmunoassay of angiotensin-I liberated by pig renin from synthetic
tetradecapeptide substrate, several of the resulting modified

peptides, *e.g.* *4 - 7* (Figure 1) were found to be specific
inhibitors of renin with a potency equal or superior to
that of the parent tetrapeptide ester *1*.

$\overset{\displaystyle CH_2CHMe_2}{|}$

4 $H_2N-CH-CH_2-Leu-Val-Phe-OMe$

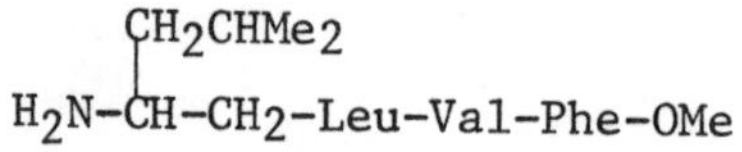

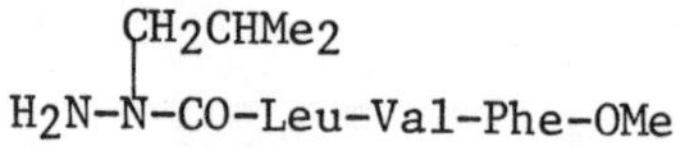

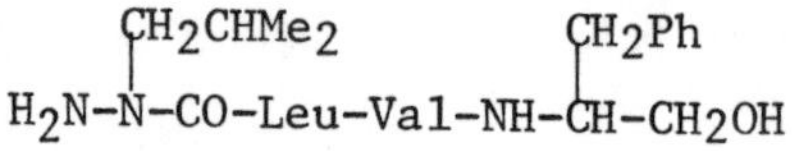

$\overset{\displaystyle CH_2CHMe_2}{|}$

5 $H_2N-N-CO-Leu-Val-Phe-OMe$

$\overset{\displaystyle CH_2CHMe_2}{|}$ $\overset{\displaystyle CH_2Ph}{|}$

6 $H_2N-N-CO-Leu-Val-NH-CH-CH_2OH$

$\overset{\displaystyle CH_2CHMe_2}{|}$ $\overset{\displaystyle CH_2CHMe_2}{|}$ $\overset{\displaystyle CHMe_2}{|}$ $\overset{\displaystyle CH_2Ph}{|}$

7 $H_2N-CH-CH_2-NH-CH-CH_2-NH-CH-CH_2-NH-CH-CH_2OH$

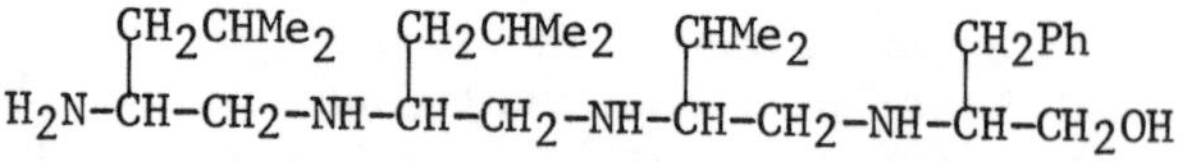

Figure 1: Bio-isosteres of peptide renin inhibitor,
H-Leu-Leu-Val-Phe-OMe

Moreover, *4* and to a lesser degree *5* and *6* inhibited
the liberation of angiotensin-I from plasma substrate by
endogenous renin in rat plasma whereas the parent peptide
1 was completely inactive in this system (see Table I).
    The modified peptides *4 - 7* were completely stable in
the presence of leucine aminopeptidase while under the
same conditions *1* was rapidly hydrolysed to the component
amino acids.
    In a group of 3 rats, made hypertensive by unilateral
renal artery constriction, the partially reduced peptide *4*
caused significant falls in blood pressure on each of 3
days when administered daily at 10 mg/kg, as compared with
a control group given saline injections. Initially there
was a short-lived rise in blood pressure of 5 mm, followed
after 2 hours by a fall lasting several hours and reaching
a maximum of 25 mm after 5 hours, which represents an
antihypertensive effect of approx. 33% (*i.e.* *4* abolished
one third of the rise in blood pressure caused by constriction of the renal artery). At higher doses (*e.g.* 20 mg/kg),
4 elicited acute cardiovascular effects (bradycardia,
hypotension) and was not investigated further.

Table I

Inhibition of Renin *in vitro* by Compounds *1 - 7*

Compound	% Inhibition		of endogenous renin cleaving natural substrate in rat plasma at 10^{-3}M
	of pig renin cleaving synthetic substrate in pH 6 phosphate buffer		
	at 10^{-3}M	at 10^{-4}M	
1	56	10	0
2	0	–	–
3	0	–	–
4	–	34	84
5	–	28	6
6	–	22	7
7	29	–	–

Modified peptides *4 - 7* were synthesized as follows:
(a) Reduced analogues. Compound *4* was obtained *via* the reductive alkylation of tripeptide ester *9* with the protected α-aminoaldehyde *8*, the latter being prepared by Pfitzner-Moffatt oxidation[2] from *N*-tosyl leucinol (Figure 2).

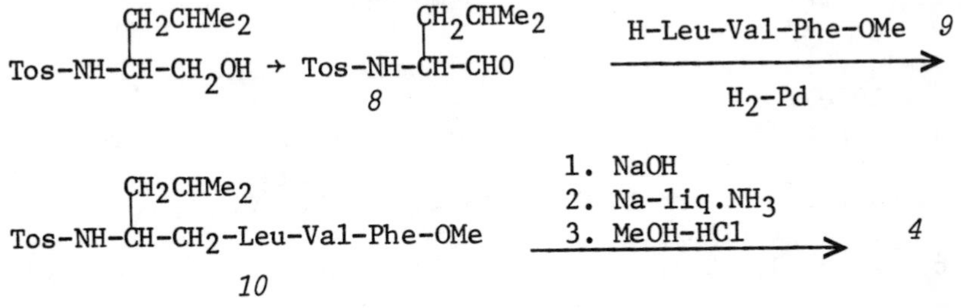

$$\underset{8}{\text{Tos-NH-CH-CH}_2\text{OH}} \underset{}{\overset{\text{CH}_2\text{CHMe}_2}{\big|}} \rightarrow \underset{8}{\text{Tos-NH-CH-CHO}} \overset{\text{CH}_2\text{CHMe}_2}{\big|} \xrightarrow[\text{H}_2\text{-Pd}]{\text{H-Leu-Val-Phe-OMe} \quad 9}$$

$$\underset{10}{\text{Tos-NH-CH-CH}_2\text{-Leu-Val-Phe-OMe}} \overset{\text{CH}_2\text{CHMe}_2}{\big|} \xrightarrow[\text{3. MeOH-HCl}]{\begin{array}{l}\text{1. NaOH}\\\text{2. Na-liq.NH}_3\end{array}} \quad 4$$

Figure 2: Synthesis of reduced analogue[4] of peptide renin inhibitor, H-Leu-Leu-Val-Phe-OMe.

Compound *6* containing a reduced C-terminal residue was prepared from the corresponding peptide ester by reduction[3] with NaBH₄. The totally reduced peptide *7* was obtained *via* reduction of *1* with LiAlH₄ in refluxing tetrahydrofuran.[4,5]

(b) Aza-peptides *5* and *6*, in which the Leu[1] residue is replaced by the isosteric carbazic acid, were synthesized *via* condensation of the protected carbazyl chloride *11*[6] with the peptide ester *9* (Figure 3).

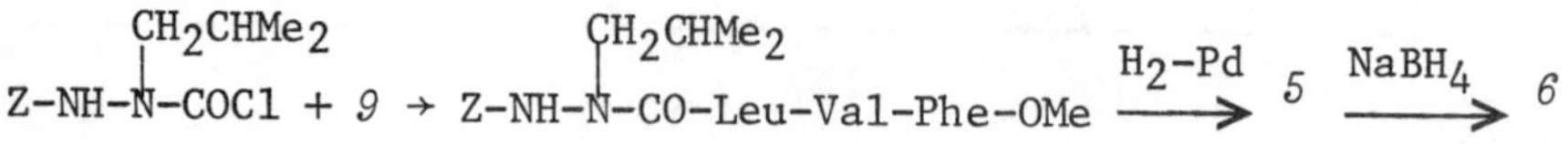

$$CH_2CHMe_2 \qquad\qquad CH_2CHMe_2$$
$$Z-NH-N-COCl + 9 \rightarrow Z-NH-N-CO-Leu-Val-Phe-OMe \xrightarrow{H_2-Pd} 5 \xrightarrow{NaBH_4} 6$$
$$\quad 11 \qquad\qquad\qquad\qquad 12$$

Figure 3: Synthesis of aza-analogues of peptide renin inhibitor, H-Leu-Leu-Val-Phe-OMe.

Acknowledgment

We acknowledge the able assistance of Mr. W. A. Million with the chemical syntheses and of Messrs. C. Butler, A. J. Carter, E. Hawkeswood and I. G. MacGregor with the biological evaluation.

References

1. Kokubu, T., E. Ueda, S. Fujimoto, K. Hiwada, A. Kato, H. Akutsu, Y. Yamamura, S. Saito, and T. Mizoguchi. Nature 217, 456 (1968).
2. Pfitzner, K. E., and J. G. Moffatt. J. Amer. Chem. Soc. 87, 5661 (1965).
3. Tanabe Seiyaku Co. Ltd. West German Patent 2,003,019 (1970).
4. Karrer, P., P. Portmann, and M. Suter. Helv. Chim. Acta 31, 1617 (1948).
5. Karrer, P., and B. J. R. Nicolaus. Helv. Chim. Acta 35, 1581 (1952).
6. Gante, J. Chem. Ber. 97, 2551 (1964).

SOME RECENT DATA ON ANGIOTENSINAMIDE II CONFORMATIONS

S. Fermandjian, D. Greff, P. Fromageot. Service
Biochimie, Département Biologie, Centre d'Etudes
Nucléaires de Saclay, 91 Gif sur Yvette, France

SUMMARY--The characteristics of Angiotensinamide II and
some of its analogs have been investigated by a variety of
techniques, under different environmental conditions. It
is concluded that one deals with monodisperse angiotensinamide II only in dilute aqueous solution or in
trifluoroethanol and hexafluoroisopropanol. In concentrated
aqueous solution, or in DMSO, the peptide is aggregated.
Data collected strongly indicate the tendency of the
molecule to fold on itself. When this tendency is enhanced by organic solvents a large proportion of the
peptide adopts a preferred overall conformation, defined
as follows: a first turn takes place at the level of
valine in position 3 and tyrosine in position 4. A second
turn at the His-Pro linkage, corresponding to a histidine
carbonyl group pointing approximately towards the same
direction as the proline carbonyl group,* brings the
phenylalanine side chain close to the valine side chains.
This delineates a hydrophobic pocket on one side of the
molecule. Arginine, tyrosine and histidine side chains
remain free of interactions.

---

*To prevent misunderstanding about the His-Pro peptide
nomenclature, we like to remind that we call *"cis"* that
conformation in which the histidine carbonyl and the
proline carbonyl are pointing to the same direction and
"trans" the opposite.

THE RAPID AND SPECIFIC PHYSIOLOGICAL responses elicited by
an intravenous injection of a variety of peptide and other
hormones are still puzzling phenomena; the more so, when
one considers the minute amounts of compounds introduced
into the circulation. The importance of specific binding
of biologically active molecules to specialized cellular
sites was early postulated and firmly stated as for instance
by von Buddenbrock[1] when he wrote: "Dem Biologen muss als
Leitfaden dienen dass die chemische Konstitution nicht das
letzte Wort ist. Das Entscheidende ist schliesslich das
Reaktionssystem."

When labelled peptide hormones of sufficiently high
specific radioactivity were obtained, the studies of their
mode of action took a new turn. The target cells could be
directly identified. Moreover, the existence of hormone-
specific binding sites on the surface of such cells could
be recognized and their affinity for the relevant hormones
measured. It is gratifying to note that the binding constants measured are in agreement with the concentrations
required to elicit a hormonal response. This progress
prompted a vivid interest in conformational relationships
between hormones and receptor sites. This is a continua-
tion of the previous efforts to correlate hormone primary
structure and biological activity. The underlying concept
is that a peptide hormone has a preferred conformation in
the prevailing environment which allows the first interaction with the specific binding site to take place.
Following this the peptide environment suddenly changes,
as well as that of the corresponding receptor site, and
additional conformational changes are envisioned, leading
to a tight binding and, conditions permitting, to a
biological response.

In the following we shall discuss some conformational
data concerning angiotensinamide II, an octapeptide of
linear primary structure corresponding to Asn-Arg-Val-Tyr-
Val-His-Pro-Phe, when of bovine origin. When extracted
from horse or hog, valine in position 5 is replaced by
isoleucine.[2]

Ideally, one should work out the conformation(s) of
such a peptide under the condition prevailing in the
vicinity of the receptor sites or *in situ*. As this is
not feasible yet, one is left with the study of angiotensinamide II under circumstances which might be far from
the physiological ones, but which should reveal the
capacities of the molecule to adopt preferential
conformations.

Infrared and Raman Spectroscopy

Between the vibration bands of peptide bonds and the structure of polypeptides, there exist empirical[3],[4] as well as theoretical relations.[5] Good agreement has been found between IR spectral analyses according to Miyazawa's theory and peptide conformations in a variety of cases, for instance with hexa-L-alanine[6] and octabenzyl glutamate.[7] The problem was to identify the peptide vibration bands in the complex spectra given by a heteropeptide like angiotensinamide II. This was accomplished through comparison with the spectra recorded for the constituant amino-acids and for peptides of increasing size.[8] We thus found in the amide I region of IR spectra an intense band close to $1635 cm^{-1}$ and a shoulder at $1685 cm^{-1}$. Their intensities relative to side chain bands decreased considerably when going from the octapeptide to the tetra and pentapeptides, suggesting that these bands corresponded to peptide bond vibrations, indeed. Their presence point to an antiparallel β structure as described for both short[9],[10] and long polypeptide chains.[11],[12] Moreover, Raman spectra showed two groups of bands between 1600 and 1700 cm^{-1}, which may contain bands at 1635, 1649, 1660, 1668 and 1689 cm^{-1}. The 1635, 1649 and 1689-1685 cm^{-1} bands correspond to the components of amide I bands according to Miyazawa's analysis of β structures, leaving open the interpretation of the 1660 and 1668 bands. Using Miyazawa's set of equations and ascribing $\nu(0.\pi) = 1685$ cm^{-1}, $\nu(\pi.0) = 1635$ cm^{-1} and $\nu(\pi.\pi) = 1649$ cm^{-1}, one gets $\nu(0.0) = 1671$ cm^{-1}. The latter, being a symetrical vibration should be IR inactive, but observed in Raman spectra and might well correspond to the 1668 cm^{-1} band. By the same computation $\nu_0 = 1660$ cm^{-1}. These values allowed an estimation of the intrachain D_1, and interchain D_1' coupling terms. The values found $D_1 = 18$ cm^{-1} and $D_1' = -7$ cm^{-1} were of the expected signs and size for an antiparalled β conformation.

The 1660 cm^{-1} frequency, attributed to an amide group that is no longer subject to vibrational interactions, may correspond to an unordered fraction in the peptide studied. This assignment is based on several arguments: the 1660 cm^{-1} amide frequency is close to values deduced from IR spectra of nylon 66 and polyglycine, which are unordered polypeptides. The contribution of the 1660 cm^{-1} band increases in the peptide fragments 1-4 and 5-8 of angiotensin, whereas the 1685 cm^{-1} band decreases. The bands localized in the amide III region (1200-1350 cm^{-1}) also support the conclusion of a significant proportion of β conformation,

i.e. the 1236 and 1274 cm^{-1} bands observed on Raman spectra
of angiotensinamide II are similar to those noted for penta-
alanine[6] (1231, 1250 and 1268 cm^{-1}) for polyserine (1235
cm^{-1}) and for polyvaline (1231, 1276, 1291 cm-1).[13] The
bands localized in the amide II (1500-1550 cm^{-1}) and amide
IV to VI regions (below 850 cm^{-1}) were not convincing and
presently of little use.

The results obtained by Raman spectroscopy of concentrated angiotensinamide II solutions were in agreement
with those reported for dry material. The multiplicity
of bands in the amide I region, the shoulder at 1685 cm^{-1}
as well as the frequencies of the amide III vibration bands
indicated a high proportion of antiparallel β conformation
as in the dry state. These results merely indicate that
under conditions which promote inter or intra peptide
interactions, an antiparallel β type of packing is more
stable than other arrangements. What does antiparallelism
stand for in the present case? IR or Raman studies did not
shed light on the problem. Rayleigh diffusion measurements
in concentrated (0.05*M* to 0.2*M*) aqueous solutions indicated
that angiotensinamide II behaved as if it were a polymer of
high molecular weight. Thus an intermolecular association
takes place, by hydrogen bonding between antiparallel peptide chains. In addition, the double strands thus generated
very likely aggregate thus mimicking the adsorption of free
angiotensin on a variety of surfaces. The conclusions
reached by IR spectroscopy on backbone conformation of solid
angiotensinamide II are supported by circular dichroism
measurements on dry angiotensin II films.[14] Reducing the
concentration of aqueous solutions of angiotensinamide II
diminishes the interpeptide interactions. Below 0.02 *M* at
pH 6, angiotensinamide II behaves as a monomer as shown by
Rayleigh diffusion experiments,[15] by sedimentation equilibrium[16] and, at still lower concentration, by thin film
dialysis.[17] Very interestingly, Printz, Williams and
Craig[18] found at 0.003-0.006 *M* peptide concentration at
least one and possibly two backbone peptide hydrogens of
Val[5]-angiotensinamide II to be involved in intramolecular
hydrogen bonding in acidic aqueous medium. This suggests
a folded conformation of the molecule. More evidence on
the ability of angiotensinamide II to adopt preferred conformations was obtained by thin film dialysis[17] and gel
filtration. Craig *et al.*[17] and de Fernandez *et al.*[19]
studied the escape time of angiotensinamide II from thin
films and observed two to four fold increases of T/2 with
rising pH (from neutral to alkaline). These results

indicate change from a compact to an expanded state. Similarly, filtration of angiotensinamide II through Sephadex
G–25 gave rise to either one or two peaks, depending on
temperature, pH and ionic strength of the eluant.[20],[21]

Circular Dichroism Studies

Circular dichroism spectra allow further insight into
the conformational behavior of angiotensin II. Comparison
of CD spectra of $0.5-1.1 \cdot 10^{-3}$ M aqueous solutions (pH 6)
of angiotensinamide II with those of analogs and lower
homologs indicated that the hormone, the [4-phenylalanine]
analog, as well as des-(8-phenylalanine)-angiotensinamide
II exhibited certain degrees of organization whereas the
shorter homologous peptides did not. On the other hand,
the [5-isoleucine, 8-alanine] analog, under identical
conditions gave a spectrum typical for an unordered conformation. These results show the conjugated importance
of chain length and amino acid composition.

Interestingly enough, heating aqueous solutions of
angiotensinamide II promoted an enhancement of the negative
CD bands located at 224 and 237 nm, Figure 1. The molar

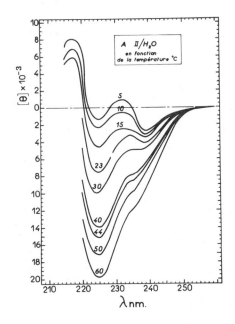

Figure 1: Circular
dichroism spectra of
angiotensinamide II in
aqueous solution at
different temperatures.[14]

ellipticity of the former changed from $[\theta] = -4 \cdot 10^3$ at 15°C to $[\theta] = -20 \cdot 10^3$ at 60°C suggesting an extensive organization of the molecule as the temperature was raised. No such dramatic changes were observed with the [5-isoleucine, 8-alanine] analog.[14] This temperature dependent folding might be a consequence of hydrophobic forces coming into play when the solute is more loosely solvated. The presence of the -Val-Tyr-Val- sequence and of phenylalanine seems to be of importance in this respect. Bringing these side chains together implies a folding of the molecule, around the tyrosine residue and a *cis* conformation of the His-Pro peptide linkage. As a matter of fact, the increase of ellipticity at 237 nm might well relate to the *cis* configuration of the His-Pro peptide bond as evidenced by the data of Legrand and Viennet[22] and those of Bovey and Hood.[23] In the absence of a terminal phenylalanine this mutual stabilization of the hydrophobic side chains would be less likely; this agrees with observed data. These results underline the importance of the surrounding medium and raise the questions (a) whether it is possible to enhance further the folding induced by a rise in temperature and (b) what is then the shape of the molecule?

A spectacular enhancement of angiotensinamide II folding is caused by organic solvents. Trifluoroethanol for instance promoted a sharpening of CD spectra leading to a more than twofold increase of $[\theta]$ at 224 nm, Figure 2. The asymmetry

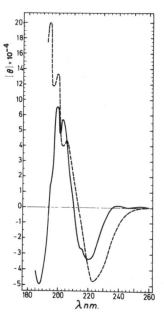

Figure 2: Circular dichroism spectra of angiotensinamide II dissolved in trifluoroethanol (-----) and of a dry film obtained from trifluoroethanol solution.

of that negative band toward the longer wave lengths suggests that the 237 nm negative band is also increased, both bands merging together. In addition positive bands become observable at 195, 198 and 205 nm. This profile is very similar to the predictions of Pysh[24] for antiparallel β conformations of peptide chromophores. We already know that in the dry state as well as in concentrated aqueous solutions, angiotensinamide II favors antiparallel β conformations. In more dilute aqueous solutions where angiotensinamide II is a monomer, we also know that a folded conformation is present, and in increasing concentrations as the temperature is raised. The situation in trifluoroethanol thus appears to be the limit of complete folding. As angiotensinamide II is not aggregated under these circumstances, antiparallel β conformation means a folding of the molecule on itself. Hydrophobic interactions, mentioned earlier, between the valine side chains represent probably one set of driving forces for a first β turn and the interactions of these valine side chains with the terminal phenylalanine side chain a second set of forces leading to another turn and together to the formation of a hydrophobic area on one side of the molecule. The roles of the different side chains of angiotensinamide II in stabilizing an overall conformation may be further studied by a variety of approaches. The availability of angiotensin analogs permitted the investigation of the influence of several amino acid substitutions. For this purpose CD spectra of various peptides dissolved in hexafluoroisopropanol were compared. This solvent was chosen for its intermediate properties between water and trifluoroethanol. Thus the subtle structural tendencies of the compounds investigated may express themselves in a more clear cut manner. First of all we examined the angiotensin II tripeptide fragment Val-Tyr-Val. It exhibits a positive band between 280 and 290 nm, Figure 3, due to the tyrosine side chain. On the other hand, angiotensinamide II shows no positive CD band between these wavelengths. Thus, the tyrosine environment in Val-Tyr-Val differs from that in angiotensinamide II. It should be pointed out that in angiotensinamide II the tyrosine side chain is free of interactions according to all available evidences.

Let us consider next [8-alanine]- and [8-isoleucine]-angiotensin II. Both give CD curves of similar shapes, different from that of the parent octapeptide, Figure 4. The [8-alanine] analog curve is reminiscent of the profile described for helical structure,[24] the more so when the

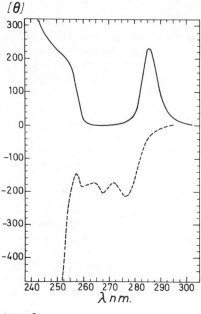

Figure 3: Circular
dichroism spectra of
Val-Tyr-Val (———) and
angiotensinamide II
(----) in hexafluoro-
isopropanol, in the
range of 240-290 nm.

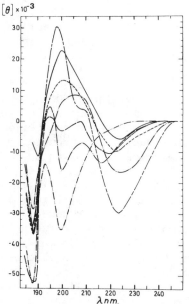
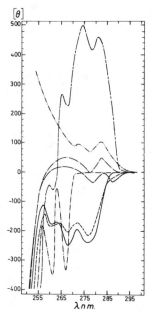

Figure 4: Circular dichroism spectra of angiotensinamide II
and some analogs in hexafluoroisopropanol. A II (-----);
[D-Arg2]-A II (———); [Ile8]-A II (———); [Pro3]-A II
(———); [Phe4]-A II (—··—); [D-His6]-A II (—···—);
[Ala8]-A II (—·—).

solvent is trifluoroethanol. One should remain aware that
CD curves may be misleading when no other information is
available, as shown by Craig[25] for some cyclic antibiotic
peptides. The [8-isoleucine] analog has less intense CD
bands, intermediate between [8-alanine]-A II and angiotensin
II. Both analogs point to the importance of the C-terminal
amino acid. When the latter is removed, as in the (1-7)-
heptapeptide, the CD curve in trifluoroethanol is similar
to that of angiotensin II. When on the other hand alanine
replaces phenylalanine, there is not only a lack of possible
interactions with the valine side chains, but another type
of association of alanine (carboxyl group ?) takes place,
leading to a different folding of the whole peptide chain.
The [8-isoleucine] analog seems to be subject to two opposite
tendencies: one is that postulated for the [8-alanine]
analog, the second may be a weak interaction between the
8-isoleucine side chain and the 3-valine side chain. The
result is a larger flexibility. Both analogs are competitive
inhibitors of angiotensinamide II,[26] but interestingly
enough, the [8-isoleucine] analog is altogether a more
potent inhibitor than [8-alanine]-A II and at high concentrations, possesses some agonistic activity. Thus the role
of a hydrophobic side chain in position 8 seems to be important both for the stability of the conformation needed
for binding as well as for the eventual triggering of the
biological response.

Further precise information about the shape of the
molecule may be gleaned from studies of CD curves. The
[2-D-arginine] analog presented a curve nearly identical
to that of angiotensinamide II suggesting that the side
chain of arginine plays no noticeable role in maintaining
the general conformation in solution. This side chain
resides, therefore, on the surface of the molecule and can
be inverted from L to D without interfering with the whole
structure. This conclusion was supported when a cyclic
derivative, obtained by condensation of the guanidine
group of angiotensinamide II with cyclohexanedione, gave
also a CD curve similar to that of the parent compound.
The cyclic derivative exhibited nearly full pressor activity,
whereas the [2-D-arginine] analog has almost none. Obviously
the spatial positioning of the basic nitrogens of arginine
is of importance for the biological response. These results
suggest that the receptor groups which interact with these
guanidine nitrogens are not embedded in a cleft as the
imidazolidinone derivative has access to them.

The location of the imidazole ring of histidine in
angiotensinamide II can be estimated by a similar approach.

Replacing L-histidine by D-histidine leads to a complete
disorganization of the structure as indicated by CD spectra.
Thus, as far as the conformation of angiotensinamide II in
organic solution is concerned, this observation eliminates
from consideration several possible locations for the L-
histidine side chain and has to be remembered when building
a model. The [3-proline] analog, on the other hand, does
present CD curves which suggest a strong enhancement of the
angiotensinamide II folded conformation; in addition, the
tyrosine environment changes, as evidenced by the positive
CD bands between 275 and 285 nm. That the rigidity introduced by the proline residue leads to a more folded conformation is in agreement with a β turn between the residues
in positions 3 and 4. This [3-proline] analog possesses half
the pressor potency of the parent hormone. This indicates,
like the CD curves, that the general features of the molecule
are not grossly modified. It shows merely a loss of
flexibility for closer adjustment reflecting probably a
restriction in the rotation of the tyrosine side chain.
It may be appropriate to note that these substitutions of
the third amino acid indicate its major importance for the
conformation at the receptor site. Finally, it is observed
that the [4-phenylalanine] analog gave CD curves similar to
that of angiotensinamide II. As already mentioned the side
chain of tyrosine appeared to be free of interactions and
no changes were expected after phenylalanine replacement.

NMR Studies

The data afforded by CD spectra have been supplemented
by others originating from NMR measurements. Proton NMR
spectra of angiotensinamide II and a variety of its constituant peptides were performed in trifluoroacetic acid[27,28]
to assign the signals corresponding to the α carbon hydrogens.
On this basis studies were made on angiotensinamide II dissolved in DMSO and in trifluoroethanol. The most clear cut
observation is related to the histidine α carbon hydrogen
in trifluoroethanol. The integration of the α carbon proton
peaks in the region of 4 ppm indicates the presence of 7
protons only, and not 8 as expected, Figure 5. Heating the
solution leads to a shift of the water peak towards high
field allowing the observation of the last C_α hydrogen
previously buried under it, Figure 6. This C_α hydrogen
has been assigned to histidine. The frequency (5 ppm)
of this histidine C_α hydrogen is more downfield than found
in control experiments or in other reports.[29] We suggest
that the histidine α carbon experiences a special environment

Figure 5   Figure 6

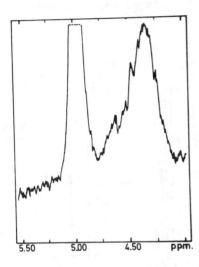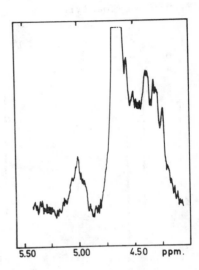

Figure 5: NMR spectrum of angiotensinamide II C_α proton region. Solvent: deuterated trifluoroethanol, concentration 3.10^{-2} M, 100 MHz, temperature 23°C.

Figure 6: NMR spectrum of angiotensinamide II C_α proton region. Same conditions as in Figure 5 but at a temperature of 60°C.

due either to the geometry of the proline residue or to specific vicinal interaction, for instance with a carboxylic group.

When angiotensinamide II is dissolved in DMSO, and the solution heated to 60°C, the C_α hydrogen peak moves downfield. This shift compares well with enhanced folding observed by CD in aqueous solution following an increase in temperature. The hydrogens carried by the C_2 and C_4 of the imidazole ring exhibit no shift or broadening in DMSO or in trifluoroethanol at room temperature. The variations of the corresponding peaks in aqueous angiotensinamide II solutions, as a function of pH indicate normal titration curves. Thus, under the conditions investigated, the imidazole ring appears to be free of interactions. This suggests that the imidazole moiety is not involved in

hydrogen bonding or located in hydrophobic regions. This
conclusion is entirely supported by the ^{13}C NMR measurements
made on the C_2 and C_4 atoms of the histidine side chain,[30]
Figure 7. The tyrosine ring gives rise to peaks corresponding

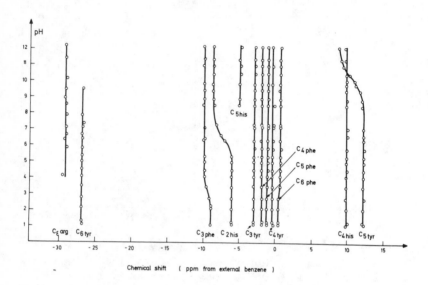

Figure 7: ^{13}C-NMR Titration curves of angiotensinamide II.
Solvent D_2O, concentration 6.10^{-2} *M*, 22.6 MHz, temperature
27°C.[30]

to the *ortho* and *meta* aromatic hydrogens. The behavior of
the tyrosine residue has been examined in DMSO, trifluoroethanol and trifluoroacetic acid, by H-NMR and in D_2O
solutions by ^{13}C-NMR. In all cases tyrosine was found to
behave normally, a conclusion, already reached by other
methods.[14] This result suggests that the tyrosine side
chain seems to be free of interaction, and not surrounded
by a hydrophobic area.

Phenylalanine aromatic side chain protons give rise to
a single peak. The frequency and the line width of this
peak appear normal in DMSO. However it is slightly
broadened in trifluoroethanol. The peak assigned to the
valine methyl group is also slightly broadened. These
changes are not dramatic but may be considered as resulting
from mutual interactions.

In this context it might be mentioned here that ^{13}C-NMR of angiotensinamide in DMSO indicates a sharpening of the proline δ carbon as the temperature increases. This gives evidence for a temperature dependent event taking place at proline. A stabilization of the proline residue in a *cis* conformation would fit with these observations, as well as with the shift noted for histidine C_α hydrogen.

NMR examination of the peptide –NH– was first carried out with the tetrapeptide Val-His-Pro-Phe dissolved in DMSO, Figure 8. Through double resonance experiments assignments of the NH and C_α protons were obtained as well as the corresponding coupling constants. It should be noted that the peaks belonging to histidine and phenylalanine NH are broadened and split, respectively. The phenomenona are observed also at the corresponding C_α proton peaks. The frequencies of the peaks in the NH region were plotted as a function of temperature in Figure 9. The NH group assigned to phenylalanine does not shift in the temperature interval studied. This behavior suggests the involvement of the phenylalanine NH group in an interaction, probably a hydrogen bond. In addition the profiles of the peaks corresponding to the C_α protons are not significantly altered. Thus the His-Pro-Phe end of the peptide appears to be stabilized.

Similar experiments were performed with angiotensinamide II, Figure 10. Here again one NH peak does not shift and another moves with a smaller frequency slope, when the temperature is increased. Unfortunately, precise assignment of peaks to individual NH has not been achieved yet because of the overlapping C_α proton peaks. The results indicate however that certainly one and most likely two hydrogen bonds involving –NH– are present in angiotensinamide II dissolved in DMSO. Similar results have been obtained by Jorgensen.[31]

Interpretation of these findings met with the following basic difficulty. In $>10^{-2}$ *M* DMSO solution angiotensinamide II is not present as a monomer but as an aggregate of at least 10 molecules as shown by Rayleigh diffusion techniques.[15] Thus the positive structural evidence obtained by NMR may correspond either to intra- or to interchain bondings. Moreover, the conformation adopted by angiotensinamide II within the aggregates may well be distinct from that in the monomeric state. In this respect, IR and Raman spectroscopy and NMR measurements of aqueous and DMSO solutions are similarly limited. Only in trifluoroethanol do the NMR data correspond to single molecules.

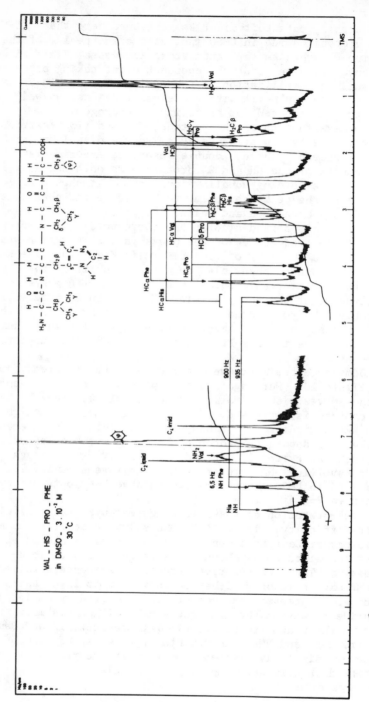

Figure 8: ¹H-NMR spectrum of Val-His-Pro-Phe. Solvent d₆DMSO, concentration 2·10⁻² M, 250 MHz, temperature 30°C. Coupled protons are indicated.

Figure 9   Figure 10

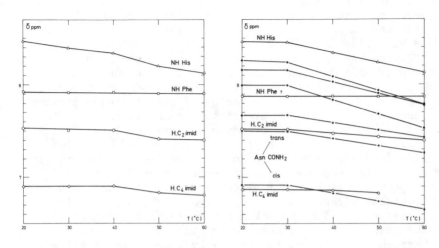

Figure 9: Plot of –NH– proton chemical shifts of Val–His–Pro–Phe as a function of temperature. Data obtained at 250 MHz.

Figure 10: Plot of –NH– proton chemical shifts of angiotensinamide II as a function of temperature. Data obtained at 250 MHz.

An Angiotensinamide II Model

Fortunately among the various techniques used there is an overlap of conclusions: the antiparallel β conformation of solid angiotensinamide II, indicated by IR spectroscopy, supports the interpretation of the CD curves obtained with dry films of the hormone and, due to their similarities, the interpretation of the CD curves of dilute solutions in organic solvents, where angiotensinamide II is monomeric.

The previously stated conclusion[32] that the principal conformation of angiotensinamide II in organic solvents is a cross β type appears thus to be based on solid grounds. The combined data suggest the presence of a turn at the level of valine in position 3 and tyrosine in position 4, and a folding of the end of the molecule due to the rotation of the histidine–proline peptide bond toward a *cis* conformation. This second turn gives rise to a hydrophobic pocket made of valine and phenylalanine side chains on one side of the molecule and is supported by the CD data, as well as by the NMR findings in trifluoroethanol solution pertaining to the C_α proton of histidine, to the phenylalanine and valine side chains. Changes at the C_α proton of histidine and of the δ carbon of proline observed by NMR when heating DMSO and concentrated aqueous solutions compare well with the heat dependent enhancement of bands seen by CD on dilute aqueous solution, and may represent the same phenomenon: a rotation of the histidine-proline bond towards a *cis* conformation. On the basis of these considerations a model of angiotensinamide II has been built[32] in which the arginine, tyrosine and histidine side chains were left free of interaction. Interestingly enough, the proposed location of the phenylalanine end is not compatible with a replacement of L-histidine by D - histidine without extensive change of the conformation, in agreement with experimental observations (Figure 4). The stabilization of the conformation is attained by hydrophobic interactions and by hydrogen bonding between peptide groups. As a matter of fact, the data of the tritium exchange technique[18] as well as NMR in DMSO solutions both indicate one strong and one weaker hydrogen bond, but under quite different conditions. The hydrogen bonds observed by Printz *et al.*[18] are certainly of intramolecular nature whereas those present in DMSO solution are of uncertain nature for the reason mentioned, even if they were assigned (probably one belongs to phenylalanine NH).

The proposed angiotensinamide II model deserves the following comments. First of all, the forces to which the peptide chain is subjected are a function of the surrounding environment conditions, solvent, pH and temperature. Under given circumstances, their combined effects lead to a higher probability for a certain type of folding. However, local arrangements may well be of nearly equal potential energy, and thus remain undefined in a range of restricted possibilities. For instance, we think that the peptide chain of angiotensinamide II assumes a turn in organic solvents starting with valine in position 3. There are several ways to undergo this turn when considering the peptide linkages.[33,34] It is not at all certain that one of the possibilities is favored to the extent that the others are excluded. We prefer the idea that in solution a variety of local transitions are going on, while the overall features are statistically preserved. This results in a molecular flexibility whose extent depends on the environment. It is by reducing this flexibility that angiotensinamide II could be crystallized.[35]

A second comment concerns the biological significance of the proposed overall conformation. Certainly, the latter originates from data collected under non-physiological circumstances. It is gratifying, however, to note that the proposed model not only fits with data obtained from amino acid substitutions and associated biological response[36] but allows further predictions of what may have happened in terms of conformational changes or triggering of the response. Future work should define more precisely the correlation between specific binding with or without biological response and refinement of the model. Presently the greater rigidity introduced in the [3-proline] analog together with the persistence of a fair amount of biological activity[37] seems to indicate that the model may not be too far from the actual conformation prevailing at the receptor sites.

Acknowledgment

We thank Dr. Riniker (Ciba, Bâle), Dr. Bumpus and Dr. Smeby (Cleveland Clinic, Cleveland, Ohio) for the angiotensin II analogs used in this work. Two of us, Serge Fermandjian and Daniel Greff were supported by the Centre National de la Recherche Scientifique, France, under the contract RCP n° 220.

References

1. Von Buddenbrock, W. In *Vergleichende Physiologie*, (Basel, Birkhäuser, 1950) vol. 4, p 16.
2. Page, I. H., and F. M. Bumpus. Physiological Reviews 41, 331 (1961).
3. Elliott, A., and E. J. Ambrose. Nature 165, 921 (1950).
4. Blout, E. R., C. De Loze, S. M. Bloom and G. D. Fasman. J. Amer. Chem. Soc. 82, 3787 (1960).
5. Miyazawa, T., and E. R. Blout. J. Amer. Chem. Soc. 83, 712 (1961).
6. Sutton, P., and J. L. Koenig. Biopolymers 9, 615 (1970).
7. Tsuboi, M., and A. Wade. J. Mol. Biol. 3, 380 (1961).
8. Fermandjian, S., P. Fromageot, A.-M. Tistchenko, J. P. Leicknam, and M. Lutz. Europ. J. Biochem. 28, 174 (1972).
9. Smith, M. A., A. Walton, and J. L. Koenig. Biopolymers 8, 29 (1969).
10. Theoret, A., Y. Grenie, and C. Garrigou-Lagrange. J. Chimie Phys. 66, 1196 (1969).
11. Miyazawa, T. J. Chem. Phys. 32, 467 (1960).
12. Rippon, W. B., J. M. Anderson, and A. G. Walton. J. Mol. Biol. 56, 507 (1971).

13. Koenig, J. L. and P. L. Sutton. Biopolymers 10, 89 (1971).
14. Fermandjian, S., J. L. Morgat, and P. Fromageot. Eur. J. Biochem. 24, 252 (1971).
15. Capitini, R. Unpublished experiments.
16. Paiva, T. B., A. C. M. Paiva, and H. A. Scheraga. Biochemistry 2, 1327 (1963).
17. Craig, L. C., E. J. Harfenist, and A. C. Paladini. Biochemistry 3, 764 (1964).
18. Printz, M. P., H. P. Williams, and L. C. Craig. Proc. Nat. Acad. Sci. U.S. 69, 378 (1972).
19. de Fernandez, M. T. F., A. E. Delias, and A. C. Paladini. Biochim. Biophys. Acta 154, 223 (1968).
20. Ferreira, A. T., O. G. Hampe, and A. C. M. Paiva. Biochemistry 8, 3483 (1969).
21. Fermandjian, S. Unpublished results.
22. Legrand, M., and R. Viennet. Bull. Soc. Chim. Fr. 679 (1965).
23. Bovey, F. A., and F. P. Hood. Biopolymers 5, 325 (1967).
24. Pysh, E. S. J. Mol. Biol. 23, 587 (1967).
25. Craig, L. C. Proc. Nat. Acad. Sci. U.S. 61, 152 (1968).
26. Khairallah, P. A., A. Toth, and F. M. Bumpus. J. Med. Chem. 13, 181 (1970).
27. Bovey, F. A., and G. V. D. Tiers. J. Amer. Chem. Soc. 81, 2870 (1959).
28. Bovey, F. A., G. V. D. Tiers, and G. Filipovich. J. Polym. Sci. 38, 73 (1959).
29. Roberts, G. C. K., and O. Jardetzky. Adv. Protein Chem. 24, 447 (1970).
30. Zimmer, S., W. Haar, W. Maurer, H. Rüterjans, S. Fermandjian, and P. Fromageot. Europ. J. Biochem. 29, 80 (1972).
31. Weinkam, R. J., and E. C. Jorgensen. J. Amer. Chem. Soc. 93. 7038 (1971).
32. Fermandjian, S., J. L. Morgat, P. Fromageot, M. Lutz, and J. P. Leicknam. *The 2nd International Symposium on Polypeptides and Protein Hormones n°2* (1971), in press.
33. Ramachandran, G. N., and V. Sasisekharan. Adv. Protein Chem. 23, 284 (1968).
34. Leach, S. J., G. Nemethy, and H. A. Scheraga. Biopolymers 4, 369 (1966).
35. Fermandjian, S., J. L. Morgat, P. Fromageot, C. Legressus, and P. Maire. FEBS Letters 16, 192 (1971).
36. Schröder, E., and K. Lübke. *The Peptides* (New York: Academic Press, 1966).
37. Khosla, M. C., N. C. Chaturvedi, R. R. Smeby, and F. M. Bumpus. Biochemistry 7, 3417 (1968).

PROTON MAGNETIC RESONANCE STUDY OF ANGIOTENSIN II (Asn[1], Val[5]) IN AQUEOUS SOLUTION

J. D. Glickson, * *W. D. Cunningham.* Division of
Molecular Biophysics, Laboratory of Molecular Biology,
University of Alabama School of Medicine, Birmingham,
Alabama 35233

G. R. Marshall. Department of Physiology and Biophysics,
Washington University School of Medicine, Saint Louis,
Missouri 63110

IN ORDER TO CLARIFY SOME OF the controversy over the conformation of angiotensinamide (AII') (Asn[1]–Arg[2]–Val[3]–Tyr[4]–
Val[5]–His[6]–Pro[7]–Phe[8]) in aqueous solution, we have analyzed
in some detail the 220 MHz proton magnetic resonance (pmr)
spectrum of this hormone in both D_2O and H_2O. Spectra of
AII' in D_2O (Figure 1) were measured near the solvent
freezing point in order to observe the αCH absorptions,
which overlapped at higher temperature with the relatively
intense solvent absorption (HDO). We assigned resonances
to specific hydrogens on the basis of chemical shift estimates for random-coil polypeptides,[1] homonuclear proton
spin decoupling experiments, peak intensities, preparation
of partially deuterated analogs of AII', and the characteristic pD_c dependence of the chemical shifts of specific
resonances (Figure 2b). The following hydrogens had
resonances with chemical shifts that differed significantly
from values reported by McDonald and Phillips[1] for random-
coil polypeptides (which appear in parenthesis): Pro
δCH_2 777, 782 Hz (2 x 736 Hz), and Val βCH 2 x 425 ± 10 Hz
(2 $\bar{x}$ 494 Hz). This suggests that except for the Pro and
the two Val residues, the sidechains of AII' experienced an
essentially solvated environment. The appearance of only

---

*Present Address: Cancer Research and Training Center,
University of Alabama School of Medicine, Birmingham,
Alabama 35233.

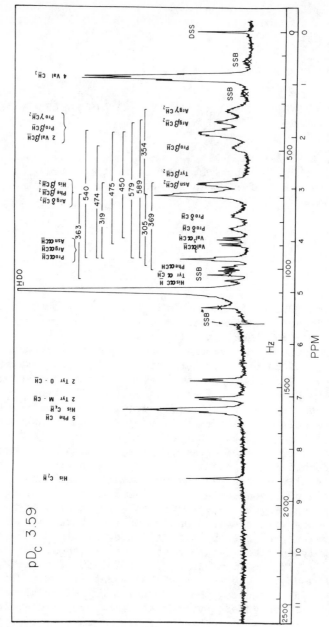

Figure 1: The 220 MHz pmr spectrum of angiotensin II (Asn¹ Val⁵) 6% (w/v) in D₂O at 5 ± 2° C and pD_c 3.59 (pH meter reading + 0.40). Chemical shifts are referred to the methyl absorption of 2,2-dimethylsilapentane-5-sulfonate (DSS), the internal standard.

one resonance of unit intensity for every proton of AII'
indicates that either the hormone assumes a rigid conformation in aqueous solution, or else transitions between the
various orientations of this hormone occur at a rate that
is "rapid on the nmr rate scale."

Further evidence consistent with a predominantly solvated
conformation was the observation that all the NH hydrogens
of AII' were completely replaced by deuterium by the time
the first spectrum was recorded (*i.e.* 6 min). However, it
must be noted that the acidic and basic sidechains of the
hormone may exert a pronounced catalytic effect on the
deuterium exchange rates,[2] which could make some of the
labile hydrogens of AII' appear much more accessible to
the solvent than they really are. In spectra of AII' in
H_2O solution, we observed six peptide NH doublets (labeled
#1-#6 in Figure 2a), two distinct Asn primary amide NH
peaks, and a broad three proton Arg-NH-C(NH)-NH_3 resonance
in the 1400-2000 Hz region of the spectrum, which also
contained the aromatic CH absorptions (Figure 2a). Broadening of NH resonances as the pH was raised results from basic
catalysis of the exchange of hydrogens between the NH groups
and water. If allowance is made for the different chemical
shifts of the various NH absorptions, a rough measure of
the relative exchange rate of a given NH hydrogen is the
pH at which its resonance broadens out--the lower this pH,
the more rapid the exchange. On this basis peptide NH #1
and the Arg guanidino protons exchange most rapidly, peptide proton (#4 = Phe) exchanges least rapidly and the
remaining peptide and carboxamide protons exchange at
intermediate rates. The C-terminal Phe NH resonance is
distinguished by the pH dependence of its chemical shift
(pK_a 3.07) (Figure 2a). The C-terminal Gly residue of
Gly-Ala-Gly behaved similarly (pK_a 3.15). Since a negatively charged C-terminal group is expected to destabilize
the amide anion transition state for proton exchange,[3]
whereas a positively charged N-terminal group is expected
to stabilize this species, anomolously slow and fast exchange rates are expected on purely inductive grounds for
the Phe and the Arg peptide NH hydrogens, respectively.
This leads us to suspect that peptide NH #1 may be Arg.

The peptide NH-αCH coupling constants: 6.5 $\pm$ 0.3,
6.0 $\pm$ 0.5, 7.2 $\pm$ 0.5, 7.3 $\pm$ 0.3 (Phe), 7.9 $\pm$ 0.3, and
8.0 $\pm$ 0.4 Hz for peptide resonances #1-#6, respectively,
exclude a right handed α-helix (about 2 Hz), and indicate
that AII' assumes a different conformation in H_2O than
the conformation assumed in dimethylsulfoxide by a related
heptapeptide investigated by Weinkam and Jorgensen.[4] This

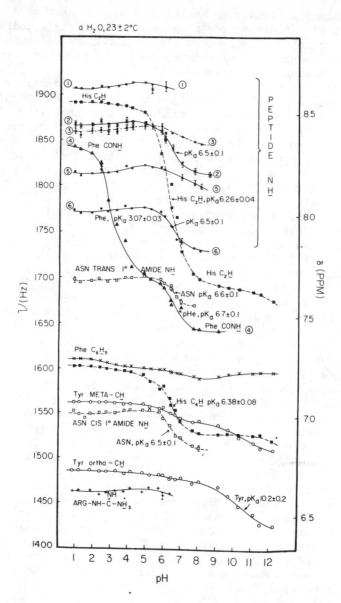

Figure 2: Chemical shifts at 220 MHz of angiotensin II
(Asn[1] Val[5] (a) resonances to low field of the solvent
absorption in H₂O solution at 23 ± 2° C as a function of
pH, and (b) of resonances to high field of the *H*DO

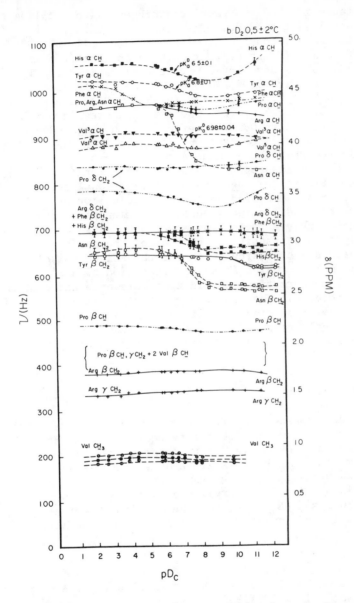

resonance in D$_2$O at 5 ± 2° C as a function of pD$_C$. The acidity of the solutions was adjusted with HCl and NaOH in H$_2$O and their deuterated analogs in D$_2$O. The sodium error was significant above pH (pD$_C$) 10.0 and has not been corrected for.

heptapeptide had two peptide hydrogens with coupling constants less than 3 Hz. The coupling constants of AII' are, however, consistent with a random-coil (6.1 Hz)[5] β-structure (7-8.5 Hz), or perhaps some other orientation.

The pK_a's associated with the titrations in Figure 2 were: carboxyl 3.07 ± 0.03 (3.0 ± 0.1), imidazole 6.26 ± 0.04 (6.82 ± 0.02), α-amino (6.98 ± 0.04), and phenol 10.2 ± 0.2 (10.5 ± 0.2), where the figures in parenthesis were obtained in D_2O and reflect the hydrogen-deuterium isotope effect. Comparison with data from small peptides[6] indicates that the α-amino group is anomolously acidic.

The simultaneous perturbation of all the amide resonances (Figure 2a) suggests a conformational change with a pK_a of 6.6 ± 0.2, indicative of involvement of both imidazole and α-amino groups. Addition of guanidine significantly affected the spectrum of AII' on the acid side of this transition, but not on the basic side. These results are consistent with the thin-film dialysis data of Craig *et al.*[7] which showed a transition of AII' from a coiled conformation in acid to a more extended conformation in base. The reported greater activity of AII' in basic solution might result from exposure of critical functional groups of AII' in basic, but not acidic, solution. The nature of the acid stable conformation is not yet clear.

Acknowledgments

The authors are indebted to Dr. Dan W. Urry for financial support, the use of laboratory equipment, and helpful discussion of the data. The assistance of William Vine in analysis of the data is gratefully acknowledged. This research was supported by grants from the Mental Health Board of Alabama (to D.W. Urry), the Public Health Service Grants AM 13025 and HE 14509, and the American Heart Association (Established Investigator Grant to G. R. Marshall).

References

1. McDonald, C. C., and W. D. Phillips. J. Amer. Chem. Soc. 91, 1513 (1969).
2. Leichtling, B. H., and I. M. Klotz. Biochemistry 5, 4026 (1966); Hvidt, A., and S. O. Nielsen. Adv. Prot. Chem. 21, 287 (1966).
3. Berger, A., A. Loewenstein, and S. Meiboom. J. Amer. Chem. Soc. 81, 62 (1959).

4. Weinkam, R. J., and E. C. Jorgensen. J. Amer. Chem.
    Soc. <u>93</u>, 7038 (1971).
5. Tonelli, A. E., and F. A. Bovey. Macromolecules <u>3</u>,
    410 (1970).
6. Cohn, E. J., and J. T. Edsall. *Proteins, Amino Acids
    and Peptides* (New York: Reinhold Publishing Corp.,
    1943) pp 84-85.
7. Craig, L. C., E. J. Harfenist, and A. C. Paladini.
    Biochemistry <u>3</u>, 764 (1964).

CONSTRAINTS ON THE RECEPTOR-BOUND CONFORMATION OF
ANGIOTENSIN II

*Garland R. Marshall, Heinz E. Bosshard, Nancy C.
Eilers, Philip Needleman.* Departments of Physiology
and Biophysics and of Pharmacology, Washington University School of Medicine, St. Louis, Missouri 63110

ONE APPROACH TO THE PROBLEM of the biologically active
conformation[1] is to reduce the inherent conformational
flexibility available to the structure. We have system-
atically investigated from a theoretical viewpoint the
restrictions on conformational freedom of the peptide
backbone introduced by increasing the steric bulk of groups
which are located on the peptide backbone. In particular,
we have investigated the replacement of the hydrogen atoms
of the peptide backbone, Figure 1, by methyl groups[2,3]
which gives an increase in the Van der Waals' radii from
1.2 angstrom to 1.85 angstrom. There are three categories
of proton replacements to be examined (Figure 1):
1) Replacement of the amide proton of the amino acid under
investigation with a methyl group to give an *N*-methyl amino
acid; 2) Replacement of the alpha proton to give an alpha
methyl amino acid; 3) Replacement of the amide proton of
the subsequent amino acid residue by a methyl group, *i.e.*
the effect of an *N*-methyl amino acid on the preceding
residue. Two simplifying assumptions are 1) the peptide
group is planar due to partial double bond character of
the amide bond, and 2) the methyl group can be treated as
a sphere, due to rapid rotation, rather than discrete atoms.
The position of all the atoms in space up to the beta carbon of the side chain are then determined by the torsional
rotations about the bond between the amide nitrogen and
alpha carbon (ϕ) and the bond between the alpha carbon and
the carbonyl carbon (ψ).

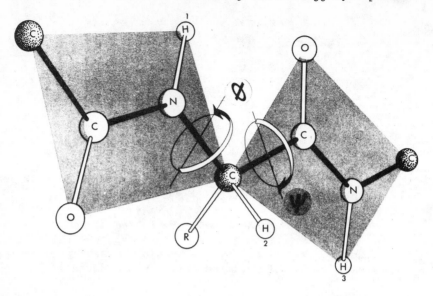

Figure 1: Two linked peptide groups whose conformation is
determined by the dihedral angles φ and ψ which refers to
torsional rotations about the single bonds passing through
the alpha carbon. Replacement of hydrogen number 1 gives
an *N*-methyl amino acid; hydrogen number 2 gives an α-
methyl amino acid; and hydrogen number 3 gives an amino
acid preceding an *N*-methyl residue. (Adapted from
Dickerson and Geis. *The Structure and Action of Proteins,*
Harper & Row, 1969).

In the calculations programs developed for a small
laboratory computer (LINC, μLINC, or PDP-12) were used
which consist of a set of general input and manipulation
programs, CHEMAST,[4] and a program for iterative searches
of conformationally dependent variables, BURLESK.[5] Such
programs were necessary because the long calculation times
made use of a computational center both too demanding and
too expensive. The CHEMAST program allows the input of a
chemical formula through a linear string of chemical
groups; *i.e.* CH3, CO, NH, *etc.* The program then calculates
the appropriate atomic co-ordinates and connectivity for the
molecule based on a dictionary which specifies the geometry
and bond distances of the appropriate chemical groups. A
stick figure representation of the model can be displayed
and manipulated on the display scope or the molecular
description handed over to more complex programs such as

BURLESK for iterating exhaustively the possible conformations and calculating variables such as potential energy.
Initial calculations[6,7] were based on the assumption of a
hard sphere for the atoms. One difficulty in interpretating
results of hard sphere calculations is that there is no
indication of the probability of a given conformation.
Agreement between conformations obtained from crystallographic studies of protein molecules and those predicted
by these calculations, while relatively good, has some
striking discrepancies.

The Kitaigorodsky potential function[8] gives an excellent
fit between the values calculated for N-acetyl-L-alanine
methyl-amide and the dihedral angle values determined for
lysozyme.[9] We have extended the comparison (Figure 2) to
six proteins: myoglobin,[10] lamprey hemoglobin,[11] insulin,[12]
carboxypeptidase A,[13] and cytochrome B5.[14] Over 89% of
the 847 non-glycine values examined are within 1 kcal/mol
of the calculated potential minima and over 95% are within
a 2 kcal/mol contour.[15]

The contours for acetyl-N-methyl-L-alanine methylamide
are shown in Figure 3. Further justification of the choice
of the Kitaigorodsky function and the contouring level is
given by crystallographic values for the only N-methyl amino
acid found in proteins, proline. Since the torsional rotation about the amide-alpha carbon bond (ϕ) in proline is
required to be approximately -60° by the cyclic imide, all
data must center around this value and are, thus, found
between the one and two kcal/mol contour for N-methyl
alanine. The contours for an amino acid preceding an N-
methyl amino acid are compared with the crystallographic
values for amino acids preceding proline in Figure 4.
Note that the allowed area is limited to positive ψ values
except for a small region between one and two kcal/mol
around the right-handed alpha helix region (ϕ = -57°, ψ =
-47°). The contours obtained for acetylaminoisobutyric
acid methylamide (alpha methyl alanine) is shown in Figure
5. Particularly striking is the marked restriction in
flexibility already indicated in the hard sphere calculations and the preference for those torsional values associated with either a right- or a left-handed helix (ϕ =
-57°, ψ = -47°; ϕ = 57°, ψ = 47°).

We are primarily interested in the interaction of a
small peptide, angiotensin II, with its macromolecular
receptor. Potential energy calculations would have to
involve the receptor as well as angiotensin to indicate a
particular receptor-bound conformation. Calculations on
the peptide alone might be useful, however, in indicating

Figure 2-4 show Kitaigorodsky potential energy plots con-
toured at X kcal/mol above the potential minimum (determined
with K_2 parameters).[7]

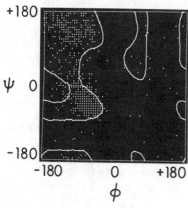

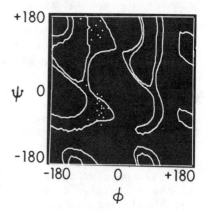

Figure 2: N–Acetyl-L-alanine
methylamide, X = 1. Points
are dihedral values determined for myoglobin, insulin,
lamprey hemoglobin, carboxypeptidase A, cytochrome B_5,
and ribonuclease S.

Figure 3: Acetyl–N–methyl-
L-alanine methylamide, X = 1
and 2. Points refer to
dihedral values determined
for proline for six proteins.

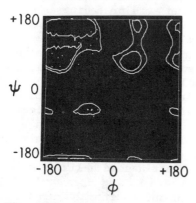

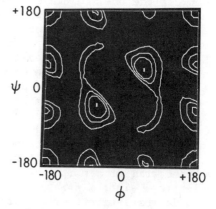

Figure 4: N–Acetyl-L-alanine
dimethylamide, X = 1 and 2.
Points refer to amino acids
which precede proline for
six proteins.

Figure 5: N–Acetylaminoisobutyric acid methylamide,
X = 0, 0.5, 1 and 2.

those peptide conformations which were so energetically
unfavorable that interaction with the receptor would be
unlikely to induce their presence. In other words, given
an estimate of the energy difference which might be overcome as a result of intermolecular interaction, one would
be able to interpret the potential energy diagrams in terms
of allowed vs. disallowed areas. A sampling of possible
intermolecular interaction energies and their likelihood of
introducing deviations from the local potential minima is
found in a protein molecule. The excellent agreement between the values of the torsional rotations (ϕ, ψ) actually
found in proteins and those calculated as being less than
2 kcal different from the potential minima for a single
amino acid derivative is striking. It implies that deviation from a local potential minimum is not common in proteins and, therefore, not likely in the interaction of a
peptide hormone with its receptor.

A variety of angiotensin analogs containing *N*-methyl
and alpha methyl amino acids have been synthesized and their
biological activities studied. Interpretation of their
biological activity in terms of conformational restraints
imposed by these unusual amino acids leads to certain conclusions regarding the receptor-bound conformation.[16] The
case for residue 5 is illustrative. A proline substitution
at this position retains 10% activity.[17] This requires a
ϕ value of approximately −60° at position 5 due to the
cyclic nature of the proline ring. Cycloleucine (1-amino-
cyclopentanecarboxylic acid) when substituted at position 5
maintains 24% activity.[17] This amino acid has the same
steric restraints as aminoisobutyric acid, and requires
ϕ_5 and ψ_5 to be either −57°, −47° or +57°, +47°, respectively.
The simplest conclusion is that the active conformation at
position 5 is that common to both proline and cycloleucine,
i.e. $\phi_5 = -57°$, $\psi_5 = -47°$, or the right-handed alpha helical
values. The proline substitution at position five also
affects the conformational freedom at position four, making
a positive value for ψ_4 likely.

The proline at position 7 of the hormone requires ϕ_7 to
be approximately −60°. The requirement for the *N*-methyl
amino acid at that position is seen by the fact that the
N-methyl alanine substitution[18] retains 22% pressor activity
while an alanine substitution[19] has less than 1% activity.
This implies a role for the *N*-methyl residue in positioning
the preceding residue, making a positive value for ψ_6 likely.

Position 3 normally contains valine, but a proline substitution retains 40-80% of the activity.[20] The torsional
rotation ϕ_3 has, therefore, a receptor-bound value near −60°.

Substitution of either aminoisobutyric acid[6] or cycloleucine[21] retains only 1% activity. Since this activity is approximately that obtained for the C-terminal hexapeptide alone[22] and residue 3 positions residues 1 and 2, this level of activity is not considered significant. It is tempting, therefore, to conclude that of the two probable conformations available to proline at position 3, $\psi_3 \simeq -47°$, or $\psi_3 \simeq 120°$, the most likely one at the receptor is that where $\psi_3 \simeq 120°$ since $\phi_3 = -57°$, $\psi_3 = -47°$ which is possible for either aminoisobutyric acid or cycloleucine is not effective. The activity of the proline substitution probably makes the value of ψ_2 positive.

Resolution of alpha methyl analogs of amino acids by deacylation with carboxypeptidase A[23] has made appropriate derivatives available such as alpha methyl-L-phenylalanine which is currently being incorporated into angiotensin analogs. This should provide additional information on positions 4 and 8.

Either the receptor conformational specificity for angiotensin is not as great as one has suspected from previous studies in that several different backbone conformations can give high levels of biological activity or the constraints derived from the analog studies apply to the receptor-bound conformation. It is of considerable interest to compare these constraints with those obtained by solution studies. In particular, the values for the receptor-bound conformation of angiotensin II for the torsional rotations ϕ_3 and ϕ_5 are of interest. Because of the activity of the two proline analogs at positions 3 and 5 and the ring constraint which limits the ϕ rotation to $-60° \pm 15°$, one can assume the active conformation has a similar rotational angle at the receptor. The value of this angle can be determined by PMR from the coupling constant between the amide hydrogen and the alpha proton. A value of $-60° \pm 15°$ would give a coupling constant of 2 ± 1 Hz.[24] In aqueous solution, all of the amide protons have been observed and have coupling constants of 6 Hz or greater.[25] This implies a conformational transition on binding of angiotensin to its receptor. One might argue that a nonpolar environment might be more comparable to that of the receptor which is assumed to be membrane-bound and whose interaction with angiotensin II may well be hydrophobic in nature. Measurement of the coupling constants for the amide protons associated with Val[3] and Ile[5] in des-Asp[1]-[Gly[2]]-angiotensin II in dimethylsulfoxide (DMSO) give values of 7.5 and 8.5 Hz respectively.[26] This implies that DMSO does not induce the conformation which is postulated at the receptor.

In conclusion, theoretical analysis of rotational freedom of the backbone has placed certain constraints on possible backbone conformations available to certain analogs of angiotensin II which show high levels of biological activity. One must, therefore, assume that these constraints are operative during binding to the receptor and subsequent processes leading to the biological action of angiotensin II. It is of particular interest that these constraints are different than those which have been determined for the solution conformation of angiotensin II. One must, therefore, conclude that a conformational transition occurs in the peptide upon binding to the receptor. Consequently, future care must be exercised before one attempts to interpret the biological activity of peptides based on their solution or even crystalline conformations.

Acknowledgments

The authors acknowledge support from the National Institutes of Health (AM 13025, HE 14509 and 14397, RR 00396) and the American Heart Association for Grants 69-722 and 70-672. In addition, acknowledgment is due the American Heart Association for Established Investigatorships (GRM, 70-111 and PN, 68-115), the European Molecular Biology Organization for a long-term fellowship (HEB), and the National Institutes of Health for a predoctoral traineeship in Neurobiology (NCE, NS-05613).

References

1. Marshall, G. R. *Peptides: Chemistry and Biochemistry* (New York: Marcel Dekker, 1970) pp 151-161.
2. Bosshard, H., C. D. Barry, and G. R. Marshall. Manuscript in preparation.
3. Marshall, G. R., N. Eilers, and H. Bosshard. Manuscript in preparation.
4. Barry, C. D., R. A. Ellis, S. M. Graesser, and G. R. Marshall. Proc. IFIP Congress 71, North-Holland Publishing Co., 1972. In press.
5. Bosshard, H. E., C. D. Barry, J. M. Fritsch, R. A. Ellis, and G. R. Marshall. Proc. 1972 Summer Simulation Conf., San Diego 1, 581 (1972).
6. Marshall, G. R., N. Eilers, and W. Vine. In *Progress in Peptide Research,* Vol 2, Lande, S., ed. (New York: Gordon and Breach, 1972).
7. Marshall, G. R. Intra-Science Chem. Reports 5, 305 (1971).

8. Venkatachalan, C. M., and G. K. Ramachandran.
   Conformation of Biopolymers, Ramachandran, G. N., ed.
   (New York: Academic Press, 1967) pp 83-105.
9. Ramachandran, G. N., and V. Sasisekharan. Adv. Protein
   Chem. 23, 284 (1968).
10. Watson, H. C. Prog. Stereochem. Vol. IV (London:
    Butterworth, 1969) pp 299-333.
11. Hendrickson, W. A., and W. E. Love. Nature New Biol.
    232, 197 (1971).
12. Blundell, T. L., G. G. Dodson, E. J. Dodson, D. C.
    Hodgkin, and M. Vijayan. Recent Prog. Horm. Res. 27,
    1 (1971).
13. Lipscomb, W. N., G. N. Reeke, J. A. Hartsuck, F. A.
    Quiocho, and P. H. Bethge. Phil. Trans. Roy. Soc. Lond.
    B 257, 177 (1970).
14. Mathews, F. S., P. Argos, and M. Levine. Symp. Quant.
    Biol. Vol. XXXVI: Structure and Function of Proteins
    at the Three-Dimension Level (Cold Spring Harbor: 1972)
    pp 387-397.
15. Marshall, G. R., J. M. Fritsch, and H. E. Bosshard.
    Manuscript in preparation.
16. Marshall, G. R., and H. E. Bosshard. Circulation Res.
    31 (Suppl. 2), 143 (1972).
17. Jorgensen, E. C., and R. J. Weinkam. *Proc. 11th European
    Peptide Symposium,* Nesvadba, H., ed. (North-Holland
    Publishing Co. In press).
18. Andreatta, R. H., and H. A. Scheraga. J. Med. Chem.
    14, 489 (1971).
19. Seu, J. H., R. R. Smeby, and F. M. Bumpus. J. Am. Chem.
    Soc. 84, 3883 (1962).
20. Khosla, M. C., N. C. Chaturvedi, R. R. Smeby, and F. M.
    Bumpus. Biochemistry 7, 3417 (1968).
21. Park, W. K., J. Asselin, and L. Berlinguet. In *Progress
    in Peptide Research,* Vol. 2, Lande S., ed. (New York:
    Gordon and Breach, 1972).
22. Schattenkerk, C., and E. Havinga. Rec. Trav. Chim.
    Pays-Bas. 84, 653 (1965).
23. Turk, J., G. T. Panse, and G. R. Marshall. 8th Midwest
    Regional ACS Meeting. Columbia, Mo. 1972.
24. Bystrov, V. F., S. L. Portnova, V. I. Tsetlin, V. T.
    Ivanov, and Yu. A. Ovchinnikov. Tetrahedron Lett. 25,
    493 (1969).
25. Glickson, J. D., W. D. Cunningham, and G. R. Marshall.
    In *Chemistry and Biology of Peptides,* Meienhofer, J.,
    ed. (Ann Arbor, Michigan: Ann Arbor Science Publ.,
    1972) pp 563-569.
26. Weinkam, R. J., and E. C. Jorgensen. J. Amer. Chem.
    Soc. 93, 7038 (1971).

SYMPOSIUM DISCUSSIONS

Summarized by Johannes Meienhofer

THE REVIEW ON THE physiological roles of angiotensin
(Peach, pp 471 to 493) was followed by an interesting discussion touching on comparisons with norepinephrin and ACTH
action, on *in vivo* stability of angiotensin, clinical testing of inhibitors and effects on the nervous system. [1-
Asparagine]-angiotensin II seems to be more active than
[1-aspartic acid]-angiotensin II on papillary muscle preparations because it is less readily degraded. When the
degradative enzymes were flushed out of the tissue before-
hand, the dose response curves of the two compounds came
close together. The degradative enzymes have never been
isolated, but most of *in vivo* angiotensin degradation seems
to occur by circulating enzymes.
    Interpretations of structure-activity correlations
become increasingly specific (pp 495 to 520). A discussant
cautioned from interpreting too categorically relationships
between affinity and intrinsic activity. With cysteine-
containing analogs (*e.g.* [Cys[8]]-angiotensin II), it is
uncertain whether the sulfhydryl form is maintained throughout an assay. More analogs with potent inhibitory activity
have been obtained, and the effects of [Sar[1], Ile[8]]- and
[Sar[1], Leu[8]]-angiotensin II also appear to be extremely
long lasting *in vivo* due to their resistance to aminopeptidase degradation. In response to an inquiry whether
1-deamino-angiotensin has been synthesized, it was pointed
out that deamino-[Sar[1]]-angiotensin II had diminished
activity (see also ref. 1, 2).
    Applauded was a short but concise description (Jorgensen)
of "syntheses of analogs of angiotensin" by the solid-phase
technique where preparations undergo the following sequential
purification: (i) countercurrent distribution, (ii) Sephadex

chromatography, and (iii) CM-cellulose chromatography.
The products are tested by: (a) NMR, (b) tlc in 6 solvent
systems, (c) high voltage electrophoresis at two different
pH values, (d) amino acid analysis, and (e) hydrolysis and
L-amino acid oxidase studies to detect racemization.
Complete homogeneity must be observed in all tests.

Synthesis of bio-isosteres of a peptide renin inhibitor
(pp 541 to 544) were complemented as the addition of another
dimension in analog synthesis. Very likely, the chemistry
will be even more complicated than in straight peptide
synthesis.

One comment alluded to multiple mechanisms of action
in potentiating effects on bradykinin through inhibition
of bradykininases.

A most enjoyable discussion developed about angiotensin
conformation following Dr. Fromageot's lecture (pp 545 to
562). Determination of the molar ellipticity and the importance of concentration in the measurements of conformation
in solution became objects of discussion. It was also
pointed out that peptides might undergo changes in certain
organic solvents, as DMSO, dichloroethanol, trifluoroethanol,
etc. to the extent that the material could sometimes not
be recovered. Therefore, peptides should, after studies
in organic solutions, always be recovered, and their integrity be checked. Trifluoroethanol was carefully redistilled
before CD measurements of angiotensin were done; finally
it crystallized from this solvent with fully intact biological potency. The question whether two forms of
angiotensin, separable by Sephadex chromatography or by
thin film dialysis represent two stable conformations or
two states of dispersity remained unsettled. Apparently,
CD spectra of the two forms appear to be indistinguishable.
An argument about which form of the His-Pro bond is to be
called *cis* and which *trans* illustrated the importance of
defining, perhaps for each individual study, the terminology
used (compare footnote, p 545). Discussions on similarities
of established peptide conformations to those at receptor
sites remain as pure conjecture, as those on the sex of an
unborn baby, but they continue to come up with the same
regularity. Similarly, the nature of the binding forces
remains anyone's guess, but topochemical studies might
provide information on spatial requirements. Thus, modification of the arginine residue in position 2 by reaction
with cyclohexane dione to give an imidazolidine derivative
does not impair the activity, indicating that the guanidine
group is not located near an aromatic residue or a narrow

cleft of the receptor. In dilute aqueous solution angiotensin is so flexible that one cannot define a conformation,
although it is far from random, and a certain degree of
folding is apparent from results of tritium exchange.

References

1. Riniker, B., H. Brunner, and R. Schwyzer. Angew. Chem.
    74, 469 (1962).
2. Riniker, B., and R. Schwyzer. Helv. Chim. Acta 47,
    2357 (1964).

SECTION VIII

HORMONAL MESSENGERS

Session Chairmen

Werner Rittel and Roderich Walter

THE HYPOTHALAMIC HORMONES--THEIR SIGNIFICANCE IN PHYSIOLOGY
AND MEDICINE

Roger Guillemin. The Salk Institute, La Jolla,
California 92037.

SUMMARY--A short review describing the concept of central
nervous system control of endocrine function through
hypothalamic pituitary afferents, the hypothalamic hypophysiotropic hormones. It discusses further the significance
of the recent characterization and total synthesis of two
of the hypothalamic hypophysiotropic hormones in terms of
current endocrinology. The use of the hypothalamic hypophysiotropic hormones in many aspects of clinical endocrinology is discussed including the possibility of the
use of synthetic analogues as a novel approach to contraceptive medication.

SINCE THE END OF THE LAST CENTURY, it has been realized
that the two major systems maintaining the homeostasis of
the organism were the central nervous system and the endocrine glands. The anterior lobe of the pituitary gland or
adenohypophysis was also shown to be the major controller
of the secretion of other (peripheral) endocrine tissues,
while the posterior lobe of the pituitary gland or neurohypophysis was shown to secrete oxytocin and vasopressin.
  Until recently, it was considered that the major, if
not sole, mechanism of the regulation of secretion of
adenohypophysial hormones was to be found in an ensemble
of simple *negative feedback* systems whereby the plasma
concentrations of peripheral hormones from the gonads, the
adrenal cortex, the thyroid, were in inverse relationship
to the amount of the corresponding pituitary hormones

secreted. The explanation for these negative feedback systems was simply in terms of a direct action of the peripheral
hormones on the pituitary gland to regulate somehow its
secretion. We will see later that if the feedback systems
are still considered an important part of the mechanism of
control of the pituitary secretions, we now have proof that
they work through the central nervous system in the case of
the gonadal and adrenal steroids or in conjunction with it,
in the case of the thyroid hormones.

A possible directing role of the central nervous system
in the control of the secretion of adenohypophysial hormones
was first suspected from numerous observations showing that
many exteroceptive stimuli triggered the secretion of pituitary hormones; exposure to cold environment stimulates secretion of TSH, nocuous stressing agents stimulate secretion
of ACTH and growth hormone, and experimentally varying amounts
of light or light-darkness ratios will in many species, from
birds to mammals modify secretions of the gonadotropins,
hence sexual receptivity and reproduction patterns.[1]

Following these early observations, it is now well established that the central nervous system participates in
the physiologic mechanisms which regulate the secretions of
the anterior lobe of the pituitary gland. For instance,
minute electrocoagulations of discrete areas of the (ventral)
hypothalamus (Figures 1, 2 and 3) specifically interfere with
secretion of ACTH,[2] TSH,[3] the gonadotropins-luteinizing
hormone (LH),[4] follicle-stimulating hormone (FSH),[5] and
growth hormone (somatotropin, STH).[6] Conversely, localized
stimulation with electrical currents in specific hypothalamic
loci will stimulate secretion of ACTH,[2] TSH,[3] gonadotropins[4,5]
and growth hormone.[6] Furthermore, secretion of pituitary
hormones such as ACTH or the gonadotropins has been shown
to be accompanied by changes in spontaneous electrical activity of the pertinent hypothalamic areas or nuclei.[7] In
all cases mentioned so far, it appears that the hypothalamus
contributes *stimulatory* inputs for acutely increasing the
secretion of a pituitary hormone over some sort of a basal
secretion which appears to take place in absence of hypothalamic stimulus.

Secretion of two other pituitary hormones, prolactin
and MSH (or melanocyte stimulating hormone) may be, on the
other hand, under some sort of a tonic *inhibition* by the
hypothalamus: Indeed, separation of the adenohypophysis
from its normal hypothalamic connections leads to increased
secretion of prolactin[8] and MSH.[9]

The negative feedback mechanism which we mentioned
earlier, in which a peripheral hormone inhibits secretion

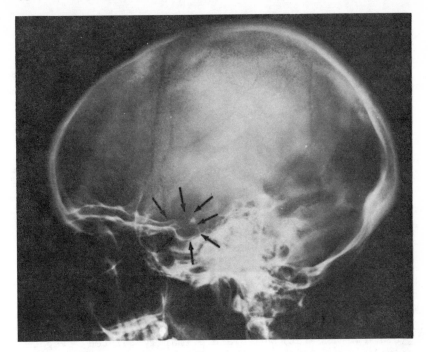

Figure 1: X-ray of the skull of a normal adult man. The pituitary gland is ensconsed in a spherical cavity of the sphenoid bone (arrows); the part of the brain immediately subjacent to it is the hypothalamus area (arrows).

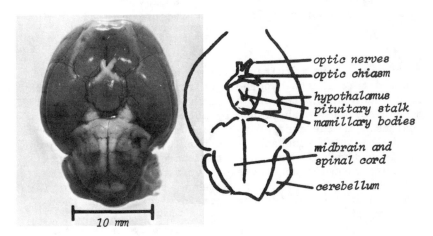

Figure 2: Ventral view of the rat brain.

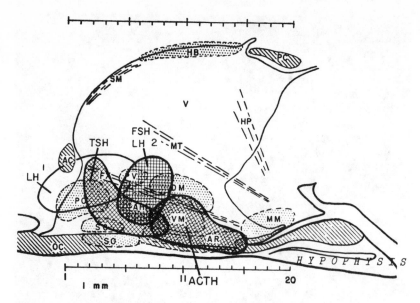

Figure 3: This simplified diagram shows four areas circled
in black lines on a sagittal section of the hypothalamus
(brain of the rat) which are primarily related, respectively, from left to right with the secretion of LH, TSH,
FSH and LH, ACTH. The diagram attempts to show that these
areas overlap considerably, one upon another; further,
that these "hypophysiotropic" areas do not coincide with
the classic nuclei of the hypothalamus described by
neuroanatomists—OC, optic chiasm; AC, anterior commissure;
PC, posterior commissure. Other abbreviations are given
as in the de Groot atlas of the rat brain.

of the pertinent pituitary factor, has been shown to be
exerted through the hypothalamus for the steroid hormones
of the adrenal cortex and of the gonads, as the feedback
mechanisms can be abolished by localized hypothalamic
lesions. In the case of thyroid hormones, the major locus
of the negative feedback is at the level of the adenohypophysis and we will see later that here again the central
nervous system is nonetheless involved also. Thus, the
evidence is now clear that, somehow, the hypothalamus is
involved in controlling the secretory activity of the
adenohypophysis.

   The hypothalamus is a phylogenetically ancient part of
the brain which receives a large number of connections from

other parts of the encephalon. Neural systems afferent
to the hypothalamus (midbrain reticular formation, limbic
system) participate also in the transhypothalamic control
of the secretion of the adenohypophysis. For instance,
localized experimental lesions of the fornix, the septum,
the amygdala and the olfactory tract, *i.e.*, fiber tracts
or nuclei connected to the hypothalamus, will interfere
with the release of the gonadotropin hormones, particularly
ovulation hormone (LH).[10],[11] Similarly, ovulation is
accompanied by characteristic changes in the electrical
activity of these centers.[10] It is known also that lesions
in the midbrain reticular formation will interfere with
secretion of ACTH and TSH.

What is the mechanism whereby the information of hypothalamic origin is conveyed to the adenohypophysis? In
contradistinction to the large tract of nerve fibers connecting the hypothalamus and the neural lobe of the
pituitary, the source of secretion of the two octapeptides
vasopressin and oxytocin, no similar nervous connection
exists between hypothalamus and adenohypophysis. There
is, however, between these two structures, a well-developed
system of capillary vessels (Figure 4): The hypothalamo-
hypophysial portal system, with its primary plexus in the
junction area between hypothalamus and adenohypophysis
(median eminence), and collecting veins around and in the
pituitary stalk and its secondary plexus throughout the
parenchyma of the anterior lobe of the pituitary; blood
flow is definitely from primary to secondary plexus, *i.e.*,
from hypothalamus to adenohypophysis. The concept thus
arose that the hypothalamic control of secretions of the
adenohypophysis might be *neurohumoral* in nature, some substance of hypothalamic origin being released in the primary
plexus of the portal vessels to be transmitted by these
vessels to the adenohypophysis where it would stimulate
secretion of pituitary hormones.

The concept of neurosecretion, *i.e.*, the ability of
some highly specialized neurons to secrete substances with
hormonal activities was long established from studies deal-
ing with invertebrates;[12] in mammalians, it is supported
by evidence for the hypothalamic origin of vasopressin and
oxytocin which are transported by axoplasmic flow to the
neurohypophysis.[13]

Investigations over the last few years have indeed
confirmed the existence in crude aqueous extracts of
hypothalamic tissue of substances capable of specifically
stimulating the secretion of ACTH, TSH, gonadotropins and
growth hormone. Similarly, it appears that some substances

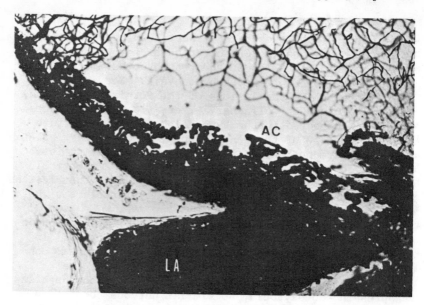

Figure 4:   Sagittal section of the ventral hypothalamus
   (median eminence) and anterior pituitary (rabbit) after
   intracarotid injection of India ink. Loops of capillary
   vessels (AC) can be seen in a network that establishes
   contact with nerve fiber terminals (not shown) of
   hypothalamic origin; these capillary vessels merge into
   larger vessels that carry blood to the anterior lobe of
   the hypophysis (LA). Photograph, courtesy of Professor
   H. Duvernoy, Department of Anatomy, School of Medicine,
   Besançon, France.

in the same crude extracts of the hypothalamus inhibit the
secretion of prolactin and MSH (and possibly also growth
hormone). These substances of hypothalamic origin, modi-
fying the secretion of pituitary hormones, have been termed
hypophysiotropic hormones or *releasing factors*; in the case
of prolactin, MSH and possibly growth hormone, we talk of
hypophysiotropic inhibitory hormones or *release-inhibiting
factors*.[14]
    Recent observations have shown that we are dealing with
different chemical entities for each one of these hypophysiotropic activities. There is now good evidence that acidic
extracts of the hypothalamus of a series of mammalian species
contain specific substances, which all appear to be relatively small polypeptides, involved in the control of the
secretion of ACTH (CRF, for corticotropin-releasing factor),

of the gonadotropins LH and FSH (LRF, for luteinizing-hormone-releasing factor, FRF, for FSH-releasing factor) and TSH (TRF, for thyrotropin-releasing factor).[14]

Regarding the control of the secretion of growth hormone, prolactin, and the melanocyte stimulating hormone (MSH), there is some experimental evidence for the existence of releasing factors (GRF, growth hormone releasing factor; PRF, prolactin releasing factor; MRF, MSH-releasing factor) as well as factors inhibiting the release of these three hormones (referred to as SRIF, somatotropin or growth hormone-release inhibiting factor; PRIF, prolactin-release inhibiting factor; MRIF, MSH-release inhibiting factor).[14]

What is known about the nature of these hypothalamic hypophysiotropic hormones?

After several years of arduous work, principally in two laboratories, and collection of several millions of fragments of sheep or pig hypothalamic fragments, two of these hypothalamic-hypophysiotropic hormones have recently been characterized. In 1969, thyrotropin releasing factor was isolated and characterized as the tripeptide <Glu-His-Pro-NH_2.[15-17] In 1971, the luteinizing hormone releasing factor, LRF, was characterized as the decapeptide <Glu-His-Trp-Ser-Tyr-Gly-Leu-Arg-Pro-Gly-NH_2.[18-21] TRF and LRF have the same respective primary sequences in the two species, ovine and porcine, from which they were isolated. TRF and LRF characterized as above, have been shown to be highly potent in stimulating respectively the secretion of TSH and LH in a variety of mammalians, including the human. Recently it has been reported that the tripeptide TRF is also highly active to stimulate the secretion of prolactin in humans;[22] from results obtained in other species, it is already apparent that there is a considerable variation in this ability of TRF to release prolactin in various species of animals. Also, it has been reported that one may be able to stimulate differentially the secretion of prolactin and TSH in humans. Thus, TRF may be closely related structurally to the still hypothetical prolactin releasing factor, PRF.

Of interest is the now well characterized observation that highly purified native LRF of porcine or ovine origin, as well as synthetic LRF-decapeptide preparations, are able to stimulate the secretion of not only the gonadotropin LH but also the other gonadotropin, FSH.[15-21] (Figure 5) The proposal has thus been made[23] that the decapeptide LRF may be the sole hypothalamic controller for the secretion of the two gonadotropins, LH and FSH. This hypothesis still leaves unanswered a series of physiological observations

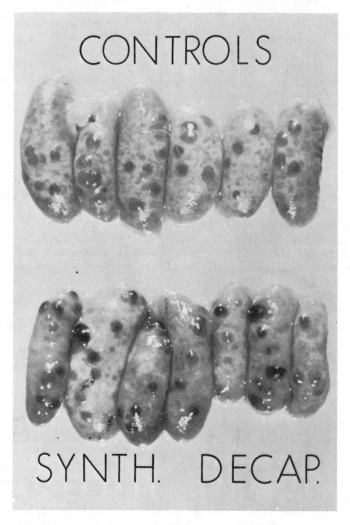

Figure 5: Induction of ovulation in rabbits by a single
intravenous injection of the synthetic decapeptide LRF.
The globular formations seen at the surface of the
control ovaries (upper row) are mature follicles
typical of the mature virgin rabbit.   24 hours after
injection of LRF, (lower row) many of these follicles
appear turgescent and very black on the photograph;
they have ovulated and their dark appearance in the
photograph is due to their being filled with blood
following ovulation (corpora haemorrhagica or Blutpunkt).

in which there is evidence for dissociated releases of FSH
and LH. Thus, at the writing of this short review, the
question of the possible existence of a specific FSH releasing factor (FRF) distinct from the decapeptide LRF,
still remains open for further investigations.

The postulated growth hormone releasing factor (GRF)
has not been characterized. A decapeptide with the primary
sequence Val-His-Leu-Ser-Ala-Glu-Glu-Lys-Glu-Ala has recently
been isolated from hypothalamic tissues of porcine origin and
proposed by Schally *et al.*[24] as being the hypothalamic
controller of growth hormone secretion, "GRH or growth
hormone releasing hormone." The significance of this
material is, at the moment, very much open to question,
since it appears to be active only in one type of bioassay
which has not been fully characterized, while it is unable
to stimulate the secretion of growth hormone in a variety
of species when following plasma growth hormone levels by
radioimmunoassays. Furthermore, it has recently been
realized that the sequence Val-His-Leu-Ser-Ala-Glu-Glu-
Lys-Glu-Ala is similar to a fragment of the beta chain of
porcine hemoglobin.[25]

Several laboratories have recently reported[26] obtaining
purified extract of hypothalamic tissues that would be endowed of growth hormone releasing activity based on radioimmunoassays, the active material being different (behavior
on G-25) from Schally's "GRH." The nature of the active
principles in these extracts is unknown at the moment. Thus,
the nature of the hypothalamic factors stimulating the
secretion of growth hormone remains to be elucidated.
Similarly, no reliable evidence is available for the nature
of the hypothalamic factors postulated in the control of
the release of prolactin (PRF) or ACTH (CRF for cortico-
tropin releasing factor).

The nature of the postulated inhibitory factors for the
release of prolactin (PRIF) or for the release of growth
hormone (SRIF, for somatotropin-release inhibiting factor)
is similarly unknown at the writing of this review.

The tripeptide Pro-Leu-Gly-NH$_2$, obviously the C-terminal
of oxytocin, has been allegedly isolated from hypothalamic
extracts and has been proposed as the MSH-release inhibiting
factor (MRIF).[27,28] Along the same lines, the pentapeptide
H-Cys-Tyr-Ile-Gln-Asn-Cys-OH or tocinoic acid has also been
reported in hypothalamic extracts and has been proposed[29]
as the primary structure of the postulated MSH releasing
factor (MRF). There is a great deal of controversy at the
moment about the significance of these two proposals regarding the biological activities attributed to these two

oligopeptides--the controversy being due to the difficulties
of the bioassays involved.   This, thus, remains for future
clarification.

Characterization of these hypothalamic hypophysiotropic
hormones has been of major significance in physiology.
Availability of the purified TRF and later of pure synthetic
TRF led to the demonstration that the well known negative
feedback mentioned above between thyroid hormone secretion
and TSH secretion is actually taking place at the level of
the pituitary tissue in some sort of antagonism between the
thyroid hormones and TRF.[30]   The molecular biology of this
negative feedback is now fairly well understood and appears
to involve a polypeptide or protein molecule not characterized
so far, induced by thyroxin or tri-iodothyronine within the
thyrotrophs of the pituitary; this polypeptide is able to
antagonize some secondary event that follows activation by
TRF at the level of the thyrotroph's plasma membrane of
some biochemical event that normally leads to the secretion
of TSH.   Several steps of this mechanism in the negative
feedback regulation of the secretion of TSH by thyroid
hormones and TRF remain to be elucidated.

There is evidence that the steroid hormones (glucocor-
ticoids or sex steroids such as estrogens, progesterone
or testosterone) participate in the negative feedback on
the secretion of the gonadotropins LH and FSH in a somewhat
different manner; while they appear to act also at the level
of the pituitary to inhibit quantitatively the pituitary
response to LRF in terms of the secretion of LH and FSH,
(probably through some mechanism similar to the one involved
in the negative feedback of thyroid hormones on TSH secretion) there is also good evidence that the steroids act
somehow at the level of the hypothalamus[10,31] to inhibit
the secretion of the hypothalamic releasing factor, LRF.
The recent availability of synthetic LRF in large quantities
will allow further investigation of this physiologically
important phenomenon.

If it is eventually and unquestionably demonstrated
that hypothalamic extracts do contain hypophysiotropic
inhibitory substances for the secretion of several of the
hypophysial hormones, the neurosecretory neurones of the
hypothalamus would thus be shown to partake in the general
physiological laws that neurophysiologists have proposed
over the last 50 years, namely, that excitatory neurones
or fibers have corresponding inhibitory neurones and fibers.
The hypothalamic neurosecretory neurones would thus correspond to a highly specialized type of neurosecretory tissue
still subjected, however, to the same types of regulatory

and integrative mechanisms that are known to be present
throughout the central nervous system. There is good and
increasing evidence that the neurosecretory neurones of the
hypothalamus respond to specific and classical neurotransmitters such as catecholamines, serotonin, which thus would
represent the neurotransmitters specifically mediating
endoceptive or exteroceptive stimuli leading to the specific
secretion of one or another of the hypothalamic hypophysiotropic hormones.[32,33]

The hypothesis of the existence of specific hypothalamic
hypophysiotropic hormones has been, over the last few years,
of major significance in the formation of concepts to explain a series of endocrine diseases. Recent availability
of TRF and LRF have dramatically confirmed these earlier
hypotheses. There are a number of diseases, particularly
in children and young adolescents, which for years have
been considered to be, possibly, of hypothalamic origin.[34]
These are usually related to clinical problems involving
deficiencies in functions of the thyroid gland, of the
gonads, of statural growth, of adrenal-cortical function.
In many cases where one could stimulate the peripheral
glands by injection of purified human pituitary hormones,
these patients could show a normal response of the peripheral
organs, thyroids, adrenal cortex, testes, ovary, or statural
growth in response to the pituitary hormones. Thus, their
primary defect was not to be found at the level of the
peripheral glands or tissues but probably at the level of
the pituitary or the hypothalamus. Direct exploration of
these two possibilities, *i.e.*, hypothalamic defect versus
pituitary defect, was not possible, except by some relatively complicated indirect means of only circumstantial
significance until availability of TRF and LRF. It is now
well recognized that these patients can be divided into
major groups in accordance to their pituitary responses to
injection of the hypothalamic releasing factors. Either
they do respond normally to injection of TRF or LRF--in
which case their primary defect is obviously not in their
pituitary but most likely at the level of their hypothalamus, or they do not respond to injection of TRF or LRF--
their primary lesion or defect being thus at the level of
the pituitary. A number of cases have thus been recently
described in which abnormality of the response of these
patients to either TRF or LRF led to further investigation
of the status of their pituitary function; in a large
number of such cases, evidence was observed for the existence in these patients of pituitary tumors which probably
would not have been suspected on the basis of clinical
observations at the time at which the patients were seen.[35]

A number of children have been known for the last few years to present what is called isolated pituitary deficiency or monotropic hypopituitarism, that is a series of syndromes in which, somehow, one pituitary hormone seems to be missing in the pituitary secretion; this has been observed for TSH, for ACTH, for LH and/or FSH and for growth hormone. Since the secretion of the other pituitary hormones appears to be normal in those children, the question was raised as to whether their defect would be pituitary or hypothalamic. There is now good evidence that, in most of these children, one can stimulate specifically the secretion of the missing pituitary hormone by administering the hypothalamic releasing factor. Obviously, in these children, the primary defect is hypothalamic, Other children with this same syndrome do not appear to respond to the hypothalamic hormone; their primary defect is thus at the level of the pituitary tissues. Replacement therapy in the human with pituitary hormones requires that only hormones of human origin be administered, with the exception of ACTH. Human gonadotropins, or growth hormone, or thyrotropin are indeed endowed of strict species specificity likely related to their complicated structures; ACTH, being a simpler polypeptide, is not species specific. Thus, when one considers the possibility of replacement therapy in these children, one must have available for years of treatment, large quantities of human pituitary hormones. Only small quantities of pituitary glands of human origin are available for preparing human pituitary hormones for clinical use. On the other hand, since the hypothalamic releasing factors can be easily synthesized in unlimited quantity and since the known sequences correspond to molecules highly active in the human, TRF and LRF are already used extensively in clinical medicine as diagnostic tools and also as therapeutic agents for chronic treatment of the pertinent deficiencies in children.[36]

Of extreme interest will be the availability of a growth hormone-releasing factor and also of the postulated growth hormone release-inhibiting factor. By far, the largest number of these children with these hypothalamic-hypophysial defects seem to suffer from deficiencies in the secretion of growth hormone. There is also a syndrome which has been described of emotionally disturbed children with stunted growth. Assuming that the pituitary secretory ability of these children for growth hormone is normal, the treatment of choice for them would be the chronic administration of the synthetic peptide with growth hormone-releasing activity while the cause of their behavioral problem is being removed. Growth hormone release inhibiting

factor would also be of major clinical significance in the
large population of adolescents suffering of what is known
as juvenile diabetes. Current concepts implicate somehow
the endogenously secreted growth hormone in those patients
in the triggering and the evolution of their usually malig-
nant diabetes.[37] Hypophysectomy is presently the only and,
unfortunately, drastic treatment for these children. It
is logical to assume that availability of a growth hormone
release-inhibiting factor should be of clinical value in
the treatment of their disease. Similarly, a not infrequent
disease known as acromegaly, in which the pituitary secretes
abnormally high amounts of growth hormone, should be alle-
viated by administration of a growth hormine release-
inhibiting factor.

The hypothalamic LH releasing factor administered acutely
in humans stimulates dramatically the secretion of gonadotropins in men and women (Figure 6).[38,39] There is early

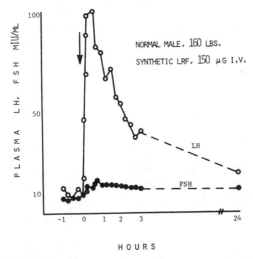

HOURS

Figure 6: Stimulation of the secretion of the gonadotropins
LH and FSH in a normal human male by a single intravenous
injection of synthetic LRF. Plasma LH and FSH measured
by radioimmunoassay. (In collab. with S. Yen, U. of
California, San Diego)

evidence that one can trigger ovulation in women who have
suffered of what has been known for years as hypothalamic
amenorrhea or infertility. The hypothalamic decapeptide
LRF is the treatment of the infertility of these women.
There is increasing evidence that chronic administration
of the hypothalamic decapeptide LRF can also reestablish

to normal the testicular function of several types of oligospermia for which no evident etiology was known and which thus, may be of hypothalamo-hypophysial origin. If LRF can stimulate ovulation in women, there is good reason to believe that it could be used as an exogenous agent to regulate the ovarian cycle and induce ovulation at a precise time in some sort of a rhythm method that could thus be used as an acceptable means of fertility control in some cultures.

There is also the possibility that analogues of LRF endowed of antagonistic activities to LRF--such molecules have recently been reported in the literature[40] and shown to be active as antagonists of LRF in experimental animals-- could be used as a novel approach to fertility control. Availability of antagonists to endogenous LRF may lead to a once-a-month type of medication that could be used as a contraceptive.

Demonstration of the existence of the hypothalamic hypophysiotropic hormones and the recent characterization of several of them have, undoubtedly, opened a new chaper in physiology and in modern medicine. I think that the physiologists and the biochemists who have made these discoveries can take added pride in the fact that once more research, originally directed at problems of purely fundamental, of "academic" significance, is leading to practical solutions in clinical medicine.

Acknowledgment

The research conducted in our laboratories is presently supported by AID (contract csd/2785), the Ford Foundation and the Rockefeller Foundation.

References

1. Harris, G. W. *Neural Control of the Pituitary Gland*, Vol. 1 (London: E. Arnold Publishers, Ltd., 1955).
2. Fortier, C. In *The Pituitary Gland*, Vol. 2 (University of California Press Publishers, 1965) pp 195-234.
3. Brown-Grant, K. In *The Pituitary Gland*, Vol. 2 (University of California Press Publishers, 1965) pp 235-269.
4. Harris, G. W., and H. J. Campbell. In *The Pituitary Gland*, Vol. 2 (University of California Press Publishers, 1965) pp 99-165.
5. Donovan, B. T. In *The Pituitary Gland*, Vol. 2 (University of California Press Publishers, 1965) pp 49-98.

6. Reichlin, S.   In *The Pituitary Gland*, Vol. 2 (University of California Press Publishers, 1965) pp 270-298.
7. Porter, R. W.   Recent Progr. in Hormone Res. 10, 1 (1954).
8. Everett, J. W.   In *The Pituitary Gland*, Vol. 2 (University of California Press Publishers, 1965) pp 166-194.
9. Landgrebe, F. W., and G. M. Mitchell.   In *The Pituitary Gland*, Vol. 3 (University of California Press Publishers, 1965) pp 41-58.
10. Beyer, C., and C. H. Sawyer.   In *Frontiers in Neuroendocrinology*, (London: Oxford University Press, 1969) pp 255-287.
11. Mason, J. W.   Psychosom. Med. 30, 565 (1968).
12. Scharrer, E., and B. Scharrer.   *Neuroendocrinology*, (New York: Columbia University Press Publishers, 1963).
13. Sachs, H., R. Portanova, E. W. Haller, and L. Share.   In *Neurosecretion* (Berlin: Springer-Verlag, 1967) pp 146-154.
14. Guillemin, R.   In *Advances in Metabolic Disorders*, Vol. 5 (New York: Academic Press, 1971) pp 1-51.
15. Burgus, R., T. F. Dunn, D. Desiderio, and R. Guillemin.   C. R. Acad. Sci. (Paris) 269, 1870 (1969).
16. Burgus, R., T. F. Dunn, D. Desiderio, D. N. Ward, W. Vale, and R. Guillemin.   Nature 226, 321 (1970).
17. Bowers, C. Y., A. V. Schally, F. Enzman, J. Boler, and K. Folkers.   Endocrinology 86, 1143 (1970).
18. Matsuo, H., Y. Baba, R. M. G. Nair, A. Arimura, and A. V. Schally.   Biochem. Biophys. Res. Comm. 43, 1334 (1971).
19. Burgus, R., M. Butcher, N. Ling, M. Monahan, J. Rivier, R. Fellows, M. Amoss, R. Blackwell, W. Vale, and R. Guillemin.   C. R. Acad. Sci. (Paris) 273, 1611 (1971).
20. Baba, Y., H. Matsuo, and A. V. Schally.   Biochem. Biophys. Res. Comm. 44, 459 (1971).
21. Burgus, R., M. Butcher, M. Amoss, N. Ling, M. Monahan, J. Rivier, R. Fellows, R. Blackwell, W. Vale, and R. Guillemin.   Proc. Nat. Acad. Sci. U. S. 69, 278 (1972).
22. Bowers, C. Y., H. G. Friesen, P. Hwang, H. T. Guyda, and K. Folkers.   Biochem. Biophys. Res. Comm. 45, 1033 (1971).
23. Schally, A. V., A. Arimura, A. J. Kastin, H. Matsuo, Y. Baba, T. W. Redding, R. M. G. Nair, L. Debeljuk, and W. White.   Science 173, 1036 (1971).
24. Schally, A. V., Y. Baba, R. M. G. Nair, and C. D. Bennett.   J. Biol. Chem. 246, 6647 (1971).
25. Veber, D. F., C. D. Bennett, J. D. Milkowski, G. Gal, R. G. Denkewalter, and R. Hirschamnn.   Biochem. Biophys. Res. Comm. 45, 235 (1971).

26. Wilber, J., T. Nagel, and W. F. White. Endocrinology 89, 1419 (1971).
27. Celis, M. E., and S. Taleisnik. Intern. J. Neuroscience 1, 223 (1971).
28. Nair, R. M. G., A. J. Kastin, and A. V. Schally. Biochem. Biophys. Res. Comm. 43, 1376 (1971).
29. Bower, S. A., M. E. Hadley, and V. J. Hruby. Biochem. Biophys. Res. Comm. 45, 1185 (1971).
30. Vale, W., R. Burgus, and R. Guillemin. Neuroendocrinology 3, 34 (1968).
31. Smith, E. R., and J. M. Davidson. Endocrinology 82, 100 (1968).
32. Fuxe, K., and T. Hökfelt. In *Frontiers in Neuroendocrinology* (London: Oxford University Press, 1969) pp 47-96.
33. McCann, S. M., P. S. Kalra, A. O. Donoso, W. Bishop, H. P. G. Schneider, C. P. Fawcett, and L. Krulich. In *Brain Endocrine Interaction*, (Basel, Switzerland: S. Karger Publishers, 1972) pp 224-235.
34. Wilkins, L. *The Diagnosis and Treatment of Endocrine Disorders* (Springfield, Illinois: Thomas Publishers, 1966).
35. Fleischer, N. J. Clin. Endoc. Metab. 34, 505 (1972).
36. Grumbach, M., and N. Kaplan. J. Clin. Investig. 28, 792 (1972).
37. Cerasi, E., and R. Luft. Acta Endocr. (Kbh) 55, 278 (1967).
38. Gual, C., A. Kastin, and A. V. Schally. Recent Progr. in Hormone Res. 28, 684 (1972).
39. Yen, S. S. C., R. Rebar, G. VanderBerg, F. Naftolin, Y. Ehara, S. Engblom, K. J. Ryan, K. Benirschke, J. Rivier, M. Amoss, and R. Guillemin. J. Clin. Endoc. Metab. 34, 1108 (1972).
40. Vale, W., G. Grant, J. Rivier, M. Monahan, M. Amoss, R. Blackwell, R. Burgus, and R. Guillemin. Science 176, 933 (1972).

STRUCTURE-BIOLOGICAL ACTIVITY RELATIONSHIPS ON THYROTROPIN AND LUTEINIZING HORMONE RELEASING FACTOR ANALOGUES

M. Monahan, J. Rivier, W. Vale, N. Ling, G. Grant, M. Amoss, R. Guillemin, R. Burgus. The Salk Institute, La Jolla, California 92037

E. Nicolaides, M. Rebstock. Parke-Davis Company, Ann Arbor, Michigan 48105

MANY ANALOGUES OF TRF (thyroid stimulating hormone releasing factor) have been synthesized and tested biologically and from this information, several generalizations can be drawn regarding structure-function relationships. Most of the data presented here on TRF have been reviewed previously[1] and we will present some ideas about the chemical interpretation of these results.

TRF analogues were examined for purity by tlc (at least four systems), amino acid analysis, mass spectral and NMR analysis. All were judged homogeneous by these criteria with the exception of III, XVII, XXII, XXIV and XXXI which exhibited trace impurities on tlc at a 20 μg load. Compounds IV and V were tested as crude unresolved mixtures [these two compounds were synthesized with the racemic pyroglutamic acid (<Glu) analogues].

Changes at the <Glu[1] position of TRF result in a considerable decrease in ability to induce secretion of thyroid stimulating hormone (TSH). Nevertheless, there are marked differences in the [R[1]]-TRF analogues tested (Table 1). These data suggest that the N^α of <Glu is involved as a nucleophile in the action of TRF. An N^α-methyl function on <Glu could introduce allosteric interference at the receptor-substrate complex level, or it could interfere with a nucleophilic process involving the N^α of <Glu. Of possible

601

Table I

Thyrotropin Releasing Factor Analogs

	Compound	% TRF Potency	pK_a^{im-His}
I	<Glu-His-Pro-NH$_2$	100	6.25
II	N^α-Me-<Glu-His-Pro-NH$_2$	1.7	6.25
III	Pro-His-Pro-NH$_2$	0.01	6.05
IV	(D,L) [thienyl]$-$C(=O)$-$L-His-L-Pro-NH$_2$	0.2	
V	(D,L) [furyl]$-$C(=O)$-$L-His-L-Pro-NH$_2$	0.01	
VI	[cyclopentyl]$-$C(=O)$-$His-Pro-NH$_2$	<0.01	
VII	<Glu-Arg-Pro-NH$_2$	0.05	
VIII	<Glu-Orn-Pro-NH$_2$	0.02	
IX	<Glu-Lys-Pro-NH$_2$	0.02	
X	<Glu-Tyr-Pro-NH$_2$	0.08	
XI	<Glu-Met-Pro-NH$_2$	1	
XII	<Glu-NH-CH(-CH$_2$-[imidazole, CH$_3$-N])-C(=O)-Pro-NH$_2$	800	5.95
XIII	<Glu-NH-CH(-CH$_2$-[imidazole, N-CH$_3$])-C(=O)-Pro-NH$_2$	0.04	6.6
XIV	<Glu-His-N[azetidine]-C(=O)-NH$_2$	1.6	

Table I Continued

	Compound	% TRF Potency	pK_a^{im-His}
XV	<Glu-His-N(CH₃)-CH₂-C(=O)-NH₂	0.32	
XVI	<Glu-His-N⟨ring,OH⟩-C(=O)-NH₂	0.14	
XVII	<Glu-His-Leu-NH₂	0.04	
XVIII	<Glu-His-Gly-NH₂	<0.02	
XIX	<Glu-His-Trp-NH₂	<0.02	
XX	<Glu-His-Pro-OMe	10	6.10
XXI	<Glu-His-Pro-OH	0.02	6.75
XXII	<Glu-His-N⟨ring⟩-CH₂OH	1.2	
XXIII	<Glu-His-Pro-NHCH₂CH₂OH	16	
XXIV	<Glu-His-Pro-NHCH₂CH	14	6.25
XXV	<Glu-His-Pro-NHNH₂	14	
XXVI	<Glu-His-Pro-NHC₆H₅	16	
XXVII	<Glu-His-Pro-Gly-NH₂	35	
XXVIII	<Glu-His-Pro-Ala-NH₂	0.5	
XXIX	<Glu-His-Pro-NMe₂	0.5	6.25
XXX	<Glu-His-Pro-NEt₂	0.05	6.45
XXXI	<Glu-His-Pro-N⟨hexagon⟩	0.2	6.45

significance to the latter interpretation are the relative
activities of compounds IV, V and VI when compared to the
relative nucleophilicities of S, O, and C. Compound III,
([Pro[1]]-TRF) ought to be more reactive than these three
compounds, except that the N^α of Pro is protonated at
physiological pH, which would reduce its utility as a
nucleophile.

Histidyl modifications involving charged groups reduce
the biological activity considerably (compounds VII, VIII
and IX). However, there appears to be a rigid requirement
for aromaticity at this position as shown by the high
activity (10%) of [Phe[2]]-TRF.[2] The marked difference in
activity between this compound and [Tyr[2]]-TRF might be explained in terms of steric interactions in the receptor-
substrate complex. [Met[2]]-TRF retains substantial activity,
although less than the aromatic substitutions, except for
[N^π-Me-His[2]]-TRF. CPK models indicate that the side chain
of methionine fills almost the same space as the side chain
of histidine with the sulfur atom able to occupy nearly the
same position as the imidazole π-nitrogen. It is possible
that both aromaticity and a general acid function are required of the imidazole of histidine. Pertinent to the
latter point is the observation that [N^π-Me-His[2]]-TRF has
a very low biological potency while [N^τ-Me-His[2]]-TRF is
eight times more active than the parent molecule. However,
other factors could also explain these results (conformational orientation, steric interactions, *etc.*).

Measurements of the pKa of the imidazole of histidine
in several TRF analogues[3] reveal that the pKa and biological
potency appear to correlate (Figure 1 and Table I). Two
points on this plot which fall outside the curve represent
compounds which are ionized at physiological pH. Formal
charges on the molecule reduce the biological potency
considerably. The relationship between pKa and potency
supports the hypothesis that the histidine side chain may
be involved as a general acid.

Prolinamide substitutions can involve retention of considerable potency. An amide substituent can easily assume
a conformation whereby it does not interfere with other
groups in the remainder of the molecule (or alternatively,
the same could be suggested for interactions in the
receptor-substrate complex). Tertiary amides are much
less active than the potent secondary amides. The presence
of a charged group (compound XXI) at the C-terminus yields
a drastic reduction in activity when compared to hydrophobic
amide substitutions.

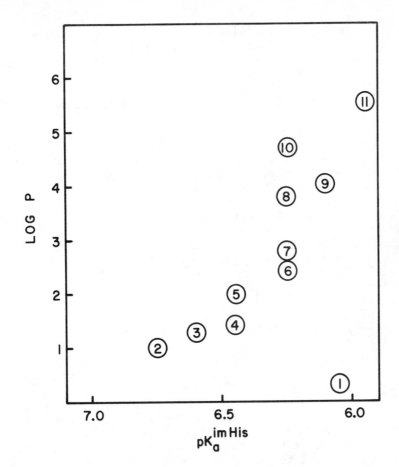

Figure 1: Logarithm of biological potency (P) *vs.* pKa of
the imidazole of histidine. Numerical designation of
points corresponds to roman numerals in Table I:
1 (III); 2 (XXI); 3 (XIII); 4 (XXX); 5 (XXXI); 6 (XXIX);
7 (II); 8 (XXIV); 9 (XX); 10 (I); 11 (XII).

    Summarizing the TRF results, we conclude that: The
molecule must be hydrophobic; rigid stereochemical or bulk
properties must be met for each residue; <Glu may act in a
nucleophilic capacity; histidine may be involved as a
general acid; and amide substitutions retain high potency
except for charge groups or groups of large bulk. It is
not surprising that TRF may exhibit poly-functionality
(*e.g.*, nucleophilic participation by <Glu and histidine

acting as a general acid), considering the remarkable
potency and specificity of such a small peptide.

   We have synthesized a variety of analogues of luteinizing
hormone releasing factor (LRF, <Glu-His-Trp-Ser-Tyr-Gly-
Leu-Arg-Pro-Gly-NH$_2$).   All LRF analogues were synthesized
on a benzhydrylamine resin by standard solid-phase techniques.
Purification, following cleavage in HF, was carried out in
two steps: Cation exchange chromatography or gel filtration
followed by partition chromatography.   Resulting products
were analyzed for purity by tlc (seven systems, 20 µg load),
amino acid analysis, NMR spectra, mass spectra (heptapeptide
or shorter), and optical rotation.   Included in these are
the peptide amides shortened from the C-terminus, Table II.
They are essentially inactive except for des-Gly^{10}-LRF (11%)
and the tripeptide <Glu-His-Trp-NH$_2$ (<0.1%). The LRF analogues
involving substitution of glycine for the other residues
exhibit potencies of *ca.* 0.1% of LRF activity except for
[Gly^4]-LRF (1.5%).   Even though [Gly^2]-LRF has a measurable
potency as an agonist, it does not exhibit the same dose-
response curves as the other analogues and is also a partial
inhibitor of LRF.   At saturation doses, it yields an intrinsic
activity of about 50% that of LRF while saturation doses of
the other analogues release LH at the same levels as the
saturation doses of LRF.   Des-His^2-LRF does not release LH
at any of the doses tested *in vitro* and also inhibits LRF
making it an even better inhibitor.   [Gly^3]- and [Ala^3]-LRF
are essentially inactive but are not antagonists.   [Phe^2]-,
[Phe^3]-, [N^τ-Me-His^2]-, and [N^π-Me-His^2]-LRF are all active
to about the same extent which suggests that aromaticity is
a requirement in these positions.   It is significant that
the methylhistidine substitutions do not yield results
analogous to those obtained with TRF.

   Even though LRF is more complicated than TRF, the outlook is encouraging in the study of structure-function
relationships of LRF.   Methods of synthesis of oligopeptide
analogues and bioassay methods for their biological activities are relatively rapid and ought to generate considerably
more data in the near future.   This will lead to a better
molecular interpretation of the physiology-biochemistry of
this biologically important molecule.

Table II

Luteinizing Hormone Releasing Factor Analogs

Compound	% LRF Potency*
<Glu–His–Trp–Ser–Tyr–Gly–Leu–Arg–Pro–Gly–NH$_2$	100
<Glu–His–Trp–Ser–Tyr–Gly–Leu–Arg–Pro–NH$_2$	10
<Glu–His–Trp–Ser–Tyr–Gly–Leu–Arg–NH$_2$	<0.01
<Glu–His–Trp–Ser–Tyr–Gly–Leu–NH$_2$	<0.01
<Glu–His–Trp–Ser–Tyr–Gly–NH$_2$	<0.01
<Glu–His–Trp–Ser–Tyr–NH$_2$	<0.01
<Glu–His–Trp–Ser–NH$_2$	<0.01
<Glu–His–Trp–NH$_2$	<0.01
<Glu–His–NH$_2$	<0.01
$[CH_3CH_2-\overset{\overset{O}{\|\|}}{C}-Gly^1]$–LRF	0.2
[Gly^2]–LRF	*
[Gly^3]–LRF	<0.001
[Gly^4]–LRF	1.5
[Gly^5]–LRF	0.1
[Gly^7]–LRF	0.2
[Gly^8]–LRF	0.1
[Gly^9]–LRF	0.2
des–His^2–LRF	<0.001
[Phe^2]–LRF	4
[N^π–Me–His^2]–LRF	2
[N^τ–Me–His^2]–LRF	6
[Ala^3]–LRF	<0.001
[Phe^3]–LRF	2
[Ala^6]–LRF	1

*Materials having special significance are routinely resynthesized and retested. We originally observed higher potencies for some of the analogues entered in Table II: The first batch of <Glu–His–Trp–NH$_2$ was observed to have 0.1% LRF potency. In a second preparation, it exhibits activity at a lower potency (*ca.* 0.002%).

Acknowledgment

This research supported by AID (Contract AID/csd 2785), Ford Foundation and Rockefeller Foundation.

References

1. Guillemin, R., R. Burgus, and W. Vale. Vitamins and Hormones 29, 1 (1971).
2. Sievertsson, H., J-K. Chang, K. Folkers, and C. Y. Bowers. J. Med. Chem. 15, 219 (1972).
3. Grant, G., N. Ling, J. Rivier, and W. Vale. Biochemistry 11, 3070, 1972.

COMPLETE AMINO ACID SEQUENCE OF PORCINE β-LIPOTROPIC
HORMONE (β-LPH)

Claude Gilardeau, Michel Chrétien. Laboratoire des
Protéines et des Hormones Hypophysaires, Institut de
Recherches Cliniques de Montréal

WE HAVE ALREADY SHOWN that porcine beta-lipotropic hormone
is a single chain protein of 91 amino acids and of molecular
weight 9650, with glutamic acid at its NH$_2$-terminus.[1]
Since then we have obtained the complete primary structure.
For this work, we have cleaved the molecule at its two
methionine residues with cyanogen bromide in 0.1 N hydrochloric acid as solvent. This reaction gave, after
chromatography on CM-cellulose, five fragments corresponding
to residues 1-47, 48-65, 66-91, 1-65, and 48-91. The Edman
degradation,[2] using a Beckman sequencer, was performed on
the entire molecule and on these fragments. The first 83
amino acid residues, except 79 and 80, could be deduced.
The remaining amino acids were placed after enzymic
degradations.

Peptide 66-91 was cleaved with chymotrypsin and yielded
two main products after purification by paper electrophoresis
in collidine acetate buffer and paper chromatography in
BPAW: one peptide (C$_1$) contains residues 66-78 and the
other (C$_2$), residues 79-91. This last peptide contains 4
lysines, and has a lysine at its NH$_2$-terminus, as determined
by dansylation. It was digested with trypsin and gave the
peptides in Table I.

Peptide C2T2 (80-84) has the sequence Asn-Ala-Ile-Val-
Lys since Ala-Ile-Val was determined by the sequencer and
since Asn was found to be the NH$_2$-terminal residue from
dansylation and leucine amino peptidase.

Table I

Amino Acid Analysis of Fragments From
Tryptic Digestion of Peptide C2

C2T1	Asp, Ala, His, Lys, Lys
C2T2	Asp, Ala, Ile, Val, Lys
C2T3	Gly, Glu

Only residues 85-91 remained unknown. These correspond
to peptides C2T1 and C2T3. Dansylation of peptide C2T1
gave aspartic acid; and leucine amino peptidase cleaved
mainly Asn and Ala after 8 hours of incubation. The sequence
proposed is: Asn-Ala-(His-Lys)-Lys.

The COOH-terminal residue of beta-LPH, as determined by
carboxypeptidase A digestion, is glutamine. Equimolar amounts
of glycine and lysine were also released in much smaller
quantities than glutamine after 24 hours of incubation.
Since peptide C2T3 contains Gly-Gln, the COOH-terminus must
be Lys-Gly-Gln; and since there is almost no histidine released by carboxypeptidase A, there must be two consecutive
lysines at positions 88-89.

From these data, the following sequence is proposed:
Glu-Leu-Ala-Gly-Ala-Pro-Pro-Glu-Pro-Ala-Arg-Asp-Pro-Glu-Ala-
                  5   10   15
Pro-Ala-Glu-Gly-Ala-Ala-Ala-Arg-Ala-Glu-Leu-Glu-Tyr-Gly-Leu
              20   25   30
Val-Ala-Glu-Ala-Glu-Ala-Ala-Glu-Lys-Lys-Asp-Glu-Gly-Pro-Tyr-
              35   40   45
Lys-Met-Glu-His-Phe-Arg-Trp-Gly-Ser-Pro-Pro-Lys-Asp-Lys-Arg-
              50   55   60
Tyr-Gly-Gly-Phe-Met-Thr-Ser-Glu-Lys-Ser-Gln-Thr-Pro-Leu-Val-
              65   70   75
Thr-Leu-Phe-Lys-Asn-Ala-Ile-Val-Lys-Asn-Ala-His-Lys-Lys-Gly-
              80   85   90
Gln.

Comparison of this sequence with the revised amino acid
sequence of sheep beta-LPH,[3] Table II, shows that twelve
amino acid residues are different; this number may change,
since the revision of sheep beta-LPH was only on the first
38 residues.

Table II

Comparison Between the Amino Acid Sequences
Of Porcine and Sheep Beta-LPHs

```
Pig     Glu-Leu Ala Gly Ala-Pro-Pro Glu Pro Ala-Arg Asp Pro-Glu-
Sheep   Glu-Leu Thr Gly Glu-Arg-Leu Glu Gln Ala-Arg Gly Pro-Glu-
                     5                      10


Pig     Ala Pro Ala-Glu Gly Ala-... Glu-... Val Lys-Asn Ala His-
Sheep   Ala Gln Ala-Glu Ser Ala-... Ser-... Lys Lys-Asn     His-
         15              20      42      83      85


Pig     Lys Lys-Gly-Gln
Sheep   Ala Lys-Gly-Gln
                 90
```

Acknowledgment

This work was supported by the Medical Research Council
of Canada.

References

1. Gilardeau, C., and M. Chretien. Can. J. Biochem. **48**,
    1017 (1970).
2. Edman, P., and G. Begg. European J. Biochem. **1**, 80
    (1967).
3. Chretien, M., C. Gilardeau, and C. H. Li. Int. J.
    Protein Chem. (in press).

STRUCTURE AND FUNCTION OF ADRENOCORTICOTROPIN: EFFECTS OF
THE *O*-NITROPHENYL SULFENYL DERIVATIVE OF THE HORMONE ON
ISOLATED RAT ADRENAL CELLS

J. Ramachandran, W. R. Moyle, Y. C. Kong. The Hormone
Research Laboratory, University of California, San
Francisco, California 94122

DURING THE COURSE OF STUDIES of the structure-function relationships of adrenocorticotropin (ACTH), it was observed
that chemical modification of the single tryptophan residue
in the molecule by reaction with *o*-nitrophenylsulfenyl (Nps)
chloride produced profound changes in the extra adrenal
actions of the hormone.[1-3] Nps-ACTH was found to inhibit
the ACTH induced stimulation of lipolysis in rat fat cells.[1]
On the other hand, Nps-ACTH was more potent than ACTH in
stimulating lipolysis in rabbit fat cells and in darkening
amphibian skins[2] (melanocyte stimulating activity). Nps-ACTH
inhibited the ACTH induced stimulation of adenylate cyclase
in rat fat cell ghosts[3] as well as in rat adrenal homogenate
fractions.[4] In order to elucidate the role of the tryptophan
residue in the adrenal stimulating actions of ACTH, the
effects of Nps-ACTH on steroidogenesis and cyclic AMP accumulation in isolated rat adrenal cells was investigated.

ACTH and Nps-ACTH were incubated with cells isolated
from rat adrenals by digestion with collagenase.[5] The production of corticosterone was monitored by a fluorometric
method[6] and cyclic AMP generation was measured by the protein
binding method of Gilman.[7]

From Figure 1 it is evident that both ACTH and Nps-ACTH
stimulate steroidogenesis to the same extent, although higher
concentrations of Nps-ACTH are required. The concentration
of Nps-ACTH required for half-maximal steroidogensis was
approximately seventy times that of ACTH (Table I) indicating

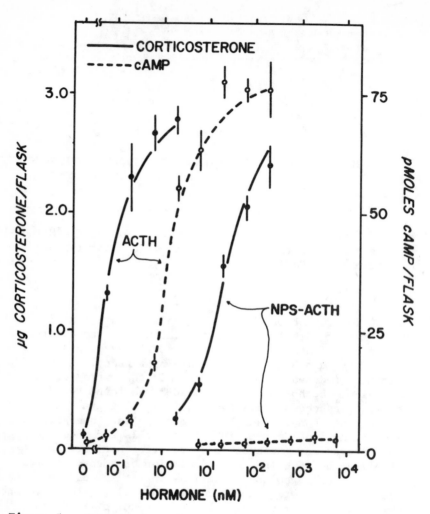

Figure 1: Stimulation of corticosterone synthesis and cyclic
  AMP accumulation in isolated rat adrenal cells in response
  to ACTH and Nps–ACTH.

an apparent decrease in the affinity of the hormone due to
modification of the tryptophan. On the other hand, Nps–ACTH
stimulated cyclic AMP accumulation marginally (3% of that
due to ACTH). In addition, Nps–ACTH inhibited competitively
the effect of ACTH on cyclic AMP accumulation. It was estimated from the inhibition studies that the apparent K_I for
Nps-ACTH was nearly equal to the concentration of ACTH

Table I

Concentrations of ACTH and Nps-ACTH Required for the
Stimulation of Steroidogenesis and Cyclic AMP in
Isolated Adrenal Cells

Hormone	Steroidogenesis*	Cyclic AMP synthesis[†]	Inhibition of cyclic AMP synthesis[#]
ACTH	0.51	3.6	
Nps-ACTH	34	39	3.8 - 5.1

The values shown refer to the average nano molar concentration
of ACTH or Nps-ACTH required to produce half-maximal stimulation of corticosterone synthesis or cyclic AMP formation.

*Both ACTH and Nps-ACTH stimulated steroidogenesis to the
same maximum.

[†]Nps-ACTH produced less than 3% of the stimulation of cyclic
AMP formation produced by ACTH at maximal effective
concentrations.

[#]Apparent K_I.

required for half-maximal stimulation of cyclic AMP production, indicating that Nps-ACTH and ACTH have the same affinity
for adrenal cells. These results suggest that the integrity
of the tryptophan residue is essential for maximal cyclic
AMP formation in adrenal cells but not for maximal steroid
synthesis. Since cyclic AMP is considered to be the intracellular messenger mediating all the actions of ACTH,[8] the
increase in intracellular cyclic AMP concentration required
for the steroidogenic response must represent a very small
fraction of the total increase in cyclic AMP concentration
produced in response to ACTH. An alternative explanation of
the results presented here would be that an unknown factor(s)
besides cyclic AMP is involved in mediating the action of
ACTH at low concentrations.

The apparently contradictory results obtained with Nps-
ACTH, namely a decrease in the affinity of the hormone
indicated by the higher concentrations required for stimulating steroidogenesis, and no change in affinity indicated
by the inhibitory effect on cyclic AMP accumulation, can be

accounted for on the basis of two types of receptors for
ACTH in the adrenal cells. Modification of the tryptophan
residue may have affected the interaction of the hormone
with the two affinity sites differently. These results
can also be explained according to the concept of spare
receptors.[9] According to this ACTH is able to stimulate
maximal steroidogenesis by interaction with only a small
per cent of the available receptors. Nps-ACTH must occupy
nearly all the available receptors to generate sufficient
cyclic AMP for the stimulation of steroidogenesis.

Acknowledgments

The authors are grateful to Professor C. H. Li for his
interest. This work was supported in part by USPHS Grant
GM-2907.

References

1. Ramachandran, J., and V. Lee. Biochem. Biophys. Res.
   Commun. 38, 507 (1970).
2. Ramachandran, J. Biochem. Biophys. Res. Commun. 41,
   353 (1970).
3. Ramachandran, J., and V. Lee. Biochem. Biophys. Res.
   Commun. 41, 358 (1970).
4. Ramachandran, J. In *Hormonal Proteins and Peptides,*
   Vol. II, Li, C. H., ed. (New York: Academic Press,
   1972).
5. Moyle, W. R., Y. C. Kong, and J. Ramachandran. Submitted
   for publication.
6. Peterson, R. E. J. Biol. Chem. 225, 25 (1957).
7. Gilman, G. G. Proc. Nat. Acad. Sci. U.S. 67, 305 (1970).
8. Robison, G. A., R. W. Butcher, and E. W. Sutherland.
   Cyclic AMP (New York: Academic Press, 1971).
9. Stephenson, R. P. Brit. J. Pharmacol. Chemotherap. 11,
   379 (1956).

CYCLIC AMP AND THE MECHANISM OF ACTION OF GASTROINTESTINAL
HORMONES

M. Samir Amer, Gordon R. McKinney. Department of
Pharmacology, Mead Johnson Research Center, Evansville,
Indiana 47721

ALTHOUGH CYCLIC AMP HAS BEEN convincingly shown to mediate
the actions of a large number of hormones, its role in the
mechanism of action of most gastrointestinal hormones is
not clear. Studies from this laboratory indicated that
gastrin, cholecystokinin and related peptides and insulin
stimulate cyclic AMP phosphodiesterase (PDE) in their target
tissues in the rabbit, Table I. Stimulation of PDE was
brought about in every case by a shift in activity from a
soluble, high Km (low affinity) to a low Km (high affinity),
and possibly particulate, form of the enzyme. The effects
of these hormones on PDE would be expected to result in a
decrease in the intracellular levels of cyclic AMP as was
shown to be the case with insulin in liver and cholecystokinin in the gall bladder. Thus it would seem that stimulation of PDE is a common mechanism for the action of a
number of gastrointestinal hormones to lower the intracellular levels of cyclic AMP. This is in contrast to the
actions of secretin and glucagon in elevating intracellular
cyclic AMP levels via stimulation of adenylate cyclase.
Thus PDE may be a site for hormonal regulation of cyclic
AMP levels rather than a switch for termination of cyclic
AMP effects.

A careful examination of the literature relating cyclic
AMP to gastric secretion indicates that this mechanism may
also mediate the effects of these hormones on gastric
secretion and that gastric secretion, if mediated by cyclic
AMP, is triggered by a decrease rather than an increase in

TABLE I

Interrelationships of Various Factors, Cyclic AMP
and Gastric Secretion

	PDE*	AC†	cAMP Levels	Gastric Secretion	Smooth Muscle
Cyclic AMP & dbAMP#				↓[1]	↓
Prostaglandin E_1		↑[2]	↑	↓[3]	↑↓
Glucagon		↑	↑	↓[4]	↓
Secretin		↑[5]	↑	↓[6]	↓
β-adrenergic stimulants		↑	↑	↓[7]	↓
Serotonin		↑	↑	↓[8]	↓
Papaverine	↓[9]		↑	↓–[10,11]	↓
Acetylcholine		↓–	↓–	↑	↑
Theophylline and methylxanthines	↓		↑	↑[12]	↓
Imidazole	↑[13]		↓↑[14]	↓[15]	↑
Histamine	↑[16,17]	↑–[18,19]	↓↑	↑	↑
Gastrin	↑[14]	–[18]	↓	↑[20,21]	↑
Cholecystokinin	↑[14]	–	↓	↑	↑
Cerulein & related peptides	↑[14]	–	↓	↑[22]	↑
Insulin	↑[23]	–	↓[24]	↑[25]	
α-adrenergic stimulants	↑[26]	–	↓[27]	↑[7]	↑[28]
Ca^{++}	↑↓	↓	↓↑	↑	↑
Balloon				↑	↑
Phenothiazines and imipramine	↓[29]		↑[30]	↓[31]	↓

*3',5'-Cyclic adenosine monophosphate phosphodiesterase
†Adenylate cyclase
#Dibutyryl cyclic AMP

↑ = increase   ↓ = decrease   – = no effect

the intracellular concentrations of the cyclic nucleotide.
Table I summarizes the effects of a large number of agents
on gastric secretion and their known or expected effects on
cyclic AMP levels. It is clear from the table that an
inverse relationship exists between cyclic AMP levels and
gastric secretion and smooth muscle tone. The notable exception is in the case of methylxanthines which do have
other effects on mucosal blood flow and permeability to
Ca^{++} that can account for their unexpected, stimulatory
effects on gastric secretion.

Stimulation of PDE may therefore mediate the gastric-
stimulatory effects of gastrointestinal hormones. It
further provides the basis of the inhibitory effects of
secretin and glucagon on gastric secretion and their
antagonism to the stimulatory effects of gastrin,
cholecystokinin and related peptides.

References

1. Levine, R. A., E. P. Cafferata, and E. F. McNally.
   Rec. Adv. Gastroenterol. 1, 408 (1967).
2. Bennett, A., and B. Fleshler. Gastroenteroloty 59,
   790 (1970).
3. Way, L., and R. P. Durbin. Nature 221, 874 (1969).
4. Lin, Tsung-Min, and M. W. Warrick. Gastroenterology
   61, 328 (1971).
5. Case, R. M., T. J. Laundy, and T. Scratcherd. J.
   Physiol. 204, 45p (1969).
6. Johnson, L. R., and M. I. Grossman. Amer. J. Physiol.
   217, 1401 (1969).
7. Misher, A., R. G. Pendleton, and R. Staples.
   Gastroenterology 57, 294 (1969).
8. Thompson, J. H. Research Comm. in Chem. Pathol. &
   Pharmacol. 2, 687 (1971).
9. Poch, G., and W. R. Kukovetz. Life Sciences 10, 133
   (1971).
10. Shanbour, L. L., C. C. Mao, D. S. Hodgins, and E. D.
    Jacobson. Gastroenterology 60, 716 (1971).
11. Blum, H., E. Mutschiller, and P. A. van Zwieten. Arch.
    Pharmak. Exp. Path. 256, 99 (1967).
12. Robertson, C. R., C. E. Rosiere, D. Blickenstaff, and
    M. I. Grossman. J. Pharmacol. Exptl. Therap. 99, 362
    (1950).
13. Robison, G. A., R. W. Butcher, and E. W. Sutherland.
    Ann. Rev. Biochem. 37, 149 (1968).
14. Amer, M. S., and G. R. McKinney. J. Pharmacol. Exptl.
    Therap. In press (1972).

15. Harris, J. B., and W. Silen. Fed. Proc. 23, 214 (1964).
16. Amer, M. S. Abstracts of A. Ph. A. Academy of
    Pharmaceutical Sciences 1, 120 (1971).
17. Honeyman, T., and H. M. Goodman. Fed. Proc. 30, 435
    (1971).
18. Perrier, C. V., and L. Laster. Clin. Res. 17, 596
    (1969).
19. Mao, C. C., L. L. Shanbour, D. S. Hodgins, and E. D.
    Jacobson. J. Lab. Clin. Med. 78, 830 (1971).
20. Gregory, H. Amer. J. Dig. Dis. 15, 141 (1970).
21. Makhlouf, G. M. Fed. Proc. 27, 1322 (1968).
22. Bertaccini, G., R. Endean, V. Erspamer, and M.
    Impicciatore. Brit. J. Pharmacol. Chemotherap. 34,
    311 (1968).
23. Senft, G., G. Schultz, K. Munske, and M. Hoffman.
    Diabetologia 4,(1968).
24. Exton, J. H., and C. R. Park. Advances Enzym. Regulat.
    5, 391 (1968).
25. Kronborg, O. Scand. J. Gastroent. 5, 481 (1970).
26. Amer, M. S. Fed. Proc. 30, 220 (1971).
27. Turtle, J. R., and D. M. Kipnis. Biochem. Biophys.
    Res. Comm. 28, 797 (1967).
28. Gagnon, D. J. Europ. J. Pharmacol. 10, 297 (1970).
29. Honda, F., and H. Imamura. Biochem. Biophys. Acta
    161, 267 (1968).
30. Abdulla, Y. H., and K. Hamadah. The Lancet 1, 378
    (1970).
31. Hano, J., and J. Bugajski. Dissert. Pharm. Pharmacol.
    21, 289 (1969).

PROPOSED MECHANISMS OF ACTION OF 3', 5'-CYCLIC AMP, SOME
CYCLIC PEPTIDES AND RELATED COMPOUNDS

G. Moll, E. T. Kaiser. Departments of Biochemistry
and Chemistry, University of Chicago, Chicago, Illinois
60637

THE SIX-MEMBERED CYCLIC PHOSPHATE DIESTER 3',5'-cyclic AMP
is thought to act as a second messenger in the action of
many hormones.[1] The possibility that cyclic AMP may owe
its effectiveness as a messenger in hormone action to its
potential ability to phosphorylate enzymes reversibly in a
manner similar to that described earlier for the "autore-
generative" interaction of the highly strained cyclic
phosphate diester, catechol cyclic phosphate,[2] with
chymotrypsin is under investigation in our laboratory.
The generalized reaction pathway we postulate for the
interaction of cyclic AMP with enzymes is illustrated in
Equation 1 below. Cyclic AMP may react with enzymes either

at their active sites or at regulatory sites with concomitant ring-opening to produce covalent phosphoryl-enzyme species.[2],[3] Under appropriate conditions where kinetic control is favored, the hydroxyl group produced by ring-opening which remains covalently bound in close proximity to the phosphoryl function can act as an intramolecular nucleophile, attacking the phosphorus, blocking the attack of water (which would cause the destruction of cyclic AMP), and causing the re-formation of cyclic AMP with the release of free enzyme. The relative concentrations of the species involved in the equilibria of Equation 1 should be affected by changes in pH, perhaps because of changes in the ionization state of important enzyme functional groups, and by other factors such as metal ion concentrations. Thus, it can be seen how cyclic AMP could regulate the action of many enzymes in a very effective way.

We have performed experiments on the denaturation of bovine skeletal muscle protein kinase in the presence of labeled cyclic AMP which indicate that a substantial amount of the labeled material remains bound to the enzyme even after denaturation. In addition, our studies on the enzymatic digestion of the complex formed between cyclic AMP and bovine brain protein kinase can be best accounted for in terms of the covalent binding of cyclic AMP to the enzyme as postulated in the scheme of Equation 1.

The analogy between the proximity effects of the newly generated hydroxyl groups produced by reactions like that shown in Equation 1 and the effect of the amino group of the newly formed amino terminal acid present in the acyl-trypsin resulting from the interaction of soybean trypsin inhibitor and trypsin has been noted.[4],[5]

Proceeding further, we have been considering the possibility that various cyclic peptides, peptides containing lactam rings, or related compounds, may exert their influence on biological reactions by processes similar to those indicated in Equation 1 above. This is illustrated in Equation 2 where the hypothetical interaction of thyrotropin-releasing hormone[6] with a protein receptor is considered. According to this hypothesis, the lactam ring present in the hormone is cleaved on reaction with a protein receptor, generating an acylated protein and releasing an amino group which remains in proximity to the acyl function and maintains the equilibrium between the intact hormone and unmodified protein species on the one hand and the modified protein on the other.

$$P + \quad \begin{array}{c} CH_2\!-\!CH_2 \\ O\!=\!C \quad CH\!-\!C\!-\!NH\!-\!CH\!-\!C\!-\!N \\ N \quad\quad CH_2 \\ H \quad HC\!=\!C \quad CONH_2 \\ N \quad NH \\ CH \end{array} \begin{array}{c} CH_2\!-\!CH_2 \\ CH\!-\!CH_2 \end{array}$$

$$\updownarrow$$

$$(2)$$

$$P\!-\!C \quad \begin{array}{c} CH_2\!-\!CH_2 \\ CH\!-\!C\!-\!NH\!-\!CH\!-\!C\!-\!N \\ H_2N \quad CH_2 \\ NC\!=\!C \quad CONH_2 \\ N \quad NH \\ CH \end{array} \begin{array}{c} CH_2\!-\!CH_2 \\ CH\!-\!CH_2 \end{array}$$

Acknowledgment

The support of this research by American Cancer Society Institutional Grant Number IN-41K, a USPHS Medical Scientist Traineeship for G. Moll, and an Alfred P. Sloan Foundation Fellowship for E. T. Kaiser is gratefully acknowledged.

References

1. Robison, G. A., R. W. Butcher, and E. W. Sutherland. *Cyclic AMP* (New York: Academic Press, 1971).
2. Kaiser, E. T., T. W. S. Lee, and F. P. Boer. J. Amer. Chem. Soc. 93, 2351 (1971).
3. Greengard, P., and J. F. Kuo. *Advances in Biochemical Psychopharmacology*, Vol. 3 (New York: Raven Press, 1970). p 287.
4. Heidema, J. H., and E. T. Kaiser. J. Amer. Chem. Soc. 92, 6050 (1970).
5. Hixson, H. F., Jr., and M. Laskowski, Jr. J. Biol. Chem. 245, 2027 (1970).
6. Bøler, J., F. Enzmann, K. Folkers, C. Y. Bowers, and A. V. Schally. Biochem. Biophys. Res. Commun. 37, 705 (1969).

Summarized by Johannes Meienhofer

HYPOTHALAMIC PEPTIDES continue to be a topic of very great interest, and the difficulties in identifying and isolating these hormonal control factors are not only due to their minute quantities in tissues, but also to bioassay problems. A discourse on the melanotropin (MSH)-release inhibiting factor,[1-3] MIRF, centered around complications with bioassays in frogs or toads itself and with comparing data from different laboratories. Both the C-terminal tripeptide of oxytocin, Pro-Leu-Gly-NH$_2$, and the ring moiety, tocinoic acid, have been implicated as inhibiting agent, and the N-terminal pentapeptide, Cys-Tyr-Ile-Gln-Asn, has been proposed as possible releasing factor. A question was raised about bioassay data on synthetic analogs of luteinizing hormone releasing factor (pp 601 to 608) which range in potency over 4-5 orders of magnitude. Concern about activities in the range of 0.001% to 0.005% pertains not only to the significance of differences between analog activities in that range, but also to the homogeneity of these decapeptides that have been synthesized by solid-phase procedures. A preparation only needs to contain 0.1% of an impurity (failure sequence, derivative, diastereoisomer, etc.) possessing 1% of the potency of LH-RH to exhibit 0.001% of activity. In reply to some questions it was pointed out that, in general, dose-response curves showed the same slope except for those analogs with inhibitory action, and that the ratios of LH-RH to FSH-RH activities remained constant in all analogs tested. The significance of the C-terminal amide, common to hypothalamic factors, along with the pyroglutamyl N-terminus, was discussed. It might be a protective factor (to get through membranes in a certain way, or to be

resistant to enzymes), or it might indicate certain possible
biosynthetic origins (Gross, pp 671 to 678).

A question was raised whether the β-lipotropic hormone
could actually be regarded as a "hormone." In reply, it
was pointed out that a related polypeptide from sheep
pituitaries had been shown some time ago to possess activity on rabbit and rat fat cells;[4] apparently no further
bioassays have yet been done on porcine β-LPH (pp 609 to 611).

The report on the effects of the *o*-nitrophenylsulfenyl
derivative of ACTH aroused a lively discussion centering
around the use of isolated adenyl cyclase, the levels of
cyclic AMP employed in this work, and the possibility of
increased metabolism of cyclic AMP by phosphodiesterase as
a cause of its disappearance. Several arguments developed
about proposed new mechanisms of action of cyclic AMP
(Kaiser, pp 621 to 623).

References

1. Celis, M. E., S. Taleisnik, and R. Walter. Proc. Nat.
   Acad. Sci. U.S. <u>68</u>, 1428 (1971).
2. Nair, R. M. G., A. J. Kastin, and A. V. Schally.
   Biochem. Biophys. Res. Commun. <u>43</u>, 1376 (1971).
3. Bower, S. A., M. E. Hadley, and V. J. Hruby. Biochem.
   Biophys. Res. Commun. <u>45</u>, 1185 (1971).
4. Li, C. H. Nature <u>201</u>, 924 (1964).

SECTION IX

ANALYTICAL TECHNIQUES

Session Chairmen

Herbert Zuber and Lyman C. Craig

THE PRECISION AND SENSITIVITY OF AMINO ACID ANALYSIS

Stanford Moore. The Rockefeller University, New York,
New York 10021

SUMMARY--Amino acid analysis bears a relationship to the
chemistry of proteins and peptides similar to that which
elementary analysis bears to the chemistry of simpler
organic molecules. The subject is reviewed in terms of
the resolution achievable by automatic ion exchange
chromatography, the variations in the speed of the analysis,
the range in sample size from 1 micromole to 1 nanomole,
methods of integration, and the precision obtainable under
practical conditions. With well-standardized instruments
and optimum resolution, 3% is a reasonable maximum deviation
from theory in the recovery of pure amino acids in a single
run; the borderline separation with some 2-hour methodologies
raises the maximum error to about 5%.

 This precision does not include the errors inherent in
the hydrolysis of the protein or the peptide to the constituent amino acids. With 6 *N* HCl at 110° the small losses
of serine and threonine can be corrected for with adequate
precision; cystine or cysteine can be most precisely determined after derivatization. Tryptophan, which is labile in
HCl, is a special case which has led to the finding that
methanesulfonic acid may be generally preferable to HCl for
acid hydrolysis. Alkaline hydrolysis is the only method
that provides quantitative recoveries of tryptophan when
carbohydrates are present in the sample. D- and L-amino
acids can be determined by chromatographic separation of
the diastereoisomeric dipeptides obtained by condensation
with L-leucine or L-glutamic acid *N*-carboxyanhydride.

 Examples are given of the approach to integral molar
ratios of amino acids obtainable in routine analyses of
chromatographically purified peptides.

INTRODUCTION--THE SUBJECT OF THIS CONTRIBUTION was suggested
by Ralph Hirschmann. In thinking about the plans for this
symposium, he felt that it would be practical to have a
discussion of how close to theory one can expect to come
in the amino acid analysis of a pure synthetic peptide. Or,
alternatively, to what extent does a given departure from
theory indicate impurity in the peptide or possible errors
in the analysis? We will attempt to answer these questions
in terms of current practice.

With a small synthetic peptide C, H, and N analysis is
a valuable criterion in the characterization of the preparation

Carbobenzoxy-leucyl-alanine methyl ester:[1]
$C_{18}H_{26}N_2O_5$. Calculated C 61.71 H 7.45 N 8.00
                Found   C 61.64 H 7.37 N 8.17

The long history of organic microanalysis provides the
chemist with guidelines on how close he can expect to come
to theory. But with large peptides, such as bovine pancreatic ribonuclease, a recent subject of synthesis,[2-4]
elementary analysis for C, H, and N remains desirable but
is less informative and the molecule is best studied in
terms of the component amino acids, in terms of the 124
amino acid residues of (in this instance) seventeen types.

Bovine pancreatic ribonuclease A:[5,6]
$C_{587}H_{909}O_{197}N_{171}S_{12}$

$Asp_{15}Glu_{12}Gly_3Ala_{12}Val_9Leu_2Ile_3Ser_{15}Thr_{10}Met_4$-
$Pro_4Phe_3Tyr_6His_4Lys_{10}Arg_4 1/2\ Cys_8Amide(NH_3)_{17}$

In elementary analysis errors can arise from inadequacies
in the combustion or the digestion and in the measurement
of the end products. Similarly, in amino acid analysis,
the sources of error are in the hydrolysis of the peptide
chain and in the determination of the resulting individual
amino acids. First let us consider the precision with
which the amino acids can be separated and measured and
then turn to the problems inherent in the preliminary step
of hydrolysis.

Automatic Amino Acid Analyzers

Present automatic amino acid analyzers owe their genesis
to the renaissance in chromatography stimulated by Martin
and Synge. If we had to follow the tradition of elementary
analysis and use specific procedures for each amino acid,
the analysis of a protein would be a very tedious and time
consuming process, as it was for Erwin Brand's[7] analysis

of β-lactoglobulin in 1946. Chromatography gave promise
of providing a physical method which would separate all of
the amino acids of a protein hydrolysate by a single technique. In the 1940's William Stein and I undertook to
develop starch columns with alcohol-water eluents[8],[9] into
a system which would give quantitative recovery of each
amino acid added to the column (Figure 1). We are not

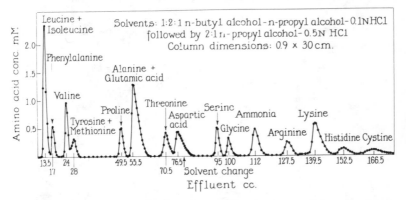

Figure 1: Chromatography of a hydrolysate of bovine serum
  albumin on a starch column with butyl alcohol-*n*-propyl
  alcohol-aqueous HCl solvent systems (1949).[10] The
  sample corresponded to about 2.5 mg of protein.

going to burden you with much ancient history in this
introduction, but it is encouraging to note how, over the
years, the results from many academic and industrial
laboratories have speeded up the process of amino acid
chromatography. When we obtained, in 1949, the type of
quantitative chromatogram[10] shown in Figure 1, with about
1 micromole of amino acid per peak, we were very happy to
be able to analyze a protein hydrolysate in about ten
days by running three such columns to resolve all overlaps.
The main features of this type of experiment were the collection of effluent fractions of precise volume[8] by an
automatic fraction collector (Figure 2) and photometric
measurement of the concentrations of the amino acids by a
quantitative version[11]-[13] of the ninhydrin reaction (Figure
3), the product from α-NH$_2$ acids being read at 570 mμ and
those from proline and hydroxyproline at 440 mμ. In the
1950's William Stein and I speeded the process up to a
five-day run (Figure 4) by turning to ion exchange

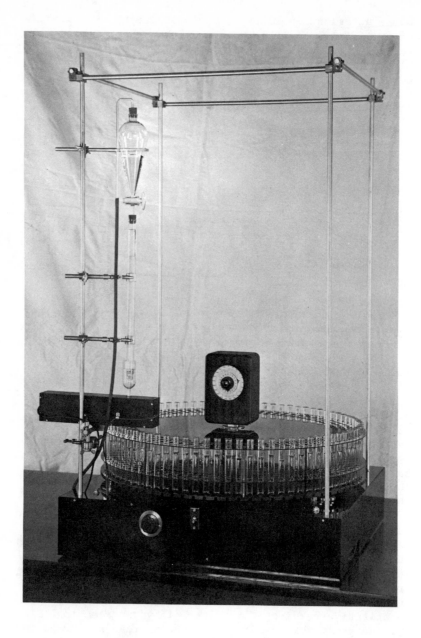

Figure 2: Automatic fraction collector for chromatographic
analysis with volume control by the counting of drops
(1948).[8]

$$R-\underset{\underset{NH_2}{|}}{\overset{\overset{H}{|}}{C}}-COOH + 2 \left[\text{Ninhydrin} \right] \xrightarrow[\substack{\text{pH 5} \\ 100^\circ C.}]{\substack{\text{Na citrate} \\ \text{buffer}}} \left[C-N=C \right] + CO_2 \uparrow + R-\overset{\overset{H}{|}}{C}=O$$

| α-NH$_2$ acid | Ninhydrin | | Diketohydrindylidene-diketohydrindamine | Carbon dioxide | Aldehyde |

Figure 3: The reaction of ninhydrin with α-amino acids as applied in the quantitative photometric analysis[11-13] of eluents from chromatograms.

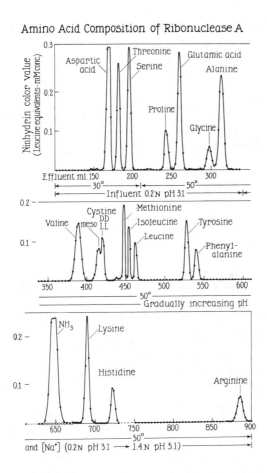

Amino Acid Composition of Ribonuclease A

Figure 4: Chromatographic analysis of a hydrolysate of 2 mg of bovine pancreatic ribonuclease A[16] on a 150 x 0.9 cm column of Dowex 50-X4 (1954).[15]

chromatography[14],[15] on 9 mm bore columns of sulfonated
polystyrene resin. By the 1960's, in cooperation with
Darrel Spackman, the process was automated[17] (Figure 5)
to give recorded curves, and (Figure 6) the speed was
increased[17],[18] to provide an overnight run with 150 and
50 cm columns.

In reference to Figure 6 we should mention the reserve
resolving power represented by the spaces between the peaks
on that chromatogram. When you analyze a synthetic peptide,
you have a fair idea of what amino acids to expect; you can
be reasonably sure that ninhydrin-positive material eluted
near the lysine position is lysine. But with naturally
occurring peptides and proteins there are additional con-
stituents to consider; for example, histones contain N^ϵ-
methyllysines,[19] actin and myosin contain 3-methylhistidine,[20]
and many microbial peptides (*e.g.* edeine[21]) have unusual
amino acids. As we turn to shorter columns, which can be
adequate for peptides or proteins of known qualitative
composition, we need to keep in mind that the analysis is
no sounder than the identification of the peaks and that
even higher resolving power, such as that developed for
analysis of physiological fluids,[17] has a role in research
on a new product.

An example of accelerated analysis [22] with 55 and 10
cm columns is given in Figure 7; the run time is about 5
hours. This particular chromatogram was obtained with
Beckman AA-15 and AA-27 spherical resins at a load of
200 nanomoles. With range cards on the recorders, the
Beckman 120-series analyzers give curves very similar to
this one at 25 nanomoles.

The manufacturers of the many commercial instruments
for this purpose have introduced their own important variations in the ion exchange resins, the ninhydrin reagent,
and the automation, with further decreases in the run times.

The latest innovation (Figure 8) is a 70-minute analysis
on a computer-operated instrument utilizing a 1.75 mm
diameter column packed with 8 micron beads of resin and
operated at a pressure of 2500 p.s.i. The instrument,
involving a number of new principles of operation, has
been designed by the Durrum Instrument Company, and draws
upon the experience of Paul Hamilton[23] and of Edgar Hare[24]
in the use of capillary columns. The narrower bore helps
to make the analysis more micro; the curve in Figure 8 was
obtained with 10 nanomoles of each amino acid. About 20
µg of protein are required for such an analysis. The
precision at the 1-nanomole level is adequate for some
purposes.

Figure 5: Automatic recording apparatus for the chromatography of amino acids (1958).[17]

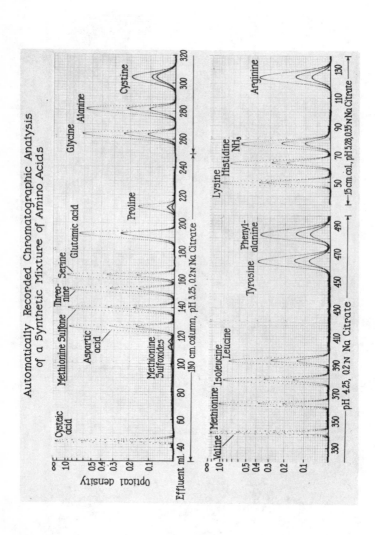

Figure 6: Chromatographic analysis of a mixture of amino acids automatically recorded in 22 hours (1958).[17] Load, 1 micromole per amino acid.

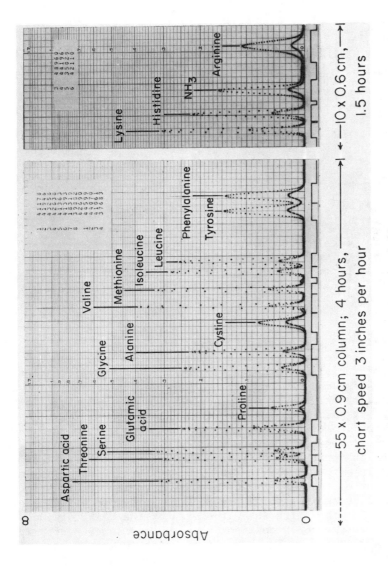

Figure 7: An accelerated 5.5 hour analysis (1963)[22] with 55 and 10 cm columns of Beckman AA-15 and AA-27 spherical resins operating at about 200 p.s.i. at 50 ml per hour. Load, 200 nanomoles per amino acid.

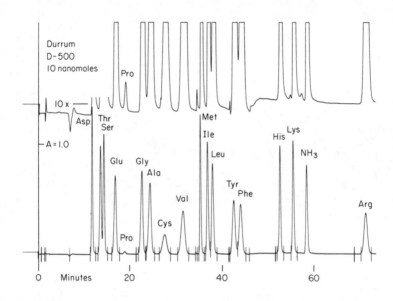

Figure 8: Accelerated 70-minute ultramicroanalysis on the
Durrum D-500 analyzer operating at 2500 p.s.i. (1972).
Load, 10 nanomoles per amino acid. The upper curve is a
10X amplification of the lower curve.

Elution is performed with three buffers from a single
column. The lower curve is for the range from zero to 1.0
absorbance units, to mid-scale. The present model does not
have a 440 mμ channel; proline is currently measured
manually (by height) at 590 mμ using the 10X-amplification
recorded by the upper curve on this figure.

Of key importance to the present subject is the need for
automatic integration when rapid analyses are performed, as
shown by the print-out in Figure 9. Integration of peaks
by hand is a burden when an analyzer is turning out a curve
every hour or so. The D-500 (reading from the right) prints
the baseline against which the peak is integrated, the color
constant for the amino acid, the area of the peak, divides
the area by the color constant to give the quantity of amino
acid (in per cent of 10 nanomoles in this example), and
records the name of the amino acid and its elution time.

With automatic integrators that plot or print the baseline against which the peaks are read and mark the beginning

PRINT-OUT FROM DURRUM D-500 ANALYZER

=
2/11/72 STD SOLUTION 10 NM
PROC=   1   RUN=   4

PEAK	MIN-SEC		NAME	QUANT	TYPE	AREA	FACTOR	BASE
1	12	2	ASP	100.5	0	27190	27046	141
2	13	50	THR	100.3	1	27618	27543	141
3	14	34	SER	101.2	2	29501	29143	141
4	17	2	GLU	100.8	0	28872	28650	141
5	22	54	GLY	100.8	0	30125	29882	140
6	24	39	ALA	101.1	0	32016	31683	140
7	27	50	CYS	101.8	0	14450	14198	140
8	31	47	VAL	101.9	0	28747	28214	140
9	35	29	MET	101.2	0	28187	27845	140
10	37	3	ILE	101.1	1	29908	29594	140
11	38	7	LEU	101.3	2	30269	29885	140
12	42	46	TYR	101.8	1	25867	25413	140
13	44	15	PHE	101.6	2	25492	25100	140
14	52	59	HIS	100.8	0	26865	26646	163
15	55	44	LYS	101.1	0	29969	29640	163
16	71	31	ARG	101.0	0	24207	23958	156

          75 50
ERR=   0

Figure 9: Print-out from the Durrum D-500 analyzer for the chromatogram illustrated in Figure 8. The third column gives the quantity of amino acid as per cent of 10 nanomoles, in this example.

and end of each integration, it is possible to inspect the curve and make rapid graphical corrections if baseline shifts have not been fully accounted for. The use of a computer to handle the data from the spectrophotometer of an amino acid analyzer offers refinements in data handling that are being explored in several laboratories and by the manufacturers of such equipment.

The Precision of the Analyses

We showed,[14] in our first manual methods, that with the stable amino acids all of the amino acid that is placed on the top of the ion exchange column comes out in the peak at pH values below 7. The precision of the analysis depends upon the reproducibility of the photometric measurement, the

constancy of the flow rates in an on-stream analyzer, the
sizes of the peaks, the steadiness of the baseline, the
precision of the standard solution used for calibration,
and the integration of the curves. With manual integration
(height times width) and the original overnight run[17] in
which resolution was optimum, the maximum deviation from
theory was about 3% over a range going down to about 1/10th
of full chart scale.

Spackman[25] has shown that this precision still holds
with 5-hour accelerated systems but increases to a maximum
deviation of about 5% for the 2-hour methodologies introduced in the 1960's. The analysis integrated in Figure 9,
however, represents the result of an accelerated, high-
pressure system that does not sacrifice resolution.

With any of these methods, if you run standards before
or after the unknowns and average duplicate or triplicate
analyses, the results can be closer to 100%, as Spackman[25]
has shown in his discussion of the standard deviation of
replicate runs.

Now let us turn to some actual results and to some of
the problems of the hydrolysis step. Table I, obtained by
Crestfield *et al.*[26] with the initial Spackman *et al.*[17]
instrument, provides a practical example of some of the
problems encountered in the analysis of a derivative of
chromatographically purified natural ribonuclease A. With
hydrolysis by 6 N HCl at 110° in evacuated tubes, it is
usually rather precise to correct serine by 10% and threonine
by 5% for the deamination occurring in 20-24 hours. For
example, here, with reduced and carboxymethylated
ribonuclease, in which the half-cystine residues are
carboxymethylcysteine, if you add 10% to 13.6 you obtain
exactly the theory of 15.0 residues for the relative molar
amount of serine. If you add 5% to 9.48 you have the 10.0
residues of threonine. In principle it is of course
sounder to determine the rate of destruction under the
precise conditions of hydrolysis, as has been done here,
by varying the hydrolysis time from 24, to 48, 73, and 102
hours and extrapolating to zero time. (For the stable
amino acids, this table also gives an example of the
reproducibility of replicate analyses.)

The small decrease in tyrosine with time, which occurred
in this instance but is not always observed, can be insured
against by including 0.2% phenol[27-29] in the 6 N HCl and
the recovery of carboxymethylcysteine is made more certain
by the inclusion of 0.1% mercaptoacetic acid. These pre-
cautions are especially important when peptides have been
eluted from paper. Such additions can complicate the

Table I

Reduced and Carboxymethylated Ribonuclease A
From Crestfield *et al.*[26]

Amino acid	Theory	Found	Hydrolysis time in hours			
			24	48	73	102
Cys(Cm)	8	8.1	7.77	8.25	8.21	8.07
Asp	15	15.2	15.4	15.5	14.8	15.1
Glu	12	12.1	12.1	11.8	12.2	12.3
Gly	3	3.1	3.05	3.22	3.09	3.09
Ala	12	12.0	12.0	12.0	12.0	12.0
Val	9	8.8	8.90	8.92	8.64	8.70
Leu	2	2.0	2.01	1.93	1.94	1.99
Ile	3	2.8*	2.18	2.54	2.69	2.81
Ser	15	15.2†	13.6	12.4	10.9	9.69
Thr	10	9.8†	9.48	9.20	8.84	8.35
Met	4	3.7	3.71	3.66	3.84	3.76
Pro	4	3.9	3.98	3.80	3.82	3.99
Phe	3	2.9	2.95	2.91	2.97	2.91
Tyr	6	5.9†	5.70	5.54	5.36	5.25
His	4	3.9	3.84	3.89	3.91	4.00
Lys	10	10.0	10.2	10.0	9.50	10.2
Arg	4	4.0	4.00	4.00	4.00	4.00

*102-hour value.

†Extrapolated to zero time.

chromatogram, however, if cystine or cysteine is present. Cysteine is eluted at the proline position but can be air-oxidized to cystine after hydrolysis;[30] mercaptoacetic acid also causes small peaks under aspartic acid and near threonine.

The longer time of hydrolysis with HCl is needed with ribonuclease for another practical reason. The protein contains an isoleucyl-isoleucyl sequence[5] which is very slow to hydrolyze; 102 hours at 110° are required to give nearly three residues. Whenever isoleucyl-isoleucyl, valyl-valyl, or isoleucyl-valyl sequences are present this problem arises.

The precision of some of the other values in this table depends upon careful deaeration of the samples to below 50 microns before they are sealed under vacuum.[30] The largest deviation from theory in this table is for methionine which is 8% low, but that result is real; the chromatograms[26] showed traces of homoserine indicating a few per cent of alkylation of methionine by iodoacetate during the derivatization.

An alternative to the determination of half-cystine or cysteine as carboxymethylcysteine is to oxidize the peptide or protein with performic acid and measure cysteic acid;[31] methionine is simultaneously determined as the sulfone. The determination as S-sulfocysteine has also recently been studied.[32]

The analysis in Table I is calculated on the basis of molar ratios, taking alanine as 12 and arginine as 4. Usually the calculation of molar ratios is best based upon the average of the results for several of the stable amino acids (including glutamic acid, aspartic acid, alanine, and leucine). When a protein of unknown composition is being analyzed, the key calculation is the correlation of the molecular weight with integral numbers of those residues present in small molar proportions.

Purity on a weight basis can be fundamental in peptide chemistry, as Johannes Meienhofer and Yoshimoto Sano[33] have emphasized. If a weight recovery is desired, the protein or peptide needs to be carefully desalted, usually by gel filtration on Sephadex G-25 in 50% acetic acid. Samples of the lyophilized product are weighed for amino acid analysis and for moisture and residual ash determinations. For example, an analysis of streptococcal proteinase,[34] referred to a sample dried to constant weight in vacuo at 100° over P_2O_5, accounted for 98% of the weight and 99% of the nitrogen of the sample.

And a discussion of the relationship of amino acid analysis and purity would not be complete without including the caution emphasized by Klaus Hofmann and his associates[35] that an amino acid analysis in agreement with theory is a necessary but not in itself a sufficient criterion for the purity of a synthetic peptide. Evidence for homogeneity by separation methods of adequate resolving power remains essential.

Most proteins contain tryptophan, which happens to be absent in ribonuclease, and which presents a special problem. It would be a great advantage to have a method of acid hydrolysis that would yield all of the amino acids, including tryptophan, which is labile under the usual conditions of hydrolysis in HCl. Hiroshi Matsubara and Richard Sasaki[36] found that the addition of 2 to 4% mercaptoacetic acid to 6 N HCl increased the recoveries of tryptophan to about 90% when carbohydrate was absent.

Teh-Yung Liu and Y. C. Chang[37] have recently made the important observation that when carbohydrate is absent, tryptophan undergoes only slight decomposition when 3 N p-toluenesulfonic acid is substituted for HCl and 0.2% tryptamine is included as a protectant. This method gave more than 90% recovery of tryptophan from ten proteins in 22 hours and multiple times of hydrolysis were used to obtain closer values by extrapolation to zero time. Teh-Yung Liu has subsequently found (Table II, personal communication) that he obtains even better results with

Table II

Hydrolysis of Proteins with 4 N Methanesulfonic Acid
Containing 0.2% 3-(2-aminoethyl)indole at 115° for 24 hours
From Teh-Yung Liu (personal communication; *cf.*[37])
(Theoretical integral residue numbers are given in parentheses)

Amino acid	Lysozyme	Chymo-trypsinogen	Myoglobin	Pepsinogen
Tryptophan	5.95(6)	7.95(8)	1.90(2)	5.13(5)
Lysine	6.00(6)	14.00(14)	19.00(19)	11.00(11)
Histidine	1.05(1)	2.04(2)	11.68(12)	3.19(3)
Arginine	11.25(11)	3.91(4)	3.88(4)	4.12(4)

4 *N* methanesulfonic acid (1 ml containing tryptamine, subsequently neutralized with 1 ml of 3.5 *N* NaOH, and made to
volume of 5 ml). At 115° he obtains 5.95 residues out of
6 in lysozyme, and similarly excellent results with
chymotrypsinogen, myoglobin, and pepsinogen. He reports
that by raising the temperature to 125° valine and isoleucine can be nearly completely liberated from their
resistant linkages in 20 hours. It may well be that
Teh-Yung Liu has arrived at the acid hydrolysis method of
the future for carbohydrate-free peptides and proteins.
One disadvantage is that the acid is not volatile and the
neutralized hydrolysate ends up as a more dilute solution
of amino acids than need be the case when HCl is used.
When an ultramicro analyzer is employed, such as the
Durrum D-500 instrument, which can take only a 40 μl
sample, only a small percentage of the hydrolyzed peptide
or protein can currently be utilized for amino acid analysis. This is only a problem when conservation of sample
is critical. An important limitation of the procedure is
that in the sulfonic acid solutions tryptophan remains
sensitive to the presence of carbohydrate; when 5% or more
of carbohydrate is present, the loss of tryptophan is
appreciable.

The surest way to obtain complete recovery of tryptophan is to perform an alkaline hydrolysis specifically for
that purpose. This year, Tony E. Hugli has succeeded in
making alkaline hydrolysis by 4.2 *N* NaOH in the presence
of carbohydrate a strictly quantitative procedure for
tryptophan;[38] it is possible to obtain 100 ± 3% recoveries
of the amino acid. The chromatography (Figure 10) is
performed on an 8 cm column of Beckman PA-35 resin in 35
minutes with 0.21 *M* sodium citrate buffer at pH 5.28. This
system, suggested by David Eaker, separates tryptophan from
lysinoalanine which Zvi Bohak[39] showed can form during
alkaline treatment of proteins. With the buffer usually
employed on a short column at pH 5.28, these two amino
acids overlap and high apparent values for tryptophan can
result.

The results with nine proteins are shown in Table III.
With sperm whale apomyoglobin, one of the tryptophan residues occurs in a valyl-tryptophan sequence and we encounter
the usual problem of the slow hydrolysis of isoleucyl- or
valyl-bonds. In this instance, 98 hours at 110° or 48
hours at 135° gave nearly the theoretical value for
tryptophan.

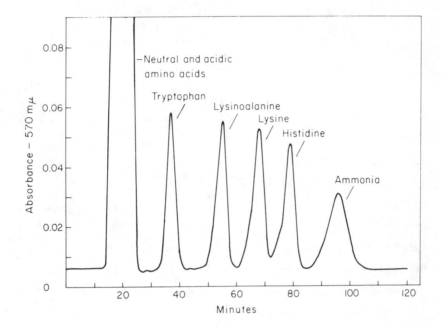

Figure 10: Chromatographic separation of tryptophan from
other amino acids by ion exchange.[38]   A 0.9 x 8.0 cm
column of Beckman PA-35 resin was developed at a buffer
flow rate of 50 ml per hour at 52° with 0.21 M sodium
citrate buffer at pH 5.4.   Ornithine, if present in an
alkaline hydrolysate is eluted just ahead of lysine.

The details of the procedure which make possible the
good recoveries by alkaline hydrolysis include the use of
polypropylene liners in glass tubes, introduced by Mark
Stahmann's laboratory[40] and the addition of starch[41] as
the most effective antioxidant tested. Michael Draper at
The Rockefeller University has been using this method to
determine the tryptophan content of synthetic peptides with
maximum precision. Spectrophotometric methods for tryptophan
can be used on the intact peptide without the problem of
hydrolysis if the peptide is soluble and the tyrosine:
tryptophan ratio is not too high. But we admit to a
preference for seeing a peak from chromatographically pure
tryptophan on a chromatogram. Enzymatic hydrolysis is an

Table III

Tryptophan Recovery from Alkaline Hydrolysates
Hydrolysis by 4.2 N NaOH (0.6 ml containing 25 mg starch)
at 110° for 16 hours unless otherwise indicated
From Hugli and Moore[38]

Protein	Residues found	Integral value
Human serum albumin	1.00	1
Bovine serum albumin	2.03	2
Bovine deoxyribonuclease	2.92	3
Bovine trypsin	3.97	4
Porcine pepsin	4.87	5
Porcine pepsinogen	5.02	5
Hen egg white lysozyme	5.75	6
Bovine α-chymotrypsinogen	7.87	8
Sperm whale apomyoglobin – 16 hrs, 110°	1.26	
98 hrs, 110°	1.89	
48 hrs, 135°	1.94	2

attractive alternative in principle, but as yet has not
proved capable of giving complete hydrolysis in all instances; when successful for tryptophan, the chromatography
is straightforward.   (In another context, enzymatic
hydrolysis is essential if chromatographic differentiation
of glutamine and asparagine is to be made with the lithium
buffer systems of Benson, Gordon, and Patterson[42]).

Analyses of Peptides

Let us consider now the precision of some analyses of
peptides smaller than ribonuclease. In Table IV are the
results of analyses of two thirteen-residue peptides isolated by gel filtration after cyanogen bromide cleavage of

Table IV

Amino Acid Compositions of Tridecapeptides Isolated After
Cyanogen Bromide Cleavage of 3-Carboxyalkyl-His-12-RNases
From Heinrikson *et al*[43]

Amino acid	*Theory*	*Found* in Carboxyethyl- derivative*	*Found* in Carboxypropyl- derivative*
Glutamic acid	3	3.04	3.00
Alanine	3	3.00	3.00
Homoserine and lactone	1	0.88	0.71
Threonine[†]	1	0.97	0.99
Phenylalanine	1	0.99	1.00
Lysine	2	2.00	2.00
Arginine	1	1.00	1.01
3-(1-carboxyethyl) His	1	0.98	
3-(1-carboxypropyl) His	1		0.97

(3-(1-carboxyethyl) His / 3-(1-carboxypropyl) His: 1 or 1)

*Amino acids present to less than 0.05 residue omitted.

†Corrected for 5% destruction.

ribonucleases alkylated at histidine-12.[43] The determination of homoserine and its lactone is not precise, because
the two are in equilibrium during chromatography. But for
the other amino acids the values for the number of residues
check theory to within 3%. The analyses of well-purified
peptides are usually reported to two decimal places. The
approximate corrections of 10% and 5% have been applied to
serine and threonine.

A single analysis of a 27-residue peptide isolated by
ion exchange chromatography from a tryptic hydrolysate of
streptococcal proteinase[44] is tabulated in Table V. The
maximum deviation is 7% for valine; a closer value on
valine would have required longer than 22 hours of hydrolysis.
This peptide was deemed pure enough to merit sequence determination which was completed by Teh-Yung Liu and which
fitted the analysis.

Table V

Amino Acid Composition of Heptaeicosapeptide Isolated From
Tryptic Hydrolysate of Carboxymethylated
Streptococcal Proteinase
From Liu *et al.*[44]

Amino acid	Found*	Integral value
Carboxymethylcysteine	0.96	1
Glutamic acid	3.90	4
Glycine	3.00	3
Alanine	5.10	5
Valine	2.79	3
Isoleucine	0.98	1
Serine[†]	1.03	1
Threonine[†]	3.12	3
Methionine	1.00	1
Proline	0.99	1
Phenylalanine	1.07	1
Histidine	1.03	1
Lysine	2.00	2

*Corrected for 10% and 5% destruction, respectively.

[†]Amino acids present to less than 0.01 residue omitted.

In sequence work we generally round off the analytical
values to one decimal place because the degree of purification of the sub-fractions and the sequentially degraded
products usually does not justify the second figure. But
a synthetic peptide of this size would merit the expression
of as much precision as the analysis can provide.

Without burdening you with more tables of this type,
we hope that we have answered Ralph Hirschmann's original
question concerning the precision with which an amino acid
analysis by ion exchange chromatography can currently be
expected to agree with theory under favorable conditions

with replicate runs or in a single rapid analysis such as
this one.

D- and L-Amino Acids

So far, we have not used the letters D- and L-. The
ion exchange columns do not differentiate between D- and
L-isomers. James Manning, at the second symposium of this
series in Cleveland, summarized the procedures through
which the chromatography can be extended to the determination of the stereochemical purity of amino acid residues
in synthetic peptides. We will use one chromatogram
(Figure 11) to cross-refer to that report.[45] The method[46]

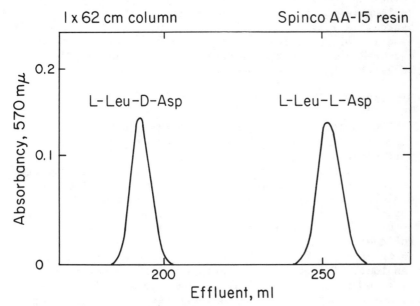

Figure 11: Separation of diastereoisomeric dipeptides on
  Beckman AA-15 resin (column 0.9 x 62 cm) at pH 3.25 and
  52°.[46]

depends upon the conversion of the D- and L-isomers into
diastereomers by condensation with L-leucine *N*-carboxyan-
hydride or L-glutamic acid NCA by the method of Hirschmann
and associates.[47] The resulting pairs of dipeptides are
separated on an amino acid analyzer. The resolution is
such that by using high loads of the L-L dipeptide, the

stereochemical purity of the starting amino acids for
peptide synthesis can be established to 99.99%. Robert
Feinberg in Bruce Merrifield's laboratory has been giving
this procedure an extensive work-out.

The determination of the stereochemical purity of an
amino acid residue in a synthetic peptide is a more difficult problem. Some means of correcting for racemization
during acid hydrolysis is needed. When the naturally
occurring L-peptide is available, its hydrolysis provides
a practical control. Bruce Merrifield suggested that when
the natural peptide is not available, hydrolysis in
tritiated HCl might provide the answer. James Manning[48]
found that tritium incorporation, as measured with a
scintillation flow cell in series with the analyzer, was
applicable to the measurement of racemization during
hydrolysis with 10 of the common amino acids. By a combination of the chromatographic experiments, the stereochemical purity of the residue in the parent peptide can
be established with an accuracy of about 1%.

Discussion

We have limited this review to the precision of amino
acid analysis by ion exchange chromatography. There are
participants in this symposium who are experts on gas
chromatography of amino acid derivatives and each advance
in that technique is a welcome addition to the tools of
protein chemistry. John Pisano[49] has written a very
thoughtful summary of the subject for Vol. XXV of Methods
in Enzymology. When derivatization is an integral part of
the chemical procedure, as in the formation of thiohydantoins
in the Edman degradation, the products are very logical
candidates for gas chromatography. For the determination
of amino acids, any derivatization process adds the variations in the yields of the derivatives to the errors inherent in the chromatographic process. Where precision
is a major objective, there are advantages in avoiding the
additional sources of error. The researches on rendering
gas chromatography an efficient method for amino acid
analysis have been stimulated by the potential speed,
convenience, and sensitivity of the process. In the meantime, the progress in liquid chromatography has essentially
matched gas chromatography in terms of speed and of sensitivity; the next contribution to this symposium, by Sidney
Udenfriend, will be touching on the latter point. The main
potential advantage of gas chromatography may then rest in

the simplicity of the separation process; any advance that makes amino acid analysis easier and cheaper will be quickly utilized.

Acknowledgments

The author is indebted to Ralph Hirschmann for suggesting the theme of this contribution, to Teh-Yung Liu for generously keeping us up-to-date on his experiments with methanesulfonic acid prior to publication, and to William H. Stein, James M. Manning, Tony E. Hugli, R. Bruce Merrifield, and Alexander R. Mitchell for reviewing this manuscript.

References

1. Bergmann, M., L. Zervas, J. S. Fruton, F. Schneider, and H. Schleich. J. Biol. Chem. 109, 325 (1935).
2. Gutte, B., and R. B. Merrifield. J. Amer. Chem. Soc. 91, 501 (1969).
3. Hirschmann, R., R. F. Nutt, D. F. Veber, R. A. Vitali, S. L. Varga, T. A. Jacob, F. W. Holley, and R. G. Denkewalter. J. Amer. Chem. Soc. 91, 507 (1969).
4. Gutte, B., and R. B. Merrifield. J. Biol. Chem. 246, 1922 (1971).
5. Hirs, C. H. W., S. Moore, and W. H. Stein. J. Biol. Chem. 235, 633 (1960).
6. Smyth, D. G., W. H. Stein, and S. Moore. J. Biol. Chem. 238, 227 (1963).
7. Brand, E. Ann. New York Acad. Sci. 47, 187 (1946).
8. Stein, W. H., and S. Moore. J. Biol. Chem. 176, 337 (1948).
9. Moore, S., and W. H. Stein. J. Biol. Chem. 178, 53 (1949).
10. Stein. W. H., and S. Moore. J. Biol. Chem. 178. 79 (1949).
11. Moore, S., and W. H. Stein. J. Biol. Chem. 176, 367 (1948).
12. Moore, S., and W. H. Stein. J. Biol. Chem. 211, 907 (1954).
13. Moore, S. J. Biol. Chem. 243, 6281 (1968).
14. Moore, S., and W. H. Stein. J. Biol. Chem. 192, 663 (1951).
15. Moore, S., and W. H. Stein. J. Biol. Chem. 211, 893 (1954).
16. Hirs, C. H. W., W. H. Stein, and S. Moore. J. Biol. Chem. 211, 941 (1954).

17. Spackman, D. H., W. H. Stein, and S. Moore. Anal.
    Chem. <u>30</u>, 1190 (1958).
18. Moore, S., D. H. Spackman, and W. H. Stein. Anal.
    Chem. <u>30</u>, 1185 (1958).
19. De Lange, R. J., A. N. Glazer, and E. L. Smith. J.
    Biol. Chem. <u>244</u>, 1385 (1969).
20. Johnson, P., C. I. Harris, and S. V. Perry. Biochem.
    J. <u>105</u>, 361 (1967).
21. Roncari, G., Z. Kurylo-Borowska, and L. C. Craig.
    Biochemistry <u>5</u>, 2153 (1966).
22. Spackman, D. H. Federation Proc. <u>22</u>, 244 (1963).
23. Hamilton, P. B. In *Methods in Enzymology*, XI, Hirs,
    C. H. W., Ed. (New York: Academic Press, 1967) pp
    15-27.
24. Hare, E. Federation Proc. <u>25</u>, 709 (1966).
25. Spackman, D. H. In *Methods in Enzymology*, XI, Hirs,
    C. H. W., Ed. (New York: Academic Press, 1967) pp
    3-15.
26. Crestfield, A. M., S. Moore, and W. H. Stein. J.
    Biol. Chem. <u>238</u>, 622 (1963).
27. Sanger, F., and E. O. P. Thompson. Biochim. Biophys.
    Acta <u>71</u>, 468 (1963).
28. Benisek, W. F., M. A. Raftery, and R. D. Cole.
    Biochemistry <u>6</u>, 3780 (1967).
29. Africa, B., and F. H. Carpenter. Biochemistry <u>9</u>,
    1962 (1970).
30. Moore, S., and W. H. Stein. In *Methods in Enzymology*,
    VI, Colowick, S. P., and N. O. Kaplan, Eds. (New York:
    Academic Press, 1963) pp 819-831.
31. Moore, S. J. Biol. Chem. <u>238</u>, 235 (1963).
32. Inglis, A. S., and T.-Y. Liu. J. Biol. Chem. <u>245</u>, 112
    (1970).
33. Meienhofer, J., and Y. Sano. J. Amer. Chem. Soc. <u>90</u>,
    2996 (1968).
34. Liu, T.-Y., N. P. Neumann, S. D. Elliott, S. Moore, and
    W. H. Stein. J. Biol. Chem. <u>238</u>, 251 (1963).
35. Beacham, J., G. Dupuis, F. M. Finn, H. T. Storey, C.
    Yanaihara, N. Yanaihara, and K. Hofmann. J. Amer.
    Chem. Soc. <u>93</u>, 5526 (1971).
36. Matsubara, H., and R. H. Sasaki. Biochem. Biophys.
    Res. Commun. <u>35</u>, 175 (1969).
37. Liu, T.-Y., and Y. H. Chang. J. Biol. Chem. <u>246</u>, 2482
    (1971).
38. Hugli, T. E., and S. Moore. J. Biol. Chem. <u>247</u>, 2828
    (1972).
39. Bohak, Z. J. Biol. Chem. <u>239</u>, 2878 (1964).

40. Oelshlegel, F. J., Jr., J. R. Schroeder, and M. A. Stahmann. Anal. Biochem. 34, 331 (1970).
41. Drèze, A. Bull. Soc. Chim. Biol. 42, 407 (1960).
42. Benson, J. V., Jr., M. J. Gordon, and J. A. Patterson. Anal. Biochem. 18, 228 (1967).
43. Heinrikson, R. L., W. H. Stein, A. M. Crestfield, and S. Moore. J. Biol. Chem. 240, 2921 (1965).
44. Liu, T.-Y., W. H. Stein, S. Moore, and S. D. Elliott. J. Biol. Chem. 240, 1143 (1965).
45. Manning, J. M., A. Marglin, and S. Moore. In *Progress in Peptide Research II*, Lande, S., Ed. (New York: Gordon and Breach, 1972) pp 173-183.
46. Manning, J. M., and S. Moore. J. Biol. Chem. 243, 5591 (1968).
47. Hirschmann, R., R. G. Strachan, H. Schwam, E. F. Schoenewaldt, H. Joshua, B. Barkemeyer, D. F. Veber, W. J. Paleveda, Jr., T. A. Jacob, T. E. Beesley, and R. G. Denkewalter. J. Org. Chem. 32, 3415 (1967).
48. Manning, J. M. J. Amer. Chem. Soc. 92, 7449 (1970).
49. Pisano, J. J. In *Methods in Enzymology*, XXV, Hirs, C. H. W., and S. N. Timasheff, Eds. (New York: Academic Press, 1972) pp 27-44.

A NEW FLUOROMETRIC PROCEDURE FOR ASSAY OF AMINO ACIDS,
PEPTIDES AND PROTEINS IN THE PICOMOLE RANGE

*Sidney Udenfriend, Stanley Stein, Peter Bohlen,
Wallace Dairman.* Roche Institute of Molecular Biology,
Nutley, New Jersey.

SUMMARY--The application of a novel reagent, fluorescamine,
to the assay of picomole quantities of primary amines is
presented. Fluorescamine which is nonfluorescent, combines
almost instantaneously with primary amines to yield highly
fluorescent products. Excess fluorescamine is rapidly
destroyed by water to yield nonfluorescent products. The
reagent has been utilized in the development of an automated
amino acid analyzer which can measure as little as 50
picomoles of each amino acid. The assay has also been
successfully applied to the automatic monitoring of peptides
from column effluents. A manual method and a semi-automated
procedure have been developed for measuring proteins in the
nanogram range. Fluorescamine has also been shown to be
more sensitive than existing procedures for monitoring the
extent of reaction in solid phase peptide synthesis.

PRIMARY AMINES COMPRISE a large number of biologically important substances and many procedures have been introduced
over the years for their quantitative assay. More and more
sensitive methods have been required as biological sophis-
tication has increased. Our laboratory has previously
reported that ninhydrin and phenylacetaldehyde combine with
primary amines in a ternary reaction to form highly
fluorescent products.[1] This reaction has been applied to
the fluorometric assay of amino acids and peptides in the
picomole range.[2] Subsequently Weigele *et al.*[3] investigated

the mechanism and determined the structure of the fluorophors obtained from several primary amines (Figure 1). This led

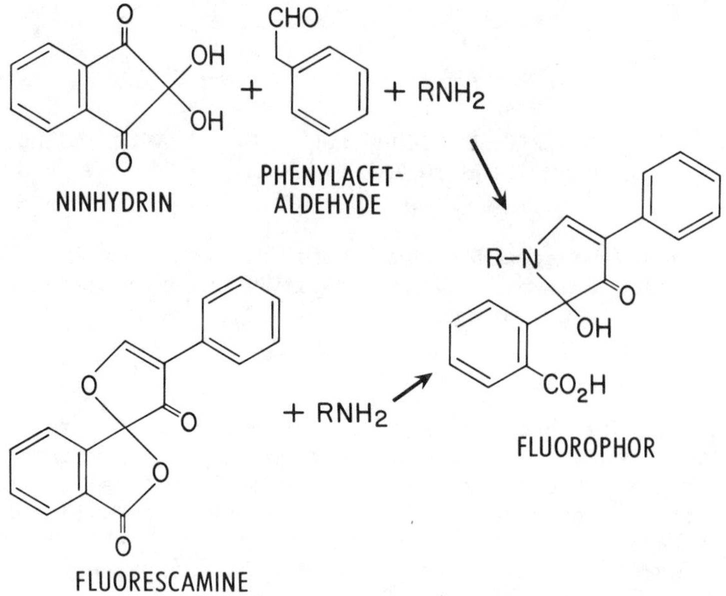

Figure 1: Reaction of primary amines with ninhydrin and phenylacetaldehyde and with fluorescamine.

Weigele *et al.*[4] to synthesize the reagent, 4-phenylspiro [furan-2(3*H*), 1'-phthalan]-3,3'-dione (fluorescamine) (Fluram)* which reacts with primary amines in a binary reaction to yield the same fluorophors as are obtained in the ternary reaction with ninhydrin and phenylacetaldehyde (Figure 1). The fluorophors in both reactions have excitation maxima at 390 nm and fluoresce maximally at 475 nm (uncorrected).

Fluorescamine has properties which offer many advantages that are unique among existing amine detecting reagents. The reagent itself is nonfluorescent. The fluorescamine

---

*Fluram--Roche Diagnostics Division, Nutley, New Jersey.

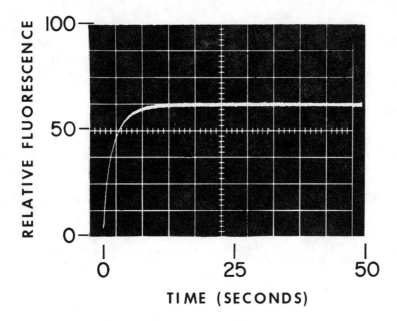

Figure 2: Oscilloscope photograph of the reaction of
fluorescamine with alanine at pH 9.0 in the Aminco-Morrow
stopped-flow fluorometer.

reaction with primary amines is extremely rapid at alkaline
pH (Figure 2) and the fluorescent product is relatively
stable. At pH 8-9 the average half-time for appearance of
maximal fluorescence is measured in hundreds of millseconds
at room temperature. Under the conditions of assay which
are presently used there is a competing reaction of
fluorescamine with water which leads to nonfluorescent
degradation products. The latter reaction has a half-time
of several seconds. The reaction has been shown to proceed
almost to completion with a large number of amino acids and
peptides (80-90% of the theoretical yield). In other words,
one adds a nonfluorescent reagent to a solution containing
an amino acid, peptide or protein and within a matter of
seconds maximal fluorescence, which is quite stable, is
obtained. Very shortly thereafter the excess reagent is
destroyed by reaction with water to yield nonfluorescent
products.
     The advantages of fluorescamine are significant. The
sensitivity is such that picomole quantities can be assayed

The rapidity of the reaction simplifies automation immensely, since there is essentially no delay and no heating apparatus is required. These properties are ideal for monitoring column effluents. In the short time since this reagent was introduced it has already found many applications.

Amino Acids

An automated apparatus has been developed for fluorometric amino acid assay. This is shown schematically in Figure 3. The column effluent mixes first with borate buffer to adjust the pH to 9, and then with fluorescamine in acetone. A coil of microtubing with a time delay of twenty seconds is sufficient to permit complete reaction and destruction of excess

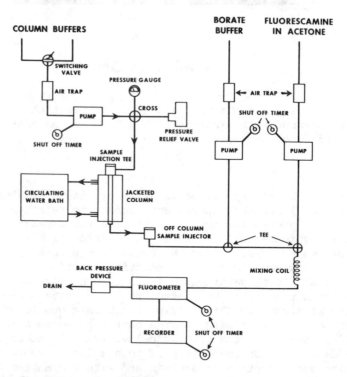

Figure 3: Schematic illustration of the flow system for automated assay with fluorescamine. Milton Roy minipumps were used for the column as well as for the 0.2 *M* sodium borate, pH 9.7 and 15 mg% solution of fluorescamine in acetone. A Beckman recorder and an Aminco filter fluorometer were employed.

reagent. Following this the solution passes through the
flow cell of an Aminco microfluorometer. The signal is
monitored on a strip chart recorder.

Chromatography of a mixture containing 250 picomoles
each of neutral, acidic and basic amino acids is shown in
Figure 4. Proline, which was present in the mixture, does

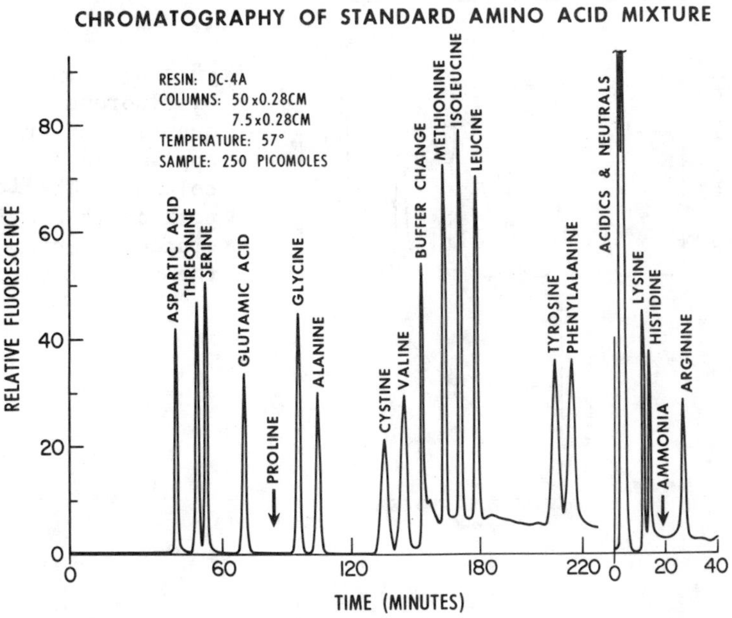

CHROMATOGRAPHY OF STANDARD AMINO ACID MIXTURE

Figure 4: Chromatography of a standard mixture of 250
picomoles of each amino acid (including proline and
ammonia) on a 50 x 0.28 cm acidics and neutrals column
and a 7.5 x 0.28 cm basics column. Durrum DC-4A resin
and Beckman concentrated citrate buffers were used.
Flow rates were approximately 7 ml/hr for the long
column and 12 ml/hr for the short column.

not appear because it is a secondary amine and therefore
nonreactive. This is a disadvantage of the method which,
it is hoped, will be overcome by additional research. A
great advantage of fluorescamine is that the reaction with
ammonia yields fluorescence which is only one-thousandth
that obtained with amino acids. Ammonia, therefore, does

not interfere in the assay of basic amino acids. Neutral, acidic and basic amino acids can be assayed on this relatively simple apparatus to at least 50 picomoles. Proportionality of the fluorescence obtained with increasing amounts of arginine is shown in Figure 5.

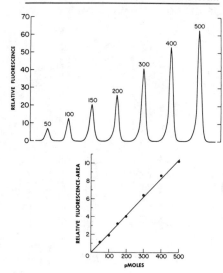

QUANTIFICATION OF ARGININE IN COLUMN EFFLUENT

Figure 5: Linearity of fluorescence with quantity of arginine applied to the short column. The lower portion is the integrated area of each of the above recordings.

The procedure as described above has been applied to the assay of amino acids in microliter quantities of serum and urine. It has also been applied to the assay of microgram quantities of protein hydrolysates.

Peptides

The fluorescence obtained with peptides has been found to be greater than that obtained with amino acids. This appears to be due to an increased quantum yield of the resulting fluorophors. Dipeptides, pentapeptides, octapeptides and even insulin and glucagon give intense fluorescence permitting assay in the 20 to 100 picomole range. Of great interest is the effect of pH on fluorescence (Figure 6). The optimal range for amino acids is between pH 8 and 9.

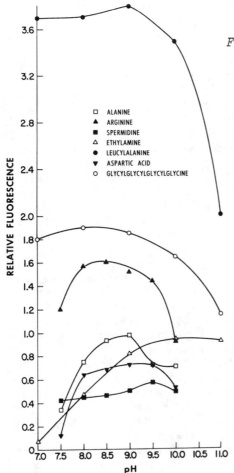

Figure 6: Influence of pH on the reaction of various primary amines with fluorescamine. One volume of fluorescamine in acetone was added to three volumes of the amine in sodium borate buffer.

Legend (in figure):
□ ALANINE
▲ ARGININE
■ SPERMIDINE
△ ETHYLAMINE
● LEUCYLALANINE
▼ ASPARTIC ACID
○ GLYCYLGLYCYLGLYCYLGLYCYLGLYCINE

(y-axis) RELATIVE FLUORESCENCE
(x-axis) pH

However, when the pH is lowered, amino acid fluorescence falls off sharply. At pH 7 the fluorescence is less than a few percent of that obtained at pH 9. In contrast, at pH 7 peptide fluorescence is almost as high as at pH 9. This differential effect of pH should make it possible to use fluorescamine to detect peptides in complex mixtures without the need for fractionation.

Fluorescamine has been used to detect peptides in urine and tissue extracts and for following the metabolism *in*

vitro of vasoactive peptides in the picomole range. Studies
on tryptic peptides of proteins for genetic comparison can
be carried out with less than 100 micrograms of protein.
Pituitary hormones have been assayed in extracts obtained
from milligram amounts of tissue.[5]

Proteins

The amino groups of proteins react readily with fluorescamine to give intense fluorescence. Two types of procedure
have been utilized for protein assay. The first is a manual
method in which protein sample, buffer and reagent solution
are mixed and fluorescence measured in a cuvette. Using the
Aminco spectrophotofluorometer the method can measure as
little as 0.5 µg per ml and the response is linear up to at
least 100 µg per ml. To increase sensitivity a semi-automated
procedure has been used. The automated apparatus shown in
Figure 3 was modified to make use of a short column of
Biogel. Each sample for assay is injected onto the Biogel
column in order to separate proteins from small molecules
which may also react with the reagent. The fluorescence in
the protein peak is compared with standards (usually
bovine serum albumin) passed over the same column. As a
result of the small bed volume of the column and the rapid
rate of reaction, each sample requires only a few minutes
for assay. The continuous flow also gives a very stable
baseline fluorescence which permits assay of as little as
50 nanograms of protein.
        Fluorescamine has been applied to the measurement of
protein in the determination of specific activities of
fractions collected from column eluates during enzyme preparation. This has been done in a discontinuous manner.
Continuous monitoring of very dilute solutions of protein
in column effluents should also be possible.
        Fluorescamine has also been applied to the detection of
amino acids and peptides on thin layer and paper chromatograms. As little as 10 to 20 picomoles of peptides can be
seen on thin layer plates.
        A potentially useful application of the new reagent is
for monitoring consecutive stages of solid-phase peptide
synthesis. A major problem in this procedure is the failure
of small amounts of the free amino group attached to the
resin to react complete with the next protected amino acid.
Even a fraction of a percent of unreacted amine at each step
can result in the accumulation of a large amount of impurity
in the desired peptide in a multi-step synthesis. Methods
for monitoring completeness of reaction have been introduced.

They are usually time consuming and not sufficiently sensitive. A. M. Felix and J. Jimenez[6] have developed a method using fluorescamine for this purpose. From their results, it is evident that the new reagent can detect much smaller amounts of incomplete coupling with greater simplicity than has heretofore been possible.

Fluorescamine yields highly fluorescent products with other types of primary amines. Quantitative assays for catecholamines and polyamines[7] such as spermine and spermidine are being developed. The effects of various parameters such as temperature, pH, organic solvent and buffer salts on the reactivity and stability of fluorescamine are still under study. A thorough understanding of their influence should result in the best possible conditions for the use of fluorescamine and lead to many additional analytical applications.

References

1. Samejima, K., W. Dairman, and S. Udenfriend. Anal. Biochem. 42, 222 (1971).
2. Samejima, K., W. Dairman, J. Stone, and S. Udenfriend. Anal. Biochem. 42, 237 (1971).
3. Weigele, M., J. F. Blount, J. P. Tengi, R. C. Czaikowski, and W. Leimgruber. J. Amer. Chem. Soc. 94, 4052 (1972).
4. Weigele, M., S. L. DeBernardo, J. P. Tengi, and W. Leimgruber. J. Amer. Chem. Soc. 94, 5972 (1972).
5. Samejima. Unpublished results.
6. Felix, A. M., and M. Jimenez, *Analytical Biochemistry*, in press.
7. Samejima, K., S. Stein, and S. Udenfriend. Unpublished observations.

THE STRUCTURE AND CONFORMATION OF THE TUBERCULOSTATIC ANTIBIOTIC VIOMYCIN

B. W. Bycroft. Department of Chemistry, University of Nottingham, Nottingham NG7 2RD, England

THE ANTIBIOTIC VIOMYCIN, first isolated[1] in 1951 from submerged cultures of certain species of *streptomyces*, has found limited clinical use in the treatment of tuberculosis. Although it possesses relatively high potency against tubercle bacilli, the marked toxic effects have prevented the widespread application of the antibiotic and its main therapeutic value lies in the treatment of cases involving drug-intolerance and bacterial resistance.

In view of the interest in the biogenesis of microbial peptide antibiotics and in structure-activity relationships in general, a knowledge of the chemical structure of viomycin was considered desirable. Initial investigations[2] had established the peptide nature of the antibiotic and shown that on acid hydrolysis it afforded the amino acids, L-serine, L-α,β-diaminopropionic acid, L-β-lysine and a new guanidine amino acid designated viomycidine (ratio 2:1:1:1 respectively) as well as urea and varying amounts of carbon dioxide and ammonia. Viomycidine, after some initial controversy, was assigned structure I[3] and this has recently been verified by several independent X-ray crystallographic analyses.[4]

I   II   III

Our own investigations supported structure I for viomycidine but further degradative studies on viomycin, in
particular mild base hydrolysis which yielded *inter alia*
glycine and 2-aminopyrimidine in place of viomycidine, as
well as an X-ray analysis of viocidic acid II, a minor
product obtained on total acid hydrolysis, clearly demonstrated that viomycidine was not present as such in the
intact antibiotic. This led us to suggest the presence of
the structural unit III in viomycin for which it was possible
to establish the absolute configuration at the α and β
centers but not at the guanidine-carbinol center.[5]   A
number of transformations were interpreted on this basis,
all of which paralleled closely those of the only other
known naturally occurring guanidine-carbinol system which
exists in the molecule of the fish poison tetrodotoxin.[6]

A further complicating feature of the chemistry of
viomycin is the presence of an extremely labile chromophoric
unit which is characterized by a strong ultraviolet absorption at 268nm (ε, 24,000) in neutral and acidic media,
shifting to 285nm (ε, 15,000) in base. The lability of
both the chromophore and the guanidine-carbinol units have
presented considerable difficulties in interpreting degradative results and earlier formulations of the chromophore
attempted to combine both these units whereas recent publications have misinterpreted results relating to the
guanidine unit and took no account of the chromophore.

It was possible by kinetic studies to demonstrate that
the urea was part of the chromophoric system and further
extensive chemical and spectral studies led us to the
formulation IV for the chromophore.[7] This was later substantiated by the synthesis of model systems which possessed
similar chemical and spectral properties.[8]

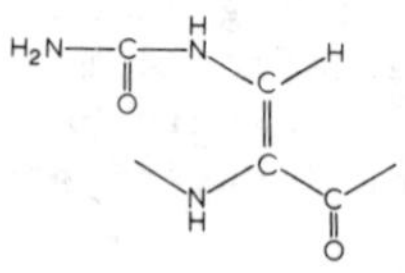

IV

With the structures of the guanidine and chromophoric
unit established it was possible, on the basis of end group
analysis and the structures of dipeptides obtained from
partial base hydrolysis, to propose tentatively the
molecular structure shown in Figure 1 for viomycin.[9]

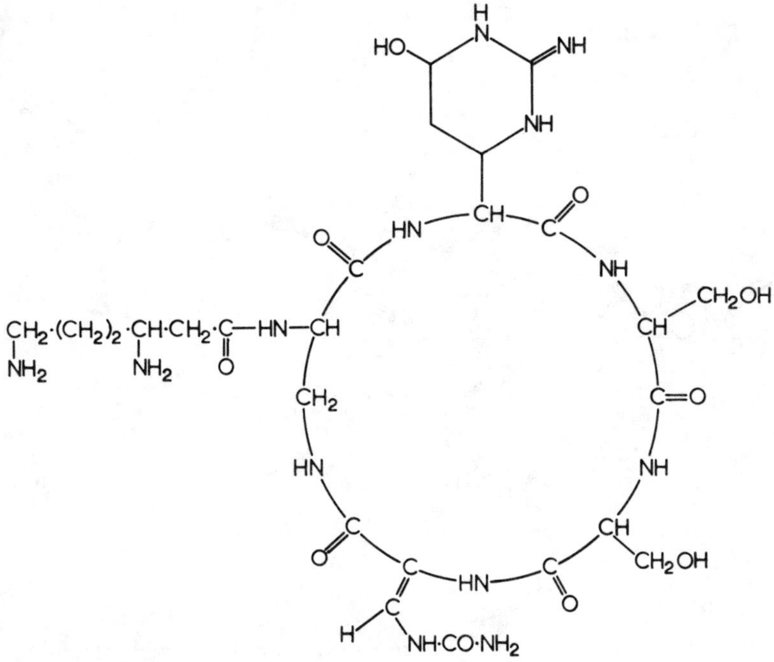

Figure 1: Structure proposed on the basis of chemical evidence

In order to substantiate this proposal the suitability of a number of derivatives of viomycin for X-ray crystallographic examination was investigated. Crystals of viomycin dihydrobromide hydrochloride $C_{25}H_{43}N_{13}O_{10} \cdot 2HBr \cdot HCl \cdot 3H_2O$, monoclinic, space group C_2 with cell dimensions a = 20.70(3), b = 15.79(3), c - 13.93(3), β = 106.17°, were subsequently prepared and the intensities of 1525 independent reflexions were recorded. Initial co-ordinates for the two bromide and chloride anions were obtained from a three dimensional Patterson synthesis and several successive rounds of structure factor calculations and Fourier syntheses phased on these ions revealed the molecular structure shown in Figure 2. The structure confirms the presence of the novel dehydro-serine ureide IV and the guanidine-carbinol system III but necessitates an amendment to the proposed amino-acid sequence. It is of interest to note that the physical and spectral properties as well as the end group analysis of viomycin are in accord with both structures. The discrepancy arose over a misinterpretation of the origin of the dipeptides

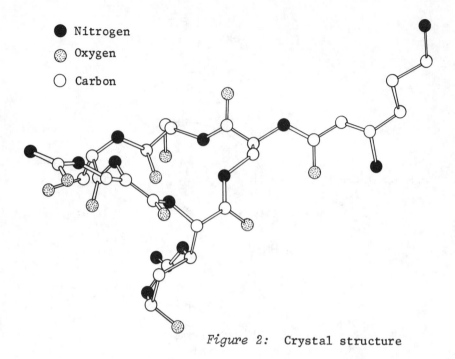

Figure 2:   Crystal structure

obtained on base hydrolysis and serves to emphasize the
caution that is necessary when determining the amino acid
sequence of peptides containing labile amino acid units.
   Since the L-configuration of the α-amino acids had
already been determined the absolute configuration of the
whole molecule is also established. A significant feature
of viomycin in the crystalline state is the hydrogen bonded
chelate ring Figure 3 the conformation of which is similar
to the β-turn structure common to many other cyclic peptides.[10] The corner positions of the β-turn in cyclic
peptides are normally occupied by the α-carbon atoms of a
D- and L-amino acid residue respectively. Conformational
energy considerations accord with the established stability
of this type of system and there is abundant evidence to
suggest that the β-turn conformation is retained in solution.[11]   Initial 100 and 220 mHz NMR studies which are as
yet incomplete provide evidence that this is also true for
viomycin.
   In viomycin these corner positions are occupied by
L-serine and the dehydroserine ureide which replaces the
usual D-amino acid residue. The presence of the same β-
turn structure in the closely related antiobiotic

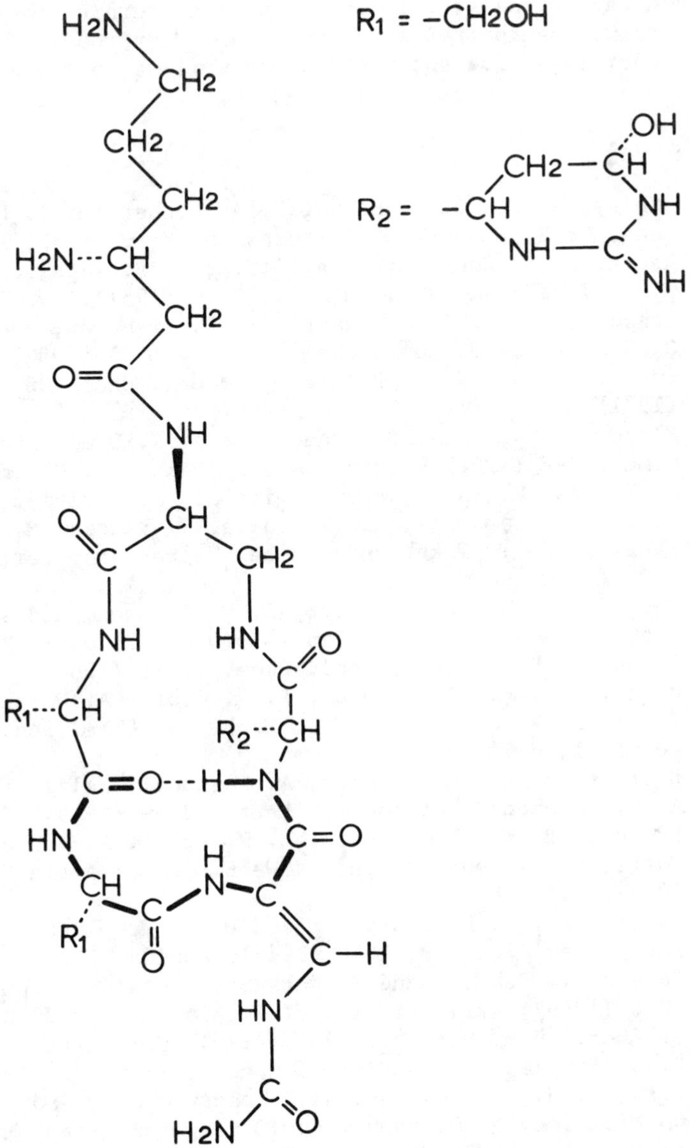

Figure 3: Correct structure showing the transannular hydrogen bond

tuberactinomycin has also been established by an X-ray
crystallographic analysis.[12] This interesting conformational
feature may be relevant to our views[13] concerning the possible relationship of D-amino acids and dehydroamino acids
in microbial peptide antibiotics, as well as to the mode
of action of this novel antibiotic.

References

1. Finlay, A. C., G. L. Hobby, F. A. Hochstein, T. M.
    Lees, T. F. Lenert, J. A. Means, S. Y. P'am, P. P.
    Regna, J. B. Routein, B. A. Sobin, K. B. Tate, and
    J. H. Kane. Am. Rev. Tuberc. 63, 1 (1951).
2. Haskell, T. H., S. A. Fusari, R. P. Frohardt, and
    Q. R. Bartz. J. Amer. Chem. Soc. 74, 599 (1952).
3. Büchi, G., and J. A. Raleigh. J. Org. Chem. 36, 873
    (1971).
4. Floyd, J. C., J. A. Bertrand, and J. R. Dyer. Chem.
    Comm., 998 (1968); Koyama, G., H. Nakamura, S. Omoto,
    T. Takita, K. Maeda, and Y. Iitaka. J. Antibiot. 22,
    34 (1969); Wakamiya, T., T. Shiba, T. Kaneko, H.
    Sakakibara, T. Take, and J. Abe. Tetrahedron Lett.,
    3497 (1970).
5. Bycroft, B. W., L. R. Croft, A. W. Johnson, and T.
    Webb. J. Chem. Soc. Perkin 1, 820 (1972).
6. Woodward, R. B. Pure Appl. Chem. 9, 21 (1964).
7. Bycroft, B. W., D. Cameron, L. R. Croft, A. Hassanali-
    Walji, A. W. Johnson, and T. Webb. J. Chem. Soc.
    Perkin 1, 827 (1972).
8. Bycroft, B. W., D. Cameron, A. Hassanali-Walji, and
    A. W. Johnson. Tetrahedron Lett., 2539 (1969).
9. Bycroft, B. W., D. Cameron, L. R. Croft, A. Hassanali-
    Walji, A. W. Johnson, and T. Webb. Experientia 27,
    501 (1971).
10. Zalkin, A., J. D. Forrester and D. H. Templeton. J.
    Amer. Chem. Soc. 88, 1810 (1966); Ashida, T., M.
    Kakado, Y. Sasada, and Y. Katsube. Acta Cryst. B25,
    1840 (1969); Karle, I. L., J. W. Gibson, and J. Karle,
    J. Amer. Chem. Soc. 92, 3755 (1970); Camilletti, G.,
    P. De Santis, and R. Rizzo. Chem. Comm. 1073 (1970).
11. Urry, D. H., and M. Ohnishi. *Spectroscopic Approaches
    to Biomolecular Conformation* (Chicago: American Medical
    Association, 1970) pp 263-300.
12. Yoshioka, H., T. Aoki, H. Goko, K. Nakatsu, T. Noda,
    H. Sakakibara, T. Take, A. Nagata, J. Abe, T. Wakamiya,
    T. Shiba, and T. Kaneko. Tetrahedron Lett. 2043 (1971).
13. Bycroft, B. W. Nature 224, 595 (1969).

STRUCTURAL RELATIONSHIPS IN AND BETWEEN PEPTIDES WITH
α,β-UNSATURATED AMINO ACIDS

Erhard Gross. Section on Molecular Structure,
Laboratory of Biomedical Sciences, National Institute
of Child Health and Human Development, National Institutes of Health, Bethesda, Maryland 20014

SUMMARY--Intramolecular structural relationships exist most
obviously in peptide molecules with α,β-unsaturated amino
acids and lanthionines. The most extensively studied
structural relationships between molecules with α,β-un-
saturated amino acids extend to nisin and subtilin, peptides
from rather different microbial origins.

THE NATURAL OCCURRENCE of the α,β-unsaturated analogs of
common amino acids has been demonstrated in at least ten
cases.[1] Thus far, most α,β-unsaturated amino acids have
been found in peptides of relatively low molecular weight
of microbial origin.

Peptides with α,β-unsaturated amino acids may be
divided into two classes: (a) those which contain a single
α,β-unsaturated amino acid having no obvious structural
relationship with other amino acids in the molecule; and
(b) those peptides which contain one or more α,β-unsaturated
amino acids that are structurally related with other amino
acids present in the molecule.

To mention but one member of the first group, telomycin
comes to mind. This peptide contains dehydrotryptophan,
which was one of the first α,β-unsaturated amino acids
found to occur naturally.[2]

A representative of the second group of peptides is
nisin[3] (Figure 1). It contains three α,β-unsaturated amino

671

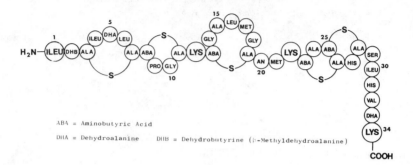

ABA = Aminobutyric Acid

DHA = Dehydroalanine   DHB = Dehydrobutyrine (β-Methyldehydroalanine)

Figure 1:   The structure of nisin.

acids, namely, two residues of dehydroalanine (I) and one residue of dehydrobutyrine (β-methyldehydroalanine, II). There are also present in the molecule on residue of lanthionine (III) and four residues of β-methyllanthionine (IV).

$$R'\!-\!HN\!-\!\underset{\underset{R\quad H}{\diagup\ \diagdown}}{\overset{\overset{\displaystyle C}{\|}}{C}}\!-\!COOH \qquad\qquad HOOC\!-\!\underset{\underset{NH_2}{|}}{CH}\!-\!CH_2\!-\!S\!-\!\underset{\overset{R}{|}}{CH}\!-\!\underset{\underset{NH_2}{|}}{CH}\!-\!COOH$$

(I)   R, H
(II) R, CH₃ } R', acyl or
           aminoacyl

(III) R, H
(IV)   R, CH₃

It is conceivable that the lanthionine and β-methyl-lanthionine residues are derived biosynthetically by the addition of cysteine to precursors in the form of the corresponding α,β-unsaturated amino acids (Figure 2). The potential of reversing the reaction via β-elimination and regeneration of the α,β-unsaturated amino acid (Figure 2) is evident.

Intramolecularly seen, another reversible reaction must be taken into consideration, namely the conversion of an α,β-unsaturated amino acid to amide and keto acid (Figure 3). In its reversible form, the reaction constitutes a peptide bond forming step which may well be operative in nature.

In addition to these relationships *within* a peptide containing α,β-unsaturated amino acids and residues of

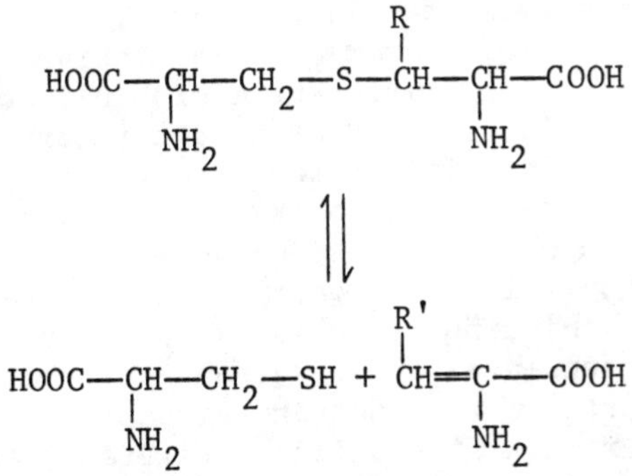

Figure 2: The structural relationship between the thioether amino acids lanthionine (R,H) and β-methyllanthionine (R,CH3) and the α,β-unsaturated amino acids dehydroalanine (R',H) and dehydrobutyrine (=β-methyldehydroalanine, R', CH3).

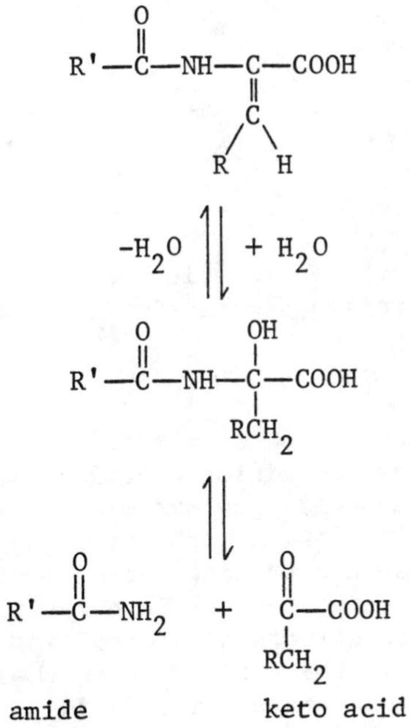

Figure 3: The reversible conversion of α,β-unsaturated amino acids to amide and keto acid.

lanthionine or substituted lanthionines, there are also
structural similarities *among* representatives of these
types of peptides from different microbial sources.

Structural relationships of this category are presently
best established for nisin (from *Streptococcus lactis*) and
subtilin (from *Bacillus subtilis*).   Both peptides contain
not only equal numbers of residues of dehydroalanine (two
each), dehydrobutyrine (one each), lanthionine (one each),
and β-methyllanthionine (four each), but also identical
COOH-terminal sequences of dehydroalanyl-lysine[4] (*cf.*
Figure 4).   Furthermore, one finds repeated within the
COOH-terminal region of subtilin (Figure 4) the unique
thirteen-membered bicyclic heterodetic structure seen in
nisin.[5]

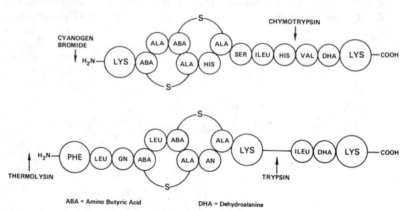

ABA = Amino Butyric Acid   DHA = Dehydroalanine

Figure 4:   Analogous heterodetic bicyclic structures in the
    COOH-terminal fragments of nisin (top) and of subtilin
    (bottom).

The extension of these studies to the lanthionine and
β-methyllanthionine containing peptides cinnamycin[6] (from
Streptomyces cinnamoneus) and duramycin[7] (from *Streptomyces
cinnamoneus* forma *azacoluta*), to date, has furnished this
information: these two peptides do no longer contain α,β-
unsaturated amino acids.   There is, however, unequivocal
evidence that this structural element was once present in
these molecules.   In the form in which these peptides are
isolated the α,β-unsaturated amino acids are masked by the
addition of functional groups across the α,β-double bond.
In cinnamycin and duramycin, the addition of the ε-amino
group of lysine to dehydroalanine (Figure 5) formed

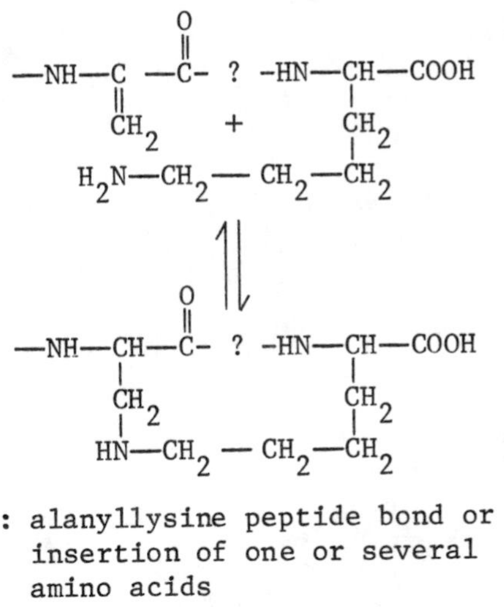

$$
\begin{array}{cccc}
 & \overset{\displaystyle O}{\underset{\displaystyle \|}{}} & & \\
-\text{NH}-\underset{\displaystyle \underset{\text{CH}_2}{|}}{\overset{\|}{\text{C}}}-\text{C}- & ? & -\text{HN}-\underset{\displaystyle \underset{\text{CH}_2}{|}}{\text{CH}}-\text{COOH} \\
 & + & \\
\text{H}_2\text{N}-\text{CH}_2- & \text{CH}_2- & \text{CH}_2
\end{array}
$$

$$
\begin{array}{cccc}
 & \overset{\displaystyle O}{\underset{\displaystyle \|}{}} & & \\
-\text{NH}-\underset{\displaystyle \underset{\text{CH}_2}{|}}{\text{CH}}-\text{C}- & ? & -\text{HN}-\underset{\displaystyle \underset{\text{CH}_2}{|}}{\text{CH}}-\text{COOH} \\
\text{HN}-\text{CH}_2- & \text{CH}_2- & \text{CH}_2
\end{array}
$$

?: alanyllysine peptide bond or
insertion of one or several
amino acids

Figure 5: The addition of the ε-amino group of lysine to
dehydroalanine.

lysinoalanine, the amino acid previously found in alkali-
treated bovine pancreatic ribonuclease.[8]
   It shall be interesting to establish the configuration
of the α-carbon atom of the alanine moiety of lysinoalanine.
The α-carbon atoms of the α-aminobutyric acid moiety in the
β-methyllanthionine residues and in one of the alanine
moieties of the lanthionine residue in nisin are all of
the D-configuration. During the biosynthesis of these
amino acids--so the hypothesis proposes--cysteine is added
across the double bond of the corresponding α,β-unsaturated
amino acid. In the absence of sequence data, the question
remains open, whether there are amino acids inserted in
lysinoalanine between the alanine and lysine moieties.
   The *in vitro* formation of lysinoalanine from dehydroalanine and lysine was demonstrated by exposing the COOH-
terminal fragment of nisin to an *N*-morpholine buffer of pH
11.4 for one hour at room temperature. Under these conditions lysinoalanine was formed rather readily (Figure 6).
   The presence of α,β-unsaturated amino acids is not
restricted to peptides of low molecular weight, nor to
those of microbial origin. Dehydroalanine, for instance,
has been identified in histidine ammonia lyase of both
bacterial[9,10] and mammalian origin[11] and in phenylalanine
ammonia lyase from a plant source.[12]

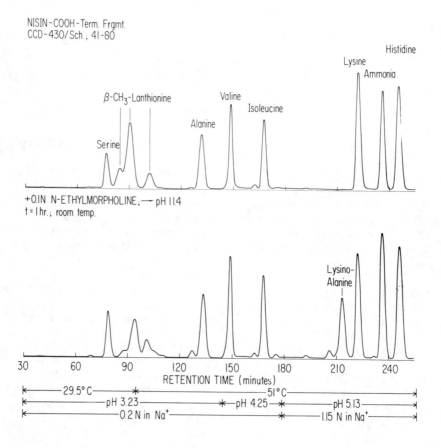

Figure 6: Lysinoalanine formation in the COOH-terminal fragment of nisin under alkaline conditions.

α,β-Unsaturated amino acids are likely to play a more important role in nature than is presently assumed. Frequently they may function in a rather cryptic way and escape detection on the grounds of their dynamic property of converting to other chemical entities (*cf.* Figures 2, 3, and 5).

The working hypothesis of such intermediary involvement of dehydroalanine was recently tested in part for the releasing factor of luteinizing and follicle stimulating hormone. Appropriate substitution of the serine residue in the releasing factor (Figure 7) should, via β-elimination and dehydroalanine formation, lead to the fragmentation of the decapeptide and result in the formation of the tripeptide amide and pyruvylhexapeptide amide.

<Glu-His-Try—NH—C—C—Tyr-Gly-Leu-Arg-Pro-Gly—C
1   5   10   NH₂

$$<\text{Glu-His-Try}-\text{NH}-\underset{\underset{\text{X}}{\overset{\displaystyle |}{\underset{|}{\text{O}}}}{\overset{\text{H}}{\underset{|}{\overset{|}{\text{C}}}}}\overset{\text{O}}{\overset{\|}{\text{C}}}-\text{Tyr-Gly-Leu-Arg-Pro-Gly}-\text{C}\underset{\text{NH}_2}{\overset{\text{O}}{\diagup}}$$

serine

β-elimination

$$<\text{Glu-His-Try}-\text{NH}-\underset{\underset{\text{CH}_2}{\|}}{\text{C}}\overset{\text{O}}{\overset{\|}{\text{C}}}-\text{Tyr-Gly-Leu-Arg-Pro-Gly}-\text{C}\underset{\text{NH}_2}{\overset{\text{O}}{\diagup}}$$

dehydroalanine

$+H_2O$

$$<\text{Glu-His-Try}-\text{C}\underset{\text{NH}_2}{\overset{\text{O}}{\diagdown}}\quad \underset{\text{CH}_3}{\overset{\text{O}}{\overset{\|}{\text{C}}}}-\overset{\text{O}}{\overset{\|}{\text{C}}}-\text{Tyr-Gly-Leu-Arg-Pro-Gly}-\text{C}\underset{\text{NH}_2}{\overset{\text{O}}{\diagup}}$$

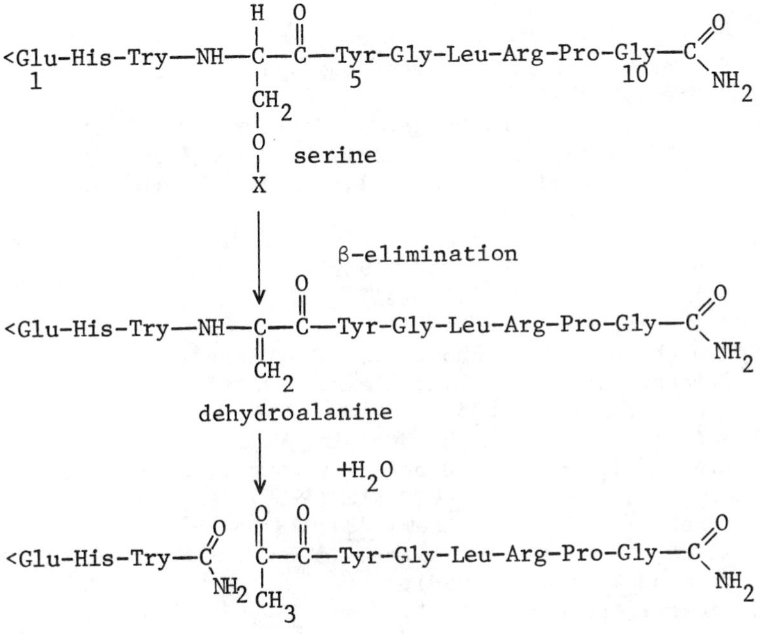

Tripeptide Amide   Pyruvylhexapeptide Amide

Figure 7: Fragmentation via β-elimination of the releasing factor for luteinizing hormone and follicle stimulating hormone.

At this stage of the investigation, the tripeptide amide has been provided by solid-phase peptide synthesis and been found to possess luteinizing hormone releasing activity and only this activity.[13]

Acknowledgment:

The able and enthusiastic collaboration of many co-workers is gratefully recorded. Mr. John L. Morell made possible the structural elucidation of nisin. Dr. H. H. Kiltz contributed much to the understanding of subtilin. Dr. Kiltz began the studies on cinnamycin which, since then, have been continued by Mr. Charles Chapin.

References

1. Gross, E. In *Handbook of Biochemistry*, 2nd Edition, Sober, H., ed. (Cleveland, Ohio: The Chemical Rubber Company, 1970) p B-50.
2. Sheehan, J. C., D. Mania, S. Nakamura, J. A. Stock, and K. Maeda. J. Amer. Chem. Soc. 90, 462 (1968).
3. Gross, E., and J. L. Morell. J. Amer. Chem. Soc. 93, 4643 (1971).
4. Gross, E., J. L. Morell, and L. C. Craig. Proc. Natl. Acad. Sci. U.S. 62, 952 (1969).
5. Gross, E., and J. L. Morell. J. Amer. Chem. Soc. 92, 2919 (1970).
6. Dvonch, W., O. L. Shotwell, R. G. Benedict, T. G. Pridham, and L. A. Lindenfelser. Antibiotics and Chemotherapy 4, 1135 (1954).
7. Shotwell, O. L., F. H. Stodola, W. R. Michael, L. A. Lindenfelser, R. G. Dvorschack, and T. G. Pridham. J. Amer. Chem. Soc. 80, 3912 (1958).
8. Bohak, Z. J. Biol. Chem. 239, 2878 (1964).
9. Givot, I. L., T. A. Smith, and R. H. Abeles. J. Biol. Chem. 244, 6341 (1969).
10. Wickner, R. B. J. Biol. Chem. 244, 6550 (1969).
11. Givot, I. L., and R. H. Abeles. J. Biol. Chem. 245, 3271 (1970).
12. Hanson, K. R., and E. A. Havir. Arch. Biochem. Biophys. 141, 1 (1970).
13. During this investigation we have enjoyed the splendid collaboration of Dr. Mortimer B. Lipsett and his associates.

A NEW TECHNIQUE FOR THE SEQUENCE DETERMINATION OF PROTEINS
AND PEPTIDES BY GAS CHROMATOGRAPHY-MASS SPECTROMETRY-
COMPUTER ANALYSIS OF COMPLEX HYDROLYSIS MIXTURES

H.-J. Förster, J. A. Kelley, H. Nau, K. Biemann.*
Department of Chemistry, Massachusetts Institute of
Technology, Cambridge, Massachusetts 02139

OVER THE PAST DOZEN YEARS considerable efforts have been
devoted to the development of new and efficient instrumental
techniques for the amino acid sequencing of peptides and
proteins.[1] Mass spectrometry offered great promise because
of its high sensitivity and structural specficity for linear
molecules. These mass spectral approaches centered around
the interpretability of mass spectra of peptide derivatives
but neglected the real problem in protein sequencing--
working with very complex mixtures of degradation peptides
on a submicromolar level. Recently, we have developed a
technique which addresses itself to this problem, and the
results obtained will be discussed below.

For various other investigations we had already developed a sophisticated gas chromatograph-mass spectrometer-
computer (GC-MS-Computer) system. A crucial aspect of this
approach is the generation of a mixture of small peptides
that represents an as complete as possible record of all
the peptide bonds present in the protein or primary degradation peptide. The identification technique must be able to
handle all peptides (with one single reaction sequence)
regardless of the nature of the amino acids present. Thus
the main emphasis is to be placed on a high degree of completeness of identification of small peptides rather than

---

*Present Address: Gerichtlich-medizinisches Institut der
Universität Zürich, Postfach 8028, Zürich, Switzerland.

the desire to handle as large a degradation peptide as possible, a process which is often limited by the presence of certain amino acids. At the outset we aimed at unambiguous identification of all possible di- and tripeptides in complex mixtures, but it turned out that the system can handle most tetra- and some pentapeptides as well; this substantially improves the confidence in the reassembly of the original structure.

Reduction of peptide derivatives produces polyamino alcohols which are well suited to both gas chromatographic separation and mass spectrometric sequencing.[2,3] Earlier difficulties with polyfunctional amino acids have now been overcome by O-trimethylsilylation of the polyamino alcohols and the efficient manipulation of peptide mixtures on a very small scale. A major effort was devoted to the generation of these peptide mixtures. Because of the diversity of the characteristics of proteins, one could not expect to find one mode of degradation applicable to all. Three different approaches were developed: (1) partial acid hydrolysis, (2) enzymatic cleavage utilizing a single enzyme or a set of enzymes; and (3) dipeptidylaminopeptidase I (cathepsin C) digestion before and after one Edman degradation step. The differences in specificity of these three govern the choice based on the amino acid composition and genesis of the original protein or peptide. Another major aspect of the problem required the development of interpretative techniques that could deal with the vast amount of data generated while analyzing an entire peptide mixture. These computer-assisted techniques are outlined later.

Extensive enzymatic and partial acid hydrolysis studies have been conducted on ribonuclease S-peptide to determine the feasibility of generating mixtures of overlapping oligopeptides which will allow reconstruction of the original peptide sequence. Mixtures have been generated by partial acid hydrolysis with 6 N HCl for 15 min to 16 hr and by employing enzymes such as chymotrypsin, papain, pepsin, pronase, subtilisin, trypsin-pepsin and trypsin-chymotrypsin. In a typical experiment 3.0 mg (1.37 µmol) of S-peptide was hydrolyzed. The hydrolyzate mixture was esterified, acetylated and reduced with $LiAlD_4$* to yield the corresponding polyamino alcohols. Silylation with pyridine-trimethylsilyldiethylamine (2:1) yields a corresponding mixture of O-silyl ethers that possess excellent gas chromatographic separability and produce easily interpretable mass spectra.

---

*$LiAlD_4$ is preferred over $LiAlH_4$ because it minimizes the occurrence of different sequence ions of the same mass.

The peptide derivative mixtures were analyzed by a
GC-MS-Computer system that consisted of a Perkin-Elmer 990
gas chromatograph - Hitachi RMU 6L mass spectrometer combination operated on line with an IBM 1800 computer for
data acquisition and control. The GC effluent is continuously scanned every 4 seconds through the mass ranges of
m/e 28-455 or *m/e* 28-743. At the termination of the GC-MS
experiment a total ionization plot, *i.e.* a computer-generated
gas chromatogram obtained by summing ion intensities over
the whole experiment (Figure 1a), mass spectra for every
scan, and mass chromatograms (plots of the abundance of
each mass during the gas chromatogram, such as Figures 1b
and 1c) are generated. These are stored on magnetic tape
for future data manipulation, and made available to the
user as microfilm copies filmed from an oscilloscope
display.[4] Thus the data is available for manual interpretation by inspection of individual mass spectra and analysis
of selected GC peaks by mass chromatograms or for partially
automatic interpretation by computer-assisted data manipulation[5] and automatic identification and sequence assembly
programs.[6] Alternatively, a technique of controlled
fractional vaporization of the sample directly into the
ion source of the mass spectrometer using the same programs
as GC-MS is available for identification of large peptides
which are not amenable to gas chromatography.

The silylated polyamino alcohols possess a repetitive
ethylenediamine backbone with carbon-carbon bonds which
cleave upon electron impact to yield ions stabilized by
neighboring nitrogen atoms and to produce mass spectra
composed almost entirely of sequence indicating ions
(Figure 2). Retention of charge on carbon atoms towards
the N-terminal end gives rise to ions representing the
sequence from the N-terminal side of the molecule (A_1, A_2,
A_3 ...), while retention of charge at the other carbon
atoms generates ions corresponding to the sequence from
the C-terminal end of the molecule (Z_1, Z_2, Z_3 ...). A
consequence of silylation is the presence of a fairly
abundant M-15 ion, (loss of methyl from the molecule)
which indicates indirectly the molecular weight. Inspection of the mass chromatograms corresponding to the *m/e*
values of possible sequence determining ions enables one
to rapidly identify the peptide derivatives. The importance
of this technique for the identification of components of
very complex mixtures cannot be overemphasized since it
allows location of amino acid and peptide derivatives containing the same N-terminal amino acid. For example, a
plot of *m/e* 150 (A_1 of Phe) as shown in Figure 1b locates

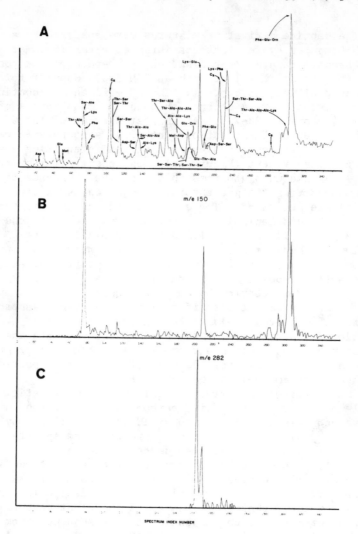

Figure 1: (a) Total ionization plot of silylated amino
alcohols obtained by digestion of ribonuclease S-peptide
by trypsin and pepsin followed by hydrazinolysis and
subsequent derivatization. C_1, 2-6-di-t-butyl-4-methyl
phenol; C_2, unidentified; C_3, C_4, Tms-sucrose; C_5, results
from elimination of TmsOH from the derivative of
Phe-Glu-Orn.

    (b) Mass chromatogram of m/e 150 of the above GC-MS
experiment.

    (c) Mass chromatogram of m/e 282 of the GC-MS experiment
in Figure 1a.

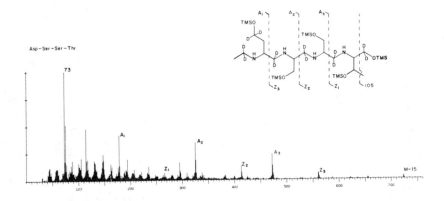

Figure 2: Mass spectrum of the silylated polyamino alcohol corresponding to Asp-Ser-Ser-Thr.

Phe, Phe-Glu and Phe-Glu-Orn. Similarly, peptide derivatives with a certain C-terminal amino acid are located by inspection of the corresponding mass chromatogram (*e.g. m/e* for the Z_1 of Glu, Figure 1c). The coincidence of maxima in mass chromatograms[7] of sequence ions makes it possible to locate and identify any peptide derivative in the gas chromatogram. For example, Phe-Glu, a relatively minor component of the mixture resulting from trypsin-pepsin degradation of S-peptide, can be readily located in the total ionization plot by mass chromatograms of *m/e* 150 and 282 (see Figures 1a-c at spectrum index number 208). Mass chromatograms also allow facile location of peptide derivatives in the presence of the unavoidable non-peptide artifacts (C_1-C_4 in Figure 1a) and the further resolution of incompletely separated components. As a further parameter the gas chromatographic characteristics, expressed by retention indices,[8] are used to great advantage. They can be calculated from values that are determined experimentally for each amino acid side chain. Since the reconstruction of the peptide structure from the sequence determining ions is a matter of simple arithmetic and the data is already residing in the computer, automation of interpretation and eventual assembly of identified overlapping peptides into the complete protein structure is possible. Several interpretive programs already exist for routine use.[6]

Table I shows the peptide derivatives that were identified by GC-MS and fractional vaporization experiments in the derivatized peptide hydrolyzate produced by trypsin and

Table I

Sequence of S-Peptide[9] with the Oligopeptide Fragments Identified by GC-MS-Computer Analysis (see Figure 1a) Tabulated Below

Lys-Glu-Thr-Ala-Ala-Ala-Lys-Phe-Glu-Arg-Gln-His-Met-Asp-Ser-Ser-Thr-Ser-Ala-Ala

Lys-Glu
Glu-Thr-Ala
Thr-Ala
Thr-Ala-Ala
Thr-Ala-Ala-Ala
Thr-Ala-Ala-Ala-Lys
Ala-Ala-Lys
Ala-Lys
Lys-Phe
Phe-Glu
Phe-Glu-Orn
Glu-Orn

His-Met-Asp*
Met-Asp
Asp-Ser
Asp-Ser-Ser
Ser-Ser
Ser-Ser-Thr
Ser-Thr
Ser-Thr-Ser
Ser-Thr-Ser-Ala
Thr-Ser
Thr-Ser-Ala
Ser-Ala
Ser-Ala-Ala

*Identified by fractional vaporization of the sample into the ion source of the mass spectrometer.

pepsin digestion of S-peptide (hydrazinolysis is employed
to transform arginine to ornithine).[10] All original peptide
bonds are represented in these fragments except those involving glutamine. Consideration of the results of one
such experiment permit a proper choice of the cleavage
method which would complete the overlap, if necessary.
Indeed, partial acid hydrolysis of S-peptide produces
(after hydrazine treatment) Orn-Glu and Glu-His-Met, among
others.

A promising third alternative is the use of cathepsin C
which sequentially cleaves dipeptides from the N-terminal
end of a peptide, except those with N-terminal Arg or Lys,
and until it encounters Pro.[11,12] This enzyme would appear
to be ideal for tryptic peptides where two cathepsin
digests, one on the intact peptide and one on the Edman-
degraded peptide, produce two sets of overlapping dipeptides.
Identification of all the dipeptides then allows complete
or partial reconstruction of the original sequence. This
strategy, which was recently also explored by others,[13-15]
was tested on glucagon and several tryptic peptides from
rabbit skeletal muscle actin.* GC-MS-Computer analysis of
silylated oligopeptide amino alcohol mixtures is an ideal
way to unambiguously characterize these dipeptides, especially since a large body of data has been accumulated
for a wide range of dipeptide derivatives examined singly
and in model mixtures.

The techniques outlined above allow sequence determination using complex mixtures of degradation peptides without
their individual separation. Furthermore, the large amount
of data obtained can be interpreted with partial assistance
of a computer which can also aid in the assembly of the
complete protein sequence. At the present time individual
interpretation is involved to a considerable extent, but
all the features necessary for a highly automated approach
exist. It is expected that during application of this
technique to more and more naturally occurring polypeptides
of known and unknown structure data and experience will be
continuously accumulated to make this approach even more
reliable, efficient, and sensitive.

---

*The tryptic peptide samples were donated by M. Elzinga,
Boston Biomedical Research Institute, Boston, Mass. and
received as lyophilized samples accompanied by amino acid
analysis results.

References

1. Biemann, K.   In *Biochemical Applications of Mass Spectrometry*, Waller, G. R., ed. (New York: Wiley Interscience, 1972) p 405.
2. Biemann, K., F. Gapp, and J. Seibl.   J. Amer. Chem. Soc. 81, 2274 (1959).
3. Biemann, K.   Chimia 14, 393 (1960).
4. Biller, J. E., H. S. Hertz, and K. Biemann.   Presented at the Nineteenth Annual Conference on Mass Spectrometry and Allied Topics, Atlanta, Ga., p 85 (1971).
5. Biller, J.E.   Ph.D. Thesis, Massachusetts Institute of Technology, Cambridge, Mass. (1972).
6. Biller, J. E., H. Nau, T. Smith, and K. Biemann. forthcoming.
7. Hites, R. A., and K. Biemann.   Anal. Chem. 42, 855 (1970).
8. Kovats, E.   Helv. Chim. Acta 41, 1915 (1958).
9. Smyth, D. G., W. H. Stein, and S. Moore.   J. Biol. Chem. 238, 227 (1963).
10. Shemyakin, M. M., Yu. A. Ovchinnikov, E. I. Vinogradova, M. Yu. Feigina, A. A. Kiryushkin, N. A. Aldanova, Yu. B. Alakhov, V. M. Lipkin, and B. V. Rosinov.   Experientia 23, 428 (1967).
11. McDonald, J. K., P. X. Callahan, S. Ellis, and R. E. Smith.   In *Tissue Proteinases*, Barrett, A. J., and J. T. Dingle, ed. (Amsterdam: North Holland Publishing Co., 1971) p 69.
12. McDonald, J. K., P. X. Callahan, B. B. Zeitman, and S. Ellis.   J. Biol. Chem. 244, 6199 (1969).
13. Valyulis, R.-A. A., and V. M. Stepanov.   Biokhimiya 36, 866 (1971).
14. Lindley, H.   Biochem. J. 126, 683 (1972).
15. Ovchinnikov, Yu. A., and A. A. Kiryushkin.   FEBS Letters 21, 300 (1972).

THE USE OF MASS SPECTROMETRY IN PEPTIDE CHEMISTRY

Piet A. Leclercq, * *P. A. White, K. Hägele, D. M. Desiderio.* The Institute for Lipid Research and Department of Biochemistry, Baylor College of Medicine, Houston, Texas 77025

IT IS NOW EVIDENT that there is an increased need for a new method to obtain the amino acid sequence of biologically active oligopeptides. Due to an increasing number of naturally occurring peptides that are extracted at the microgram level and/or that contain a blocked *N*-terminus, it has been necessary to employ sequencing techniques other than wet chemical techniques such as the Edman method. Examples of both of the above difficulties have been reported.[1,2]

For these reasons, the thrust behind the research done in this laboratory has been towards the sequencing of biologically active oligopeptides on the microgram level by means of mass spectrometry. The limitations of this newer instrumental technique include volatility of the sample and an upper limit of the mass of the peptide of 1000-1500 mass units.

This report will discuss the derivatization of peptides that is necessary to provide sufficient volatility for mass spectrometry (as a rough estimate, it is necessary to have at least 10^{-5} mm vapor pressure at 300°C).

Based upon the pioneering work of Lederer,[3] Biemann,[4] Lande[5] and others, we found that we had to address our efforts to three areas of derivative formation.   First,

*Present Address: Department of Chemistry, Eindhoven University of Technology, Eindhoven, The Netherlands

acetylation had to be performed on the nanomolar level in a reasonable amount of time. Second, the permethylation reaction had to be made applicable at the nanomolar level for all of the amino acids. Third, arginine had to be included in any analysis. While it is true that enzymatic hydrolyses would furnish *C*-terminal Arg peptides, the isolation of a small amount of an Arg-containing peptide would possibly not be amenable to such a technique.

Therefore, the following laboratory procedure has been developed and applied to a wide variety of oligopeptides. The two (or three) step derivatization process has been found to be quite useful:

Step 1.   Convert the arginine side chain to dimethyl-pyrimidyl ornithine by a reaction with acetylacetone.[4]

$$
\begin{array}{ccc}
\text{------ NHCHCO ------} & & \text{------ NHCHCO ------}\\
(CH_2)_3 & \xrightarrow{\;CH_3COCH_2COCH_3\;} & (CH_2)_3\\
NH & & NH\\
C{=}NH & & \text{pyrimidine ring with CH}_3\text{ groups}\\
NH_2 & &
\end{array}
$$

This reaction has been found to proceed smoothly on the 100 nanomolar level with a nonapeptide such as bradykinin.[6]

Step 2.   In order to avoid quaternization during the permethylation reaction, the free amine groups are acetylated. The peptide is dissolved in methanol, dissolution aided with ultrasonic treatment,[7] and acetic anhydride added. The reaction mixture sits at room temperature for three hours. Because such a wide range of reaction times had been reported,[8] we investigated the optimum for this parameter.[9] Reagents are removed under vacuum.

Step 3.   *N,O,S* permethylation replaces all hydrogen atoms not bound to carbon with a methyl group. This is accomplished by reacting the peptide with the methylsulfinylmethide carbanion (DMSO⁻) to produce a peptide polyanion, then adding methyl iodide to effect methylation. The reaction proceeds at room temperature for one hour and is terminated by adding water. The permethylated peptide is extracted with chloroform.

The derivatization reactions above have been employed on a wide variety of oligopeptides. Lys-Tyr-Glu has been sequenced by mass spectrometry with no difficulties.[9] The

N^α and N^ϵ groups were acetylated and the carboxyl groups and hydroxyl groups methylated. <Glu–Pro–Tyr–His–NH$_2$ was investigated.[10] The N-terminal pyroglutamic acid residue obviates many enzymatic and chemical methods. Also, the proline residue inhibits many enzymes. Histidine was difficult to derivatize[10,11] and the carboxamide group inhibits carboxypeptidase. The O,N permethylated peptide yields all sequence information. Methionine was specifically investigated due to the ease of oxidation of the sulfur. No difficulties were experienced with acetylated permethylated Thr–Met and Met–Gly–Met–Met.[12]

Finally arginine-containing peptides were investigated.[6] This is necessary because some natural peptides[13] contain Arg, and sufficient peptide is not always available for enzymatic hydrolysis. All three steps above were necessary to derivatize Arg, Arg–Arg, Ser–Arg–His–Pro and Arg–Pro–Pro–Gly–Phe–Ser–Pro–Phe–Arg (bradykinin). The sequence was obtained in all cases.

With each mass spectrum discussed above, the amino acid sequence was obtained by means of a computer program adapted from Biemann's program.[14] The input data consist of all peaks in the low resolution mass spectrum and the quantitative amino acid analysis. (In a high resolution mass spectrum, this analysis is not necessary.) All possible sequences are exhaustively searched for and the output lists the more probable sequences. In all cases, the most probable sequence is the correct one.

In conclusion, all of the chemical and instrumental methodology discussed here has been developed out of the necessity of obtaining the amino acid sequence of natural oligopeptides on submilligram quantities. In addition, these peptides usually have a blocked N-terminus. As usual, an unequivocal sequence must be obtained in order to synthesize the sequence and to verify the biological activity. Concurrent with the discussed techniques, ancillary methods are being developed. These include peptide synthesis, computer programs to enable finding synthetic model compounds from the literature,[15] and chemical ionization methods to enhance the overall mass spectrometric sensitivity.[16]

Acknowledgment

The authors gratefully acknowledge financial support from the National Institutes of Health (GM–13901, 69–2161, 71–2303, 2055 and RR 254) and the Robert A. Welch Foundation (Q–125). Technical assistance was furnished by

Misses C. Weise and P. Crain. D. M. Desiderio was a fellow
of the Intra-Science Research Foundation (1971-1975).

References

1. Burgus, R., T. F. Dunn, D. M. Desiderio, D. N. Ward,
   W. Vale, and R. Guillemin. Nature 226, 321 (1970).
2. Ungar, G., D. M. Desiderio and W. Parr. Nature 238,
   198 (1972).
3. Vilkas, E., and E. Lederer.   Tet. Lett. 3089 (1968).
4. Vetter-Diechtl, H., W. Vetter, W. Richter, and K.
   Biemann. Experientia 24, 340 (1968).
5. Polan, M. L., W. J. McMurray, S. R. Lipsky, and S.
   Lande. Biochem. Biophys. Res. Commun. 38, 1127 (1970).
6. Leclercq, P. A., L. C. Smith, and D. M. Desiderio.
   Biochem. Biophys. Res. Commun. 45, 937 (1971).
7. Björndal, H., C. G. Hellerqvist, B. Lindberg, and S.
   Svensson. Angew. Chem. Int. Ed. 9, 610 (1970).
8. Thomas, D. W., B. C. Das, S. D. Géro, and E. Lederer.
   Biochem. Biophys. Res. Commun. 32, 519 (1968).
9. Leclercq, P. A., and D. M. Desiderio. Anal. Lett. 4,
   305 (1971).
10. White, P. A., and D. M. Desiderio. Anal. Lett. 4,
    141 (1971).
11. Polan, M. L., W. J. McMurray, S. R. Lipsky, and S.
    Lande. J. Amer. Chem. Soc. 94, 2847 (1972).
12. Leclercq, P. A., and D. M. Desiderio. Biochem. Biophys.
    Res. Commun. 45, 308 (1971).
13. Matsuo, H., Y. Baba, R. M. G. Nair, A. Arimura, and
    A. V. Schally, Biochem. Biophys. Res. Commun. 43,
    1334 (1971).
14. Biemann, K., C. Cone, B. R. Webster, and G. P.
    Arsenault. J. Amer. Chem. Soc. 88, 5598 (1966).
15. Weise, C. L., and D. M. Desiderio. Internat. J. of
    Computers in Biology and Medicine, in press.
16. Leclercq, P. A., K. Hagele, B. S. Middleditch, R. M.
    Thompson, and D. M. Desiderio. *Twentieth Annual
    Conference on Mass Spectrometry and Allied Topics*,
    Dallas, June, 1972.

THERMAL DEGRADATION OF PEPTIDES TO DIKETOPIPERAZINES AND
ITS APPLICATION TO SEQUENCE DETERMINATION

A. B. Mauger. Research Foundation of the Washington
Hospital Center, Washington, D.C. 20010.

THE FORMATION OF DIKETOPIPERAZINES during thermal degradation of peptides has various analytical applications,
particularly since their gas chromatographic properties
have been described.[1,2] The technique reported here involves
pyrolysis-gas chromatography, P G C, in which the peptide
is held at 400° for 20 sec in the flowing carrier gas. The
pyrograms so obtained are reproducible and quantitation can
be introduced by addition to the sample of a suitable internal standard. Thus far the only diketopiperazines produced
result from adjacent pairs of amino acids in the peptide.

Since dipeptides cyclize readily at elevated temperatures,
the P G C method has potential for their quantitative analysis.
For example, L-prolyl-L-proline, which is difficult to measure
using the amino acid analyzer, was converted by pyrolysis to
the diketopiperazine in quantitative yield. It was therefore
possible to measure this dipeptide using P G C with an internal standard.

Thermal degradation of tripeptides under comparatively
mild conditions was investigated in 1938.[3] Tripeptides with
the general sequence ABC generated the diketopiperazine AB
and the amino acid C. This reaction can even proceed at
room temperature in the case of D-valyl-L-prolyl-sarcosine.[4]
Several tripeptides were submitted to the P G C technique.
The results are summarized in Table I. With one exception
the tripeptides in which glycine is *N*-terminal produced both
AB and BC diketopiperazines, with AB predominating. For
example, glycyl-leucyl-alanine gave glycyl-leucyl and
leucyl-alanyl diketopiperazines in yields of 67 and 23%

Table I

Pyrolysis–Gas Chromatography   (P G C) of Tripeptides

Tripeptide ABC	Diketopiperazines Produced	
	AB	BC
H–Gly–Gly–Ala–OH	+ +	+
H–Ala–Gly–Gly–OH	+ +	–
H–Gly–Leu–Ala–OH	+ +	+
H–Ala–Leu–Gly–OH	+ +	–
H–Gly–Phe–Ala–OH	+ +	+
H–Ala–Phe–Gly–OH	+ +	–
H–Gly–Pro–Ala–OH	+ +	–

respectively, for a total of 90% conversion to one or other
diketopiperazine. In such cases the data establish the
sequence. However, those tripeptides in which glycine was
C–terminal produced only the AB diketopiperazine, as did
the case in which B was proline. In the latter instance,
as in the aforementioned room–temperature reaction, the AB
peptide bond can adopt the *cis* conformation, permitting
more rapid cyclization.

    In early studies on the sequence of actinomycin C_1,
valyl–prolyl and sarcosyl–N–methylvalyl diketopiperazines
were isolated and identified after thermal degradation in
hydrazine hydrate.[5] Assuming that threonine cannot be
C–terminal in view of the lactone structure, 16 of 48 possible sequences remained. With the far simpler P G C
procedure prolyl–sarcosyl diketopiperazine was identified
in addition to the above, reducing the number of possible
sequences to two and utilizing only micrograms of actinomycin.
Application of this method to novel actinomycins was reported
recently.[6] It is noteworthy that N–terminal degradative
sequencing procedures cannot be applied to cyclopeptides
such as actinomycin, and that in this series mass spectrometry
has also been unproductive thus far. In these circumstances
the P G C approach to sequence determination is useful.

    In extending this technique to peptides in general,
various limitations are apparent. Theoretically, if a

diketopiperazine is produced from each adjacent pair of amino acids in the peptide, a minimum of two possible sequences (unless the sequence is symmetrical) result, and the number may be higher where sequential degeneracy exists. In practice, there are limits to the length of peptide chain which will produce all possible diketopiperazines, and there are problems with those amino acids which possess a third functional group. Prior permethylation[7],[8] of the peptide provides an approach to solving some of these problems, since polar groups are protected and *cis* peptide bond formation is promoted. Furthermore, gas chromatography of the resulting methylated diketopiperazines[1],[9] is far more satisfactory than that of the underivatized compounds.

For the identification of peaks in pyrograms of peptides, comparison of retention times with those of all the possible diketopiperazines is obviously cumbersome. Present studies include the use of gas chromatography–mass spectrometry for this purpose.

Acknowledgment

This investigation was supported by Public Health Service Research Grant No. CA-11627 from the National Cancer Institute.

References

1. Mauger, A. B. J. Chromatog. 37, 315 (1968).
2. Westley, J. W., V. A. Close, D. E. Nitecki, and B. Halpern. Anal. Chem. 40, 1888 (1968).
3. Lichtenstein, N. J. Amer. Chem. Soc. 60, 560 (1938).
4. Meienhofer, J., Y. Sano, and R. P. Patel. In *Peptides: Chemistry and Biochemistry*, Weinstein, B., and S. Lande, eds. (New York: Marcel Dekker, 1970) pp 419–434.
5. Brockmann, H., G. Bohnsack, and C. H. Süling. Angew Chem. 68, 66 (1956).
6. Mauger, A. B. Chem. Comm. 39 (1971).
7. Das, B. C., S. D. Gero, and E. Lederer. Biochem. Biophys. Res. Comm. 29, 212 (1967).
8. Coggins, J. R., and N. L. Benoiton. Canad. J. Chem. 49, 1968 (1971).
9. Mauger, A. B., R. B. Desai, I. Rittner, and W. J. Rzeszotarski. J. Chem. Soc. In press.

AUTOMATED EDMAN DEGRADATION MONITORING OF SOLID-PHASE
PEPTIDE SYNTHESIS

Hugh D. Niall, G. W. Tregear, J. Jacobs. Endocrine
Unit, Massachusetts General Hospital, and Department
of Medicine, Harvard Medical School, Boston,
Massachusetts.

MONITORING OF THE COUPLING and deprotection steps in solid-
phase peptide synthesis is of the greatest importance in
insuring completeness of reaction at each cycle and thus
avoiding or reducing the extent of the formation of deletion
peptides. No single method yet devised is satisfactory for
this purpose. New approaches are needed for both "real
time" monitoring during the progress of the synthesis, and
"retrospective" monitoring of the purity of the final product, whether still attached to the resin or after purification. We have examined the use of Edman degradation for
these purposes, using an automated procedure with radioactive
(^{35}S) phenylisothiocyanate, in the Beckman model 890
"Sequencer."

Completeness of Coupling

The method is illustrated in Figure 1. Note that the
deprotection step is conveniently carried out in the
sequenator cup, using trifluoroacetic acid or heptafluoro-
butyric acid in 1 chloro-butane. The same reagent can be
used for the cleavage step in the Edman degradation.
Special programs are necessary, particularly with the
Merrifield-type resins, to avoid washout of polymer particles from the cup. Here the "undercut" cup of the most
recent Beckman Model 890 "Sequencer" is especially helpful,
since the undercut tends to keep the resin particles from

COMPLETENESS OF COUPLING

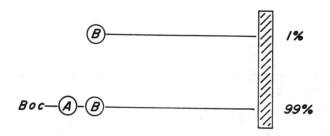

1 Add aliquot of resin to sequenator

2 Deprotect

3 React with $C_6H_5NCS^{35}$

4 Measure ratio $\dfrac{[PTH - \circledB]}{[PTH - \circledA]}$

Figure 1: Procedure for monitoring completeness of coupling reaction using automated Edman degradation.

rising up the cup wall. This problem does not arise with the graft copolymer resins developed for solid-phase synthesis[1] since their high density prevents any washout.

Figure 2 shows the results of a degradation cycle on a peptide from bovine parathyroid hormone of sequence Phe-Met-His-Asn-Leu-Gly-Resin, after addition of the phenylalanine. The gas chromatographic tracing of the phenylthiohydantoin (PTH) amino acids obtained shows that *N*-terminal Met is still present at the 3% level, with trace amounts of Leu, Gly and possibly Asn. The ninhydrin test[2] in this case was negative, indicating the greater sensitivity of the end-group approach.

Completeness of Deprotection

Methods for specific monitoring of the deprotection reaction using automated Edman degradation are currently being developed, but will not be discussed here.

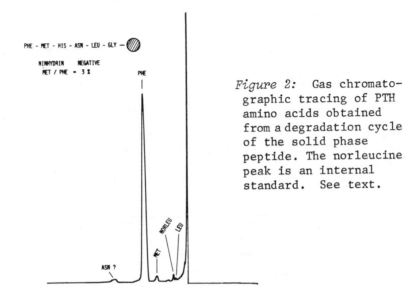

Figure 2: Gas chromatographic tracing of PTH amino acids obtained from a degradation cycle of the solid phase peptide. The norleucine peak is an internal standard. See text.

Purity of the Final Product

The rationale of this approach is illustrated in Figure 3. If there is a 1% deletion error at each step in the synthesis of a pentapeptide A-B-C-D-E, then when repetitive cycles of Edman degradation are carried out, the ratio of "wrong" to "correct" amino acid found at cycle 1 will be 1/99, at cycle 2, 2/98, and so on. It can be seen that the ability to detect deletion errors is greatly improved by an "amplification factor" inherent in the Edman method. Thus if the method could not detect an incomplete reaction of, say, 0.5% at any one cycle, the cumulative effect of successive Edman cycles on the final product would soon make the problem obvious.

We have applied this approach to a variety of synthetic peptides produced by the solid-phase approach and have found that deletion errors of the order of 0.5-1.0% per cycle are common even when the ninhydrin test has suggested that the coupling reaction was complete. Incompleteness of deprotection could of course also be responsible.

This approach assumes that the Edman degradation does not cleave off more than one residue at any one cycle. This seems to be true except perhaps for histidine residues for which a premature cleavage reaction has been described.[3]

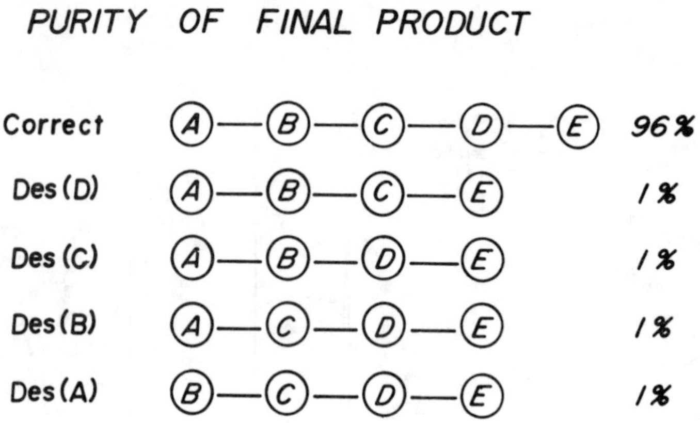

PURITY OF FINAL PRODUCT

Correct   (A)—(B)—(C)—(D)—(E)   96%

Des(D)   (A)—(B)—(C)—(E)   1%

Des(C)   (A)—(B)—(D)—(E)   1%

Des(B)   (A)—(C)—(D)—(E)   1%

Des(A)   (B)—(C)—(D)—(E)   1%

EDMAN CYCLE	1	2	3	4
RATIO $\frac{Error}{Correct}$	$\frac{1}{99}$	$\frac{2}{98}$	$\frac{3}{97}$	$\frac{4}{96}$

Figure 3: Application of Edman degradation in assessing purity of a synthetic pentapeptide. Note amplification of deletion errors with successive degradation cycles.

Conclusions

The automated Edman degradation can be used as an effective monitoring procedure for solid-phase peptide synthesis. A single cycle of degradation with identification can be completed in about 2 hours; this time can almost certainly be reduced by appropriate program changes. The method has several advantages over existing approaches. It is quantitative, allowing kinetic measurements. It is direct, since the amino acids with free *N*-terminals are identified directly. It is particularly sensitive in the detection of amino acid deletions in the final product since the errors at individual steps are additive.

References

1. Tregear, G. W. This volume, see page 175.
2. Kaiser, E., Colescott, R. L., Bossinger, C. D., and Cook, P. I. Anal. Biochem. $\underline{34}$, 595 (1970).
3. Schroeder, W. A. *Methods in Enzymology*, Volume XI (New York: Academic Press, 1967) p 445.

CHEMICAL METHOD OF SEQUENTIAL DEGRADATION OF POLYPEPTIDES
FROM THE CARBOXYL ENDS

*Saburo Yamashita,** *Nobuo Ishikawa*. Biochemistry
Division, Hoshi College of Pharmacy, 2-4-41 Ebara,
Shinagawa-ku, Tokyo, Japan

CHEMICAL SEQUENTIAL DEGRADATIONS of peptides from the
carboxyl ends have been one of the interests in peptide
chemistry for almost half a century.[1-7] Formation of
thiohydantoin rings[8] to selectively cleave peptide bonds
at the carboxyl end has been studied. Recently milder
conditions have been introduced to improve hitherto
employed rather drastic conditions.[5,6] Attention was
also directed to minimize the required quantity of polypeptide. This communication deals with the mechanism of
thiohydantoin ring formation at C-terminal amino acids,
minimization of the quantity of polypeptide, and initial
rate of hydrolysis of peptidyl thiohydantoins.

The intermediate C-terminal activated compound must
be either II or III, or both. Starting from optically
active L-aspartic acid, L-tyrosine, L-arginine, and L-
valine, it was found that the optical activity of the
resulting thiohydantoins was zero, which might indicate
that the thiohydantoin formation proceeds through compound
III. Considering the above mechanism and the suggestion
of Waley and Watson[2] that the direct formation of acyl
isothiocyanate (IV) is desirable to avoid occasional side
reactions by the labile oxazolone (III), the use of acetyl
chloride[6] to form the mixed anhydride (II) seems to be
advantageous. In fact, the yield of thiohydantoins by
way of (II)[6] is better than that through (III).[4,5]

---

*To whom inquiries should be directed.

$$\underset{\text{(I)}}{\text{Dns-NH-CH-C-N-CH-C-OH}} \quad \xrightarrow[\text{(CF}_3\text{CO)}_2\text{O}]{\text{Ac}_2\text{O}} \quad \left[\underset{\text{(II)}}{\text{Dns-NH-CH-C-N-CH-C-O-C-CH}_3} \right.$$

with R_2, O, H, R_1, O groups on (I); R_2, O, H, R_1, O, O groups on (II)

$$\left. \underset{\text{(III)}}{\text{Dns-NH-CH}-\overset{R_2}{\text{C}} \begin{matrix} \text{O} ----- \text{C=O} \\ | \qquad\qquad | \\ \qquad \text{CH-R}_1 \\ \diagdown \text{N} \diagup \end{matrix}} \right]$$

$$\xrightarrow{\text{NCS}^-} \quad \underset{\text{(IV)}}{\text{Dns-NH-CH-C-N-CH-CONCS}} \quad \longrightarrow \quad \underset{\text{(V)}}{\text{Dns-NH-CH-C}-\begin{matrix} \overset{R_1}{\text{CH}} ----- \text{C=O} \\ | \qquad\qquad | \\ \text{N} \qquad\quad \text{NH} \\ \diagdown \underset{\|}{\text{C}} \diagup \\ \text{S} \end{matrix}}$$

with R_2, O, H, R_1 groups on (IV); R_2, O groups on (V)

$$\xrightarrow[\text{(H}_2\text{O)}]{\text{H}^+} \quad \underset{\text{(VI)}}{\text{Dns-NH-CH-C-OH}} \quad + \quad \underset{\text{(VII)}}{\begin{matrix} \overset{R_1}{\text{CH}} ----- \text{C=O} \\ | \qquad\qquad | \\ \text{HN} \qquad\quad \text{NH} \\ \diagdown \underset{\|}{\text{C}} \diagup \\ \text{S} \end{matrix}}$$

with R_2, O groups on (VI)

Amino acid sequence analysis of polypeptides was successful up to around 10 amino acids from the C-termini,[7] but a rather large quantity (10^{-8} to 10^{-7} mole) of starting material was required, since during the repeated cycles the recovery of the remaining peptide from Sephadex column chromatography was difficult when monitored by the optical density at 280 nm.[7] Thus, the amino terminus of the original peptide was marked by dansylation (I), and the eluate containing the Dns-peptide could be recovered almost quantitatively, and 10^{-8} to 10^{-9} mole of the peptide can be used for the sequence analysis. Dansylation of thiohydantoins (VII) was not successful.

The initial rate of hydrolysis of peptidyl and acetyl thiohydantoin (V) to give VI and VII was studied spectrophotometrically[5] [shift of peak absorption from 280 nm (V)

to 260 nm (VII)]. It was found that the half life of
peptidyl thiohydantoin (V) was generally approx. 2 min in
0.1 N HCl (comparable to H^+ form of Dowex resin[5] at 25°.
Peptidyl thiohydantoins were more rapidly hydrolyzed than
acetyl thiohydantoins. The findings may contribute to
set up experimental conditions for the sequential analysis
of peptides.

References

1. Schlack, P., and W. Kumpf. Hoppe-Seyler's Z. Physiol.
   Chem. <u>154</u>, 126 (1926).
2. Waley, S. G., and J. Watson. J. Chem. Soc. 2394 (1951).
3. Stark, G. R. Biochemistry <u>7</u>, 1796 (1968).
4. Cromwell, L. D., and G. R. Stark. Biochemistry <u>8</u>,
   4735 (1969).
5. Yamashita, S. Biochim. Biophys. Acta <u>229</u>, 301 (1971).
6. Kubo, H., T. Nakajima, and Z. Tamura. Chem. Pharm.
   Bull. (Tokyo) <u>19</u>, 210 (1971).
7. Yamashita, S., and N. Ishikawa. Proc. Hoshi Pharm. <u>13</u>,
   136 (1971).
8. Johnson, T. B., and W. H. Scott. J. Amer. Chem. Soc.
   <u>35</u>, 1136 (1913).

CHEMISTRY AND BIOLOGY OF PEPTIDES

THE PRIMARY STRUCTURE OF AN APOLIPOPROTEIN FROM THE HUMAN
HIGH DENSITY LIPOPROTEIN FAMILY: ApoLp–Gln–II (ApoA–II)

H. B. Brewer, Jr., S. E. Lux, R. Ronan, K. M. John.
Molecular Disease Branch, National Heart and Lung
Institute, National Institutes of Health, Bethesda,
Maryland 20014

THE PLASMA LIPOPROTEINS can be classified by ultracentrifugal
flotation or paper electrophoresis into four major classes
which include chylomicrons, very low density, low density,
and high density lipoproteins (HDL).[1,2] HDL isolated by
ultracentrifugal flotation (densities 1.063 to 1.21) consists of an approximately equal proportion of protein and
lipid.[3] Approximately 90% of the delipidated protein
moiety of HDL is composed of two major apoproteins; the
remaining 10% is composed of the three major apoproteins
associated with the very low density lipoprotein family.
The two major apoproteins have been designated as apoA–I
and apoA–II, and by their carboxyl terminal residues as
apoLp–Gln–I and apoLp–Gln–II.[4-10] Scanu *et al.* have previously reported that the molecular weight of apoLp-Gln-II
(designated as fraction IV) can be reduced from 16,000–
17,000 to 7,500–8,500 by cleavage of a disulfide bond
indicating that the intact apoprotein is composed of two
polypeptide chains.[11]
    The purpose of this report is to summarize our recent
studies on the covalent structure of apoLp–Gln–II.

Materials and Methods

    The HDL utilized in these studies was isolated from a
single healthy volunteer by ultracentrifugal flotation,
delipidated, and fractionated by chromatography on DEAE

cellulose in 6*M* urea.[12]   Analytical methods have been previously reported in detail.[12,13]   Enzymatic digestion of
apoLp-Gln-II with trypsin, cyanogen bromide cleavage, and
isolation of the individual polypeptide fragments by ion
exchange and gel permeation chromatography has been described
elsewhere.[14]   Isolated fragments were shown to be homogeneous
by disc gel electrophoresis, tlc, amino acid analysis, and
Edman amino terminal analysis.   Enzymatic digestion of
apoLp-Gln-II with pyrrolidonecarboxylyl peptidase (E:S,1:6)
was performed in 0.05*M* NH$_4$HCO$_3$ (pH 7.8).   Following the
incubation the sample was lyphilized, and then extracted
with acetone or dimethylformamide to solubilize the pyrrolidone carboxylic acid.   The extract was evaporated to dryness
with nitrogen, redissolved in a small volume of dimethylformamide and identified by mass spectroscopy (Finnigan
Quadropole, isobutane carrier gas) or as the trimethylsilyl
derivative by gas-liquid chromatography on the CFC blended
column at 105°C.[15]   Carboxypeptidase A and B digestion
(E:S, 1:50) was performed in 0.2*M* NH$_4$HCO$_3$ (pH 7.6), and the
released amino acids were identified by amino acid analysis.

    Manual phenylisothiocyanate degradations were performed
by the three stage procedure of Edman.[16]   Automated degradations were performed on the Beckman Sequencer (Model 890B).
Selected peptide fragments were treated with 4-sulfophenyl-
isothiocyanate prior to automated degradation to increase
their hydrophilicity.[17]   The phenylthiohydantoin amino acids
(PTH amino acids) were identified by tlc,[16] gas-liquid
chromatography,[15,18] and mass spectroscopy.[19]

Results

    Human apoLp-Gln-II has been isolated to homogenity by
ultracentrifugation followed by DEAE cellulose chromatography.[12]   The purified apoprotein gave a single band on
disc gel electrophoresis (Figure 1A). Following reduction
of the cystine residue with 2-mercaptoethanol and carboxymethylation with iodoacetic acid the molecular weight of
apoLp-Gln-II (SCMC apoprotein) was reduced from approximately 18,000 to 9,000 by gel filtration in 6*M* guanidine
confirming the report of Scanu *et al.*[11] that the apoprotein
is composed of two polypeptide chains.   The SCMC apoprotein
could not be separated into distinct chains by disc gel
electrophoresis (Figure 1A), gel filtration, or DEAE-
cellulose chromatography.[12]   Edman degradation of the
intact apoprotein or SCMC apoprotein revealed no detectable
amino terminal residue. Digestion of the SCMC apoprotein
with pyrrolidonecarboxylyl peptidase revealed the amino
terminal amino acid to be pyrrolidone carboxylic acid

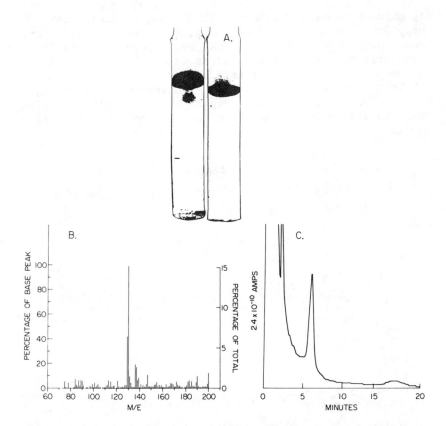

Figure 1: Disc gel electrophoresis (A) of native apoLp-
Gln-II (left) and reduced and carboxymethylated (SCMC)
apoLp-Gln-II (right); (B) Chemical-ionization mass
spectrum and (C) gas-liquid chromatogram of the amino
acid residue, pyrrolidone carboxylic acid, released
following digestion of SCMC apoLp-Gln-II with
pyrrolidonecarboxylyl peptidase.

(pyroglutamic acid) by mass spectroscopy (Figure 1B) or
gas-liquid chromatography (Figure 1C). Carboxypeptidase A
digestion revealed glutamine as the carboxyl terminal and
threonine as the penultimate residue. These combined results which are reported in detail elsewhere indicated
that apoLp-Gln-II is composed of two identical polypeptide
chains connected by a single disulfide bridge.[12]
   The amino acid composition of the SCMC apoprotein was
determined by timed acid hydrolysis, and total enzymatic

digestion, and is as follows: $SCMC_1$, $<Glu_1$, Asp_2, Asn_1, Thr_6, Ser_6, Glu_8, Gln_7, Pro_4, Gly_3, Ala_5, Val_6, Met_1, Ile_1, Leu_8, Tyr_4, Phe_4, Lys_9. The SCMC apoprotein contains 77 amino acid residues. There was no detectable carbohydrate on the purified SCMC apoprotein.

Tryptic digestion of the SCMC apoprotein revealed eight major and one minor peptide (Figure 2). The major peptides range in size from a single residue, lysine, T-7, to a 22

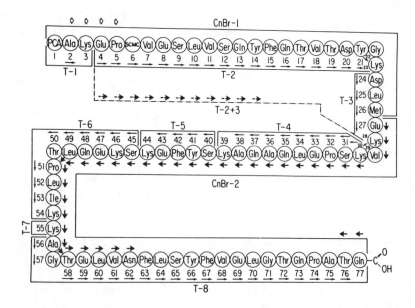

Figure 2: Amino acid sequence of reduced and carboxymethylated apoLp-Gln-II. *Large open arrows* refer to Edman
degradations on the apoprotein following digestion with
pyrrolidonecarboxylyl peptidase; *small arrows* to degradations on the isolated tryptic peptides; *small dashed arrows*
to degradations on tryptic peptide T-2 + 3; *large solid
arrows* to degradations on cyanogen bromide peptide CNBr-2;
and *large solid arrows* of residues 76 and 77 to carboxypeptidase A digestion of the apoprotein.

amino acid peptide, T-8. The minor peptide was a peptide
composed of two tryptic peptides, T-2 + 3. T-1, a tripeptide, contained pyrrolidone carboxylic acid as the amino
terminal residue and was designated the first tryptic

peptide; T-8 was devoid of a basic residue, and contained glutamine as the carboxyl terminal residue indicating that it was the carboxyl terminal peptide (Figure 2). Cyanogen bromide cleavage of the SCMC apoprotein produced two peptides, CNBr-1 and CNBr-2. CNBr-1 contained homoserine-homoserine lactone by amino acid analysis, amino terminal pyrrolidone carboxylic acid, and was assigned to the amino terminal position (Figure 2). The amino terminal and carboxyl terminal residue of CNBr-2 were glutamic acid and glutamine respectively indicating that CNBr-2 was the carboxyl terminal cyanogen bromide peptide (Figure 2).

The complete covalent structure of apoLp-Gln-II was determined by manual and automated Edman degradations on the intact SCMC apoprotein, and tryptic and cyanogen bromide fragments. Peptide T-1 was initially digested with pyrrolidonecarboxylyl peptidase, and then degraded by the Edman procedure with PTH alanine being identified as the second residue (Figure 2). Each of the remaining tryptic peptides T-2 to T-7 were degraded both by the manual as well as the automated technique (small arrows, Figure 2). Peptide T-8, the carboxyl terminal peptide, contains no basic residue and was degraded only by the manual procedure (small arrows, Figure 2). The carboxyl terminal residue of each of the tryptic peptides was determined by amino acid analysis following the appropriate number of cycles for each peptide. The alignment of the tryptic peptides T-1 and T-2 was established by digestion of 1.0 μmol of the intact SCMC apoprotein with pyrrolidonecarboxylyl peptidase followed by four Edman degradations (open arrows, Figure 2). In addition, the overlap tryptic peptide T-2 + 3, which contained the amino acids present in peptides T-2 and T-3, was degraded with results consistent with the alighment of the peptides as T-2 followed by T-3 (dashed arrows, Figure 2). An automated degradation of 1.2 μmol of CNBr-2 was performed in order to align peptides T-3 to T-8. This degradation (35 cycles) aligned the tryptic peptides (T-3 to T-8), and confirmed the sequence of the manual degradations on tryptic peptides T-4 to T-7 (large solid arrows, Figure 2). The carboxyl terminal sequence provided by the Edman degradation of T-8 was confirmed by carboxypeptidase digestion of the SCMC apoprotein with the carboxyl terminal residue being glutamine, and the penultimate residue threonine (Figure 2).

These combined results provided a single unique sequence for the SCMC apoLp-Gln-II (Figure 2). The structure for the intact apoprotein, apoLp-Gln-II, is shown in Figure 3.

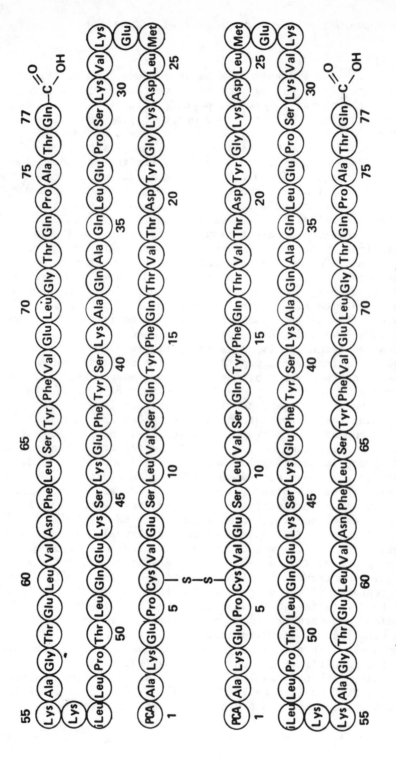

Figure 3: Covalent structure of apoLp-Gln-II (apoA-II).

Discussion

The covalent structure of apoLp-Gln-II is unique in that it contains two identical polypeptide chains. Each chain contains 77 residues connected by a single disulfide bridge at position 6 in the sequence. The apoprotein contains a large number of glutamic acid functions, serine, threonine, and lysine residues, however no tryptophan, histidine, or arginine. The amino terminal residue is pyrroline carboxylic acid, and the carboxyl terminal residue is glutamine. No unique or repeating sequence is present in the molecule, however at a number of positions in the sequence (positions 3,4; 23,24; 27,28; 43,44; and 46,47) a dicarboxylic acid residue is adjacent to a basic residue. ApoLp-Ala, an apoprotein isolated from the very low density lipoprotein family, also has an association of basic and dicarboxylic acid residues (positions 24,25; 51,52; 58,59,60).[20]

The lipid binding site(s) of the plasma lipoproteins is as yet undefined. It is interesting to speculate that the high frequency of paired basic and dicarboxylic acid residues is important in this binding. In this regard, molecular models are compatible with the alignment of the negatively charged dicarboxylic acid with the positively charged choline or ethanolamine, and the negatively charged phosphate with the positively charged amino group of the basic amino acid. Further studies will be necessary to test this hypothesis, and to extend these studies of the plasma lipoproteins to a study of the lipid-protein interactions in membrane proteins in general. In addition, the elucidation of the covalent structure of the plasma lipoproteins will now enable more detailed studies to be performed on the role of this unique class of proteins in lipid transport in normal as well as disease states.

References

1. deLalla, O. F., and J. W. Gofman. Methods Biochem. Anal. 1, 459–478 (1959).
2. Scanu, A. J. Lipid Res. 7, 295–306 (1966).
3. Fredrickson, D. S., R. I. Levy, and R. S. Lees. New England J. Med. 276, 32 (1967).
4. Shore, B., and V. Shore. Biochemistry 7, 2773–2777 (1968).
5. Shore, V., and B. Shore. Biochemistry 7, 3396–3403 (1968).
6. Shore, B., and V. Shore. Biochemistry 8, 4510–4516 (1969).

7. Scanu, A., J. Toth, C. Edelstein, S. Koga, and E. Stiller. Biochemistry 8, 3309-3316 (1969).
8. Rudman, D., L. A. Garcia, and C. H. Howard. J. Clin. Invest. 49, 365-372 (1970).
9. Camejo, G., Z. M. Suarez, and V. Munoz. Biochim. Biophys. Acta 218, 155-166 (1970).
10. Kostner, G., and P. Alaupovic. FEBS Lett. 15, 320-324 (1971).
11. Scanu, A. M., C. Edelstein, and C. T. Lim. FEBS Lett. 18, 305-307 (1971).
12. Lux, S. E., K. M. John, and H. B. Brewer, Jr. J. Biol. Chem. (1972) in press.
13. Brewer, H. B., Jr., H. T. Keutmann, J. T. Potts, Jr., R. A. Reisfeld, R. Schueter, and P. L. Munson. J. Biol. Chem. 243, 5739-5747 (1968).
14. Lux, S. E., K. M. John, R. Ronan, and H. B. Brewer, Jr. J. Biol. Chem. (1972) in press.
15. Pisano, J. J., T. Bronzert, and H. B. Brewer, Jr. Anal. Biochem. 45, 43-59 (1972).
16. Edman, P. *Protein Sequence Determination* (New York: Springer-Verlag, 1970) pp 211-255.
17. Braunitzer, G., B. Schrank, and A. Ruhfus. Hoppe-Seyler's Z. Physiol. Chem. 351, 1589-1590 (1970).
18. Pisano, J. J., and T. Bronzert. J. Biol. Chem. 244, 5597-5607 (1969).
19. Fales, H. M., Y. Nagai, G. W. A. Milne, H. B. Brewer, Jr., T. J. Bronzert, and J. J. Pisano. Anal. Biochem. 43, 288-299 (1971).
20. Brewer, H. B., Jr., R. Shulman, P. Herbert, R. Ronan, and K. Wehrly. Advan. Exp. Med. Biol. 26, 280 (1972).

STRUCTURAL STUDIES ON BOVINE PEPSINOGEN AND PEPSIN

Beatrice Kassell, John Kay, Joseph P. Marciniszyn, Jr.
Department of Biochemistry, Medical College of Wisconsin,
Milwaukee, Wis. 53233

BOVINE PEPSINOGEN AND PEPSIN have proved to be sufficiently
different in composition and specificity from the better
known porcine zymogen and enzyme to be of considerable
interest.[1,2,3] This report describes the sequence information that has been obtained for the C-terminus of the enzyme
and for the peptides released from the N-terminus of the
zymogen on activation.

Bovine pepsinogen was isolated as described previously[1]
and freed from a trace of extraneous peptides by affinity
chromatography[4] on polylysine-Sepharose 4B in 0.05 M
phosphate, pH 6.5. The pure zymogen was activated at pH
2.0 and 0° for 4 min. The activation mixture was adjusted
to pH 3.5 and immediately applied to the affinity column
(2.6 x 43 cm), equilibrated in 0.05 M formate, pH 3.5.
The activation peptides were not retarded and were eluted
as one fraction whereas the bovine pepsin was retained by
the resin and was eluted with a linear gradient of NaCl.

The peptide mixture was fractionated by column chromatography on Bio-Gel P2 and by high voltage paper electrophoresis. Sequence analysis gave the results shown in
Figure 1, with porcine pepsinogen[5] and bovine prorennin[6]
included for comparison.

Thus, the N-terminal portion of bovine pepsinogen is
much closer in structure to porcine pepsinogen than to
bovine prorennin.

Bovine pepsin contains 3 residues susceptible to attack
by trypsin.[3] In porcine pepsin, the corresponding susceptible residues are present among the 20 amino acids at the
carboxyl terminus, the sequence of which is known.[7]

713

	1	5	10	15

A. Leu-Val-Lys — Val-PRO-Leu-Val-Arg-Lys-LYS-SER-LEU-ARG-Gln-Asn-LEU
B. Ser-Val-Lys-Leu-Ile-PRO-Val-Val-Lys-Lys-LYS-SER-LEU-ARG-Gln-Asn-LEU
C. Ala-Glu-Ile-Thr-Arg — Ile-PRO-Leu-Tyr-Lys-Gly-LYS-LYS-SER-LEU-ARG-Lys-Ala-LEU

	20	25	30	35

A. -Ile-Lys-Asp-GLY-Lys-LEU-Lys-Asp-PHE-LEU-LYS-THR-HIS-Lys — His-ASN-Pro-Ala
B. [Ile-Glu-Asn-GLY-Lys-LEU-Lys-Glu]PHE][LEU-LYS-THR-HIS(Lys,Val)Arg-ASN-Met-Gly
C. -Lys-Glx-His-GLY-Leu-LEU-Glu-Asp-PHE-LEU-LYS

	40	45

A. -SER-LYS-TYR-Phe-Pro-Ala-Glu
B. -SER-LYS-TYR-Leu][Ile-Arg-Glu][Ala-Ala-Thr-Leu]
C.

Figure 1: The N-terminal sequences of porcine (A) and bovine (B) pepsinogens and of bovine prorennin (C).

Bovine Val-Glu(Glu,Glu,Thr,Ser,Pro,Gly,Ala,Leu,ILE,LEU,GLX,ASX,ASX,ASX,SER,THR,GLY)Tyr-PHE-

Porcine   ILE-LEU-GLN-ASP-ASP-ASP-SER(THR,GLY,Ser)PHE-

Porcine Glu-Gly-Met-Asn-Val-Pro-Thr-Ser-Ser-Gly-Glu-Leu-Trp-Ile-Leu-Gly-Asp-Val-Phe-Ile-Arg-

Bovine [GLN-TYR-Phe-THR-VAL-PHE-ASP-ARG-Gly-ASN-ASN-Gln-ILE-GLY-LEU-ALA-PRO-VAL-ALA-COOH]

Porcine GLN-TYR-Tyr-THR-VAL-PHE-ASP-ARG-Ala-ASN-ASN-Lys-ILE-GLY-LEU-ALA-PRO-VAL-ALA-COOH

Figure 2: Partial C-terminal sequences of bovine and porcine pepsins.

Bovine pepsin, prepared as described previously,[4] was denatured[8] at pH 11.2 and 37.5° for 15 min and was digested with TPCK-trypsin at pH 8 for 2 hr at 37.5°. The digestion products were separated on Sephadex G-75 into 4 fractions. The carboxyl terminal portion of 19 amino acids, the sequence of which was determined by Rasmussen and Foltmann[9] was present in the fraction of lowest molecular weight, and resembles the structure of this area of porcine pepsin (Figure 2).

A fraction of intermediate size was purified by ion exchange chromatography. This peptide of 53 amino acids, containing a single arginine residue and corresponding approximately in composition to the penultimate carboxyl terminal region of porcine pepsin,[7,10] was further digested with chymotrypsin at pH 7.8 and 42° for 5 hours. One chymotryptic peptide, containing 21 amino acids, was purified. Part of this peptide resembles a portion of the sequence of porcine pepsin (Figure 2); the substitution of tyrosine for serine involves a single base change. Other peptides are being purified for sequence investigation.

Acknowledgments

This work was supported by USPHS grant AM-09826 and by a grant from the Patrick & Anna M. Cudahy Fund. We are indebted to Dr. Foltmann for sending a copy of his paper before publication.

References

1. Chow, R. B., and B. Kassell. J. Biol. Chem. 243, 1718 (1968).
2. Meitner, P. A., and B. Kassell. Biochem. J. 121, 249 (1971).
3. Lang, H. M., S. S. J., and B. Kassell. Biochemistry 10, 2296 (1971).
4. Nevaldine, B., and B. Kassell. Biochim. Biophys. Acta 250, 207 (1971).
5. Ong, E. B., and G. E. Perlmann. J. Biol. Chem. 243, 6104 (1968).
6. Foltmann, B. Seventh FEBS meeting, 1971, abstract no. 137.
7. Dopheide, T. A. A., S. Moore, and W. H. Stein. J. Biol. Chem. 242, 1833 (1967).
8. Huang, W.-Y., and J. Tang. J. Biol. Chem. 245, 2189 (1970).
9. Rasmussen, K. T., and B. Foltmann. Acta Chem. Scand. 25, 3873 (1971).
10. Kostka, V., and L. Morávek. Eighth FEBS Meeting, 1972, Abstract No. 365.

SYMPOSIUM DISCUSSIONS

Summarized by Johannes Meienhofer

THE LUCID SURVEY about amino acid analysis (S. Moore) and
the disclosure of a new fluorometric reagent (S. Udenfriend)
were followed by an animated discussion. The audience was
reminded of a simple tryptophane determination in hydrolysates by fluorometry after dilution, since tryptophane is
highly fluoroescent (excited at 285 nm, fluorescence at 350
nm). No chromatographic separation is required.

The probability was discussed whether a mixture of
different components could give the same amino acid analysis
as that expected for the pure protein under study, and indeed this was experienced with synthetic trypsinogen activation peptide (Val-Asp$_4$-Lys); and it is also known to occur
with analyses of preparations synthesized by the solid-phase
method. As a discussant stated, it is essential for a
meaningful evaluation of amino acid analyses of solid-phase
synthetic products that (a) all steps of the synthesis are
monitored, *e.g.* by titration, and (b) step by step analysis
can be done *during* the synthesis instead of afterwards.

As larger and larger peptides are being synthesized a
correct amino acid analysis does not necessarily prove
homogeneity, if taken as the only criterion. However, that
must not be used as an excuse for not determining amino
acid analysis at all; rather, this as well as elemental
analysis are *"necessary but not sufficient"* data.

The question was raised whether any of the various
published procedures for hydrolysis of resin-bound peptides
has by now been found to be fully satisfactory, but apparently that seemed not yet to be the case. It was suggested
that data obtained by peptide-resin hydrolysis should be
checked by data obtained after removal of the peptide from
the solid phase by alcoholysis.

A discourse developed on precision of amino acid analysis (to improve on the present 3% maximum deviation would appear to be very difficult) and also on the benefits and dangers of very high sensitivity. Not only do the demands for purity increase when smaller and smaller samples are analyzed, but also natural environmental levels of amino acids start to become a serious source of potential contamination. In current practice (S. Moore) analyses are routinely done at the 10 nanomole level (0.2 ml HCl or CH_3SO_3H and small tubes for hydrolysis) not requiring special purification of acids or blank runs (or stopping to breathe!). Only samples of short supply are analyzed at the 1 nanomole level, and then HCl and H_2O need to be glass-distilled, glassware to be especially well rinsed, and blank runs are required.

Fluorescamine assays, however, can be done at the picomole level (pp 655 to 663). The reagent[1] is also being applied to monitor solid-phase synthesis for completeness of peptide coupling.[2] A discussant wondered whether ninhydrin-negative peptides, as Ile-Arg, would react with fluorescamine; this was not yet known but just under investigation. A number of questions and comments dealt with properties of fluorescamine. The reagent is said to be stable in dry form, not acutely toxic, and gives good color yields even with β-alanine, aspartic acid, also with spermine and spermidine, but not with proline, N-methylamino acids, or any other secondary amine. Quenching appears not to be a problem. Excess reagent hydrolyses fast and there is no interfering absorption at 390 nm.

The interesting structure of viomycin (pp 665 to 670) was discussed with respect to its high extinction at 268 nm which has been ascribed to the molecule through synthesis and spectrophotometric examination of model compounds.

The suggestion that α,β-unsaturated amino acids might play a role in the generation of peptides from larger precursors (pp 671 to 678) led to a lively discussion touching on the source of thyrotropin releasing factor formation, nisin oligomerization, addition of amides across keto functions, and a possible separation of LH and FSH releasing activities in LH-RH/FSH-RH.

Reference

1. Weigele, M., S. L. DeBernardo, J. P. Tengi, and W. Leimgruber. J. Amer. Chem. Soc. **94**, 5927 (1972).
2. Felix, M. A., unpublished.

SECTION X

FORECAST

AN ATTEMPT TO FORECAST THE FUTURE OF PEPTIDE CHEMISTRY

Ralph Hirschmann. Merck Sharp & Dohme Research Laboratories, Division of Merck & Co., Inc., Rahway, New Jersey 07065.

SUMMARY--The problems in forecasting scientific developments in peptide chemistry are briefly discussed. The discussion about plausible future developments includes a brief discussion of the current role of peptides in medicine, possible future application of peptides for ecological purposes and a brief discussion of the impact on peptide chemistry of the recently discovered releasing hormones. The problems in the design of an antagonist of a peptide hormone are briefly considered.

In the field of protein chemistry the synthetic problems are assessed as are problems related to the folding of proteins. An attempt is made to forecast some future developments involving membrane proteins. The lecture concludes with a brief analysis of the unique properties of peptides.

Problems of Forecasting

"Then came the magicians, the astrologers, the Chaldeans and the soothsayers." *Daniel*, IV, 7, *c.* 165 B.C.

Attempting to predict the future of peptide chemistry is at the same time a fascinating undertaking and a hopeless assignment. As scientists engaged in peptide research we find prognostication about peptide chemistry difficult because we tend to predict the future by extending the present. Most of us find it hard to envisage and foresee those developments which are unexpected and which are therefore

721

often most dramatic in their impact. Thus I believe that
my comments this afternoon are likely to fail to predict
the most exciting developments of the next 30 years, and
are most apt to predict fulfillment of endeavors which have
already been initiated. This is not a new dilemma. No one
prior to Sumner's isolation of urease in 1926[1] could have
predicted that enzymes would prove to be proteins. Similarly the enormous progress in peptide synthesis during
the past 40 years became predictable only after Professor
Leonidas Zervas, an Honorary President of this symposium,
and the late Max Bergmann introduced the carbobenzoxy
protecting group in 1932.[2] The effect on peptide chemistry
of the synthesis of oxytocin by Professor Vincent du
Vigneaud,[3] our other Honorary President, and of the elucidation of the structure of insulin by Sanger[4] is well known
to all of us. The point is that once the unexpected has
been discovered or accomplished, nearly everyone can predict the next step, but it is unlikely that these developments would have been predictable *a priori*.

Some Plausible Future Developments

Ecology

Since we are living in a society which is becoming
increasingly ecology minded, growing cognizance is likely
to be taken of the fact that our peptides are biodegradable
substances. The fact that Zaoral and Sláma[5] have succeeded
in synthesizing an active analog of insect juvenile hormone
containing two amide bonds may be viewed as a step in this
direction.

Current Role of Peptide Hormones in
Medicine. Synthetic Advances.

It is reasonable to assume, however, that the most important developments in peptide chemistry will continue to
be tied to biological phenomena. The importance of insulin
as a life-saving agent has been recognized for many years
and a limited number of other peptide hormones such as ACTH
are also enjoying a steadily growing role in medicine. It
may take many years before biologists and physicians can
determine the clinical value of recently discovered hormones
such as calcitonin and of such other biologically-active
peptides as nerve growth factor and gastric inhibitory
polypeptide, some of which have been discussed at this
symposium. It is noteworthy and reassuring that a polypeptide

such as ACTH, or at least one of its fragments, can be
synthesized in kilogram amounts. The synthesis of peptides
in the molecular weight range of 3000 to 4000 daltons can
now be approached with confidence. Often the synthesis
entails the combined use of solution chemistry and synthesis
on a solid support. Since advances in sequence determination
have easily equalled those in synthesis, we have reached the
point where *chemistry* is no longer the rate determining step
on the road from the isolation of a biologically-active
peptide to its clinical use, at least for those compounds
with a molecular weight which does not greatly exceed 4,000
daltons. Compared with hurdles such as safety assessment,
detailed evaluation by biologists, pharmacologists and
clinical investigators, the problems of synthesis are now
beginning to be regarded as being among the easiest to
solve. We may view this development with some pride, although I suspect that this state of affairs is likely to
become also the source of growing frustration on the part
of the synthetic peptide chemist, who hopes for rapid
clinical acceptance of the compounds which he has been
able to synthesize.

Impact of the Releasing Hormones

At recent peptide symposia in this country and in Europe
it seemed to some of us that although everybody was working
very hard, it was not clear where we were really moving.
This year we seem to be more confident about the immediate
future of peptide chemistry, and this is due largely, I
believe, to the impact of recent developments in the field
of the releasing hormones, which have received much attention
at this symposium. Compounds such as TRH and LH-RH are
active at exquisitely low dose levels, they are of low
molecular weight and can be prepared by synthesis in pure
form. This is important because, by contrast, the proteins
released by these hormones are macromolecules. For example,
LH-RH is a decapeptide, but LH is a glycoprotein which has
not been sequenced, much less synthesized. Thus a glycoprotein becomes biologically accessible *via* the synthesis
of a small peptide. For the same reason the isolation and
synthesis of GH-RH--especially human GH-RH--will hold promise
for those who are now dependent for normal growth on injections of isolated GH, the supply of which is discouragingly
low and the large-scale synthesis of which still seems out
of reach. Admittedly, those dwarfs whose pituitaries do
not function would not find GH-RH a replacement of GH. It
also seems likely that biological surprises will emerge from

research with the releasing hormones. The fact that TRH
induces the release of prolactin in man is a case in point.
Finally, I should like to suggest that the hypothalamus
will not prove to be the last source of novel peptides of
biological significance.

Hormone Antagonists. Opportunities and Problems

Peptide hormones are important not only in their own
right, but often also because they can be expected to guide
us in the design of an antagonist. In some cases the
antagonists are likely to be of greater interest than the
hormones themselves. Examples of this situation include
gastrin, angiotensin II and LH-RH.

I should point out, however, that in general we have
been more effective in finding inhibitors of enzymes than
antagonists of hormones. The explanation, I believe, lies
in the fact that in the design of enzyme inhibitors we can
often be guided by knowledge about both the enzyme and its
substrate. This is not the case with hormones where our
knowledge about the receptor is generally very limited.
Indeed, our probing the shape of a receptor *via* synthetic
hormone analogs has been compared with the use of a screw-
driver by a blind man who seeks to discover the shape of a
grand piano. As medicinal chemists we are skillful in
systematizing and rationalizing the information about
structure-activity relationships which has been accumulated.
However, to be able to say that we can truly understand
these data, we must be in a position to predict. This is
clearly much more difficult and will, in my opinion, require
a breakthrough in concept as yet unforeseen. Let me illustrate the complexity of the problem with a recent observation.
From the elegant and extensive studies with gastrin analogs
by J. S. Morley[6] and his associates at ICI we thought we
"knew" that an aspartic acid residue in the penultimate
position of the carboxy-terminal end of the molecule is
indispensable for gastrin-like activity. The fact that a
tetrazolyl analog was fully active nicely supported the
proposed receptor model. How startling then was the observation by H. H. Trout and M. I. Grossman[7] that an octapeptide
in which alanine replaces the aspartic acid residue stimulates
gastric acid secretion. In his effort to gain understanding
about receptors, the peptide chemist is no worse off than
other medicinal chemists who seek to "see" the receptor with
the aid of our currently available tools, both chemical and
physical. These tools are likely to seem quite primitive to
those who will attend peptide symposia 30 years from now.

Still we must make the most of the methods now at our disposal and it is indeed encouraging that several papers at this symposium reported exciting progress with the synthesis of angiotensin antagonists possessing *in vivo* activity. I believe, therefore, that in the years ahead we shall witness the synthesis of peptides that are hormone antagonists possessing clinical value. This will be a notable advance. I stress "of clinical value" because we in the pharmaceutical industry are acutely aware that compounds possessing only *in vitro* activity have no value in medicine.

Proteins

Challenges of Protein Synthesis

Let us now turn to larger molecules. Enzymes and other proteins will continue to challenge the peptide chemist for many years to come. More than ten years ago Dr. Crick wrote the following about proteins: "In biology, proteins are uniquely important. They are not to be classed with polysaccharides, for example, which by comparison play a very minor role. The main function of proteins is to act as enzymes." The synthesis of molecules possessing enzymatic activity was first accomplished in 1968 and announced simultaneously by two groups in 1969.[8,9] Since that time the syntheses of other enzymes and of growth hormone have been described. These studies have shown that proteins *can* be synthesized but they have also served to focus attention on the shortcomings of existing methodology.

In spite of important advances which have been described since 1969, it is apparent that the synthesis of a pure crystalline protein is still a very formidable undertaking. It seems safe to predict that the synthesis of crystalline enzymes will be announced in the near future. It is possible that this will initially be accomplished through ingenuity in the purification of the synthetic product and its precursors and that the synthesis of a pure protein without excessive recourse to resourceful purification techniques will follow only later on. I have no doubt that advances in synthesis will be forthcoming which will bring about these developments.

Other Challenges of Protein Chemistry

Enzymes will continue to challenge us and our colleagues in other disciplines also from a theoretical point of view. There may be several who feel that they have succeeded in

unraveling the mysterious catalytic powers of enzymes, but
I sometimes have the feeling that no two people share quite
the same explanation. Surely much remains to be done.
Peptide chemists have already contributed significantly to
our present understanding of enzyme catalyzed reactions.
Semi-synthetic approaches have proven highly rewarding in
giving us information about the role in binding of individual
amino acids of the protein. It is likely that our understanding will increase rapidly once the *total* synthesis of
analogs of enzymes becomes less of an ordeal than it is at
present. Another aspect of protein chemistry in which the
sea of the unknown is presently very great concerns the
folding of proteins. One is confident that the future will
see notable advances in this area, although I wonder whether
the techniques with which we are familiar today will suffice
to permit this breakthrough. We are all aware of the tremendous triumphs which X-ray crystallography has achieved
in elucidating the tertiary structure of proteins, but I
have always been disappointed that no one seems to be able
to look at a three-dimensional model of a protein and say,
"Yes, of course! Now I understand the folding of the
enzyme, its specificity and its catalytic activity."

Membrane Proteins

In discussing peptides and proteins even as recently as
in the late 1960's, the view was generally accepted that
nothing would ever be discovered which could rival enzymes
in their basic biological importance. After all, the enzymes combine the so-called twin miracles of specificity
and catalytic activity and they are required even for
nucleic acid synthesis. However, it seems that a rival to
the enzymes may be emerging in the form of membranes. It
is interesting, I think, that we tend to think of the
structural proteins of membranes with the same distaste
today with which the majority of chemists--in fact just
about everybody except Emil Fischer--regarded peptides 70
years ago. In an informative editorial review published
in *Nature* in 1969[10] the proteins of membranes were described
as "disagreeable materials." *Nature* went on to say: "Their
solubility properties are intractable, and because their
native environment is essentially non-aqueous and because
they exist in a complex state with lipids, there is no
reason to suppose that their physical properties when extracted bear any very close relation to those in the native
state." These very appropriate comments should be enough
to discourage anyone, but it is apparent that membrane

chemistry is in fact an active, exciting and rapidly expanding field, albeit an extremely difficult one. A recent achievement was the demonstration by Professor Hofmann[11] that a fragment of ACTH which was biologically active also showed high affinity for an adrenal cortical particulate fraction. What molecules could be better suited to interact selectively with cell membranes than peptides? That affinity chromatography, already an important tool of the protein chemist, will continue to play a critical role in the isolation of membrane receptors seems almost a certainty.

Outlook

Those of us who have witnessed the rise and decline of adrenocorticoid steroid synthesis from one of the most active fields of hormone research to one which is at least at present relatively quiescent might well ask whether the same fate might be in store for peptide chemistry as well. I do not think so. There is only so much even nature can do with the cyclopentanoperhydrophenanthrene ring system and there are only so many alkyl, halo and other substituents which even the most imaginative medicinal chemist can attach to this skeleton. On the other hand, using the 20 coded amino acids nature has found a way to build molecules of all sizes and shapes which display a high capacity to interact with each other. The future looks bright, not because we are so clever, but because the building blocks which are at our disposal permit the synthesis of compounds with seemingly unlimited versatility of chemical, physical and biological properties.

References

1. Sumner, J. B. J. Biol. Chem. <u>69</u>, 435 (1926).
2. Bergmann, M., and L. Zervas. Ber. <u>65</u>, 1192 (1932).
3. du Vigneaud, V., C. Ressler, J. M. Swan, C. W. Roberts, P. G. Katsoyannis, and S. Gordon. J. Amer. Chem. Soc. <u>75</u>, 4879 (1953).
4. Ryle, A. P., F. Sanger, L. F. Smith, and R. Kitai. Biochem. J. <u>60</u>, 541 (1955).
5. Zaoral, M., and K. Sláma. Science, <u>170</u>, 92 (1970).
6. Morley, J. S. Proc. Roy. Soc. B. <u>170</u>, 97 (1968).
7. Trout, H. H., and M. I. Grossman. Nature New Biology <u>234</u>, 256 (1971).
8. Gutte, B., and R. B. Merrifield. J. Amer. Chem. Soc. <u>91</u>, 501 (1969).

9. Hirschmann, R., R. F. Nutt, D. F. Veber, R. A. Vitali,
   S. L. Varga, T. A. Jacob, F. W. Holly, R. G. Denkewalter.
   J. Amer. Chem. Soc. $\underline{91}$, 507 (1969).
10. Nature $\underline{221}$, 614 (1969).
11. Hofmann, K., W. Wingender, and F. M. Finn. Proc. Natl.
    Acad. Sciences (USA) $\underline{67}$, 829 (1970).

The lack of cooperation and even communication has rendered much of this effort sterile. It seems to me--and I hope this is not merely wishful thinking--that in the coming second phase we shall have much better integration of the chemical and biological effort, and a much more rational approach to the whole problem both in the design and in the evaluation of the products. In such an approach I would also see the answer to Ralph Hirschmann's anticipatory bit of gloom about the frustration of the peptide chemist: If the chemist is able to participate vicariously in the thrills of the biological exploration, he need not fear a dull time.

Protein Models

Providing models of proteins has always been an important function of peptide chemistry, whether these were model substrates in the early studies of proteolytic enzymes, or the polymeric models which have taught us so much about the rules of the conformational game and, more recently, about immunogenicity. The model structures are becoming ever more sophisticated; today, as we have seen, the unique advantages of cyclic peptides as conformational models are being vigorously exploited. I believe that this function of peptide chemistry will continue to be important, though some of the ways in which we have contributed in the past have become obsolete through more direct approaches to the structure and properties of the proteins themselves. Looking today at the knowledge of catalytic mechanisms contributed by organic chemists interested in enzymology; at the information on enzyme structures available from x-ray crystallography; and extrapolating the growing confidence with which the conformations of cyclic peptides may be predicted, I think that we shall see renewed activity in the field of highly simplified enzyme models. There has already been some exploration of this field on what one might call the "let's-see-how-far-we-can-get" principle but I suggest that in the next few years we shall see more earnest and perhaps more successful efforts in this direction.

Methods

Isolation

In the work on the smaller biologically active peptides, isolation is to my mind still the most difficult and critical phase and its practitioners have my sincere respect and

admiration. Isolation is still largely an art, but surely
there are ways in which it can be made more of a science.
One of the most important of these is likely to be affinity
chromatography, coupled with immunology since antibodies
can in principle provide affinity sorbents which are made
to measure. The usual difficulty is the shortage of the
isolated material, of sufficient purity, which is required
to raise the antibodies in the first place; and when such
material does become available, there is always so much
that one would want to do with it that one is reluctant to
sacrifice it for this one purpose.

Perhaps the dilemma could be avoided by a sort of
iterative approach: Use a relatively crude extract, prepared by a sequence of conventional purification procedures,
to raise correspondingly heterogeneous antibodies; use and
reuse these in the form of an affinity sorbent to accumulate,
more simply, further material of the same degree of purity;
carry the purification of this material further by conventional techniques; use the more highly purified product to
raise a more specific antiserum; discard the first antibody,
and use the new one to extract purer material directly from
the crude extract; purify this material further... and so
on to final purification. This is the sort of procedure I
would expect to develop from the interaction with immunology.

Dr. J. Pisano has pointed out to me, and I entirely agree
with him, that in isolation work (as in the study of structure-activity relations) we shall have to develop more
communication and closer cooperation between the chemist
and the biologist: Very often, the bioassay forms the
bottleneck in the isolation of a biologically active material
and the development of a rapid and convenient bioassay may
be decisive for the rate of progress and, indeed, for the
success of a project.

Structure Determination

I would not dare to predict the outcome of the competition between ion-exchange chromatography, gas chromatography,
and mass spectrometry. I do remember William H. Stein once
telling me what a healthy stimulus it had been for the
further refinement of ion-exchange chromatography to have
mass spectrometry treading on its heels; by the same token,
however the race over the next lap may go--and it is going
to be neck and neck most of the way--we may expect that it
will bring accelerated progress and this can only profit us.

Synthesis

I am often shocked to realize to what extent we are
still working essentially by rule of thumb, in spite of the
highly advanced state of contemporary organic chemistry in
its kinetic and mechanistic aspects. We know very little
about the kinetics of coupling even for simple peptides
and Klaus Hofmann has pointed out how ignorant we are about
the behavior of larger peptides. It is perhaps a hope
rather than a prediction that peptide chemists, and perhaps
others who appreciate the importance of peptide synthesis,
will increasingly contribute to the store of mechanistic or
even merely quantitative information on which we can draw
to make our syntheses more rational. Normally, the synthetic chemist is of course impatient to get to his product,
but I would suggest that occasionally he could save a lot
of time by investing a little more of it in systematic
study and intellectual effort instead of proceeding by mere
empirical experimentation.

One synthetic approach which I confidently expect to
develop is the partial synthesis, or relay synthesis, of
proteins. It is almost inevitable that we should try to
use fragments obtained from readily available natural proteins in combination with synthetic fragments to get modifications of such proteins without having to go through
the complete synthesis of the whole peptide chain each time.
However, this approach will not be fully effective if we
pursue it merely with the methods which happen to be on
hand from analytical protein chemistry and classical peptide synthesis. Special degradative and synthetic techniques will have to be developed, or modified for this
specific purpose. We have seen a similar development,
which is still far from complete, in the techniques of
solid-phase synthesis; and indeed, even in classical synthesis tasks of increasing complexity may force us, or
should at least stimulate us, to take more trouble over
developing methods adequate to a given purpose rather than
trying to make do with the most commonly used standard
procedures.

On the other hand, the confrontation of conventional
peptide synthesis with "non-classical" techniques such as
solid-phase synthesis has shown drastically how large a
share of our manual labor is taken up by workup and isolation, because of the conditions imposed on us by the varying
properties of our products. It is only natural that in
"conventional" synthesis, too, we are consciously trying
to develop more stereotyped, universal procedures for

manipulating our intermediates, so as to gain some of the
advantages of the solid-phase and related techniques while
retaining the important assets of solution chemistry. I
would expect this development to continue with even greater
emphasis.

There is one type of synthetic approach to structure-
activity relations which would be anathema to the classical
organic chemist but which has some pragmatic justification:
What I would call the "subtractive" approach. By this I
mean the synthesis of a product, however ill-defined, in
which a particular amino-acid residue has been omitted or
changed, preferably by omission of a side-chain or functional group, and examination of this product for biological
activity. Such a procedure is valid as long as it is used
to give a "yes-or-no" answer, or strictly only a "yes" answer:
If the product *does* show biological activity, we may fairly
conclude that whatever has been left out is *not* essential
for activity. Provided we confine ourselves merely to
this statement and do not start drawing quantitative conclusions, this seems a perfectly legitimate method of
gathering limited information. We might expect to see it
increasingly used as suggested by Dr. N. Izumiya, for a
preliminary exploration preparatory to the design and
more rigorous synthesis of a defined, simplified structure.
So much in defense (if defense it needs) of the "subtractive
approach."

Purification and Characterization

We have repeatedly spoken about the need for new methods
of purifying and characterizing synthetic peptides, particularly large peptides. Such a need does, indeed, exist and
the methods will no doubt be developed; but we shall see a
good deal of improvement even if those methods which are
already available are more extensively and effectively used.
To give but one example, immunological criteria are now
generally applied by examining the reaction of a synthetic
product with antibodies to the purified natural material.
This is standing the method on its head. We should be using
the synthetic materials as antigens and test by e.g.
immunodiffusion if the synthetic product contains antigenic
molecules *other* than the natural product. Used properly,
I am sure immunological techniques will become increasingly
powerful and important in determining the purity of synthetic
materials.

I hope we shall be able to develop some degree of con-
sensus about the sort of evidence we shall require before

using terms such as "pure material," "fully identical," or
"the synthesis of such-and-such an enzyme." Unless we do
this, we shall be misleading others, who take these expres-
sions at their face value, about the state and prospects of
our field.

Peptide Symposia

One prediction I can make with complete confidence:
In spite of glamorous newcomers such as the prostaglandins,
peptides and proteins will continue to be an important and
challenging group of substances to the chemist and biologist
alike; and for that reason, Peptide Symposia such as this
will continue to be exciting and stimulating occasions.

CLOSING REMARKS

F. Merlin Bumpus. Research Division, Cleveland Clinic
Foundation, Cleveland, Ohio 44106

DR. FRUTON AFTER HEARING the presentation on scotophobin
has pointed out that he, too, has a new biologically active
peptide. He has named this "peptidophobin" and threatened
that if I talked too long he would synthesize it to "determine
its structure."
     But more seriously, comments on this meeting should
include comparisons of the three peptide symposia which
have been held in this country. The first one, which was
in 1968 at Yale,[1] was rather limited in scope; dealt largely
with sequence determination and synthesis of peptides. Very
little work on conformational analysis was presented. The
next, in 1970 in Cleveland,[2] was expanded considerably--
there was more discussion of conformational analysis; some
little time was given to the biologists. This lack of participation by pharmacologists was certainly a weakness of
the second meeting. I believe that this has been overcome
here; and if I am allowed to star-gaze along with my colleagues, I would suggest that we continue to join with the
biologists to obtain a better understanding of the peptide
hormone mechanisms.
     Another great advance in this meeting was the comment
made by Dr. Klaus Hofmann. He recognized solid-phase synthesis. The next step now is for Dr. Bruce Merrifield to
recognize fragment condensation. This would almost equal
Nixon's recent trip to China.
     During the meeting we heard much about conformational
analysis, energy minimization techniques, C-13 NMR, CD
studies, X-ray crystallography; and to many of us, it was
quite a surprise to learn of the "looseness of structure,"

737

especially of a diketopiperazine. Much was made of the
solvent and salt effects, and rightfully so. At times it
seemed that with the proper choice of solvents we can each
have our own peptide conformation. There will be enough
to go around this way.

Fortunately, those in the field of conformational
analysis are applying their talents to the simpler molecules; especially to cyclic peptides, and those that are
involved in ion binding are extremely interesting. From
the papers given at the meeting one must conclude that
this interest has led to much work on new methods of
cyclization, both by solid-phase and solution methods.
We learned about these cyclizations, both when they're
wanted or unwanted. We also have been informed about new
and better protecting groups, different condensation
methods, and polymer modification which may either reduce
or increase the degree of cyclization.

Problems related to cross-linking and swelling of
resins used for solid-phase synthesis have been discussed.
With continued effort along these lines increased yields
will result from the solid-phase methods. Methodology for
the determination of the completeness of condensation reactions has been discussed. This has not been stressed
enough, and I am sure that great progress will be made in
this area in the future.

Several new biologically active peptides have been
characterized; some have been obtained by modification of
structure. One of the interesting things presented was
the isolation of a "T" shaped phagocytosis stimulating
peptide (Tuftsin); there aren's many chemists able to have
a conformation directly from the name of their isolated
compounds.

There was much discussion on angiotensin, including
the presentation of a new group of competitive antagonists.
They'll have tremendous usefulness; I think, possibly, not
so much clinically as we had hoped for. New inhibitors to
renin, to converting enzyme have been synthesized.

New protecting groups for amines and other functional
groups were reported. Many of these will find great use
and, fortunately, they are useful for solid-phase as well
as for the conventional methods. The new catalysts for
condensation are going to be helpful, and I am sure that
many of us will adapt these readily.

The previous speakers have discussed the importance of
the recent accomplishments in the field of the hypothalamic
hormones. It may have been interesting if they had told us
more how these are being used on the population problem;
possibly that's for our next meeting.

If I'm allowed to make one comment concerning predictions for the future, I would predict that the receptor mechanisms are going to be studied a lot more than in the past. Many people feel that all hormones must act *via* the cyclic AMP mechanism. This may not be necessarily true. Present research in this area of mechanism analysis, only in its infancy, should certainly lead to some very interesting results in the near future.

The Program Committee was very wise in saving the discussion on analytical techniques for the last session. As we constantly heard throughout the meeting there are many problems concerning purity of synthetic peptides. Conclusions concerning the identity of the synthetic product(s) are often too broad and based upon too few criteria of homogeneity.

The methodology that we have heard about here should lend guidance during the next few years and should remind us to limit our claims to that which can be justified by sufficient evidence. I think we're all somewhat guilty from time to time when we are anxious to obtain a biologically active substance synthetically. The use of synthetic peptide analogs for studies on structure-activity relationship has become commonplace. A high degree of homogeneity and exact identity is of utmost importance. As research progresses on interaction of hormone with receptor site, use of pure and properly characterized hormone preparations will become even more important.

And finally, on behalf of all here attending, I think that we owe a debt of great gratitude to Dr. Meienhofer and his Program Committee for planning this meeting. It was well organized. It's been informative and I think we've all had a good time and we'll all be looking forward to the next meeting.

References

1. The proceedings of the First American Peptide Symposium, Yale University, New Haven, Connecticut, August 12-15, 1968, have been published in: *Peptides: Chemistry and Biochemistry*, Weinstein, B., and S. Lande, eds. (New York: Marcel Dekker, 1970).
2. The proceedings of the Second American Peptide Symposium, Cleveland Clinic Foundation, Cleveland, Ohio, August 17-19, 1970, have been published in: *Progress in Peptide Research*, Lande, S., ed. (New York: Gordon and Breach, 1972).

LIST OF PARTICIPANTS

H. L. Aanning, University of Nebraska, Omaha, Neb. 61805.
K. Agarwal, Massachusetts Inst. of Tech., Cambridge, Mass. 02139
B. M. Altura, Yeshiva University, Bronx, N.Y. 10461
M. S. Amer, Mead Johnson Research Center, Evansville, Ind. 47721
G. W. Anderson, Lederle Laboratories, Pearl River, N.Y. 10965
R. Anderson, Pierce Chemical Co., Rockford, Ill. 61105
Ruth H. Angeletti, Wash. Univ. School of Medicine, St. Louis, Mo.
E. Atherton, Children's Cancer Res. Found., Inc., Boston, Mass.
S. Bajusz, Res. Inst. for Pharmaceutical Chemistry, Budapest
L. Balaspiri, Medical College of Ohio, Toledo, Ohio 43614
L. E. Barstow, University of Arizona, Tucson, Ariz. 85721
C. Bayley, Cyclo Chemical, Los Angeles, Cal. 90001
C. A. Benassi, University of Ferrara, Ferrara, Italy
C. Bennett, Merck Sharp & Dohme Res. Lab., Rahway, N.J. 07065
N. L. Benoiton, University of Ottawa, Ottawa, Canada
Zmira Bernstein, Whitehall, Pa. 18052
H. C. Beyerman, Delft Univ. of Tech., Delft 8, The Netherlands
K. Biemann, Mass. Inst. of Technology, Cambridge, Mass. 02139
E. R. Blout, Harvard Medical School, Boston, Mass. 02115
Agnes Bodanszky, Case Western Reserve Univ., Cleveland Ohio
M. Bodanszky, Case Western Reserve Univ., Cleveland, Ohio
D. Borovas, Cyclo Chemical, Los Angeles, Cal. 90001
H. E. Bosshard, Washington University, St. Louis, Mo. 63110
C. D. Bossinger, Armour Pharmaceutical Co., Kankakee, Ill. 60901
F. A. Bovey, Bell Laboratories, Murray Hill, N.J. 07974
R. A. Bradshaw, Wash. Univ. School of Medicine, St. Louis, Mo.
H. B. Brewer, National Heart & Lung Inst., Bethesda, Md. 20014
E. Bricas, Université de Paris, 91 - Orsay, France
H. Brown, Harvard Medical School, Boston, Mass. 02115
T. Brown, University of Colorado, Denver, Colo.
K. Brunfeldt, Danish Inst. of Protein Chem., Copenhagen
F. M. Bumpus, Cleveland Clinic, Cleveland, Ohio 44106
R. Burgus, The Salk Institute, San Diego, Cal. 92112
J. A. Burton, Mass. General Hospital, Boston, Mass. 02114
B. W. Bycroft, Univ. of Nottingham, Nottingham NG72RD England
F. M. Callahan, Lederle Laboratories, Pearl River, N.Y. 10965
R. Camble, Imperial Chemical Indus. Ltd., Macclesfield England
S. F. Cernosek, Jr., Brandeis Univ., Waltham, Mass. 02154

I. M. Chaiken, Natl. Inst. of Health, Bethesda, Md. 20014
K.-K. Chan, Hoffmann-La Roche Inc., Nutley, N.J. 07110
R. K. Chawla, Emory Univ. School of Medicine, Atlanta Ga. 30322
R. Chow Cheng, Dow Chemical Co., Midland, Mich. 48640
M. Chrétien, Clinical Res. Inst. of Montreal, Montreal, Canada
R. Ciabatti, Dow Chemical Co., Midland, Mich. 48640
R. L. Colescott, Armour Pharmaceutical Co., Kankakee, Ill. 60901
A. Cosmatos, Mt. Sinai Sch. of Med. of CUNY, New York, NY 10029
L. C. Craig, Rockefeller University, New York, NY 10021
W. Danho, University of Baghdad, Baghdad, Iraq
C. M. Deber, Harvard Medical School, Boston, Mass. 02115
G. S. Denning, Jr., Norwich Pharmacal Co., Norwich, NY 13815
D. M. Desiderio, Baylor College of Medicine, Houston, Tex. 77025
J. Don, Harvard University, Cambridge, Mass. 02138
L. C. Dorman, Dow Chemical Co., Midland, Mich. 48640
M. S. Doscher, Wayne State Univ. Sch. of Med., Detroit, Mich.
M. W. Draper, Rockefeller University, New York, NY 10021
V. du Vigneaud, Cornell University, Ithaca, NY 14850
D. F. Dyckes, Cornell University, Ithaca, NY 14850
B. Elpern, Schwartz/Mann, Orangeburg, NY 10962
B. W. Erickson, Rockefeller University, New York, NY 10021
A. Failli, Ayerst Research Labs., Montreal, Canada
A. M. Felix, Hoffmann-LaRoche Inc., Nutley, NJ 07110
D. C. Fessler, Norwich Pharmacal Co., Norwich, NY 13815
F. M. Finn, University of Pittsburgh, Pittsburgh, Pa. 15213
E. Flanigan, Brookhaven National Lab., Upton, LI, NY 11973
G. Flouret, Northwestern University, Chicago, Ill. 60611
B. F. Floyd, The Procter & Gamble Co., Cincinnati, Ohio 45239
H.-J. Förster, Universität Zürich, Zürich, Switzerland
R. Freer, University of Colorado, Denver, Colo. 80220
M. Fridkin, Mass. Institute of Tech., Cambridge, Mass. 02139
O. M. Friedman, Collaborative Res., Inc., Waltham, Mass. 02154
P. Fromageot, Cen. d'Études Nucl. de Saclay, Gif-Sur-Yvette, Fr
J. S. Fruton, Yale University, New Haven, Conn. 06520
D. Gallagher, Schwarz/Mann, Orangeburg, NY 10962
M. Garff, Beckman Instr., Inc., Palo Alto, Calif. 04304
H. Garg, Mass. General Hospital, Boston, Mass. 02178
T. Gerritsen, Univ. of Wis. Medical Center, Madison, Wis. 53706
H. J. Giallombardo, New England Nuclear Corp., Boston, Mass.
C. Gilardeau, Clinical Res. Inst. of Montreal, Montreal, Canada
L. Gill, Children's Cancer Res. Found., Inc., Boston, Mass.
B. F. Gisin, The Rockefeller University, New York, NY 10021
C. B. Glaser, University of California, San Francisco, Cal.
J. Glickson, Univ. of Alabama in Birmingham, Birmingham, Ala.
M. Goodman, University of California, San Diego, Cal. 92037
H. J. Goren, The University of Calgary, Calgary, Alberta, Canada
R. I. Gregerman, Baltimore City Hospital, NIH, Baltimore, Md.
J. H. Griffin, National Inst. of Health, Bethesda, Md. 20014
C. M. Groginsky, The University of Arizona, Tucson, Ariz. 85721
E. Gross, National Inst. of Health, Bethesda, Md. 20014
M. A. Gross, National Inst. of Health, Bethesda, Md. 20014

M. Guarneri, University of Ferrara, Ferrara, Italy
M. A. Guiducci, Schwartz/Mann, Orangeburg, NY 10962
R. Guillemin, The Salk Institute, San Diego, Cal. 92112
F. R. N. Gurd, Indiana University, Bloomington, Ind. 47401
F. S. Guziec, Jr., Mass. Inst. of Tech., Cambridge, Mass. 02139
M. Halperin, Children's Cancer Res. Found., Inc., Boston, Mass.
B. Hansen, Steno Memorial Hospital, Gentofte, Denmark
J. F. Harbaugh, Beckman Instruments, Inc., Palo Alto, Cal. 94304
C. H. Hassall, Roche Prod. Ltd., Welwyn Garden City, England
E. Hayon, U.S. Army Natick Laboratories, Natick, Mass. 01760
R. F. Hirschmann, Merck Sharp & Dohme, Rahway, NJ 07065
R. G. Hiskey, Univ. of North Carolina, Chapel Hill, NC 27514
R. S. Hodges, Rockefeller University, New York NY 10021
K. Hofmann, University of Pittsburgh, Pittsburgh, Pa. 15213
T. R. Hollands, Dept. of Natl. Health & Welfare, Ottawa, Canada
D. A. Hoogwater, Delft Univ. of Technology, Delft 8, Netherlands
V. J. Hruby, University of Arizona, Tucson, Arizona 85721
J. L. Hughes, Armour Pharmaceutical Co., Kankakee, Ill. 60901
H. Immer, Ayerst Research Labs., Montreal, Quebec, Canada
J. K. Inman, National Inst. of Health, Bethesda, Md. 20014
M. Itoh, Case Western Reserve Univ., Cleveland, Ohio 44106
D. A. J. Ives, Univ. of Toronto, Willowdale, Ontario, Canada
N. Izumiya, Kyushu University, Fukuoka, Japan
J. Jacobs, Massachusetts General Hospital, Boston, Mass. 02114
P. Jacobs-Brooks, Northeastern University, Boston, Mass. 02115
E. Jay, Massachusetts Institute of Tech., Cambridge, Mass. 02139
W. H. Johnson, Laboratory of Molecular Biophysics, Oxford, Engl.
D. A. Jones, G. D. Searle & Co., Chicago, Ill. 60680
W. C. Jones, Cornell University, Ithaca, N.Y. 14850
E. C. Jorgensen, University of California, San Francisco, Cal.
H. Joshua, Merck Sharp & Dohme Res. Labs., Rahway, NJ 07065
E. Kaiser, Armour Pharmaceutical Co., Kankakee, Ill. 60901
E. T. Kaiser, University of Chicago, Chicago, Ill. 60637
A. Kapoor, St. John's University, Jamaica, NY 11432
I. Karle, Naval Research Laboratory, Washington, DC 20390
B. Kassell, Medical Col. of Wisconsin, Milwaukee, Wis. 53233
E. Katz, Georgetown University, Washington, DC 20007
J. C. Kauer, DuPont Experimental Station, Wilmington, Del. 19899
J. M. Kauffman, New England Nuclear Corp., Watertown, Mass.
K. Kawasaki, University of Pittsburgh, Pittsburgh, Pa. 15213
J. A. Kelley, Massachusetts Inst. of Tech., Cambridge, Mass.
D. S. Kemp, Massachusetts Inst. of Tech., Cambridge, Mass. 02139
H. T. Keutmann, Massachusetts General Hosp., Boston, Mass. 02114
P. A. Khairallah, Cleveland Clinic Found., Cleveland, Ohio 44106
M. C. Khosla, Cleveland Clinic Found., Cleveland, Ohio 44106
L. Kisfaludy, Gedeon Richter, Budapest X, Hungary
Y. S. Klausner, Case Western Reserve Univ., Cleveland, Ohio
J. E. Klingbeil, Abbott Laboratories, No. Chicago, Ill. 60064
W. König, Farbwerke Hoechst AG, 6230 Frankfurt(Main)-Höchst, Germ.
K. D. Kopple, Illinois Inst. of Technology, Chicago, Ill. 60616
D. Kotelchuck, Mt. Sinai School of Medicine, New York, NY 10029

J. Kovacs, St. John's University, Jamaica, NY 11432
K. Juromizu, Children's Cancer Res. Found., Inc., Boston, Mass.
H. Lackner, Universität Göttingen, 3400 Göttingen, West Germany
S. Lande, Yale University, New Haven, Conn. 06510
C. Lantz, Pierce Chemical Co., Rockford, Ill. 61105
A. E. Lanzilotti, Lederle Laboratories, Pearl River, NY 10965
D. A. Laufer, University of Massachusetts, Boston, Mass. 02116
J. Lecocq, University of California, Berkeley, Cal. 94720
K.-Y Lee, Children's Cancer Res. Found., Inc., Boston, Mass.
C. H. Li, University of California, San Francisco, Cal. 94122
N. Ling, The Salk Institute, San Diego, Cal. 92112
K. Lübke, Schering AG, Berlin 65, Germany
J. A. MacLaren, CSIRO Wool Res. Labs., Parkville N2, Australia
V. Madison, Harvard Medical School, Boston, Mass. 02115
R. Makineni, Bachem, Inc., Marina Del Rey, Cal. 90291
M. Manning, Medical College of Ohio, Toledo, Ohio 43614
A. Marglin, Tufts University, Boston, Mass. 02111
D. L. Marshall, Battelle Memorial Inst., Columbus, Ohio 43201
G. R. Marshall, Washington University, St. Louis, Mo. 63110
G. Matsueda, University of Colorado, Denver, Colo. 80220
A. B. Mauger, Washington Hospital Center, Washington, DC 20010
J. McDermott, University of Ottawa, Ottawa, Ontario, Canada
W. H. McGregor, Wyeth Lab., Inc., Philadelphia, Pa. 19101
J. Meienhofer, Children's Cancer Res. Found., Inc., Boston, Mass.
R. B. Merrifield, Rockefeller University, New York, NY 10021
R. A. Mikulec, G. D. Searle & Co., Chicago, Ill. 60680
A. R. Mitchell, Rockefeller University, New York, NY 10021
E. J. Modest, Children's Cancer Res. Found., Inc., Boston, Mass.
M. W. Monahan, The Salk Institute, San Diego, Cal. 92112
S. Moore, Rockefeller University, New York, NY 10021
C. Moriarty, Children's Cancer Res. Found., Inc., Boston, Mass.
J. D. Morrisett, Baylor College of Medicine, Houston, Texas 77025
T. Mukaiyama, Tokyo Institute of Technology, Tokyo, Japan
V. A. Najjar, Tufts University, Boston, Mass. 02110
H. Nau, Massachusetts Inst. of Tech., Cambridge, Mass. 02139
P. Needleman, Washington Univ. Medical School, St. Louis, Mo.
V. Nelson, Ayerst Research Laboratories, Montreal, Canada
J. J. Nestor, Jr., Cornell University, Ithaca, NY 14850
H. Nesvadba, Sandoz Forschunginstitut, A-1235 Wien, Austria
H. D. Niall, Massachusetts General Hospital, Boston, Mass. 02114
E. D. Nicolaides, Parke, Davis & Co., Ann Arbor, Mich 48105
D. E. Nitecki, University of California, San Francisco, Cal.
R. Noble, University of California, San Francisco, Cal. 94122
K. Noda, National Inst. of Health, Bethesda, Md. 20014
E. G. Nybell, Ohio State University, Columbus, Ohio 43210
Y. Okada, Mt. Sinai Sch. of Med. of CUNY, New York, NY 10029
R. Oliver, Pierce Chemical Co., Rockford, Ill. 61105
M. A. Ondetti, The Squibb Inst. for Med. Res., New Brunswick, NJ
D. Ontjes, University of North Carolina, Chapel Hill, NC 27514
Yu. A. Ovchinnikov, Shemyakin Inst. for Chem. of Nat. Prod., Moscow

A. C. M. Paiva, Escola Paulista de Medicina, São Paulo, Brazil
K. N. Parameswaran, Yale Univ. Sch. of Med., New Haven, Conn.
W. K. Park, Laval University, Quebec 10e, Quebec, Canada
W. Parr, University of Houston, Houston, Texas 77004
A. Patchornik, The Weizmann Inst. of Science, Rehovoth, Israel
D. Patel, Bell Laboratories, Murray Hill, N.J. 07974
R. P. Patel, Tufts University, Boston, Mass. 02111
J. Patterson, Schwartz/Mann, Orangeburg, NY 10962
M. J. Peach, University of Virginia, Charlottesville, Va. 22903
L. G. Pease, Harvard Medical School, Boston, Mass. 02115
J. F. Pechére, Centre de Recherches Biophysiques et Biochimiques
    du C.N.R.S., 34-Montpellier, France
W. J. Peterson, Peninsula Laboratories, San Carlos, Cal. 94070
I. Photaki, The University of Athens, Athens 144, Greece
J. Pisano, National Inst. of Health, Bethesda, Md. 20014
D. Ponzi, Massachusetts Inst. of Tech., Cambridge, Mass. 02139
J. Porath, Biokemiska Inst., 751 21 Uppsala 1. Sweden
J. T. Potts, Jr., Massachusetts General Hosp., Boston, Mass.
J. Ramachandran, University of California, San Francisco, Cal.
D. Regoli, Centre Hospitalier Univ., Sherbrooke, Quebec, Canada
C. Ressler, Institute for Muscle Disease, Inc., New York NY 10021
D. H. Rich, University of Wisconsin, Madison, Wis. 53706
W. Rittel, Ciba-Geigy Ltd., 4002 Basle, Switzerland
J. Rivier, The Salk Institute, San Diego, Cal 92112
J. Roberts, Sloan-Kettering Inst. for Cancer Res., New York NY
R. Rocchi, Universita di Ferrara, Ferrara, Italy
R. Rodin, Boyce Thompson Inst. for Plant Res., Yonkers, NY 10701
R. W. Roeske, Indiana Univ. Sch. of Med., Indianapolis, Ind.
J. D. Rosamond, Cornell University, Ithaca, NY 14850
R. Roskoski, Jr., Rockefeller University, New York NY 10021
M. Rothe, University of Mainz, 65 Mainz, Germany
R. Roy, Robroy Scientific, Marshfield, Mass. 02050
J. Rudinger, Eidg. Tech. Hochschule Zürich-Hönggerberg, Zürich
M. E. Safdy, Miles Laboratories, Inc., Elkhart, Ind. 46514
S. I. Said, University of Texas, Dallas, Texas 75235
S. Sakakibara, Protein Research Found., Minoh, Osaka, Japan
T. S. A. Samy, Children's Cancer Res. Found., Boston, Mass.
B. Sarkar, The Hospital for Sick Children, Toronto, Canada
J. P. Scannell, Hoffmann-La Roche Inc., Nutley, NJ 07110
P. W. Schiller, Johns Hopkins University, Baltimore, Md. 21218
C. Schwabe, Medical Univ. of South Carolina, Charleston, SC 29401
R. Schwyzer, Eidg. Tech. Hochschule Zürich-Hönggerberg, Zürich
R. Shapira, Emory University, Atlanta, Georgia 30322
J. C. Sheehan, Massachusetts Inst. of Tech., Cambridge, Mass.
J. E. Shields, Eli Lilly & Co., Indianapolis, Ind. 46206
M. Shimizu, Beckman Instruments, Inc., Palo Alto, Cal. 94304
F. Sipos, Squibb Inst. for Medical Res., New Brunswick, NJ 08903
E. R. Simons, Harvard Medical School, Boston, Mass. 02115
R. R. Smeby, Cleveland Clinic, Cleveland, Ohio 44106
I. C. P. Smith, National Research Council, Ottawa, Canada

E. L. Smithwick, Jr., The Lilly Res. Labs., Indianapolis, Ind.
C. R. Snell, Brandeis University, Waltham, Mass. 02154
G. L. Stahl, Case Western Reserve Univ., Cleveland, Ohio 44106
C. H. Stammer, University of Georgia, Athens, Ga. 30601
W. Steglich, Technische Universität Berlin, Berlin, West Germany
J. M. Stewart, University of Colorado, Denver, Colo. 80220
M. Szelke, Balaton, Hawksdown, Walmer, Deal, Kent, England
I. Teplan, Isotope Institute Budapest, Budapest, Hungary
D. A. Tewksbury, Marshfield Clinic Found., Marshfield, Wis. 54449
M. A. Tilak, The Lilly Research Labs., Indianapolis, Ind. 46206
K. Titlestad, Universitetet i Oslo, Oslo 3, Norway
A. A. Torellas, Schwarz/Mann, Orangeburg, NY 19062
G. W. Tregear, Massachusetts General Hosp., Boston, Mass. 02114
B. Tye, Massachusetts Inst. of Technology, Cambridge, Mass.
S. Udenfriend, Hoffmann-La Roche Inc., Nutley, NJ 07110
J. Van Rietschoten, Massachusetts General Hosp., Boston, Mass.
D. F. Veber, Merck Sharp & Dohme Res. Labs., Rahway, NJ 07065
P. Von Dreele, Cornell University, Ithaca, NY 14850
R. Walter, Mt. Sinai Sch. of Med. of CUNY, New York, NY 10029
S. S. Wang, Hoffmann-La Roche Inc., Nutley, NJ 07110
A. M. Warner, Northeastern University, Boston, Mass. 02115
B. Weinstein, University of Washington, Seattle, Was
T. Wieland, Max-Planck-Institut, 69 Heidelberg 1, West Germany
E. B. Williams, University of California, San Francisco, Cal.
R. E. Williams, Natl. Res. Council in Canada, Ottawa, Canada
T. C. Wuu, Medical College of Ohio, Toledo, Ohio 43614
D. Yamashiro, University of California, San Francisco, Cal.
N. Yanaihara, Shizuoka College of Pharmacy, Shizuoka-shi, Japan
N. Yoshida, Shionogi & Co., Ltd., Osaka, Japan
J. Young, University of California, Berkeley, Cal. 94720
P. E. Young, Harvard Medical School, Boston, Mass. 02115
R. L. Young, New England Nuclear Corp., Boston, Mass. 02118
H. Zahn, Deutsches Wollforschungsinstitut an der Technischen
     Hochscule, 51 Aachen, West Germany
A. R. Zeiger, Thomas Jefferson Univ., Philadelphia, Pa. 19107
L. Zervas, The University of Athens, Athens (144), Greece
G. Zimmerman, Harvard Medical School, Boston, Mass. 02115
H. Zuber, Eidg. Tech. Hochschule Zürich-Hönggerberg, Zürich

SPONSORS

Abbott Laboratories
Armour Pharmaceutical Company
Ayerst Research Laboratories
Beckman Instruments, Inc.
CIBA-GEIGY
Hoffmann-La Roche Inc.
Merck & Company, Inc.
Norwich Pharmacal Company
Pierce Chemical Company
Sandoz Pharmaceuticals
Schering Corporation
G. D. Searle and Company
The Dow Chemical Company

SUBJECT INDEX

AUTHOR INDEX

D 2